2010 International Conference on Advanced Optoelectronics and Lasers

(CAOL 2010)

Sevastopol, Ukraine
10-14 September 2010

Editors:

I. A. Sukhoivanov   O. V. Shulika

IEEE Catalog Number: CFP10814-PRT
ISBN: 978-1-4244-7043-3

Copyright © 2010 by the Institute of Electrical and Electronic Engineers, Inc
All Rights Reserved

Copyright and Reprint Permissions: Abstracting is permitted with credit to the source. Libraries are permitted to photocopy beyond the limit of U.S. copyright law for private use of patrons those articles in this volume that carry a code at the bottom of the first page, provided the per-copy fee indicated in the code is paid through Copyright Clearance Center, 222 Rosewood Drive, Danvers, MA 01923.

For other copying, reprint or republication permission, write to IEEE Copyrights Manager, IEEE Service Center, 445 Hoes Lane, Piscataway, NJ 08854. All rights reserved.

***This publication is a representation of what appears in the IEEE Digital Libraries. Some format issues inherent in the e-media version may also appear in this print version.**

IEEE Catalog Number: CFP10814-PRT
ISBN 13:   978-1-4244-7043-3

Additional Copies of This Publication Are Available From:

Curran Associates, Inc
57 Morehouse Lane
Red Hook, NY 12571 USA
Phone: (845) 758-0400
Fax: (845) 758-2633
E-mail: curran@proceedings.com
Web: www.proceedings.com

2010 International Conference on Advanced Optoelectronics and Lasers (CAOL 2010)

Sevastopol, Ukraine
10-14 September 2010

IEEE Catalog Number: CFP10814-POD
ISBN: 978-1-42447-043-3

CAOL*2010 program committee

M. Marciniak	CAOL PC Co-Chair, National Institute of Telecommunications, Warsaw, Poland
V. A. Svich	CAOL PC Co-Chair, V. N. Karazin National University, Kharkov, Ukraine
E. Alvarado Mendez	University of Guanajuato, Mexico
J. A. Andrade Lucio	University of Guanajuato, Mexico
G. Belenky	State University, New York, USA
T. Benson	University of Nottingham, Nottingham, UK
I. V. Blonskiy	Institute of Physics, National Academy of Science of Ukraine, Kyiv Ukraine
P. Chamorro-Posada	Universidad de Valladolid, Valladolid, Spain
R. De La Rue	University of Glasgow, UK
A. Desyatnikov	Nonlinear Physics Centre, Australian National University, Canberra, Australia
S. Donati	University of Pavia, Italy
I. V. Dzedolik	Taurida V.I. Vernadsky National University, Simferopol, Crimea, Ukraine
N. N. Elkin	State R&D Center TRINITI, Troitsk, Russia
W. Freude	University Karlsruhe, Germany
A. F. Glova	State Research Center of the Russian Federation Troitsk Institute for Innovation and Fusion Research, Troitsk, Russian Federation
H. Guo	Laboratory of Light Transmission Optics, South China Normal University, Guangzhou, PR China and School of Electronics Engineering & Computer Science, Peking University, Beijing, PR China
C. Jagadish	Australian National University, Canberra, Australia
A. Joulie	Laboratory CEM2, University Montpellier, France
H. Kawaguchi	Yamagata Univ., Japan
A. P. Kiselev	Steklov Math Institute RAS, St.Petersburg, Russia
V. K. Kononenko	Institute of Physics BAS, Minsk, Belarus
M. Koshiba	Hokkaido University, Japan
A. V. Kudryashov	Moscow State Open University, Moscow, Russia
A. Lapucci	Istituto Nazionale di Ottica Applicata, Firenze, Italy
V. V. Lysak	Gwangju Institute of science and technology, Gwangju, Republic of Korea
V. A. Makarov	International Teaching and Educational Laser Centre of the Moscow State University
V. A. Maslov	V. N. Karazin National University, Kharkov, Ukraine
V. K. Miloslavsky	V. N. Karazin National University, Kharkov, Ukraine
A. I. Nosich	Institute Radiophysics and Electronics of National Academy of Sciences of Ukraine, Kharkov, Ukraine
M. F. Pereira	Sheffield Hallam University, UK
S. Ponomarenko	Dalhousie University, Halifax, Canada
R. Rojas-Laguna	University of Guanajuato, Mexico
B. Sahraoui	Laboratoire POMA CNRS UMR 6136, Universiteé d'Angers, France
R. Schatz	KTH, Stockholm, Sweden
Y. S. Shmaliy	University of Guanajuato, Mexico
O.V. Shulika	National University of Radio Electronics, Kharkov, Ukraine
I. A. Sukhoivanov	University of Guanajuato, Mexico
S. V. Svechnikov	Institute of Semiconductor Physics of National Academy of Sciences of Ukraine, Kiev, Ukraine
S. I. Tarapov	Institute of Radiophysics and Electronics, Kharkov, Ukraine
R. Vlokh	Institute of Physical Optics, Ukraine

CAOL*2010 organizing committee

I. A. Sukhoivanov	General Chair, University of Guanajuato, Mexico
J. A. Andrade-Lucio	University of Guanajuato, Mexico
V. N. Berzhansky	Local Organization, Taurida National V.I.Vernadsky University, Simferopol, Ukraine
I. V. Dzedolik	Local Chair, Taurida National V. I. Vernadsky University, Simferopol, Crimea, Ukraine
A. V. Degtyarev	Local Organization, V. N. Karazin National University, Kharkov, Ukraine
E. Eres'ko	Local Organization, Taurida National V.I.Vernadsky University, Simferopol, Ukraine
V. I. Fesenko	Local Organization, National University of Radio Electronics, Kharkov, Ukraine
A. I. Filipenko	National University of Radio Electronics, Kharkov, Ukraine
T. B. Gryschenko	Social Programme, National University of Radio Electronics, Kharkov, Ukraine
S. G. Gryschenko	Local Organization, National University of Radio Electronics, Kharkov, Ukraine
O. V. Gurin	Local Organization, V. N. Karazin National University, Kharkov, Ukraine
I. V. Guryev	Web-site Support, National University of Radio Electronics, Kharkov, Ukraine
O. Ibarra-Manzano	University of Guanajuato, Mexico
S. O. Iakushev	Local Organization, National University of Radio Electronics, Kharkov, Ukraine
A. V. Kublik	Coordinator, Registration, National University of Radio Electronics, Kharkov, Ukraine
M. V. Klymenko	Local Organization, National University of Radio Electronics, Kharkov, Ukraine
A. Levchenko	Local Organization, V. N. Karazin National University, Kharkov, Ukraine
V. A. Maslov	Co-chair, Local Management, V. N. Karazin National University, Kharkov, Ukraine
Y. P. Machekhin	National University of Radio Electronics, Kharkov, Ukraine
J. Mikulska	Local Organization, Taurida National V.I.Vernadsky University, Simferopol, Ukraine
S. I. Petrov	Co-chair, National University of Radio Electronics, Kharkov, Ukraine
T. F. Ruban	Local Organization, V. N. Karazin National University, Kharkov, Ukraine
A. Rubass	Local Organization, Taurida National V. I. Vernadsky University, Simferopol, Crimea,
I. M. Safonov	Local Organization, National University of Radio Electronics, Kharkov, Ukraine
A. N. Shaposhnikov	Local Organization, Taurida National V. I. Vernadsky University, Simferopol, Ukraine
O. V. Shulika	Coordinator, Publication Chair, Web-designer, National University of Radio Electronics, Kharkov, Ukraine
A. N. Topkov	Local Organization, V. N. Karazin National University, Kharkov, Ukraine
G. V. Tkachenko	Local Organization, National University of Radio Electronics, Kharkov, Ukraine
A. V. Vasyanovich	National University of Radio Electronics, Kharkov, Ukraine

Contents

Plenary Papers

(Invited) Micro-/Nano-Photonic Device Structures Applied to Communications, Sensing and Consumer Optoelectronics
R. M. De La Rue, A. Samarelli, D. Logan, A. Khokhar, R. Dylewicz,
P.Velha, D. Macintyre, N. Johnson, F. Rahman, M. Sorel 1

(Invited) Optical Soliton Pulses with Relativistic Characteristics
G. S. McDonald, J. M. Christian, T. F. Hodgkinson 4

(Invited) Trends and Challenges in Optical Packet Networking: The Network Layer Perspective
M. Klinkowski, M. Jaworski, M. Marciniak 7

(Invited) Tunable Nonlinear Metamaterials and Plasmon Nanofocusing
Y. Kivshar 12

(Invited) Strong Turbulence Analysis of Sinusoidal and Hyperbolic Gaussian Beams
H. T. Eyyuboglu 15

(Invited) Evolution of Nanophotonics from Semiconductor Photonic Crystal Device to Metal/Semiconductor Plasmonic Device
K. Asakawa, Y. Sugimoto, N. Ikeda, D. Tsuya, Y. Koide, Y. Watanabe, N. Ozaki,
D. Kumar V., T. Nomura, D. Inoue, A. Miura, H. Fujikawa, K. Sato 18

(Invited) New Coupled Micro-Ring Resonator Structures and Laser Arrays
P. Chamorro-Posada, F.J. Fraile-Pelaez 25

(Invited) Recent Progress on Deep-ultraviolet Coherent Light Sources
Qinjun Peng, Zhimin Wang, Shenjin Zhang, Fengfeng Zhang, Feng Yang, Dafu Cui, Zuyan Xu 28

(Invited) Effective Mode Volume of a Natural Mode of an Open Dielectric Resonator with an Active Region
A. I. Nosich, E. I. Smotrova, V. O. Byelobrov, T. M. Benson 33

(Invited) Pneumatic Photonic Crystals: Properties and Multiscale Indication of Pressure
E.Ya. Glushko 36

(Invited) Non-Stationary Coherent Population Trapping in Frequency-Modulated Fields
V. L.Derbov, A. S.Lyubimov 38

(Invited) Mechanisms of Holographic Recording in Photopolymers
V.Obukhovsky, T.Smirnova 40

(Invited) Superradiant Heterolasers
V. V. Kocharovsky, M. A. Garasoyv, P. A. Kalinin, E. R. Kocharovskaya, A. A. Belyanin, V. V. Kocharovsky 43

(Invited) Selected Fundamental and Applied Problems of Laser mode Formation
V.G. Niziev 46

(Invited) Properties of the Interaction of Laser Radiation with a Gaseous Dust Medium
A. F. Glova, A. Yu. Lysikov, M. M. Zverev 52

(Invited) Diffraction Feature and Applications of Holographic Bragg Gratings
E. A.Tikhonov 55

(Invited) (110) Quantum Well Based Spin VCSELs
H. Kawaguchi, S. Koh 58

(Invited) Dark Resonances in Rubidium Atoms Excited by Optical Comb
A. M. Negriyko, R. A. Malitskiy, V. M. Khodakovskiy,
V. I. Romanenko, I. V. Matsnev, L. P. Yatsenko 61

(Invited) 589 nm High Power Sum-Frequency Generation with Quasi-Continuous-Wave Diode-Pumped Nd:YAG Lasers
Bo Yong, Xie Shiyong, Wang Zhichao, Shen Yu, Xu Jialin, Cui Dafu, Peng Qinjun, Xu Zuyan 64

(Invited) High Power Wavegude CO_2 Lasers for Technologies and Medicine
V. Ya. Panchenko, V. S. Golubev, V. V. Vasiltsov, E.N.Egorov 67

(Invited) 3D Model of Vertical Cavity Surface Emitting Laser with Resonant Array of Quantum Wells and External Mirror
N. N. Elkin, A. P. Napartovich, D. V. Vysotsky 71

(Invited) Light Fields with Complex Polarization Structure
V.G. Volostnikov, S.P. Kotova, O. Yu. Moiseev, A.V. Volkov, E.N. Vorontsov, D.M. Yakunenkova 74

II

(Invited) Shack-Hartmann Wavefront Sensor – Advantages and Disadvantages
A. Kudryashov, V. Samarkin, A. Alexandrov, J. Sheldakova, V. Zavalova 76

Nonlinear Photonics

(Invited) Nonlinear Solitary Waves in Positive-Negative Refractive Media
A. Maimistov 78

(Invited) Multipole Model for Metamaterial Homogenization
A. Chipouline, J. Petschulat, C. Menzel, C. Rockstuhl, T. Pertsch, F. Lederer 81

(Invited) Nonlinear Coupling in Plasmonic Structures
J. R. Salgueiro, Y.S.Kivshar 83

(Invited) Nonlinear Dynamics of Spiraling Laser Beams
A. S. Desyatnikov 86

(Invited) Rotating Three-Dimensional Solitons
S. Skupin, F. Maucher, W. Krolikowski 89

(Invited) Symmetry-Induced Forces on Phase Singularities
A. Ferrando, M.A. García-March 92

(Invited) Optical Nanoprobing via Spin-Orbit Interaction of Light
K. Y. Bliokh, O. Rodríguez-Herrera, D. Lara, E. A. Ostrovskaya, C. Dainty 95

Masses of Polaritons
I. V. Dzedolik 97

Refraction of Grey Solitons at Defocusing Kerr Interfaces
J. Sanchez-Curto, P. Chamorro-Posada, G.S. McDonald 100

Localized States and Oscillations Induced by Coherent Interaction of Waves in Nonlocal Media
S. Bugaychuk, R. Conte, E. Kozij, O. Kolesnyk, G. Klimusheva 103

Soliton based DPSK Encoded Sequences: Stability and Dynamical Properties
J. E. Prylepskiy, S. A. Derevyanko, S. K. Turitsyn 106

Magneto-Optical Kerr Effect in a Dielectric Film with Cubic Optical Nonlinearity on a Magneto-Electric Slab
Yu. S. Dadoenkova, I. L. Lyubchanskii, Y.P. Lee, Th.Rasing 109

Direct Quantitative Estimation of the Spatial Energy Migration Between Ions Yb^{3+} in $Gd_3Ga_5O_{12}$ Matrix
E. V. Ivakin, I. G. Kisialiou 112

(Invited) Liquid Crystal Defects: Nonlinear Optics and Optical Vortices
E. Brasselet 115

(Invited) Nonlocal Interactions of Solitons in Bias-Free Nematic Liquid Crystals
Ya. V. Izdebskaya 118

(Invited) Reversible Optical Nonreciprocity in Periodic Structures with Liquid Crystals
A. E. Miroshnichenko, E. Brasselet, Y. S. Kivshar 121

Fast Near-Infrared CdHgTe:V:Mn Photorefractive Material for Optical and Biomedical Applications
Y. P. Gnatenko, P. M. Bukivskij, Y. P. Piryatinski, I. O. Faryna, M. S. Fur'yer,
O. A. Shigiltchoff, R. V. Gamernyk, N. Kukhtarev, and T. Kukhtareva 124

Nonlinear LBO and BBO Crystals for Extreme Light Sources
A. E. Kokh, T. B. Bekker, K. A. Kokh, N. G. Kononova,
V. A. Vlezko, Ph. Villeval, S. Durst, D. Lupinski 127

Laser Resonators and Beam Propagation

Kaleidoscope Laser Properties and New Optical Fractal Contexts
J. M. Christian, G. S. McDonald, A. S. Heyes, J. G. Huang 130

Modification of the Power Radiated by an Electrical Dipole in the Presence of a Thin Dielectric Disk
M. V. Balaban, R. Sauleau, A. I. Nosich 133

Adaptive Optical System with Water-Cooled Bimorph Deformable Mirror
V.V.Samarkin, M.E.Dryagin, A.V.Kudryashov, P.N.Romanov 136

A Theorem for the Propagation of Electromagnetic Waves Through a Multilayered Structure Consisting of Left Handed Material and Dielectric
M. F. Ubeid, M. M. Shabat 137

Theoretical and Experimental Investigations Compact Slab Guide Resonators
V.A. Saetchnikov, E. A. Tcheriavskaia 139

The Contribution of Evanescent Waves into the Power Transfer
A.B. Katrich 141

Simulation of Lasing Modes in a Kite-Shaped Microcavity Laser
E. I. Smotrova, A. I. Nosich 144

Space Confinement of Random Lasing
V. P.Yashchuk, O. A. Prygodiuk 147

Fast Adaptive Optical System for Laser Beam Correction
A.L.Rukosuev, A.G. Alexandrov, A.V.Kudryashov, V.V.Samarkin 150

Optimizing Energy Performance of «Zig-Zag» Slab Lasers in Form of Flat Truncated Prisms
I. S. Manak, V. V. Zukowski, M. S. Leanenia 152

Microchip Laser with Active Output Mirror
V. V. Kiyko, V. I. Kislov, E. N. Ofitserov 155

Hill-Climbing Algorithm for Adaptive Optical System with Shack-Hartmann Sensor
J. Sheldakova, A. Rukosuev, P. Romanov, A. Kudryashov, V. Samarkin 157

Design of the Interference Structures Taking Into Account Absorption by Thin-Film Materials
Ya. Bondarchuk, H. Petrovska 159

Resonance Oscillations in Quasiperiodic Layered Structures with Lumped Inhomogeneities
V. F. Borulko, D. V. Sidorov 162

Disc and Microdisc Resonators with Strip-Like Diffraction Gratings
O. I. Belous, A. I. Fisun, V. L. Pasynin 165

Spiral Resonators with Whispering Gallery-like Modes
O. V. Goroshko, R. V. Golovashchenko, V. N. Derkach 167

Metal Waveguide Resonator with a Tilted Mirror
A. V. Volodenko, O. V. Gurin, A. V. Degtyarev,
V. A. Maslov, V. A. Svich, A. N. Topkov, T. F. Ruban 169

Optical Measurements

Measurement of Linear Polarization of Laser Radiation with Two-Grating Thin-Wire Bolometer
V.M. Kuzmichov, S.V. Pogorelov, B.V. Safronov, V.P. Balkashin, I.A. Priz 173

Novel Semiconductor Optical Amplifier with Tapered Active Channel
Yu.O. Kostin, A.A. Lobintsov, S.D. Yakubovich 175

Effect of Laser Beam Shape on Differential-Phase Measurements
Yu. V. Pilgun, E. N. Smirnov 177

The Relations Between the Basic Characteristics of Scattered Laser Irradiation Field and the Location Objects Surface Parameters
Arakcheev P. V., Semerenko D. A., Maryanina A. S., Nazarov S. I., Buryi E. V. 179

Diagnostics of Laser Radiance Penetration into Material by Multi-Channel Pyrometer
V. S. Golubev, A. V. Dubrov, Yu. N. Zavalov, V. D. Dubrov, N. G. Dubrovin 182

Optical Measurements of Ferroelastics ($Cs_3Bi_2I_9$): New Phenomena and Optical Devices
F. V. Motsnyi 185

Optical Interferometry and Data Analysis of Laser-Produced Plasmas
A. A. Aliverdiev, D. Batani, R. Dezulian, T. Vinci, A. Benuzzi-Mounaix, M. Koenig,V. Malka 188

Measuring Surface Distribution of Narrowband Radiation Wavelength by Colorimetric Method
A. V. Kraiski, T. V. Mironova, T. T. Sultanov, V. A. Postnikov 191

IV

Basing of the Television-Picture Generator Application for Distance Monitoring of Turbulent Atmosphere
G. N. Dolya, A. N. Katunin, A. N. Bulay, E. S. Chudovskaja 193

Experimental Investigations of Influence of Additionally Induced Polarization of the Ground State of
Atoms on Multiphoton Transitions
I. I. Bondar', V. V. Suran 196

Influence of Semiconductor Photoreceiver Spectral Responsivity at Different Temperature on Optical Measurements
A. V. Polyakov, M. A. Ksenofontov 199

Research of Bolometer Parameters of the Trellised Receiver of Laser Radiation
Kokody N. G., Pak A. O., Safronov B. V., Balkashin V. P., Priz I. A., Perepechai M. P. 202

Measurement of Intensity Distribution of Radiation in a Laser Beam by Shear Mechanism
N.G. Kokodiy, Li Zhenhua 205

Holographic Shack-Hartmann Wavefront Sensor
D. V. Podanchuk, V. P. Dan'ko, M. M. Kotov 208

The Commercially Available Version of Shack-Hartmann Wavefront Sensor
P. N. Romanov, V. E. Zavalova, A. V. Kudryashov, A. L. Rukosuev 211

High Intensity Laser Beam Wavefront Diagnostics and Correction at the Advanced Laser Light Source Facility
A. Alexandrov, S. Fourmaux, J. C. Kieffer, A. Kudryashov, F. Martin, T. Ozaki, S. Payeur 213

Optical Investigation of InGaAsP/InP Double Heterostructure Wafers
V. Rakovics 216

Determination of Active Region Overheating Temperature of GaN-based Light Emitting Diodes
Promising for Laser Media Pumping
E. V. Lutsenko, V. N. Pavlovskii, A. V. Danilchyk, M. V. Rzheutski,
A. G. Vainilovich, V. Z. Zubialevich, A. V. Muravitskaya, G. P. Yablonskii 219

Locating the Waist of a Gaussian Laser Beam
K. I. Muntean 222

High power laser beam position stabilization system by means of adaptive optics
M. E. Dryagin, A.G. Alexandrov, P. N. Romanov, A. L. Rukosuev, V. V. Samarkin 225

Enhanced interferometric technique for nondestructive characterization of crystalline optical materials: automated express
refractive index measurements
I. D. Karbovnyk, N. A. Andrushchak, Ya.V. Bobitskii 226

Lasers and Applications

Photophysical, Photochemical and Lasing Properties of Dipyrrolylmethene Complexes in Solutions and Solid Matrices
R. Kuznetsova, L. Samsonova, T. Kopylova, G. Mayer, O. Sikorskaya, Yu. Aksenova,
N. Zulina, E. Antina, S. Yutanova, L. Antina, T. Pavich, K. Solovyov, S. Arabei 228

Measurements and Modeling of Optical Distortions Relaxation in High Power Nd:Glass Lasers
V. E. Zavalova, A. V. Kudryashov, A. L. Rukosuev 231

Application of $Sm_2Ti_2O_7$ in Technology of Mirrors for He-Ne Laser
V. F. Zinchenko, V. I. Maksimenko, V. P. Sobol', L. V. Sadkovska, Ye. V. Timukhin 233

Modeling of XeCl Excilamps, Taking Into Account of Process of Halogen Regeneration
S. S. Anufrik, A. P. Volodenkov, K. F. Znosko 236

XeCl-Excilamp with the Capacitance Discharge
S. S. Anufrik, A. P. Volodenkov, K. F. Znosko 239

Point of Care Fiber Optical Sensor for Non-Invasive Multi Parameter Monitoring of Blood and Human Tissue Biochemistry
V. A. Saetchnikov, E. A. Tcheriavskaia, G. Schweiger 242

The Damage of DNA Induced by UV Nanosecond Laser Excitation at 193 nm
N. N. Vtyurina, S. L. Grokhovski, I. V. Filimonov, O. I. Medvedkov,
D. Yu. Nechipurenko, S. A. Vasiliev, Yu. D. Nechipurenko 244

Real Time Monitoring of Micro and Nano Particles, Blood Phantoms in Situ by Optical Micro Resonance Methods
V. A. Saetchnikov, E. A. Tcheriavskaia, G. Schweiger, A. Ostendorf 247

The Phenomenon of Laser-Induced Photodissosiation of Hemoglobin Complexes in Cutaneous Blood Vessels and its Biomedical Application
M. M. Asimov 249

Some Problems of Modelling Processes of Relaxed Optics in the Regime of Saturation the Excitation
P. P. Trokhimchuck 252

Laser-Microwave Spectroscopy of Au I Atoms in F Rydberg State
S. F. Dyubko, V. A. Efremov, V. G. Gerasimov, M. N. Efimenko, M. P. Perepechay, K. B. MacAdam 255

Holographic Sensors of Glucose in Model Solution and Serum
V. A. Postnikov, A. V. Kraiski, T. T. Sultanov, V. V. Deniskin 257

Thermosensor Diagnostics (TSD) of Laser Welding Process
A. F. Keremzhanov , P. P. Arkhipov, A. G. Lazarenko 259

About the Influence of Metallic Additives On Copper Vapor Laser Output Parameters
E. A. Svitlichniy, V. A. Kelman, Yu. V. Zhmenyak, V. V. Zvenihorodskiy, Yu. O. Shpenik 262

Influence of InGaN/GaN/Si Electroluminescent Heterostructure Design and Quantum Well Thickness on their Luminescent and Laser Properties
V. N. Pavlovskii, E. V. Lutsenko, A. V. Danilchyk, V. Z. Zubialevich,
A. V. Muravitskaya, G. P. Yablonskii, H. Kalisch, R. H. Jansen, B. Schineller, M. Heuken 265

Modelling of THz-Laser Radiation Pulse Shape
V. K. Kiseliov, V. P. Radionov 268

Reparative Regeneration Processes of Thyroid Gland and Timus at Experimental Autoimmune Thyroiditis after Laserotherapy
I. Gopkalova, V. Dubovik 270

Micro-/Nano-Photonic Device Structures Applied to Communications, Sensing and Consumer Optoelectronics

(Invited Paper)

Richard M. De La Rue, *IEEE Fellow*, Basudev Lahiri, Antonio Samarelli, Dylan Logan, Ali Khokhar, Rafal Dylewicz, Philippe Velha, Douglas Macintyre, Nigel Johnson, Faiz Rahman and Marc Sorel,
Optoelectronics Research Group, School of Engineering,
University of Glasgow, Glasgow G12 8QQ, Scotland, U.K.

Abstract: Device structures in high refractive index materials, notably semiconductors such as silicon, will form the basis for high-density integrated photonics. Nanometre-scale structuring - in conjunction with photonic crystal, photonic wire and plasmonic/metamaterial principles - will be required in order to realise the full potential for a variety of applications.

Photonic devices that involve nanometre-scale structuring - and that satisfy the unavoidable requirement for precision on a nanometric scale in their realisation - will be used in a variety of applications in the near future. Compact and energy efficient devices have become possible through exploitation of strong confinement - so that delivery, for example, of all-optical functionality at power levels that are compatible with 'green' photonics is a realistic goal. Targets such as making coarse wavelength-division multiplexed (CWDM) communications systems available - with multi-Gigabit per second channel capacity and at acceptable cost levels - to the individual user, fiber-to-the-home (FTTH) - and in FTTx situations more generally, have become credible.

Such compact 'nano'-photonic devices will provide all of the basic functionalities required, such as generation, detection, modulation, switching and wavelength selection - and they will greatly enhance the exploitable processing capability in a variety of different optoelectronic/photonic systems. For example, the availability of efficient four-wave mixing processes will enable phase-accurate translation of data channel frequencies at much lower optical and electrical power levels than currently required. A new wave of photonic device technology will result from the routine use of sub-micrometre scale waveguide cross-sections in silicon and other high refractive index materials, including both GaAs-based and the common InP-based epitaxial quantum well and quantum dot structures. Slow light structures will surely play a role, but will require carefully tailored dispersion properties - while both temporal and spatial soliton behaviour could find a useful role. The combined effect of current trends in guided-light photonics technology implies that high-density integration will become a reality, particularly when the compact optical isolator devices are realised that will result from incorporation of slow-light behaviour.

Devices that efficiently deliver the various desirable functionalities now possible are potentially suitable for

exploitation in fibre-optical communications systems. Data communication at seriously high bit-rates like 100 GBit/s, over short distances down to inter-chip level, and at low cost and power consumption levels will become routinely available to the individual consumer. Nanophotonic devices based on photonic-wire (PhW), photonic-crystal (PhC) and plasmonic or metamaterial principles are likely to play a central role in delivering the functionality required. Combinations of these principles will increase versatility and make it possible to choose, in a locally selective manner, between the behaviour required, for example, in fast-switching or rapid wavelength selection and the behaviour required in low-loss signal distribution and the interconnection of electronic devices. Serious technological challenges, in terms of both precision and reproducibility, will need to be addressed if the potential benefits of the nano-photonics approach are to be exploited successfully. Reducing the characteristic dimensions of light guiding structures fabricated in high refractive index semiconductors such as silicon or indium phosphide is required for enhancement of device performance. But the reduced cross-sectional dimensions involved, e.g., in photonic wire waveguides and photonic crystal channel waveguides, also imply greater sensitivity to changes in the material properties or the optical modal distribution.

One-dimensional periodic waveguide structures based on a silicon photonic wire geometry [1, 2] are potentially applicable for both fibre-optical communications and bio-medical sensing applications. New versions of high quality factor PhC/PhW microcavities device structures based on silicon-on-insulator have emerged recently [3, 4] - involving coupled cavities in one case and suspended structures in the other. Planar metamaterial structures formed by regular arrays of metallic split-ring resonators (SRRs) have been revisited, with interesting new results [5-7]. From the point of view of bio-sensing (or organic sensing more generally) such resonant arrays can be activated using surface-enhanced resonant Raman scattering (SERS) when illuminated by light with suitable wavelengths in the visible part of the spectrum, but illumination at wavelengths in the mid infra-red enables the alternative approach known as surface-enhanced infra-red absorption (SEIRA) spectroscopy to be used. We have also documented the impact of using a titanium metallisation layer, initially driven by the technological requirement for adequate adhesion to the substrate, below the gold layer that is standard for fabrication of planar metamaterial structures.

In the domain of 2D waveguide photonic crystal structures, aspects of the general problem that continue to be worthy of exploration are the form of the basic crystal lattice and the shape and orientation of the 'atoms' or 'molecules' from which the crystal lattice is formed. Different combinations of lattice symmetry - e.g. kagome - hole/pillar shape and orientation continue to be worth exploring, as a recent paper [8] demonstrates. Finally, the important potential application of photonic Quasi-crystal (PQC) structures for enhancing the light extraction efficiency and controlling the emission pattern of blue LEDs should be mentioned [9]. A stamp (imprint tool) for heat-and-pressure nano-imprint lithographic (NIL) production of PQC structures is shown in Fig. 1. Moving away from waveguide format PhC lattices that only use circular hole shapes promises to produce significantly superior performance [10] in some situations. The use of elliptical holes in lattice blocks that define channel waveguides has been predicted to make it possible to obtain a desirable combination of very slow light and negligible higher-order dispersion effects.

Research in silicon-based photonics has continued to increase in extent and importance [11]. Various forms of resonant cavity structure are likely to play a role in silicon waveguide-based photonic systems of the future. All-optical [12] switching at low energy levels is one example of a potentially important application - while account may need to be taken of the fact that ring-resonator cavities can exhibit significant levels of coherent back-scattering because of multiple transit effects [13]. Slow light behaviour can be obtained both in suitably designed PhC channel waveguides and in ring-to-ring, photonic wire-based, coupled resonator optical waveguides (CROWs). An example of the basic structure is shown in the micrograph of Fig. 2.

It is of interest to compare alternative approaches to meeting the challenges of realising compact slow-light structures based on planar waveguide technologies. In particular, the alternatives of using slow-light in PhC channel guides and ring-resonator based CROW structures should be considered. From a communications systems point-of-view, what emerges from a proper comparison of the two technological approaches is a 'horses-for courses' result [14]. The PhC approach will score where bit-rates that are above one Terabit per second are required, while the ring-based CROW structure scores decisively at bit rates below one hundred gigabits per second. It is arguable that the demonstrated capabilities for individual cavity switchability and buffered storage - and the overall delay tunability of ring-based CROW structures - also provide them with a competitive edge. The high finesse that can be obtained in a compact geometry by using a single silicon photonic wire ring resonator has been exploited to obtain usefully large responsivity in ion-implanted photodiodes at the standard fibre-telecommunications wavelengths [15]. This compactness is illustrated in the scanning electron micrograph shown as Fig. 3.

This brief review has concentrated on recent research to which the author has been more-or-less directly connected. It ranges over several domains under the general heading of micro-/nano-photonics - and these domains may be labelled as metamaterials, photonic (quasi-)crystals and photonic wire structures. Several different metals have been featured within the domain of metamaterials, while the words 'silicon photonics' are appropriate for much - but not all - of what has been covered. Applications in areas that include high-performance LEDs for consumer optoelectronics, silicon photonic ICs and biomedical-sensing could all benefit from the results of this research.

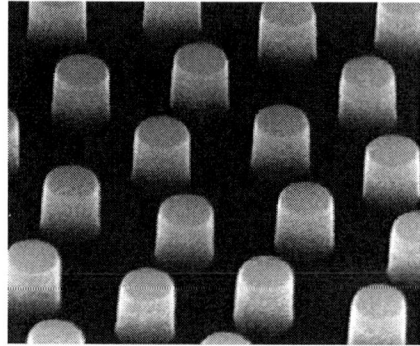

Fig. 1: Scanning electron micrograph of silicon stamp for NIL fabrication of PQC structures.

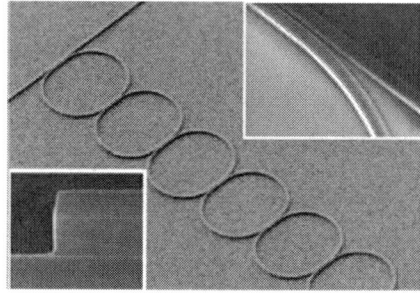

Fig. 2: Scanning electron micrographs of coupled ring (CROW) slow light structure.

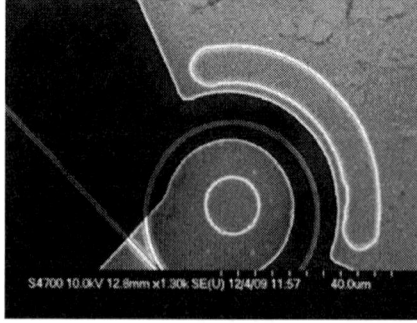

Fig. 3. SEM image of photonic wire waveguide photodiode integrated with ring-resonator structure.

REFERENCES

[1] A.S. Jugessur, J. Dou, J. S. Aitchison, R.M. De La Rue, and M. Gnan, 'A photonic nano-Bragg grating device integrated with micro fluidic channels for bio-sensing applications', Microelectron. Eng. **86**(4-6), 1488–1490, (2009).

[2] M. Gnan, W.C.L. Hopman, G. Bellanca, R.M. de Ridder, R.M. De La Rue and P. Bassi, 'Closure of the stop-band in photonic wire Bragg gratings', *Optics Express*, **17**(11), pp. 8830-8842, 25th May (2009).

[3] A.R. Md Zain, N.P. Johnson, M. Sorel and R.M. De La Rue, 'Coupling strength control in photonic crystal/photonic wire multiple cavity devices', *Electronics Letters*, **45** (5), pp. 283 - 284, (26th February 2009).

[4] Ahmad Rifqi Md Zain, Nigel P. Johnson, Marc Sorel, Richard M. De La Rue, 'High Quality-Factor 1-D-Suspended Photonic Crystal/Photonic Wire Silicon Waveguide Micro-Cavities', *IEEE Photonics Technology Letters*, **21**(24), December 15, 2009, pp.1789 - 1791.

[5] Basudev Lahiri, Ali Z. Khokhar, Richard M. De La Rue, Scott G. McMeekin and Nigel P. Johnson, 'Asymmetric split ring resonators for optical sensing of organic materials', *Optics Express*, **17**(2), pp. 1107-1115, 19th January (2009).

[6] Basudev Lahiri, Scott G. McMeekin, Ali Z. Khokhar, Richard M. De La Rue and Nigel P. Johnson, 'Magnetic response of split ring resonators (SRRs) at visible frequencies', *Optics Express*, **18**(3), pp. 3210 - 3218, 1st Feb (2010).

[7] Basudev Lahiri, Rafal Dylewicz, Richard M. De La Rue, and Nigel P. Johnson, 'Impact of titanium adhesion layers on the response of arrays of metallic split-ring resonators (SRRs)', *Optics Express,* **18**(11), pp. 11202 - 11208, 24th May (2010).

[8] A. V. Dyogtyev, I. A. Sukhoivanov, and R. M. De La Rue, 'Photonic band-gap maps for different two dimensionally periodic photonic crystal structures', *Jour. Applied Physics*, **107**, 013108 (1-7), 12th Jan (2010).

[9] Ali Z. Khokhar, Keith Parsons, Graham Hubbard, Faiz Rahman, Douglas S. Macintyre, Chang Xiong, David Massoubre, Zheng Gong, Nigel P. Johnson, Richard M. De La Rue, Ian M. Watson, Erdan Gu, Martin D. Dawson, Steve J. Abbott, Martin D. Charlton and Martin Tillin, 'Nanofabrication of gallium nitride photonic crystal light-emitting diodes', *Microelectronic Engineering*, **87**, 2200 - 2207, (2010).

[10] Swati Rawal, R.K. Sinha and Richard M. De La Rue, 'Slow Light Miniature Devices with Ultra-Flattened Dispersion in Silicon-on-Insulator Photonic Crystal', *Optics Express*, pp. 13315-13325, **17**(16), 3rd Aug (2009).

[11] Richard M. De La Rue, 'Breakthroughs in Silicon Photonics 2009', *IEEE Photonics Journal*, **2**(2), pp. 233 - 236, April (2010).

[12] Michele Belotti, Matteo Galli, Dario Gerace, Lucio Claudio Andreani, Giorgio Guizzetti, Ahmad R. Md Zain, Nigel P. Johnson, Marc Sorel and Richard M. De La Rue, 'All-optical switching in silicon-on-insulator photonic wire nano–cavities', *Optics Express*, **18**(2), pp. 1450 - 1461, 18th Jan (2010).

[13] F. Morichetti, A. Canciamilla, M. Martinelli, A. Melloni, A. Samarelli, R.M. De La Rue, and M. Sorel, 'Coherent backscattering in optical micro-ring resonators', *Applied Physics Letters*, **96**, pp. 081112-1 - 081112-3, (2010).

[14] A. Melloni, A. Canciamilla, C. Ferrari, F. Morichetti, L. O'Faolain, T. Krauss, R.M. De La Rue, A. Samarelli and M. Sorel, 'Tunable Delay Lines in Silicon Photonics: Coupled Resonators and Photonic Crystals, a Comparison', *IEEE Photonics Journal*, **2**(2), pp. 181 - 194, April (2010).

[15] D. Logan, P. Velha, M. Sorel, R.M. De La Rue, A.P. Knights and P.E. Jessop, 'High Sensitivity Defect-enhanced Silicon Ring-resonator Photodetectors at Telecom Wavelengths', IPR Topical Meeting, Monterey, paper IWF6, 26th - 28th July (2010).

Optical soliton pulses with relativistic characteristics

G. S. McDonald, J. M. Christian, T. F. Hodgkinson
Materials & Physics Research Centre, University of Salford, U.K.

(Invited Paper)

Abstract: The slowly varying envelope approximation and the ensuing Galilean boost to a local time frame are near-universal features of conventional scalar pulse models. Here, we will give an overview of our recent progress with a new approach to nonlinear pulse modelling, which is based on a Helmholtz-type formalism.

I. INTRODUCTION

It can be safely said that optical soliton pulses are one of the most thoroughly investigated and well-understood phenomena in nonlinear photonics. Since the seminal works of Hasegawa and Tappert [1], and later the experiments of Mollenauer et al. [2], the cornerstone of many investigations has been the slowly-varying envelope approximation (SVEA). The SVEA, in combination with a subsequent Galilean boost to a local time frame, tends to reduce the complexity of the longitudinal (spatial) part of wave operator, with temporal effects (such as higher-order dispersion and Raman scattering) left unchanged. While this approach has some clear-cut advantages [by replacing the elliptic (or hyperbolic) governing equation with a parabolic one], there are some physical effects that fall outside its remit. One such effect is spatial dispersion [3], recently discussed by Biancalana and Creatore in the context of pulse envelopes in semiconductor planar waveguides [4].

The SVEA and subsequent Galilean boost have enjoyed unbridled longevity in the literature over the past forty years for two main reasons. Firstly, they often provide an adequate description of the phenomena being observed. Secondly, a large body of knowledge exists on how to solve the resultant parabolic governing equations. In this Invited presentation, we report on our new Helmholtz approach to nonlinear pulse modelling, whereby these two classic simplifications are omitted. By deploying similar methods to those used over the past 12 years to study nonlinear beams, we have been able to derive several classes of exact analytical soliton solution to a new governing equation. Our results are fundamentally different from those already published [4] because of the frame of reference in which we analyze the pulses. They have a simple physical interpretation, and some surprising connections to special relativity have been uncovered.

II. HELMHOLTZ PULSE MODEL

We begin by considering a scalar electric field $E(t,z)$ that is traveling down the longitudinal axis z of a waveguide,

$$E(t,z) = A(t,z)\exp\left[i\left(k_0 z - \omega_0 t\right)\right] + A^*\left(t,z\right)\exp\left[-i\left(k_0 z - \omega_0 t\right)\right], \quad (1)$$

where t is the time coordinate. Here, $A(t,z)$ is the envelope that modulates a carrier wave with optical frequency ω_0 and propagation constant $k_0 = n_0\omega_0/c$, where n_0 is the linear refractive index of the core medium at ω_0 and c is the vacuum speed of light. The transverse spatial variation of the electric field is controlled by the structure of the waveguide itself. By substituting Eq. (1) into the corresponding Maxwell equations and Fourier transforming to the temporal frequency domain, it can be shown that [5]

$$\frac{\partial^2 \tilde{A}}{\partial z^2} + i2k_0 \frac{\partial \tilde{A}}{\partial z} + \left(k^2 - k_0^2\right)\tilde{A} = 0, \quad (2)$$

where $\tilde{A} \equiv \tilde{A}(\omega,z)$ denotes the Fourier transform of the pulse envelope. The parameter k^2 that appears in Eq. (2) is the mode eigenvalue – it is obtained by solving Maxwell equations for the transverse part of the confined field. The factor $(k^2 - k_0^2)$ is often approximated by $2k_0(k - k_0)$, and the remaining linear term $k \equiv k(\omega)$ is expanded around ω_0 according to

$$k(\omega) - k_0 = \sum_{j=1}^{\infty} \frac{k_j}{j!}\left(\omega - \omega_0\right)^j + \Delta k_{\mathrm{NL}}, \quad (3)$$

where $k_0 \equiv k(\omega_0)$, and $k_j \equiv (\partial^j k/\partial \omega^j)_{\omega_0}$ for $j = 1, 2, 3, \ldots$ The last term on the right-hand side of Eq. (3) is the nonlinear correction to k, taken to be $\Delta k_{\mathrm{NL}} = n_2 I \omega_0/c$, where n_2 is the Kerr coefficient and I is the intensity. The summation in Eq. (3) is truncated by assuming that terms beyond $j = 2$ make only a negligible contribution; the two expansion coefficients of interest are thus $k_1 = (\partial k/\partial \omega)_{\omega_0}$ and $k_2 = (\partial^2 k/\partial \omega^2)_{\omega_0}$, which parameterize the (inverse) group velocity and (inverse) group-velocity dispersion, respectively. By Fourier transforming back to the time domain, it can be shown that A must satisfy

$$\frac{1}{2k_0}\frac{\partial^2 A}{\partial z^2} + i\left(\frac{\partial A}{\partial z} + k_1 \frac{\partial A}{\partial t}\right) - \frac{k_2}{2}\frac{\partial^2 A}{\partial t^2} + \gamma |A|^2 A = 0. \quad (4)$$

where the coefficient of the nonlinear term is $\gamma = n_2/2n_0$. At this juncture, one should recognize that the double-z derivative, $\partial^2 A/\partial z^2$, appears naturally in the governing equation [5]. It is this term that is routinely neglected in analyses of pulse propagation phenomena.

Recently, it has been shown for the first time that spatial dispersion in some semiconductor materials can provide a second contribution to the coefficient of $\partial^2 A/\partial z^2$ [4]. This field-exciton coupling augments the propagation contribution $1/2k_0$ to yield a lumped coefficient,

978-1-4244-7043-3/10 $26.00 © 2010 IEEE

$$\frac{1}{2k_0} + \frac{n_0 \Gamma \Delta \tilde{\omega}_0}{2\delta\omega^2 c} .\qquad(5)$$

Here, $\Gamma \equiv \hbar/2M_x^*$, M_x^* is the effective exciton mass, $\tilde{\omega}_0$ is a resonant frequency, Δ is a dimensionless parameter related to the oscillator strength for the coherent exciton-photon interaction, and $\delta\omega$ is a frequency detuning). A salient point is that the coefficient of $\partial^2 A/\partial z^2$ can, in principle, become negative when $M_x^* < 0$ [4].

After rescaling, the following governing equation for the dimensionless envelope u may be derived:

$$\kappa \frac{\partial^2 u}{\partial \zeta^2} + i\left(\frac{\partial u}{\partial \zeta} + \alpha \frac{\partial u}{\partial \tau}\right) + \frac{s}{2}\frac{\partial^2 u}{\partial \tau^2} + |u|^2 u = 0 .\qquad(6)$$

The normalized space and time coordinates are $\zeta = z/L$ and $\tau = t/t_p$, respectively, where t_p is the duration of a reference pulse and $L = t_p^2/|k_2|$. The sign of the group velocity dispersion is flagged by $s = \pm 1 = -\mathrm{sgn}(k_2)$ (+1 for anomalous; –1 for normal), and $\alpha \equiv k_1 t_p/|k_2|$. The spatial dispersion parameter is $\kappa = \kappa_0 + D$, where $\kappa_0 \equiv 1/2k_0 L = c|k_2|/2n_0\omega_0 t_p^2$ and $D \equiv n_0\Gamma\Delta\omega_0/2\delta\omega^2 cL = |k_2|n_0\Gamma\Delta\omega_0/2\delta\omega^2 ct_p^2$. Finally, $u = A/A_0$, where $A_0 = (\gamma L)^{-1/2} = (2n_0|k_2|/n_2 t_p^2)^{1/2}$.

III. Models in the Local Time Frame

When considering problems involving pulse propagation, one typically follows a prescribed route to get from the more general nonlinear-Helmholtz governing equation (6) to the more straightforward nonlinear-Schrödinger (NLS) model. Firstly, one typically invokes the SVEA by arguing that the term in $\partial^2 u/\partial \zeta^2$ is small. A Galilean boost to a frame moving at the group velocity $1/\alpha$ is then implemented by defining

$$\tau_{\mathrm{loc}} = \tau - \alpha\zeta \quad \text{and} \quad \zeta_{\mathrm{loc}} = \zeta .\qquad(7a,b)$$

The coordinates $(\tau_{\mathrm{loc}}, \zeta_{\mathrm{loc}})$ are typically referred to as the "local time frame". They define a unique frame of reference in which a pulse centered on the carrier frequency ω_0 is *at rest*. In this rest frame, the field u satisfies the familiar NLS equation with a cubic nonlinearity,

$$i\frac{\partial u}{\partial \zeta_{\mathrm{loc}}} + \frac{s}{2}\frac{\partial^2 u}{\partial \tau_{\mathrm{loc}}^2} + |u|^2 u = 0 ,\qquad(8)$$

for which many different types of solution are known.

The natural question to ask is, "what happens if one keeps the $\partial^2 u/\partial \zeta^2$ term in Eq. (6) when implementing the Galilean boost?" In that case, the governing equation takes on a cross-derivative operator:

$$\kappa\frac{\partial^2 u}{\partial \zeta_{\mathrm{loc}}^2} + i\frac{\partial u}{\partial \zeta_{\mathrm{loc}}} + \frac{1}{2}\left(s + 2\kappa\alpha^2\right)\frac{\partial^2 u}{\partial \tau_{\mathrm{loc}}^2}$$
$$- 2\kappa\alpha\frac{\partial^2 u}{\partial \zeta_{\mathrm{loc}}\partial \tau_{\mathrm{loc}}} + |u|^2 u = 0 .\qquad(9)$$

To proceed, one might, for instance, consider only those families of solutions where $\kappa\alpha^2 \ll O(1)$, which enables the coefficient of the temporal dispersion term to be simply $s/2$. One could also present a case, based on order-of-magnitude considerations, for omitting the cross-derivative term. In so doing, one ends up with

$$\kappa\frac{\partial^2 u}{\partial \zeta_{\mathrm{loc}}^2} + i\frac{\partial u}{\partial \zeta_{\mathrm{loc}}} + \frac{s}{2}\frac{\partial^2 u}{\partial \tau_{\mathrm{loc}}^2} + |u|^2 u = 0 .\qquad(10)$$

Biancalana and Creatore have recently analysed model (10) [4], which is a temporal analogue of the well-known spatial nonlinear Helmholtz equation [6,7]. They performed a linear stability analysis [8] of its continuous-wave solutions, and derived exact analytical bright and dark solitons. The work of Biancalana and Creatore [4] is particularly important since it appears to be the first inclusion of Helmholtz-type effects in scalar pulse modelling.

IV. The Laboratory Frame

Despite the excellent progress made with model (10), the above approximations to eliminate inconvenient terms is not always going to be a fully satisfactory approach [4]. For instance, the mapping between normalized and dimensional quantities is no longer exact. With this in mind, it is instructive to compare how the mathematical structure of the two simplified models differ from the envelope equation in the laboratory frame – Eq. (8) drops two terms ($i\alpha\partial u/\partial\tau$ and $\kappa\partial^2 u/\partial\zeta^2$), while Eq. (10) drops only a single term ($i\alpha\partial u/\partial\tau$). Importantly, both simplified models admit exact analytical solutions, and this is undoubtedly one of the reasons for their success.

The nub of the problem is that if one wishes to keep the $\kappa\partial^2 u/\partial\zeta^2$ term (which is essential, for instance, when describing spatial dispersion [4]), then Galilean boost (7) results in a local governing equation that is more complicated than the original equation! So at the outset, the conventional coordinate transformation serves no useful purpose. This leaves one with a fairly stark choice. One could either work with the approximate models (8) and (10), with all their inherent advantages and disadvantages; or one can abandon the near-universal Galilean transformation and remain in the laboratory frame.

It turns out that, if the latter option is chosen, a huge amount of progress can be made. So much, in fact, that it is surprising that analyses of optical pulses *in the laboratory frame* (i.e., in the frame where experiments are always performed and measurements made) seem to be almost completely absent from the literature. Rather than seeing how many terms can be dropped from Eq. (6), one can instead approach the problem in the reverse sense: Eq. (6) contains just *one* extra term compared to the spatial Helmholtz equation. This term even has a particularly tractable form – a linear operator $i\alpha\partial u/\partial\tau$ (i.e., there are no higher-order or cross-derivatives, and no complicated nonlinearities to consider). Hence, one expects to be able to deploy the same mathematical methods [6–8] and computational [9] tools that have been used extensively over the past decade to study Helmholtz spatial solitons.

*CAOL*2010 International Conference on Advanced Optoelectronics & Lasers, 10-14 September, 2010, Sevastopol, Ukraine*

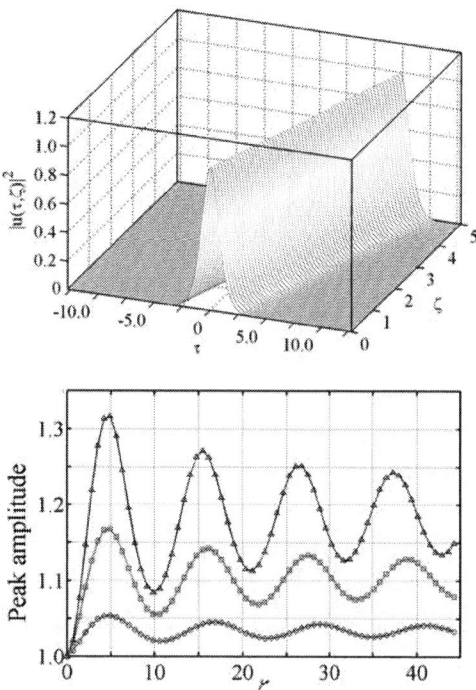

Fig. 1. Top: propagation of an exact bright soliton of Eq. (6). Bottom: typical self-reshaping oscillations for a perturbed bright soliton with $\kappa < 0$.

V. HELMHOLTZ SOLITON PULSES

In this presentation, we will give a comprehensive overview of our investigations into Eq. (6). This will include a linear stability analysis of the continuous-wave solutions, and establishing where those solutions are modulationally stable against both long- and short-wave periodic perturbations.

Exact analytical bright and dark soliton solutions will also be reported, and their space-time geometry considered in detail. New parameter regimes will be considered that have no counterpart in the spatial domain – namely, $\kappa < 0$ (in the spatial domain, κ is the nonparaxial parameter, which must be positive [4,6,7]). A wide range of generic features will also be identified. Crucially, the properties of these new Helmholtz *temporal* solitons are found to depend on the sign of the product $s\kappa$, rather than $\text{sgn}(s)$ or $\text{sgn}(\kappa)$ separately.

One of the key results – and also one of the most interesting – is the velocity combination rule for Helmholtz solitons. This law, which is geometric in nature, is strongly reminiscent of the way velocities add together in relativistic particle mechanics. In fact, when $\text{sgn}(s\kappa) = -1$, there is a one-to-one mapping with special relativity theory. Deeper insight can be gained by considering the transformation laws for Eq. (6), which show that the velocity combination rule is an intrinsic property of the model itself, rather than a property of particular (e.g., soliton) solutions.

Analysis has also uncovered *why* the Galilean transformation (7) is inherently incompatible with the Helmholtz governing equation (6). In conventional pulse theory, there exists a single unique reference frame where *all* pulses (centred on the carrier frequency ω_0) are stationary. No such frame exists for the solutions of Eq. (6) because the transformation to the rest frame of any pulse is parameterized by the characteristics (e.g., the intensity) of the pulse itself.

Finally, we will provide supporting evidence of the stability of the new soliton pulses that has been obtained from extensive computer simulations [9]. For example, bright solitons subjected to an initial perturbation to their local temporal shape tend to exhibit monotonically-vanishing oscillations, relaxing back toward an invariant solution (see figure 1).

VI. CONCLUSION

We have taken the first steps toward understanding nonlinear optical pulses from a new perspective by studying their behaviour from the laboratory frame. In the course of our work, we have found that this frame is the natural frame from which to describe pulses. Furthermore, the internal inconsistencies introduced by the classic Galilean boost (7) can be quite subtle (indeed, some of them remained hidden until the more general Helmholtz model was investigated). We have discovered, what we believe to be, a compact and elegant framework for describing optical pulses. The framework is exact [in the sense that no further approximation beyond Eq. (6) is required] and self-consistent. The existence of bright and dark soliton solutions with novel relativistic characteristics is also reported.

REFERENCES

[1] A. Hasegawa and F. Tappert, "Transmission of stationary nonlinear optical pulses in dispersive dielectric fibres. I. Anomalous dispersion," Appl. Phys. Lett. vol. 23, no. 3, pp. 142-144, 1973.

[2] L. F. Mollenauer, R. H. Stolen, J. P. Gordon, and W. J. Tomlinson, "Extreme picosecond pulse narrowing by means of soliton effect in single-mode optical fibers," Opt. Lett. vol. 8, no. 5, pp. 289-291, 1983.

[3] V. M. Agranovich and V. L. Ginzburg, *Crystal Optics with Sptial Dispersion, and Excitons*, Berlin, Springer, 1984.

[4] F. Biancalana and C. Creatore, "Instabilities and solitons in systems with spatiotemporal dispersion," Opt. Exp. vol. 16, pp. 14882-14893, 2008.

[5] K. J. Blow and N. J. Doran, "Solitons in optical fibres," in *Nonlinear Waves in Solid State Physics* A. D. Boardman et al. Eds. Plenum 1990, pp. 325-369.

[6] P. Chamorro-Posada, G. S. McDonald, and G. H.C. New, "Nonparaxial solitons," J. Mod. Opt. Vol. 45, no. 6, pp. 111-1121, 1998.

[7] P. Chamorro-Posada and G. S. McDonald, "Helmholtz dark solitons," Opt. Lett. vol. 28, no. 10, pp. 825-827, 2003.

[8] J. M. Christian, G. S. McDonald, and P. Chamorro-Posada, "Helmholtz bright and boundary solitons," J. Phys. A: Math. Theor. vol. 40, pp. 1545-1560, 2007.

[9] P. Chamorro-Posada, G. S. McDonald, and G. H. C. New, "Non-paraxial beam propagation methods," Opt. Commun. vol. 192, pp. 1-12, 2001.

Trends and Challenges in Optical Packet Networking: The Network Layer Perspective

(Invited Paper)

Mirosław Klinkowski, Marek Jaworski, and Marian Marciniak

Department of Transmission and Optical Technologies, National Institute of Telecommunications
1 Szachowa Str., 04-894 Warsaw, Poland, email: mklinkow@itl.waw.pl

Abstract—The immense growth of the Internet which is accompanied by the emergence of new communication technologies, applications, and devices, results in the evolution from voice-centric toward data-centric optical networks. In this paper we overview some of the trends in optical packet networking and identify several related research problems. Our particular focus is on the network layer aspects of future optical networks.

I. INTRODUCTION

During the past couple of years, optical networking has undergone tremendous changes which have been governed by developments of networking capabilities and emerging applications. This evolution can be translated to denser WDM transmission systems (i.e., $80-160$ wavelengths per fiber) operating at higher line rates (e.g., $10/40/100$ Gbit/s) with coarser granularities at switching level (optical circuit/burst/packet switching, OCS/OBS/OPS) [1], as well as the emerging of new technologies such as Optical Orthogonal Frequency Division Multiplexing (O-OFDM). At the same time, we observe the integration of diverse network technologies (wireless and wired networks) and disparate communication paradigms (mobile and fixed communications), as well as the convergence of different applications and services (data and multimedia) within a unified Internet Protocol (IP) packet-oriented network infrastructure. The observed trends are accompanied by rapid development of wired and wireless access technologies (FTTx, WiFi, WiMAX, LTE) as well as new applications (grid, P2P, gaming, 3DTV) which results in a tremendous increase of network traffic generated by users and network devices. All these factors have an impact on the requirements put on optical networks both in the backbone, where the aggregated traffic is transported, and in the access, where different communication technologies meet together. These requirements concern, among others, large transmission capacities, low costs, quality guarantees, energy efficiency, high granularity, dynamic resource reservation, flexible network reconfiguration, and interoperability with diverse network technologies.

The purpose of this paper is to overview current trends in optical networking and identify some research challenges that have to be addressed in order to enable future progress in development of next-generation packet-oriented optical networks. Our particular interest is in the network layer aspects.

The reminder of this paper is organized as follows. In Section II we focus on next-generation optical transport networks (OTN). In Section III we address the issue of fiber-wireless (FiWi) convergence in access networks. In Section IV we discuss two new challenges in optical networking, namely, the physical layer impairment-aware network design and the energy efficiency. Finally, in Section V we conclude our discussion.

II. NEXT-GENERATION OPTICAL TRANSPORT NETWORKS

The main role of future optical networks is to provide a transport infrastructure for legacy and new IP services. Nowadays, transport networks are based mainly on static point-to-point and ring connections, and they involve an excessive protocol stack (IP/ATM/SDH) The role of optics in communication networks is limited to the realization of transmission functions only, mainly due to still costly all-optical devices (tunable lasers, wavelength converters, fast switching elements, etc.) and the lack of viable technologies (optical RAM). Nonetheless, anticipating future advances in the optical technologies, the next generation optical transport networks are expected to perform either some or all the switching and control functions in the optical domain.

In this Section, we discuss several architectural and technological choices for next-generation packet-oriented OTNs and we present some commonly accepted solutions.

A. Architectural and Technological Choices

1) Communication Modes: The fact that the Internet is a connectionless packet-based network is the main driver in the development of data-centric OTNs. There are two candidate operational models considered by the research community for future data-centric OTNs, namely, the circuit switching (CS) model and the packet/burst switching (PS/BS) model. CS networks are connection-oriented and, in particular, the transmission of data packets through the network is realized on pre-established circuits. On the contrary, in PS/BS networks the data stream consists of statistically multiplexed packets, or bursts of packets, which belong to different connections and share common transmission link resources.

2) Frequency Division Multiplexing Technologies: At present, the most advanced technology that can deal with the huge capacity requirements of OTNs is the Dense WDM (DWDM) technology. DWDM transmission systems multiplexes multiple optical carrier signals on a single optical fiber by using different wavelengths (colors) of laser light to carry different signals.

978-1-4244-7043-3/10 $26.00 © 2010 IEEE

Recently, the concept of Optical Orthogonal Frequency Division Multiplexing (O-OFDM) emerged [2]. The key idea behind O-OFDM is to split a high-rate data-stream into a number of low-rate data-streams that are transmitted simultaneously over a number of subcarriers.

In light of the emerging demand for 100 Gb/s and higher data rates in future optical transport systems, optical OFDM is considered to be a promising enabling technology. In contrast with DWDM systems, OFDM introduces significantly higher spectrum-efficiency.

B. OTN Architectures

1) Wavelength-Routed OCS Networks: Taking into account the current status of optical technologies the short-term OTNs will apply the CS model in DWDM networks, resulting in the so called wavelength-routed optical circuit switching (WR-OCS) network. The transmission of data packets through the WR-OCS network is realized on pre-established optical circuits called the *lightpaths*. In WR-OCS, each lightpath is identified with a physical route, i.e., a subset of network links, and the DWDM wavelength channels assigned in consecutive links. The establishment of lightpaths in the network is performed by OCS nodes, also called the optical cross-connects (OXC). The switching capability of OXCs can be achieved by means of commercially available technologies such as, for instance, micro-electro-mechanical (MEMS) elements. The relatively long switching times of MEMS are sufficient for rather slow OCS operation.

The switching granularity of the OXC is a wavelength. It means that no further aggregation and drop of data is possible before the lightpath reaches its destination. In order to allow the transmission of data packets between nodes that do not have direct lightpath connections, electronic routers are used on top of the OXC nodes. In such a multi-layer architecture, electronic routers serve both as network ingress/egress nodes and traffic aggregators. They route packets over existing lightpath connections and intermediate routers towards their destination. It is worth noting that the packets undergo optical–electronic–optical (OEO) conversion every time they come across a router.

The cost of electronic devices, their high energy consumption, the heat dissipation and device dimensioning issues are the main drawbacks of the OCS-based packet transport networks.

From the network-layer perspective, the design of packet-oriented WR transport networks requires dedicated models and algorithms so that to take into account the multi-layer network architecture, which consists of the optical network layer and the electronic router network layer. Efficient methods for the cross-layer optimization in network planning, routing, resilience, etc. are up-to-date research problems.

2) Wavelength-Routed OBS/OPS Networks: OBS and OPS technologies have arisen as an alternative to the OCS network model. The principal objective of OBS/OPS is the provisioning of sub-wavelength switching granularity in the optical domain so that the wavelength resources could be used temporarily and shared between different connections. With respect to DWDM-based OCS networks, where wavelength resources are limited, this feature allows to increase the network scalability as well as its adaptability to the bursty characteristics of IP traffic.

In OBS/OPS, the data packets from the client networks are assembled into optical bursts/packets at the edge of the network and send towards their destination on wavelength channels. The bursts/packets are switched over the network by means of intermediate optical switching nodes and they remain in the optical domain all the way. In order to perform selective switching of arriving bursts/packets, the intermediate nodes have access to the control information that is generated and carried with the bursts/packets. The control information is separated from the user data either in time (OPS) or in space (OBS). A dynamic character of OBS/OPS requires fast switching elements. Thanks to the aggregation of a number of data packets into the burst in OBS, this requirement is slightly relaxed with respect to OPS. Indeed, in OPS the length of the optical packet is considered to correspond to the length of the electronic data packet.

In the perspective of network optimization the implementation of burst/packet switching techniques directly in the transport network will bring more statistical sharing of physical resources and will reduce the connection costs. Besides, there is no need for the energy-consuming electronic routers and expensive OEO conversion in the network.

Nowadays, the main challenge for OBS/OPS networks is the lack of low-cost all-optical processing devices. For this very reason, OBS/OPS is considered as a long-term solution for OTNs. Apart from technological difficulties, OBS/OPS needs a special treatment so that to solve problems typical of data-centric networks. The problem of routing with quality of service (QoS) guarantees [3] and the development of control plane protocols [4] are, among other things, key designing issues in OBS/OPS networks. Besides, for network planning and routing purposes, new modeling approaches are required to take into account the wavelength-continuity constraints in OBS networks. So far, models based on the Erlang B loss formula have been considered under rather ideal assumption of full-wavelength conversion capability in switching nodes [5]. For our recent, more complete surveys on the QoS and routing methods, the reader is referred to [6] and [7].

3) O-OFDM Networks: O-OFDM is identified as a promising solution due to its intrinsic characteristics of spectrum efficiency and scalability achieved by means of flexible bandwidth provisioning [8]. Due to the finer granularity of an OFDM-based network compared to a traditional WDM network just enough bandwidth can be allocated to a given traffic demand. In this context, O-OFDM enables a new network architecture, where any two nodes can be connected with the amount of bandwidth required, either providing a sub-wavelength service or super-channel connectivity.

O-OFDM is still at an early stage of development and its application in optical transport networks is unlikely in the near future. Besides technological challenges, new research problems arise in the network layer. Indeed, the flexible bandwidth allocation of O-OFDM-based networks introduces new constraints in network design [9]. In fact, since bandwidth allocation is not fixed but can vary, traditional RWA algorithms of WDM networks are not applicable anymore. As a consequence, a problem of Routing and Spectrum Allocation have to be addressed [9].

III. Covergent Fiber-Wireless (FiWi) Networks

The integration of wired and wireless networks has recently gained considerable research interest. The Fiber-Wireless (FiWi) architectures [10] are very attractive since they combine the best attributes of both worlds. The optical networks offer almost limitless amount of bandwidth and wireless networks provide ubiquitous access and support for user mobility. Among promising scenarios of FiWi integration we can discern:

- Wireless-Optical Broadband Access Network (WOBAN) [11][12],
- Radio over Fibre (RoF) [13],
- Free-Space Optics (FSO) [14].

Below we address the first two scenarios. Both are dedicated for the deployment in access networks.

A. Wireless-Optical Broadband Access Network (WOBAN)

The idea behind WOBAN is to run fiber as far as possible from the telecom central office (CO) toward the end-user and then to provide wireless access. This concept is very attractive since in some situations it may be costly to connect the end user to the CO with the fiber and, at the same time, it may be impossible to provide wireless access from the CO because of limited spectrum. Consequently, the WOBAN network architecture consists of an optical backhaul, such as e.g., a passive optical network (PON), and wireless access in its front end (e.g., WiFi or WiMAX). The communication in the network is achieved by means of optical line terminals (OLT) and optical network units (ONU), in the backhaul part of the network, and wireless gateways and wireless routers, in the wireless part of the network. The wireless gateways serve as an interface with the backhaul network. The wireless routers provide end-user connectivity and may form a multihop wireless mesh network (WMN) that further increases the coverage of the network.

An interesting engineering design and optimization problem is how far the fiber should penetrate before wireless takes over [15]. Other challenges concern the problem of routing and network resilience [11].

B. Radio over Fibre (RoF)

RoF networks allow an optical link to transmit a modulated radio frequency (RF) signal, by these means, moving the processing functionalities to the Central Station (CS) and reducing the radio Access Point (AP) complexity. In some deployment scenarios, such as e.g. indoor picocellular access networks, hundreds of antennas need to be supported and the RoF approach can result in a simplified overall system design

due to the aggregation of RF signal generation and network management at a central location [16].

As more applications spring up, overcrowding and interference at high end of the microwave region pushes operating frequencies toward millimeter-wave (mm-wave) band, such as 60 GHz. Higher frequency operation provides larger instantaneous bandwidth for greater transfer of information with reduced dimensions for antennas and other components. As a consequence, at the front end of the RoF network, mm-waves are potentially useful RF resources to cope with Gb/s data transmission [16]. It is worth noting that 60 GHz millimeter waves are highly attenuated in the air (approx. 10 dB/km) and, therefore, are useful only in the short distance communication. Thanks to the extremely low attenuation in fibres, below 0.2 dB/km around 1550 nm wavelength, the RoF transmission will extend significantly the reach and utility of mm-wave band systems.

A pasive optical network (PON) is a short-term candidate for the optical back end of the RoF network. The wavelength division multiplexed PON (WDM-PON) is considered to be the next evolutionary solution which increases system capacity and scalability as well as it improves network security [17]. The introduction of wavelength routing capabilities [18] will allow the reconfigurable DWDM-RoF networks to accommodate a huge number of APs with a flexible allocation of resources according to traffic demands and the movement of nomadic users.

Among research topics related to RoF networks is the problem of the impairments in RoF transmission, which occur in an optical link in a mm-wave band, e.g. for 60 GHz carrier frequency [19]. From the network design perspective, an exemplary problem is the resource allocation problem in reconfigurable DWDM-RoF networks [20] under dynamic scenarios with mobile and temporally active users.

IV. New Challenges in Optical Networking

A. Physical Layer Impairment-Aware Network Design

Optical network architectures are evolving from traditional opaque networks, in which optical signal undergoes an OEO regeneration, toward all-optical (i.e., transparent) networks. The lack of practical all-optical regeneration gives rise to the so called semi-transparent network architectures, in which a set of sparsely but strategically placed OEO regenerators is used to maintain the acceptable level of signal quality.

In this context, the common problem in optical networks is the problem of Routing and Wavelength Assignment (RWA). So far, in the literature the assumption was that the optical layer is a perfect transmission medium and therefore all outcomes of a RWA algorithm are considered valid and feasible. The reality is that the actual performance of the system may be unacceptable for some RWAs due to the optical signal degradation, which is particularly harmful in high-rate and long-distance transmission systems. Indeed, as optical signals traverse the optical fiber links and also propagate through passive and/or active optical components, they encounter many impairments that affect the signal intensity level, as well as its temporal, spectral and polarization properties. For this reason the incorporation of physical layer impairments (PLI) in (semi-)transparent optical network planning and operations has recently received increasing attention from the research community [21].

The problems that still require effective solutions are the PLI-aware RWA (PLI-RWA) problem and the Regeneration Placement (RP) problem. Both PLI-RWA and RP are NP-complete problems [22][23], i.e., there is no algorithm known to solve them efficiently (in polynomial time). Most of the solutions proposed in the literature to address these problems employ analytical PLI estimation models and use heuristic algorithms. The few works that are based on network optimization methods make use of PLI models that estimate only static impairments, which do not involve actual network state information. A particularly challenging problem is the modelling of dynamic PLIs and their incorporation into efficient design methods.

Although the PLI-aware methods presented in the literature focus mainly on OCS networks, still there are issues specific to OBS/OPS, such as short connection holding times and the effect of optical signal dispersion on the transmission offset time, which require dedicated solutions. Some RP algorithms incorporating a static PLI model have been proposed in [24]. Similarly as in OCS, in OBS there is need for the development of computationally efficient and applicable for the optimization purposes the analytical models of dynamic PLIs.

B. Energy Efficiency and Awareness

One of the most important problems facing our civilization nowadays is the problem of increasing (exponentially) energy consumption and the associated problem of carbon dioxide emission. Consequently, many activities are carried out all around the world aiming both at the reduction of energy consumption and the replacement of existing energy sources with renewable ones. Among these activities, there are new initiatives, such as the Green Touch started by scientists of Alcatel-Lucent Bell Labs and the Green Communications initiated by ITU, which aim at the development of the energy efficient communication devices and networks.

New energy efficient technologies and techniques should aim at the reduction of energy consumption in the network but without negative impact on its performance. An important concept concerns the energy-awareness in the network and it characterizes a technology that adjusts its behaviour or performance according to the energy sources (either renewable or fossil) which supply the network and its components.

In the context of optical networking, it is expected that energy-efficiency and energy-awareness might be achieved through dedicated energy-driven design solutions and optimization schemes. [9]. One of the possible solutions is the reduction of the use of electronic routers in the network by keeping the traffic in the optical domain as far as possible. Indeed, optical devices are characterized by much smaller energy consumption per bit than electronic devices and, therefore, minimizing the number of potential IP hops can bring

significant power savings. To this aim, the traffic might bypass intermediate IP routers by means of direct lightpath connections [25]. Also, putting underutilized network components in low-energy consumption modes and re-routing the traffic over already awaken devices might help to save power in the network [26].

V. Conclusions

The main purpose of this paper was to overview current trends in optical networking taking into account the emergence of new communication technologies and applications. In the next future, we should observe further development of the WR-OCS networks, however, new technologies, such as e.g. O-OFDM, sound very promising and they may attract research interests as well. The OBS/OPS technology is still immature and further advances in the development of inexpensive and functional devices is required. Among new challenges we can mention the development of integrated optical-wireless network technologies enabling ubiquitous access and high bandwidth available for end users. Another problem concerns network modelling and design methods incorporating, among others, the impact of physical layer impairments on the signal degradation, so that to support efficient network operation and provide quality guarantees. Last but not least is the issue of energy efficiency and awareness which requires dedicated technologies and techniques and which introduces new objectives and constraints on the optical network design problem.

Acknowledgment

This work has been supported by the Polish Ministry of Science and Higher Education under the contract 643/N-COST/2010/0.

The authors express their appreciation for very fruitful interactions with the members of COST Action 2100 - Pervasive Mobile & Ambient Wireless Communications.

References

[1] J. Berthold *et al.*, "Optical networking: past, present, and future", *IEEE JLT*, vol. 26, no. 9, pp. 1104–1118, 2008.

[2] I. B. Djordjevic and B. Vasic, "Orthogonal frequency division multiplexing for high-speed optical transmission", *Opt. Express* 14, pp. 3767-3775, 2006.

[3] M. Klinkowski *et al.*, "Virtual Topology Design in OBS Networks", in *Proc. IEEE ICTON 2010*.

[4] P. Pedroso *et al.*, "An interoperable GMPLS/OBS Control Plane: RSVP and OSPF extensions proposal", in *Proc. PCSN2008*, Graz, Austria, Jul. 2008.

[5] M. Klinkowski *et al.*, *Graphs and Algorithms in Communication Networks - Studies in Broadband, Optical, Wireless, and Ad Hoc Networks.* Springer-Verlag, 2009, pp. 165-181.

[6] M. Klinkowski *et al.*, *Current research progress of optical networks*, Springer-Verlag, 2009, pp. 1-20.

[7] M. Klinkowski *et al.*, "An Overview of Routing Methods in Optical Burst Switching Networks", *Elsevier OSN Journal*, vol. 7, no. 2, pp. 41-53, Apr. 2010.

[8] M. Jinno *et al.*, "Spectrum-efficient and scalable elastic optical path network: Architecture, benefits, and enabling technologies", *IEEE Comm. Mag.*, vol. 18, pp. 66-73, Nov. 2009.

[9] I. Tomkos, "New challenges in next generation dynamic optical network planning", in *Proc. IEEE ICTON 2010*.

[10] N. Ghazisaidi, M. Maier, and Ch. M. Assi, "Fiber-wireless (FiWi) access networks: A survey", *IEEE Comm. Mag.*, Feb. 2009.

[11] S. Sarkar, S. Dixit, and B. Mukherjee, "Hybrid Wireless-Optical Broadband-Access Network (WOBAN): A Review of Relevant Challenges", *IEEE JLT*, vol. 25, no. 11, 2007.

[12] W.-T. Shaw *et al.*, "Hybrid Architecture and Integrated Routing in a Scalable Optical–Wireless Access Network", *IEEE JLT*, vol. 25, no. 11, 2007.

[13] Ch. Lim *et al.*, "Fiber-wireless networks and subsystem technologies", *IEEE JLT*, vol. 28, no. 4, Feb. 2010.

[14] A. K. Majumdar and J. C. Ricklin, "Free-Space Laser Communications, Principles and advantages", *Springer Science LLC*, 2008.

[15] S. Sarkar *et al.*, "Hybrid Wireless-Optical Broadband Access Network (WOBAN): Network Planning Using Lagrangean Relaxation", *IEEE/ACM Trans. on Netw.*, vol. 17, no. 4, 2009.

[16] M. Sauer, A. Kobyakov, and J. George, "Radio over fiber for picocellular network architectures", *IEEE JLT*, vol. 25, no. 11, Nov. 2007.

[17] G.-K. Chang *et al.*, "Key technologies of WDM-PON for future converged optical broadband access networks", *J. Opt. Commun. Netw.*, vol. 1, no. 4, Sep. 2009.

[18] T. Kuri *et al.*, "Reconfigurable dense wavelength division multiplexing millimeter-wave-band radio-over-fiber access system technologies", *IEEE JLT*, vol. 28, no. 12, 2010.

[19] M. Jaworski and M. Klinkowski, "Optical Transmission Impairments in 60 GHz Radio-over-Fiber System", in *Proc. IEEE ICTON 2010*.

[20] M. Klinkowski, M. Jaworski, and D. Careglio, "Channel Allocation in Dense Wavelength Division Multiplexing Radio-over-Fiber Networks", in *Proc. IEEE ICTON 2010*.

[21] S. Azodolmolky *et al.*, "A Survey on Physical Layer Impairments Aware Routing and Wavelength Assignment Algorithms in Optical Networks", *Computer Networks*, vol. 53, no. 7, pp. 926-944, 2009.

[22] R. Ramaswami and K. N. Sivarajan, "Routing and Wavelength Assignment in All-Optical Networks", *IEEE/ACM Trans. on Netw.*, vol. 3, no. 5, pp. 489-500, 1995.

[23] M. Flammini, "On the Complexity of the Regenerator Placement Problem in Optical Networks", in *Proc. ACM SPAA 2009*.

[24] O. Pedrola *et al.*, "Modelling and Performance Evaluation of a Translucent OBS Network Architecture", in *Proc. IEEE Globecom 2010*, accepted.

[25] G. Shen and R. S. Tucker, "Energy-minimized design for IP over WDM networks", *J. Opt. Commun. Netw.*, vol. 1, no. 1, Jun. 2009.

[26] M. Gupta and S. Singh, "Greening of the Internet", in *Proc. ACM SIGCOMM 2003*.

Tunable Nonlinear Metamaterials and Plasmon Nanofocusing

Yuri S. Kivshar

Nonlinear Physics Centre, Research School of Physics and Engineering,
Australian National University, Canberra ACT 0200, Australia
E-mail: ysk124@rsphyse.anu.edu.au

(Invited Paper)

Abstract—We discuss tunability and nonlinear properties of metamaterials operating at microwave and optical frequencies. We also study the fundamental nonlinear effects in metal-dielectric waveguides supporting surface plasmon polaritons, and discuss nanofocusing of a plasmon beam and soliton formation in a tapered waveguide in the presence of losses.

Index Terms—Nonlinear plasmonics, metamaterials

I. INTRODUCTION

Properties of hypothetical left-handed media with both $\epsilon < 0$ and $\mu < 0$ were first discussed by V. Veselago in 1968 [1]. However, it took more than thirty years until these ideas became realistic. The experimental realization of left-handed media was achieved in composite structures which possess features smaller than the wavelength of incident light that behave as artificial atoms making the structures to exhibit properties not available in natural materials. This is why such structures are called metamaterials. A rapid advance in nanofabrication allowed the fabrication of metamaterials in optics. One of the goals driving this research field is tunability of metamaterial structural properties enabling novel functionalities and applications.

In this talk we overview the activities of our group in Canberra in the study of tunable nonlinear metamaterials and related plasmonic nanophotonic structures. We summarize our recent achievement in experimental nonlinear metamaterials. Then, we demonstrate the electro-optic and all-optical tunability of left-handed optical metamaterial structure - trilayer fishnet structures infiltrated with a nematic liquid crystal and operating in the infra-red optical range. In particular, such tunability allows switching of the material refractive index from positive to negative by the applied voltage across the material.

Because left-handed materials at optical frequencies are closely related to plasmonics, we study the fundamental nonlinear effects in plasmonics and, as the first step, analyze the families of guided modes of a nonlinear slot waveguide and revealed that the symmetric mode undergoes the symmetry breaking and becomes primarily localized near one of the interfaces. We study propagation of surface plasmon polaritons in a nonlinear tapered slot waveguide and demonstrate taper-induced plasmon nanofocusing and the formation of a plasmon soliton. In addition, we analyze phase matching in planar metal-dielectric nonlinear waveguides supporting highly

localized plasmon polariton modes and reveal that quadratic phase matching between the plasmon modes of different spatial symmetries becomes possible in the planar waveguide geometry.

II. TUNABLE AND NONLINEAR METAMATERIALS

Metamaterials, which are typically regular arrays of sub-wavelength resonant particles or metal-dielectric patterns, offer us a new degree of freedom in controlling the electromagnetic response of matter. Thus we are no longer completely constrained by the properties of existing materials, but can tailor the response in an almost arbitrary fashion, for example achieving very high [2], very low [3], and negative [4] values of refractive index, permittivity and/or permeability. Because of the inherently strong dispersion of resonant metamaterials, they must be modified in order to operate in a different frequency band. Therefore, there is a significant push to have a further degree of control over these materials - tunability of their response. Fortunately, the engineered nature of metamaterials allows their properties to be controlled externally, either by dynamically modifying their structure, or by adding some nonlinear inclusion and controlling with external fields [5]. Examples of the latter approach include the introduction of varactor diodes, ferroelectrics and photoconductive semi-conductors. On the other hand, even without resorting to such exotic (and often lossy) constituents, there is a great deal of freedom to manipulate the structure itself, and this is the approach we propose here. For theoretical model, we consider specifically the split ring resonator as one of the most important metamaterial elements, noting that whilst the details of near-field interaction are structurally specific, our approach can be applied to a wide variety of structures. We will show experimental results on the tuning of the chiral properties of the twisted double-wire particles. Moreover, since the nonlinear properties of metamaterials strongly depend on linear characteristics of the structure, we can utilize the same approach for tuning the nonlinear response of the metamaterials.

Theory of nonlinear metamaterials [6] predicted that the hysteresis-type dependence of magnetic permeability on the field intensity may allow dramatic changes of the material properties. As the first step towards creating tunable nonlinear metamaterials we studied dynamic tunability of the magnetic

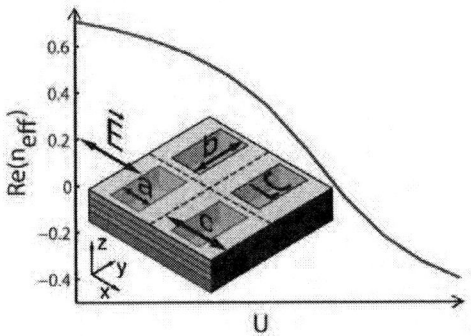

Fig. 1. Dependence of the effective refractive index on the voltage applied between the gold film and an electrode. Inset shows schematically a fishnet metamaterial structure infiltrated by a liquid crystal.

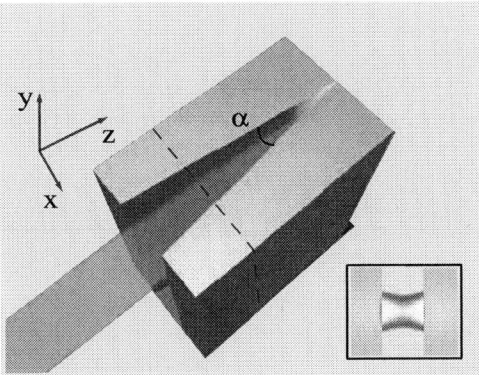

Fig. 2. Schematic of a three-dimensional plasmon mode focusing in a tapered metal-dielectric-metal slot waveguide. Focusing occurs in the horizontal plane, due to a taper, and in the vertical plane, due to the nonlinear self-action.

resonance of a single nonlinear split-ring resonator [7] and revealed different tuning regimes of metamaterial. At higher powers the nonlinear response of the split-ring resonator becomes multi-valued, indicating that the memory effect can be potentially observed in nonlinear metamaterials. Recently, we fabricated the first tunable nonlinear magnetic [8] and nonlinear electric [9] metamaterials by placing varactors in each of the split-ring resonators or each of the electric resonators of the structure. We measured a very pronounced shift of the resonance, and observed a change of the transmission through the nonlinear metamaterial with split-ring resonators for different power levels [8]. We observe experimentally the intensity-suppressed and intensity-induced transparency, when the frequency of the incoming wave is at the left edge of the resonance. The achieved nonlinear suppression of the beam transmission is 20 dB for magnetic metamaterial [8], and almost 50dB for electric metamaterial [9].

III. TUNABLE FISHNET STRUCTURES

It is proven that one of the most effective designs of a left-handed medium in the optical spectral range is a fishnet structure. The structure represents a periodic array of holes milled in a multilayer metal-dielectric-metal stack (see inset in Fig. 1). It was shown that the fishnet design can exhibit broadband negative permittivity and permeability accompanied by a high transmission [10]. Optical properties of a standard fishnet structure are defined by geometrical parameters and relative permittivity of the materials composing the structure. However, the infiltration of the structure with liquid crystals would allow controlling the refractive index of the materials inside the holes of the fishnet. This control is achieved by the reorientation of its pre-aligned molecules by an external electric or optical field.

Fishnet structure exhibits negative effective permittivity below the plasma frequency, which matches the cutoff frequency of the holes i.e. the transmission peak of the hole mode ($\lambda_c \approx 2b$, where b is a hole size along the direction of magnetic field). Negative magnetic response is related to the excitation of surface plasmon polaritons in the dielectric gap between metal plates. If the surface plasmon polaritons are excited below the plasma frequency, it results into simultaneously

negative effective permittivity and permeability and therefore into negative refractive index.

The cutoff frequency depends also on the dielectric permittivity (refractive index) of a substance located inside the holes. liquid crystals possess high birefringence, for example, E7 has $\Delta n \approx 0.2$ in near-infrared. Thus, if one infiltrates the fishnet holes with the liquid crystals, it would be possible to shift the cutoff frequency of the structure reorienting the liquid crystal molecules by an external electric or magnetic field. In addition, if an surface plasmon polaritons resonance located close enough to λ_c, it is possible to achieve switching of the effective index from positive to negative values as shown in numerical calculations presented in Fig. 1.

In our work we find the necessary geometrical parameters for gold-ZnO-gold trilayer fishnet and demonstrate the switching process of the refractive index of the order of 1.1 (five times larger than the tuning of the liquid crystal index). The liquid crystal is reoriented by applying voltage U between the gold film and an electrode placed above the sample. Experimental studies are underway to confirm our theoretical predictions. Our results to dynamically tune the refractive index of an artificial metamaterial represent an important milestone in the development of such materials and open novel applications for adaptive control of the metamaterial parameters.

IV. NONLINEAR PLASMONIC TAPERED WAVEGUIDES

Tapering of plasmonic waveguides has been suggested for high energy concentration in a deep subwavelength regime [11], [12]. Recent studies of tapered plasmonic slot waveguides revealed that at certain tapering angles the amplitude of a plasmon mode at the metal-dielectric interface may increase, thus effectively overcoming losses in the system [13]. In this work we study nonlinear transverse self-action of plasmon beams propagating in tapered metal-dielectric-metal waveguides with the Kerr-type nonlinear dielectric.

First, we consider linear two-dimensional waveguides. Assuming an adiabatic change of the waveguide width along plasmon propagation, we apply the slowly-varying envelope approximation and describe analytically the evolution of the plasmon amplitude at the metal-dielectric interface. We find the optimal shape of a taper, for which the amplitude of the plasmon mode remains constant, corresponding to the effective loss compensation [14]. It appears that the shape of this optimal taper is well described as a linearly tapered waveguide with some optimal angle α_c. For the tapering angles larger than α_c, we observe the plasmon nanofocusing close to the taper tip, as was originally predicted in Ref. [13].

In the three-dimensional case, nanofocusing is not achieved due to diffraction of plasmonic beams in the transverse dimension. It is well known that the beam diffraction can be compensated by the beam self-action in Kerr nonlinear media, and at certain conditions the spatial solitons may be formed [15], [16].

For the study of nonlinear tapered plasmonic slot waveguides, we apply the slowly varying approximation and derive the effective nonlinear Schrödinger equation for the slowly varying plasmon amplitude $A(y, z)$ with the coefficients varying with plasmonic beam propagation, depending on the tapering angle [17],

$$2i\sigma\frac{\partial A}{\partial z} + I\frac{\partial^2 A}{\partial y^2} + i\left(\frac{\partial \sigma}{\partial h}\frac{dh}{dz} + \Gamma\right)A + N_{\mathrm{nln}}|A|^2 A = 0, \quad (1)$$

where the $I = <E_{x0}^2 + E_{z0}^2>_x$ is the effective beam intensity, $\sigma = <E_{x0}H_{y0}>_x$ is proportional to the overall energy flow in the plasmon propagation direction per unit length,

$$\Gamma = <\varepsilon''(E_{x0}^2 + E_{z0}^2)>_x$$

is the effective dissipation in the system defined with the mode structure, and

$$N_{\mathrm{nln}} = <\nu(E_{x0}^2 + E_{z0}^2)^2>_x$$

is the effective nonlinear coefficient, where $< \cdot >_x$ corresponds to the integration over the transverse coordinate x.

We solve Eq. (1) by the beam propagation method, and study the soliton propagation in the taper for three different tapering angles, i.e. below optimal angle, near optimal angle, and larger than the optimal angle. For the soliton launched in a tapered waveguide with the angles below the optimal tapering angle (see Ref. [14]), we observe the dynamics similar to the beam propagation in planar nonlinear waveguides where the plasmon beam broadens due to strong losses present in the system [15], [16].

For the tapering angles close to the optimal angle, we find that the losses can be effectively compensated due to tapering, so that the plasmon beam launched into the system propagates practically without a change of its shape. We find that in this way the soliton beam can propagate for up to 20 μm.

Finally, we consider the beam propagation in the tapers with the tapering angles larger than the optimal angle. In this case, the beam width narrows with propagation, and the energy becomes focused in the lateral dimension as well, causing the plasmon amplitude to increase even more than being expected in the two-dimensional case [13]. Therefore, the energy focusing due to tapering and spatial beam narrowing with nonlinear amplitude concentration can be observed simultaneously.

V. CONCLUSIONS

We have analyzed tunability of nonlinear metamaterials operating at microwave and optical frequencies. More specifically, we We have analyzed the near-field coupling within metamaterials, considering both the relative orientation and the offset between the centers of two neighboring resonators. Using a pair of split-ring resonators as a simple model, we have shown the coupling mechanisms at work in our recently proposed tuning scheme, based on the direct calculation of the interaction energy. We have confirmed that these mechanisms can predict qualitatively the performance of a realistic metamaterial structure. This paves a road towards a reliable design and development of tunable metamaterials for various applications We have studied also fundamental nonlinear effects in plasmonic structures including guided modes of nonlinear slot waveguides, and discuss self-focusing of a plasmon beam and soliton formation in a taper in the presence of losses.

ACKNOWLEDGEMENTS

I acknowledge numerous contributions of my colleagues and students at the Nonlinear Physics Center, as well as my collaborators. This work was supported by the Australian Research Council.

REFERENCES

[1] V. G. Veselago, "The electrodynamics of substances with simultaneously negative values of ϵ and μ", Sov. Phys. Uspekhi **10**, 509-514 (1968).

[2] M. G. Silveirinha and C. A. Fernandes, Nonresonant structured material with extreme effective parameters, Phys. Rev. B **78**, 033108 (2008).

[3] R. W. Ziolkowski, Propagation in and scattering from a matched metamaterial having a zero index of refraction, Phys. Rev. E **70**, 046608 (2004).

[4] 13.J. B. Pendry, A. Holden, D. Robbins, and W. Stewart, "Magnetism from Conductors and Enhanced Nonlinear Phenomena," IEEE Trans. Microwave Theory Tech. **47**, 2075-2084 (1999).

[5] M. Gorkunov and M. Lapine, "Tuning of a nonlinear metamaterial band gap by an external magnetic field", Phys. Rev. B **70**, 235109 (2004).

[6] A.A. Zharov, I.V. Shadrivov, and Yu.S. Kivshar, "Nonlinear properties of left-handed metamaterials," Phys. Rev. Lett. **91**, 037401 (2003).

[7] D. Powell, I. Shadrivov, Yu. Kivshar, and M. Gorkunov, "Self-tuning mechanisms of nonlinear split-ring resonators," Appl. Phys. Lett. **91**, 144107 (2007).

[8] I. Shadrivov, A.Kozyrev, D.van der Weide, Yu.Kivshar, "Tunable transmission and harmonic generation in nonlinear metamaterials", Appl. Phys. Lett. **93**, 161903 (2008).

[9] D. Powell, I. Shadrivov, and Yu. Kivshar, "Nonlinear electric metamaterials", Appl. Phys. Lett. **95**, 084102 (2009).

[10] J. Valentine, S. Zhang, T. Zentgraf, E. Ulin-Avila, D.A. Genov, G. Bartal, and X. Zhang, "Three-dimensional optical metamaterial with a negative refractive index", Nature **455**, 376-379 (2008).

[11] A.J. Babadjanyan, N.L. Margaryan, and Kh. V. Nerkararyan, "About the nature of increase of the nonlinear optical response of a rough surface", J. Appl. Phys. **87**, 3785-3788 (2000).

[12] M.I. Stockman, "Nanofocusing of optical energy in tapered plasmonic waveguides", Phys. Rev. Lett. **13**, 137404 (2004).

[13] D.K. Gramotnev, "Adiabatic nanofocusing of plasmons by sharp metallic grooves: geometrical optics approach," J. Appl. Phys. **98**, 104302 (2005).

[14] A.R. Davoyan, I.V. Shadrivov, Yu.S. Kivshar, and D.K. Gramotnev, "Optimal tapers for compensating losses in plasmonic waveguides", Phys. Status Solidi RRL 1-3 (2010).

[15] E. Feigenbaum and M. Orenstein, "Plasmon-soliton", Opt. Lett. **32**, 674-676 (2007).

[16] A.R. Davoyan, I.V. Shadrivov, and Yu.S. Kivshar, "Self-focusing and spatial plasmon-polariton solitons", Opt. Express **17**, 21732-21736 (2009).

[17] A.R. Davoyan, I.V. Shadrivov, A.A. Zharov, D.K. Gramotnev, and Yu.S. Kivshar, "Nonlinear nanofocusing in tapered plasmonic waveguides", submitted to Phys. Rev. Lett (2010).

Strong turbulence analysis of sinusoidal and hyperbolic Gaussian beams

(Invited *Paper*)

H. T. Eyyuboğlu

Çankaya University, Ankara, Turkey

Abstract: We formulate the on-axis asymptotic scintillations of sinusoidal and hyperbolic Gaussian beams, namely cos and cosh Gaussian beam, under strong turbulence conditions. Our results are displayed in the form of plots of the scintillation index against propagation distance and wavelength at several source and propagation parameters. It is revealed that at small source sizes, cosh Gaussian beams will exhibit the lowest scintillations, but at large source sizes, this will reverse and cos Gaussian beams will attain such characteristics. It is further found that in strong turbulence, longer wavelengths will offer less scintillations, a behavior similar to the one, encountered in weak scintillation conditions.

INTRODUCTION

Turbulence is created by the random refractive index variations in the atmosphere due to wind and temperature changes. This in turn causes the intensity fluctuations in the received optical beam, leading to scintillations whose scale is measured by a quantity called scintillation index, basically corresponding to the statistical variance. Strong turbulence arises in cases when the Rytov variance largely exceeds unity. This particular variance is given by $\sigma_R^2 = 1.23 C_n^2 k^{7/6} L^{11/6}$, where C_n^2 is refractive index structure constant, $k = 2\pi / \lambda$ is the wave number with λ standing for the wavelength, and L is the propagation distance. Hence it is possible to create strong turbulence circumstances by appropriately adjusting C_n^2, λ, L so that $\sigma_R^2 \gg 1$. In practical cases however, we are concerned with strong turbulence, if long distances are involved in optical communications.

Some literature exists for strong turbulence. In particular a separate chapter in [1] is devoted to this subject. The other sources that may be listed are; in [2], over a horizontal path of 11.8 km, scintillation measurements, including the strong scintillation cases were made and with the data that was collected experimentally, it was proven that probability density function of the normalized intensity nicely fitted the theoretically expected form of negative exponential. It was shown in [3] that the saturation value of scintillation index approached unity for any degree of source coherence in strong fluctuation conditions. Reference [4] reports the results of scintillation measurements in moderate to strong turbulence conditions, also assessing the effects of inner and outer scale of turbulence parameters. Several other researchers may be added to this list, but these works almost entirely cover plane, spherical and Gaussian beam waves. The present work makes an original contribution in this sense, that it extends the strong turbulence theory to other beam types such as sinusoidal and hyperbolic Gaussian beams.

FORMULATION

A sinusoidal and hyperbolic Gaussian beam placed on a transverse input plane having the radial and angular coordinates s and θ can be constructed from the summation of two beams such that

$$x(s,\theta) = \sum_{i=1}^{2} \exp\left(-s^2 / \alpha_s^2\right) \exp\left[js\left(\cos\theta + \sin\theta\right)D_i\right] \quad (1)$$

where α_s is named the Gaussian source size, $j = (-1)^{0.5}$, D_i is known as the displacement parameter whose adjustment (together with the amplitude factor, excluded here) will produce all types sinusoidal and hyperbolic Gaussian beams [5]. But since the source fields of sine and sinh beams pass though zero around on-axis and this situation does not change even upon propagation in free space, thus creating ambiguities in the evaluation Rytov based scintillation index, we restrict our attention in this study to cos and cosh types of sinusoidal and hyperbolic Gaussian beams. To get to the strong turbulence scintillation index expression for the beams defined in (1), first we have to derive the weak turbulence version. The procedure for this is described in several previously published works [6-8]. Below we pursue the same steps.

The free space field on an output plane which is separated from the input plane with an axial distance of L, can be found by applying Huygens Fresnel integral to (1) which will result in

$$y(r,\phi,L) = \sum_{i=1}^{2} \frac{k\alpha_s^2}{k\alpha_s^2 + 2jL}$$

$$\times \exp\left[-\frac{kr^2 - jk\alpha_s^2 r\left(\cos\phi + \sin\phi\right)D_i + j\alpha_s^2 D_i^2 L}{k\alpha_s^2 + 2jL}\right] \quad (2)$$

where r and ϕ denote the output plane coordinates. A function $Q(\kappa)$ is obtained from (2) via the following expression

$$Q(\kappa) = \frac{k^2}{2\pi(L-\eta)y(r,\phi,L)}\exp\left[\frac{jkr^2}{2(L-\eta)}\right]$$

$$\times \int_0^\infty dr_1 \int_0^{2\pi} d\phi_1 r_1 y(r_1,\phi_1,\eta)\exp\Big\{j\kappa r_1 \cos(\phi_1 - \psi)$$

$$+ \frac{jk}{2(L-\eta)}\left[r_1^2 - 2r_1 r\cos(\phi_1 - \phi)\right]\Big\} \quad (3)$$

where κ and ψ indicate the magnitude and the angular orientation of spatial frequency. Upon substituting in (3) for the output plane field $y(\ )$ from (2) and performing the required integrations, $Q(\kappa)$ will develop into

$$Q(\kappa) = \exp\left\{\frac{j\kappa\left(k\alpha_s^2 + 2j\eta\right)\left[kr\cos\left(\psi - \phi\right) - 0.5\kappa\left(L - \eta\right)\right]}{k\left(k\alpha_s^2 + 2jL\right)}\right\}$$

$$\times jk\sum_{i=1}^{2}\exp\left\{-j\frac{\alpha_s^2 D_i}{k\alpha_s^2 + 2jL}\left[\kappa\left(L - \eta\right)\left(\cos\psi + \sin\psi\right)\right.\right.$$

$$\left.- kr\left(\cos\phi + \sin\phi\right) + D_i L\right]\right\}$$

$$\times\left\{\sum_{i=1}^{2}\exp\left[j\frac{kr\alpha_s^2\left(\cos\phi + \sin\phi\right)D_i - \alpha_s^2 D_i^2 L}{k\alpha_s^2 + 2jL}\right]\right\}^{-1} \quad (4)$$

Finally, by utilizing $Q(\kappa)$ in the following manner,

$$b_{wt} = 4\pi\int_0^\infty d\kappa\int_0^L d\eta\int_0^{2\pi}d\psi\kappa\Phi(\kappa)$$

$$\times\left\{\left|Q(\kappa)\right|^2 + \mathrm{Re}\left[Q(\kappa)Q(-\kappa)\right]\right\} \quad (5)$$

we arrive at the scintillation index, b_{wt} for weak turbulence conditions. In (5), $\Phi(\kappa)$ signifies the spectrum function, $|\ |$ means the absolute value, $\mathrm{Re}[\ ]$ is the real part of the enclosed expression. From (5), it is possible to write the strong turbulence scintillation index, b_{st} as shown below [1],

$$b_{st} = 1 + 16\pi\int_0^\infty d\kappa\int_0^L d\eta\int_0^{2\pi}d\psi\kappa\Phi(\kappa)$$

$$\left\{\left|Q(\kappa)\right|^2 + \mathrm{Re}\left[Q(\kappa)Q(-\kappa)\right]\right\}\exp(-I) \quad (6)$$

Here the argument in the newly added exponential, i.e. I acts as a spatial low pass filter incorporating the integration of the plane wave structure function of phase over the normalized distance variable. The related details are provided in [1]. By setting $\Phi(\kappa) = 0.033C_n^2\kappa^{-11/3}\exp\left(-0.0285l_0^2\kappa^2\right)$, that is Tatarskii spectrum with l_0 indicating the inner scale of turbulence, I comes out to be

$$I = 1.0933C_n^2\kappa^2\eta^3 l_0^{-1/3}\left(1 - \frac{\eta}{R_L}\right)^2$$

$$\times {}_2F_1\left[1/6, 1.5, 2.5, -2.033\left(\frac{\kappa\eta}{kR_L}\right)^2\left(\frac{R_L - \eta}{l_0}\right)^2\right]$$

$$-1.0933C_n^2\kappa^2\eta^2 l_0^{-1/3}R_L\left\{\left(1 - \frac{L}{R_L}\right)^3\right.$$

$$\times {}_2F_1\left[1/6, 1.5, 2.5, -2.033\left(\frac{\kappa\eta}{kl_0}\right)^2\left(1 - \frac{L}{R_L}\right)^2\right]$$

$$-\left(1 - \frac{L}{R_L}\right)^3 {}_2F_1\left[1/6, 1.5, 2.5, -2.033\left(\frac{\kappa\eta}{kl_0}\right)^2\left(1 - \frac{\eta}{R_L}\right)^2\right]\right\} \quad (7)$$

where ${}_2F_1(\ )$ is the hypergeometric function, R_L represents the free space radius of curvature for hyperbolic and sinusoidal beams given in (1). R_L was formulated in [9] for such beams under atmospheric turbulence, from there, we deduce the free space version by letting $C_n^2 \to 0$, thus

$$R_L = \sum_{i_1=1}^{2}\sum_{i_2=1}^{2}\left\{0.25k^2\alpha_s^2\left[1 - 0.25\left(D_{i_1} - D_{i_2}^*\right)^2\alpha_s^2\right]\right.$$

$$+\left[\alpha_s^{-2} + 0.25\left(D_{i_1} + D_{i_2}^*\right)^2\right]L^2\right\}\exp\left[-0.25\left(D_{i_1} - D_{i_2}^*\right)^2\alpha_s^2\right]$$

$$\times\left\{\sum_{i_1=1}^{2}\sum_{i_2=1}^{2}\left[\alpha_s^{-2} + 0.25\left(D_{i_1} + D_{i_2}^*\right)^2\right]L\right.$$

$$\times\exp\left[-0.25\left(D_{i_1} - D_{i_2}^*\right)^2\alpha_s^2\right]\right\} \quad (8)$$

Finally, by inserting in (6) from (4), (7) and (8), the strong turbulence index, b_{st} can be found for beams of (1). Note that in this process, the triple integral of (6) can be reduced to double integral by analytically solving the integration over ψ. Despite these rearrangements however, the end result becomes too lengthy, hence not presented here for space considerations.

RESULTS AND DISCUSSIONS

From [1], we know that (6) gives an asymptotic representation of scintillation index in the strong (saturation) regime. Thus, in the accompanying plots C_n^2, k and L are selected such that $\sigma_R^2 \gg 1$ is always satisfied. To this end, C_n^2 is commonly set to 10^{-13} m$^{-2/3}$ and λ is taken as 1.55 µm except the last figure where the variation of scintillation index against wavelength is analyzed. For the spectrum function $\Phi(\kappa)$, l_0 is taken to be 1 mm.

Fig.1 demonstrates the behavior of scintillation index at a source size value of $\alpha_s = 1$ cm. The falling trend in this figure against propagation distance is in conformity with the existing literature [1, 4]. Furthermore, Fig. 1 illustrates that at shorter propagation distances compared with the Gaussian beam, cosh Gaussian beam will have slightly less scintillations, while the reverse will be valid for cos Gaussian beams. For both cosh and cos Gaussian beams, this deviation will become more pronounced with rising values of the displacement parameter D. At longer propagation distances however, the scintillations of all beams merge toward the same value as seen in Fig. 1.

Switching to a source size of $\alpha_s = 4$ cm as done in Fig. 2, we note that the falling trend of the scintillation index against propagation distance continues almost unchanged, but in the case of larger source size, firstly the scintillation levels have been raised, and secondly the respective positions of cosh and cos Gaussian beams have shifted. That is, now cosh Gaussian beams have more scintillations than the Gaussian beams and the scintillations of cos Gaussian beams are below the scintillations of Gaussian beams.

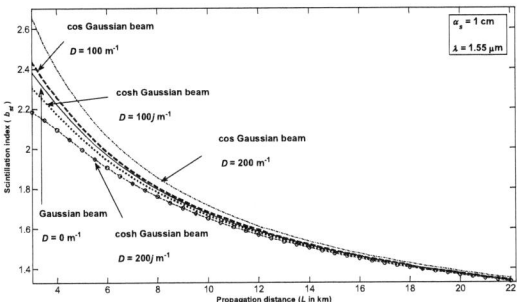

Fig. 1. Variation of scintillation index against propagation distance at source size of $\alpha_s = 1$ cm.

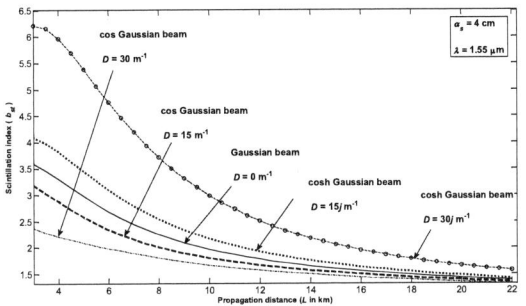

Fig. 2. Variation of scintillation index against propagation distance at source size of $\alpha_s = 4$ cm.

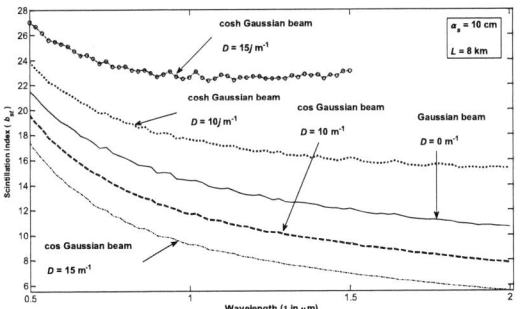

Fig. 3. Variation of scintillation index against wavelength at source size of $\alpha_s = 10$ cm and $L = 8$ km.

Finally we examine the variation of scintillation index against wavelength in Fig. 3, where we set the source size and propagation distance to 10 cm and 8 km respectively. From the collective assessment of Figs. 1, 2 and 3, it is concluded that at such large source sizes, scintillations have increased considerably. Furthermore, according to Fig. 3, the status of cosh Gaussian beams possessing the highest amount of scintillations at large source sizes, has not been altered. From the curves of Fig. 3, we gather that scintillations in strong turbulence are reduced at longer wavelengths regardless of beam types. Such a behavior appears to be in conformity with the one found in weak turbulence regime [6].

CONCLUSION

In this study, we have investigated the scintillation behavior of sinusoidal and hyperbolic Gaussian beams in strong turbulence. For this purpose we have derived the relevant formulation and presented our numeric evaluations in the form of graphs. Our results show that at small source sizes, cosh Gaussian beams have the smallest scintillations, whereas at large source sizes, this advantage is taken over by cos Gaussian beams. But at large source sizes, all beams produce higher levels of scintillations than those at smaller source sizes. When plotted against wavelength, we find that, scintillation index falls with increasing wavelength, a behavior inline with the one observed in weak turbulence conditions.

ACKNOWLEDGMENT

This work was performed within the framework of the Scientific and Technological Research Council of Turkey (Tübitak) grant number 108E130. The author gratefully acknowledges the support provided by Çankaya University and Tübitak.

REFERENCES

[1] L. C. Andrews, R. L. Phillips, *Laser Beam Propagation through Random Media*, 2nd Ed., SPIE Optical Engineering Press, Bellingham, Washington, 2005.

[2] Y. Jian, J. Ma, L. Tan, S. Yu, W. Duf, "Measurement of optical fluctuation over an 11.8 km turbulent path" Opt. Express, vol. 16, no. 10, pp. 6963-6973, 2008.

[3] S. C. H. Wang, M. A. Plonus, C. F. Ouyang, "Irradiance scintillations of partially coherent source in extremely strong turbulence", Appl. Opt. vol. 18, no. 8, pp. 1133-1135, 1979.

[4] F. S. Vetelino, B. Clare, K. Corbett, C. Young, K. Grant, L. Andrews, "Characterizing the propagation path in moderate to strong optical turbulence" Appl. Opt. vol. 45, no. 15, pp. 3534-3543, 2006.

[5] Ç. Arpali, C. Yazıcıoğlu, H.T. Eyyuboğlu, S. A. Arpali, Y. Baykal, "Simulator for general-type beam propagation in turbulent atmosphere", Opt. Express, vol. 14, no. 20, pp. 8918-8928, 2006.

[6] H. T. Eyyuboğlu, Y. Baykal, "Scintillations of cos-Gaussian and annular beams", J. Opt. Soc. Am. A, vol. 24, no. 1, pp. 156-162, 2007.

[7] H. T. Eyyuboğlu, Y. Baykal, E. Sermutlu, Y. Cai, "Scintillation advantageous of lowest order Bessel-Gaussian beams", Appl. Phys. B, vol. 92, no. 2, pp. 229-235, 2008.

[8] H. T. Eyyuboğlu, Y. Baykal, X. Ji "Scintillations of Laguerre Gaussian beams", Appl. Phys. B, vol. 98, no. 4, pp. 857-863, 2010.

[9] H. T. Eyyuboğlu, X. Ji, "An analysis on radius of curvature aspects of hyperbolic and sinusoidal Gaussian beams", Appl. Phys. B, DOI 10.1007/s00340-010-4039-1.

CAOL*2010 International Conference on Advanced Optoelectronics & Lasers, 10-14 September, 2010, Sevastopol, Ukraine

Evolution of nanophotonics from semiconductor photonic crystal device to metal/semiconductor plasmonic device

(Invited Paper)

K. Asakawa[1], Y. Sugimoto[2], N. Ikeda[2], D. Tsuya[2], Y. Koide[2], Y. Watanabe[2], N. Ozaki[3],
D. Kumar V.[4], T. Nomura[5], D. Inoue[5], A. Miura[5], H. Fujikawa[5] and K. Sato[5]

[1] University of Tsukuba, Tsukuba, Japan

[2] National Institute for Materials Science, Tsukuba, Japan

[3] Wakayama University, Wakayama, Japan

[4] Indian Institute of Information Technology Design and Manufacturing, Jabalpur, India

[5] Toyota Central R&D Labs. Inc., Nagakute, Japan

Abstract:. This paper reviews our recent activities on nanophotonics based on a photonic crystal (PC)/quantum dot (QD)-combined structure for an all-optical device and a metal/semiconductor composite structure using surface plasmon (SP) and negative refractive index material (NIM). The former structure contributes to an ultrafast signal processing component by virtue of new PC design and QD selective-area-growth technologies, while the latter provides a new RGB color filter with a high precision and optical beam-steering device with a wide steering angle.

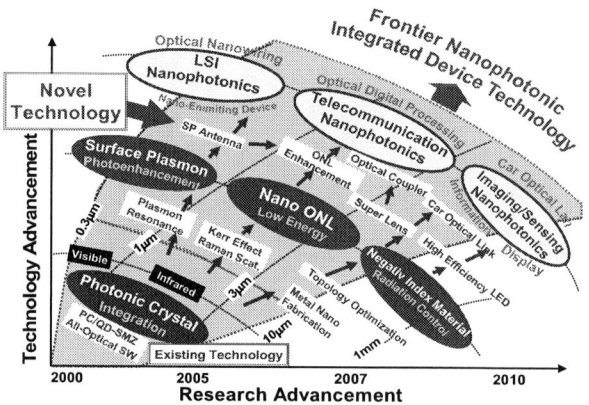

Fig. 1 Key roles, their relevance and road map in our nano-photonics research.

1. Introduction

For several decades, photonic technologies have devoted to optical signal transmission, storage and processing in the optoelectronic community. Their integration technologies have also been developed since 1970's as represented by opto-electronic integrated circuits (OEICs) in 1980's, planar light wave circuits (PLCs) in 1990's, and photonic crystal (PC)-based waveguide in 2000's [1]. Research on a quantum dot (QD) was also intensively advanced [2]. Recently, appearance of surface plasmon (SP) [3] and negative refractive index material (NIM) [4] have stimulated such tend towards a nano-photonics containing integrated circuit. Figure 1 shows our nano-photonics roadmap which predicts potential abilities of PC, QD, SP and NIM technologies towards new fields such as silicon photonics in LSI nano-electronics and novel imaging/ sensing applications as well as advanced photonics for next generation photonic network system. Up to date, we have clarified that optical nonlinearity (ONL) in the QD contributes to a key role of a PC-based all-optical switch, while SP and NIM have potential abilities for our original optical nano-antenna application, novel RGB color filter and optical beam-steering device. This paper reviews such recent nano-photonic technologies and their applications.

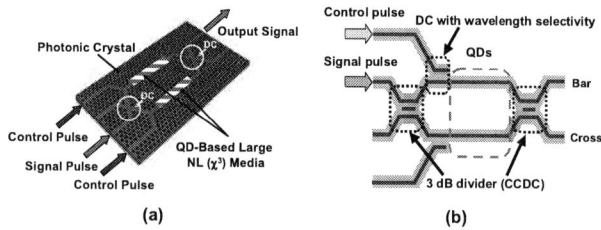

Fig.2 (a) Schematic diagram of the PC-SMZ. (b) Waveguide configuration in the PC-SMZ.

2. Photonic crystal waveguide

The PC, having a wide band gap and strong dispersion [1], exhibits strong optical confinement and light/matter interaction in the defect waveguide. The QD, having a high density-of-state [2], exhibits large ONL under low pumping energy. We

have proposed all-optical switches with a PC and QD combined symmetrical Mach-Zehnder (SMZ) configuration [5], named PC-SMZ (analogue switch) and PC-FF (digital flip flop switch) [6]. Figures 2 (a) and (b) show a schematic of the PC-SMZ and its PC patterns with several directional couplers (DCs), respectively. The PC waveguide is composed of air-bridge type slab waveguides with single line defect, as shown

978-1-4244-7043-3/10 $26.00 © 2010 IEEE

in Figs. 3 (a) and (b). A solid line circle in Fig. 3 (c) shows a cross section of about 0.7-μm-wide and 0.25-μm-thick single-line defect PC waveguide [7, 8].

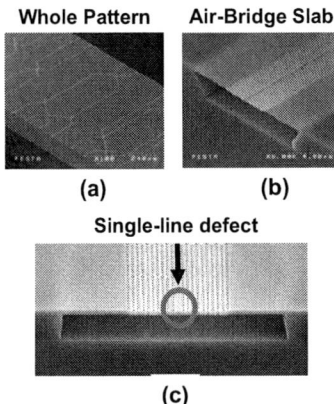

(a)   **(b)**

Single-line defect

(c)

Fig. 3 (a) SEM photograph of the whole PC-SMZ pattern. (b) Cleaved edge of the air-bridge PC waveguide. (c) Magnified photograph of the cleaved PC waveguide.

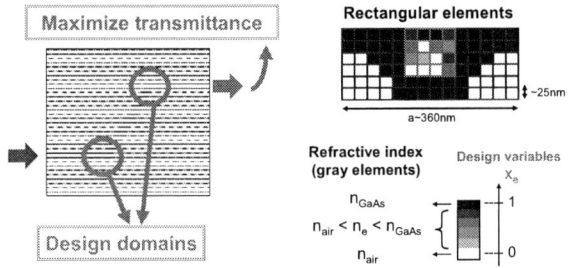

Fig. 4 Computational model of the TO design including the design domain.

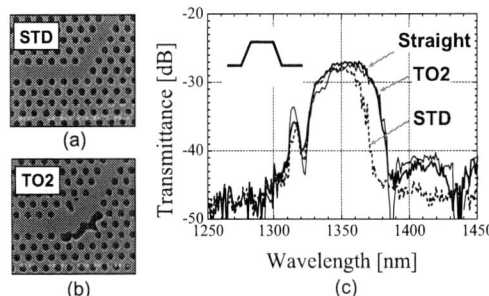

Fig. 5 (a) and (b) SEM images of fabricated bend waveguides designed with standard (STD) and TO (TO2) methods, respectively. (c) Their measured transmittance spectra compared with that for the straight waveguide.

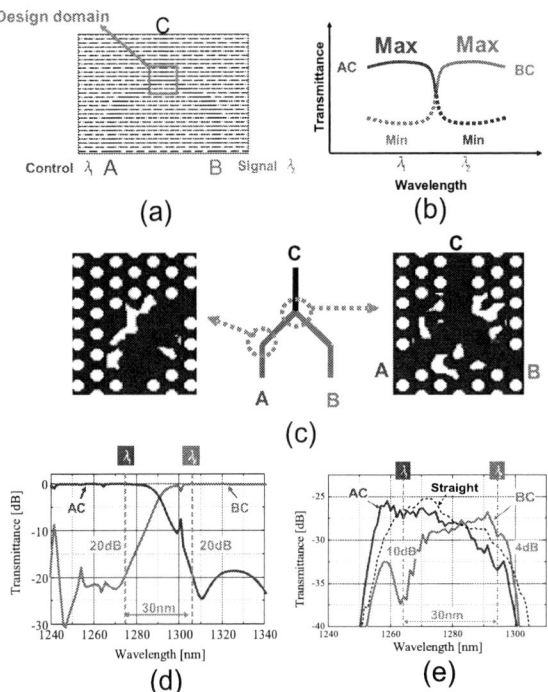

Fig. 6 (a) Computational model of the TO-designed Y-junction. (b) Schematic picture of the required asymmetric transmission spectra. (c) SEM images of the bend (left) and Y-junction (right) parts in the fabricated TO-designed waveguides. (d) and (e) Respective calculated and measured transmission spectra of the channels AC and BC in the TO-designed Y-junction.

For design of the PC-SMZ with wide/flat-band waveguides, a new design method called topology optimization (TO) has been developed [9,10], as shown in Fig. 4. The TO procedure is performed to maximize transmittance in a straightforward line (indicated in arrows) by modifying refractive index distributions in the design domain indicated by solid circles. The TO design has so far been applied to intersection [11], bend [12], wavelength-selective Y-junction [13] and DC waveguides. Figures 5 (a) and (b) show SEM photographs of fabricated bends designed by the standard (STD) and TO (TO2) methods. As shown by the measured transmittance spectra in Fig. 5 (c), the TO bend shows quite the same transmittance spectra as a straight waveguide, as shown in a thin solid line. In addition, an abrupt decrease of the transmittance in the long wavelength, observed in the STD bend, is not seen in the TO bend. These results verify the dramatic improvement in the TO bend.

The TO method is effective for designing the wavelength-sensitive Y-junction also, as shown in Fig. 6. As shown in Figs. 6 (a) and (b), the TO method was adopted to (1) maximize the transmittances along AC and BC channels, (2) minimize the transmittance along unwanted AB and BA channels, and (3) optimize them at two separate (by 30 nm in this case) wavelengths. After the TO design, unique air-hole patterns appear, as shown in Fig. 6 (c). Comparison of calculated transmission spectra between AC and BC channels at 30-nm-apart wavelengths result in 20 dB isolation, as shown in Fig. 6 (d). Figure 6 (e) shows corresponding measured transmission spectra along the channels AC and BC. The results almost reproduce the calculation mentioned above, although the isolation is not enough. The result is found to be effective for the wavelength-selective Y-junction used in the PC-FF.

3. Quantum dot ONL materials
3.1 Selective area growth of quantum dot

Recent development of self-assembled InAs QDs has attracted a great deal of attention because of its potential for use in telecommunication systems. As indicated in Fig. 2 (a), the PC-SMZ requires InAs QDs partially embedded in the GaAs PC waveguide. In addition, a digital all-optical switch, PC-FF, promising in the future optical digital processing system, as shown in Fig. 7, requires two separate QD sites, that is, QD ensemble with the wavelength $\lambda_1= 1.29$ μm in SW-1 and QD ensemble with wavelength $\lambda_2= 1.31$ μm in SW-2 [6]. For these purposes, selective-area-growth (SAG) of QDs has been developed by using a metal-mask (MM) method [14, 15]. The MM method enables us to grow the QD ensemble selectively in a required area of several tens- to several hundreds-μm in size. Wavelength of the SAG-QD can be controlled by the thickness of an inserted InGaAs layer as a stress reducing layer (SRL) between the InAs-QD and GaAs-spacer layer [16]. Figure 8 (a) shows a typical atomic force micrograph (AFM) image of the QD ensemble on the unmasked regions grown by the MM method. The QD density is as high as $4 \times 10^{10} cm^{-2}$. QD uniformity is estimated to be 28 meV in term of FWHM (full width at half maximum) in the PL peak, as shown in Fig. 8 (b). Figure 8 (c) shows a cross-sectional SEM photograph of the PC GaAs air-bridge waveguide embedded with four-InAs-QD layers.

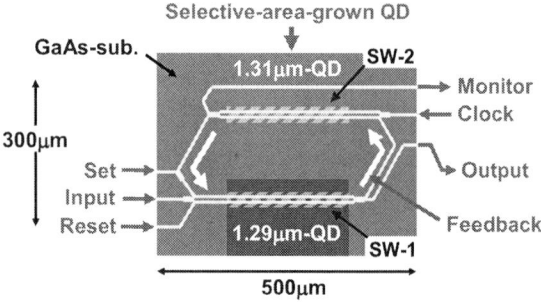

Fig. 7 Configuration of the PC waveguide and areas of QDs with different absorption wavelengths in the PC-FF

The MM is designed as to be 180° rotatable on the MBE wafer holder. After the QD1 is grown, an $In_{0.2}Ga_{0.8}As$ SRL is deposited on the QD and capped with a GaAs layer. Then, the MM is rotated by 180° and the QD2 is grown in the same manner except for the thickness of the SRL. For instance, the SRL thickness is set to 2.7 nm for QD1 and 4.5 nm for QD2. Figure 9 (a) shows a view of the PL intensity mapping for the neighboring QD1 and QD2, while Fig. 9 (b) shows PL spectra from the corresponding QD1 and QD2 [17]. Thus, wavelength difference ($\Delta\lambda$) by ~ 20 nm can be realized as designed. These techniques are suitable for implementation of the PC-FF. By using these technologies, design and fabrication of an actual PC-FF device is under implementation. Optical flip flop operation requires improvement of both the PC waveguide design and QD growth-area size as well as a practical coupling method to an external optical fiber and easy packaging. However, this technique will pave the road to a key optical digital node device in a future ultra-fast and energy-saving photonic network system.

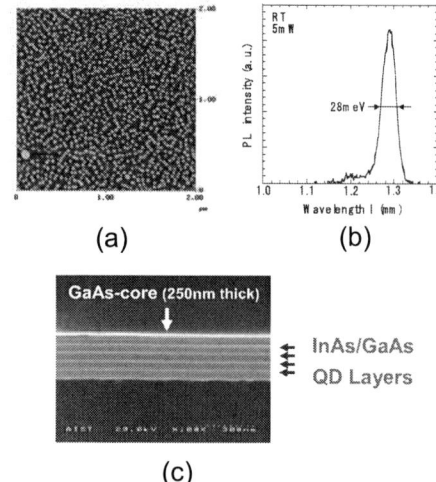

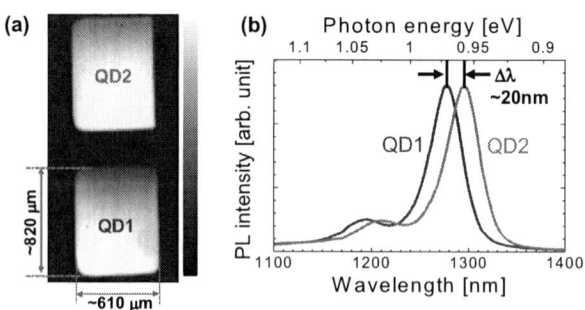

Fig. 8 (a) AFM image showing high density QDs. (b) PL spectrum showing highly uniform QDs. (c) SEM photograph of a cross-sectional view of GaAs PC waveguide embedded with four InAs QD layers.

Fig. 9 (a) PL intensity mapping of QD1 and QD2 grown separately before and after 180° rotation of the MM. (b) PL spectra of QD1 and QD2 exhibiting different emission peaks at 1276 and 1296 nm.

3.2 Ultra-fast operation of PC/QD all-optical switch

An optical switching response of the PC-SMZ is shown here. In 2004, a successful switching operation of the PC-SMZ pumped by a set of on-pulse and off-pulse was reported [18, 19, 20]. Resultant rise and fall times were as fast as ~2ps and input pulse energy was as low as 100fJ. Here, modulation (switching ratio) was 50%, half the desired value (100%) due to the deficient phase shift $\Delta\phi$ of $\pi/2$ and not $\tilde{\pi}$. Nevertheless, high repetition-rate operation of the PC-SMZ has been recently demonstrated by using a four-pulse train with a period of 25ps, as shown in Fig. 10 (a). Operations with repetition frequencies of 40 and 20 GHz were demonstrated, as shown in Figs. 10 (b) and (c), respectively [21]. A switching ratio was as low as 16 %. A possible improvement of the switching ratio is discussed next.

In Table 1, the left-hand column shows summarized performances of the PC-SMZ at 40 GHz operation. The middle column shows design parameters necessary for 100% modulation available by the current techniques. Here, $\Delta\phi$ means phase shift over the QD arms induced by the ONL in QDs, while n_g means a group index as defined by the reciprocal of the group velocity in the PC waveguide. The waveguide was designed as to achieve the low group velocity for enhancing light/matter interaction, thus increasing the ONL at the QD arms [22]. In the new design, 100% modulation is to be achieved at 40 GHz by increasing the QD layer, n_g and QD arm. If the ONL of QDs will be enhanced furthermore by other means in nano-scale, 100% switching will be possible with much reduced size. The right-hand column shows expected values of the above-mentioned items in the PC-SMZ when the ONL in QDs is enhanced with the aid of surface plasmonic optical nano-antenna. The possibility of this scheme is discussed in the next section.

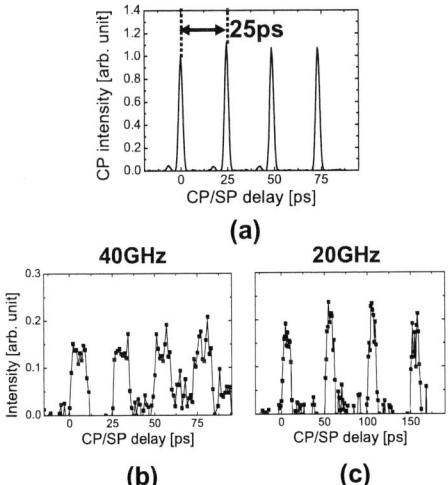

Fig. 10 (a) Input four-pulse train with the period of 25 ps used for sequential operation. (b) ~ (c) Output pulses for sequential switching operation of the PC-SMZ with repetition frequencies of 40 and 20 GHz, respectively.

Table 1 Summarized performances and future design parameters of the PC-SMZ.

QD state	QD in PC		QD with SP
PC/QD structure	Result @ 40GHz	New Design @ 40GHz	Optical nano-antenna
Modulation	16%	100%	100%
Δf	30°	180°	180°
QD layer	3	5	New Criteria
n_g	7	21	
QD arm	400μm	500μm	~20μm

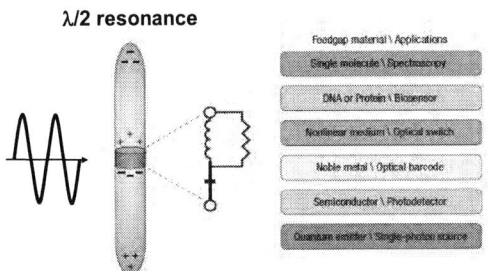

Fig. 11 Fundamental structure (left) of an optical nano-antenna and several items for possible applications (right).

4. Surface-plasmonic material
4.1 Plasmonic optical nano-antenna

Surface plasmon (SP) is a light wave propagating at the metal/dielectric boundary as a result of coupling to the electron motion in the metal. Since energy of the SP is confined in the interface, enhanced ONL in nano-scale and other interesting phenomena and applications have been reported recently. One of the applications of the SP is an optical nano-antenna (ONA), as shown in Fig. 11 [23]. When the light is incident on the ONA, a split metal bar with a narrow feed gap and total length of a half-wavelength plays a dipole antenna like the Yagi antenna. At the resonance condition, incident light extremely enhances e-field in the vicinity of the feed gap, thus resulting in strong light/matter interaction and produces possible unique applications. Insertion of DNA of protein or semiconductor produces a highly sensitive biosensor or photo-detector, while insertion of nonlinear medium or quantum emitter enables a low-energy optical switch or high-efficiency single-photon source, as indicated in Fig. 11.

Figure 12 (a) shows a proposed scheme of mounting an ONA on the surface of the PC waveguide with QDs embedded near the ONA. It is expected that the propagating light wave, incident into the ONA, causes strong e-field at the ONA feed gap and enhance ONL of the QDs in the PC waveguide. For quantitative estimation of such a scheme, an e-field enhancement factor at the edge of a metal ONA was calculated, resulting in Fig. 12 (b). Calculation was done at the normal incident configuration. Figures 12 (c) and (d) show calculated e-field distributions at the outer and center planes, respectively. In this way, e-field enhancement at the edge of the feed gap is estimated to exceed 1,200 and 300 in the center and outer planes, respectively, as shown in Fig. 12 (e) [24]. The high e-field will be expected to enhance the ONL of the QD near the ONA, thus positively supporting the scenario of enhancing the phase shift in the PC-SMZ under the low pumping energy, as suggested in Table 1.

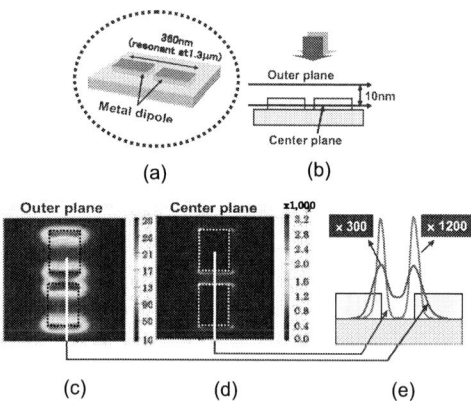

(a)   (b)

(c)   (d)   (e)

Fig. 12 (a) ONA on the surface of the PC waveguide. (b) Cross section of the ONA on the waveguide. (c) and (d) Calculated e-field distributions at the outer and center planes in (b), respectively. (e) E-field enhancement at the edge of the feed gap in the center and outer planes.

4.2 Surface-plasmonic optical color filter

Another unique feature of the SP is an extra-ordinary transmission of light incident on a nano-scale via-hole on a metal/dielectric material [3, 25, 26]. This feature can be applied to an RGB color filter with high precision. Recent experimental results of such a device in an aluminum-film hole-array is shown here [27-29]. Figure 13 is a targeting picture. Usage of the aluminum enables RGB color filters covering a shorter wavelength as a result of a high plasma frequency than in use of silver and gold. In addition, nanofabrication of aluminum materials has been matured by the conventional semiconductor technologies.

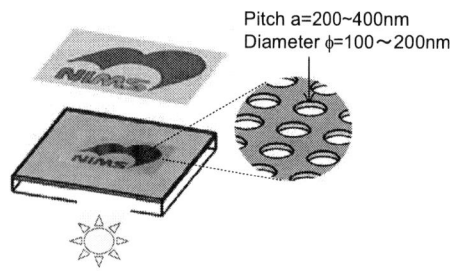

Fig. 13 Schematic picture of an RGB color filter comprising an aluminum-film hole-array.

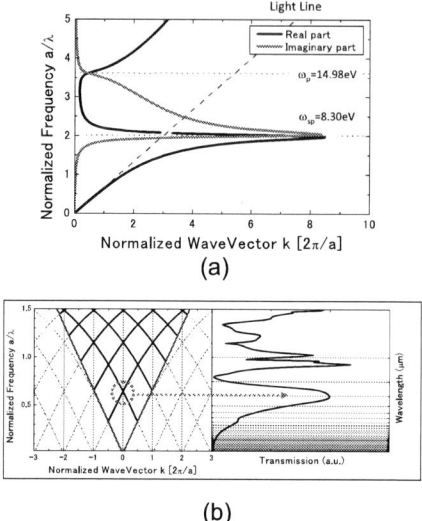

(a)

(b)

Fig. 14 (a) Dispersion relation of the SP in an aluminum/SiO$_2$ system. (b) Dispersion relation of the SP in a periodic structure (left) and transmission spectrum at the Γ point (right).

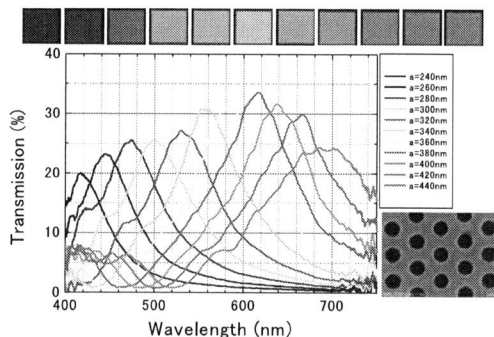

Fig. 15 Measured optical transmission spectra for samples with different lattice constants in an air-hole array

Figure 14 (a) shows a dispersion relation of the SP in an aluminum/SiO$_2$ system calculated by the Lorentz-Drude model. The SP generates in the frequency range below ω_{sp} (8.30ev), while $\omega_{sp} < \omega < \omega_p$ (14.98eV) and $\omega_p < \omega$ correspond to absorptive and transparent regions for the light, respectively. In the periodic structure, on the other hand, the SP is subject to a band diagram as a result of Bloch function, as shown at the left hand side in Fig. 14 (b). In the figure, the transmission spectrum at the Γ point is shown also. The large peak indicated by the dashed arrow corresponds to the energy of the SP used in the experiment. Samples were triangle-lattice air-hole arrays with a unit circle hole pattern, where lattice constant a = 240~420nm (a/λ=0.6, λ =400~700nm). Final aluminum hole-array patterns on the glass were covered with SiO$_2$ layer. Figure 15 shows measured optical transmission spectra for samples with different lattice constants in an air-hole array. The light source was a halogen lamp (white light), while a detector was a CCD camera. A transmitted color (wavelength) has been found to agree with a calculated result in the band diagram. In this way, an SP induced RGB color filter with high color (wavelength) selectivity has been demonstrated for the first time.

5. Negative refractive index material with metal/dielectric composites

A negative refractive index material (NIM) is one of the recent topics in the nano-photonic field as indicated in Fig. 1 [4, 30-36]. Recently we proposed a planar-prism type beam steering device using NIM which is available in an infrared wavelength region (1.55 μm) [37]. The negative refractive index effect generates at a composite structure of multi-stacked aluminum (Al) metal layers embedded in the host SiO₂ medium. An air-hole array with sub-wavelength shapes in size is formed in the plane. The beam steering is caused by the air-hole shape changing side by side from circle to square in the steering direction. Figure 16 shows a schematic diagram of the device. The detailed structure is as follows. In the thickness direction, the structure has five stacked aluminium/SiO₂ composite layers. Each unit layer has 20nm-thick Al film and 80nm-thick SiO₂, thus forming alternative Al/SiO₂ layers with a period (d) of 100nm. Dispersion effect of this structure exhibits the negative refractive index behaviour. The secret of this device is that the dispersion is gradually controlled side by side due to the spatially changing air-hole shape. Figure 17 shows dispersion curves for stacked composite with 1 μm × 1 μm square and single air-hole in it. In order to form a graded index distribution, a curvature, R, of the square corner in the air-hole is gradually changed from 0 to 250nm, as shown by the different dispersion curves in Fig. 17. In the figure, when the normalized frequency is changed from f1 (0.99) to f3 (0.94), a radiated angle of the output beam is changed from 0° to 12.5°. In the lateral direction on the plane, such an element is arranged 10 by 10 with 1 μm pitch over the entire plane. For this structure, simulated results of the electromagnetic wave propagation at the corresponding frequencies are shown in Fig. 18. Beam steering behaviour in this structure is clearly shown. For practical application, output beam angle is changed by changing the refractive index of the liquid crystal infiltrated in the air-hole by applying the external electric of magnetic field. The planar prism is achieved in this way.

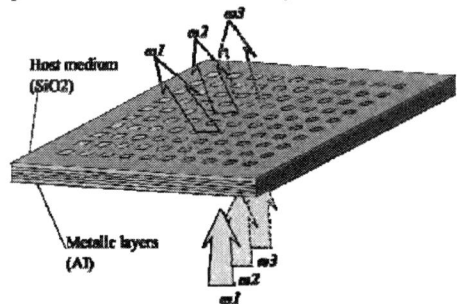

Fig. 16 Schematic diagram of the planar prism with graded NIM for near infrared beam steering

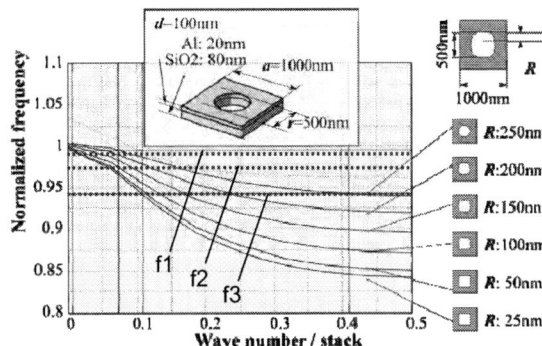

Fig. 17 Calculated band diagrams of the planar prism with varied curvatures of corners at the square holes.

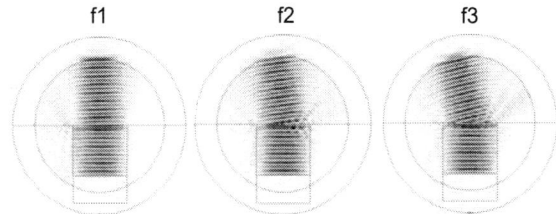

Fig. 18 Simulated electromagnetic wave propagations at the corresponding frequencies, f1, f2 and f3 in Fig. 17.

6. Conclusion

Our resent research on nano-photonics involving the PC, QD, SP and NIM has been reviewed. For PC/QD-combined devices, new topology optimization method for design of high performance PC waveguides and novel selective area growth method for QD-containing advanced photonic intergraded circuits have been developed and applied to ultrafast and low-energy all-optical switches using ONL in the QD-embedded waveguide. On the other hand, SP has been investigated for a potential application to an optical nano-antenna which enhances the ONL in the PC/QD waveguide and a novel plasmonic RGB color filter. Finally, negative refractive index in the stacked metal/dielectric composite with an air-hole array has been applied to an optical beam-steering device with a wide range of steering angle. These nano-photonics will pave the road to novel sensing/imaging components as well as new integrated signal processing components/circuits in a next generation photonic network system.

Acknowledgement

Authors would like to acknowledge kind contributions of A. Watanabe, S. Nakamura, S. Ohkouchi, Y. Nakamura, H. Nakamura, K. Kanamoto, Y. Tanaka, Y. Takata and Y. Kitagawa. Research works reviewed here were partly supported by the NEDO projects, Japan.

References

1. E. Yablonovitch, Phys. Rev. Lett., **58**, 2059-2062 (1987).
2. Y. Arakawa and H. Sakaki, Appl. Phys. Lett., **40**, 939 (1982).
3. T. W. Ebbesen, H. J. Lezec, H. F. Ghaemi, T. Thio and P. A. Wolff, Nature, **391**, 667-669 (1998).
4. J. B. Pendry, A. J. Holden, W. J. Stewart, I. Youngs, Phys. Rev. Lett. **76**, 4773 (1996). J. B. Pendry, A. J. Holden, D. J. Robbins and W. J. Stewart, IEEE Trans. Microwave Theory and Techniques, 47, 2075 (1999).
5. K. Tajima, Jpn. J. Appl. Phys., **32,** pp. L1746-L1748 (1993).
6. K. Asakawa, Y. Sugimoto, Y. Watanabe, N. Ozaki, A. Mizutani, Y. Takata, Y. Kitagawa, H. Ishikawa, N. Ikeda, K. Awazu, X. Wang, A. Watanabe, S. Nakamura, S. Ohkouchi, K. Inoue, M. Kristensen, O. Sigmund, P. I. Borel, and R. Baets, New Journal of Physics **8**, 1-26 (2006).
7. Y. Sugimoto, Y. Tanaka, N. Ikeda, H. Nakamura, K. Kanamoto, S. Ohkouchi, Y. Watanabe, K. Inoue, and K. Asakawa, IEEE Journal on Selected Areas in Communications, **23**, 1308-1314 (2005).
8. N. Ikeda, Y. Sugimoto, Y. Tanaka, K. Inoue, Y. Watanabe, and K. Asakawa, Semiconductor Science and Technology, **22**, 1-9 (2007).
9. P.I. Borel, L.H. Frandsen, A. Harpøth, J.B. Leon, H. Liu, M. Kristensen, W. Bogaerts, P. Dumon, R. Baets, V. Wiaux, J. Wouters and S. Beckx, Electronics Letters, **40**, 1263-1264 (2004)
10. J. S. Jensen, O. Sigmund, L.H. Frandsen, P.I. Borel, A. Harpøth, and M. Kristensen, IEEE Photonics Technology Letters, **17**, 1202-1204 (2005).
11. Y. Watanabe, Y. Sugimoto, N. Ikeda, N. Ozaki, A. Mizutani, Y. Takata, Y. Kitagawa, and K. Asakawa, Optics Express, **14**, 9502-9507 (2006).
12. Y. Watanabe, N. Ikeda and Y. Sugimoto, Y. Takata, Y. Kitagawa, A. Mizutani, N. Ozaki, and K. Asakawa, J. Appl. Phys., **101**, 113108 (2007).
13. Yoshinori Watanabe, Naoki Ikeda, Yoshiaki Takata, Yoshinori Kitagawa, Nobuhiko Ozaki, Yoshimasa Sugimoto and Kiyoshi Asakawa, J. Phys. D: Appl. Phys. **41**, 175109 (2008).
14. S. Ohkouchi, Y. Nakamura, H. Nakamura, N. Ikeda, Y. Sugimoto and K. Asakawa, J. Crystal Growth, **293**, 57-61 (2006).
15. N. Ozaki, Y. Takata, S. Ohkouchi, Y. Sugimoto, Y. Nakamura, N. Ikeda, K. Asakawa, J. of Crystal Growth, **301–302**, 771–775 (2007).
16. S. Ohkouchi, Y. Nakamura, H. Nakamura, and K. Asakawa, Jpn. J. of Appl. Phys., **44**, 5677–5679 (2005).
17. N. Ozaki, Y. Takata, S. Ohkouchi, Y. Sugimoto, N. Ikeda, K. Asakawa, Appl. Surf. Sci., **254**, 7968–7971 (2008).
18. H. Nakamura, S. Nishikawa, S. Kohmoto, K. Kanamoto, and K. Asakawa, J. Appl. Phys., **94**, 1184-1189 (2003).
19. H. Nakamura, K. Kanamoto, Y. Nakamura, S. Ohkouchi, H. Ishikawa and K. Asakawa, J. Appl. Phys., **96**, 1425-1434 (2004).
20. H. Nakamura, Y. Sugimoto, K. Kanamoto, N. Ikeda, Y. Tanaka, Y. Nakamura, S. Ohkouchi, Y. Watanabe, K. Inoue, H. Ishikawa, and K. Asakawa, Optics Express, **12**, 6606-6614 (2004).
21. Yoshinori Kitagawa, Nobuhiko Ozaki, Yoshiaki Takata, Naoki Ikeda, Yoshinori Watanabe, Yoshimasa Sugimoto, and Kiyoshi Asakawa, IEEE J. Light Wave Technology, **27**,1241-1247 (2009).
22. K. Inoue, N. Kawai, Y. Sugimoto, N. Carlsson, N. Ikeda, and K. Asakawa, Phys. Rev. B **65**, 121308 (2002).
23. M. L. Brongersma, Nature photonics, **2**, 270-272 (2008).
24. Dinesh Kumar V. and Kiyoshi Asakawa, Photonics and Nanophotonics – Fundamentals and Applications (PNFA), **6**, 148-153 (2008).
25. T. Ishi, J. Fujikata, K. Makita, T. Baba, and K. Ohashi, Jpn J. Appl. Phys. **44**, L364–L366 (2005).
26. C. Genet1 and T. W. Ebbesen, Nature, **445**, 39-46 (2007).
27. N. Ikeda, D. Tsuya, Y. Sugimoto, Y. Koide and K. Asakawa, Proceedings of PECS-VIII, Sidney, Australia, April, 2009.
28. N. Ikeda, D. Tsuya, Y. Sugimoto, Y. Koide, A. Miura, D. Inoue, T. Nomura, H. Fujikawa and K. Sato, IEICE Technical Report, 2009 (in Japanese).
29. Daisuke Inoue, Tsuyoshi Nomura, Atsushi Miura, Hisayoshi Fujikawa, Kazuo Sato, Naoki Ikeda, Daiji Tsuya, Yoshimasa Sugimoto, Yasuo Koide, Kiyoshi Asakawa, Proceedings of Optical MEMS & Nanophotonics 2009, Florida, USA, August, 2009.
30. J. B. Pendry, Phys. Rev. Lett., **85**, 3966-3969 (2000).
31. D. R. Smith, J. B. Pendry, M. C. K. Wiltshire, Science, **305**, 788-792 (2004).
32. A. Sanada, C. Caloz and T. Itoh, IEEE Trans. on Microwave Theory Tech., **52**, 1252-1263 (2004).
33. V. M. Shalaev, W. Cai, U. K. Chettiar, H-K. Yuan, A. K. Sarychev, V. P. Drachev, and A.r V. Kildishev, Optics Letters, **30**, 3356- 3358 (2005).
34. S. Zhang, W. Fan, N. C. Panoiu, K. J. Malloy, R. M. Osgood, and S. R. J. Brueck, Phys. Rev. Lett. **95,** 137404 (2005)
35. G. Dolling, C. Enkrich, M. Wegener, C. M. Soukoulis, S. Linden, Science, **312**, 892-894 (2006).
36. S. Linden, C. Enkrich, G. Dolling, M. W. Klein, J. Zhou, T. Koschny, C. M. Soukoulis, S. Burger, F. Schmidt, and M. Wegener, IEEE J. Selected Topics in Quantum Electronics, **12,** No. 6, November/ December (2006).
37. T. Nomura, D. Inoue, A. Miura, H. Fujikawa, K. Sato, N. Ikeda, D. Tsuya, Y. Sugimoto, Y. Koide, K. Asakawa, Proceedings of 4th International Conference on Nanophotonics, Tsukuba, Japan, May-June, 2010.

CAOL*2010 International Conference on Advanced Optoelectronics & Lasers, 10-14 September, 2010, Sevastopol, Ukraine

New coupled micro-ring resonator structures and laser arrays

P. Chamorro-Posada[1], F.J. Fraile-Pelaez[2]

[1]Dpto. de Teoría de la Señal y Comunicaciones e Ingeniería Telemática, Universidad de Valladolid, ETSI Telecomunicación, Campus Miguel Delibes s/n, 47011 Valladolid, Spain

[2]Dpto. de Teoría de la Señal y Comunicaciones, Universidad de Vigo, ETSI Telecomunicación, Campus Universitario s/n, 36310 Vigo, Spain

(Invited *Paper*)

Abstract: We have recently put forward a new type of zig-zag micro-ring resonator chain. This device has been demonstrated, for passive structures, using SiN technology. But the active structures are the ones exhibiting the most interesting behaviour and, for them, new fast and slow light phenomena have been predicted. For the stable operation of such systems, we have proposed and analysed a laser array configuration.

Coupled micro-ring resonator devices have attracted much attention for the realisation of complex linear and nonlinear photonic signal processing. They can find application, for instance, as optical filters [1], in group velocity dispersion compensation [2] or in optical sensing [3]. Particularly appealing is their use for the control of the group velocity of light in slow and fast light applications.

There exists a range of micro-ring structures offering different types of behaviour. Previous analyses have been mostly based on side-coupled integrated spaced sequences of resonators and coupled-resonator optical waveguides or a mixture of both [4, 5]. We have recently put forward a novel type of structure named zig-zag micro-ring resonator chain [6] (or relay ring resonator structure [7]). The basic building block is a two-coupler micro-ring resonator, as shown in Fig. 1; a structure conventionally employed as an add-drop filter. The zig-zag chain is obtained as the cascaded connection of N such stages, as shown in Fig. 2. As it can be appreciated in the figure, one remarkable property of our structure is the existence of N output signals for an N-th order device.

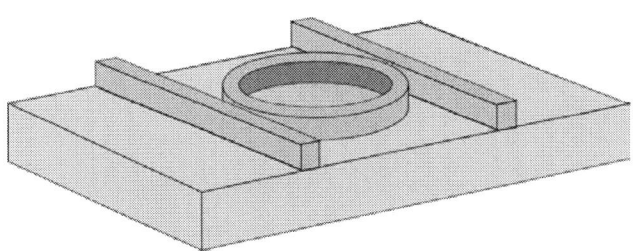

Fig. 1. Basic building block of the zig-zag microring resonator chain.

In the following analysis, $L=2\pi R$ is the total ring length, with R the ring radius. We neglect any dispersive effect of the short

transmission sections, which will be much smaller than the dispersive effects induced by the zig-zag structure itself. We will consider either passive structures, with $g\leq1$ the propagation loss along half a ring, or active devices with $g>1$. $\tau_0=L/(2v)$ is the propagation delay in half a ring, which we use for reference, and v is the propagation velocity in the waveguide sections. We assume equal evanescent couplings between resonators and waveguides for all the structure with coupling parameters r and $t=\sqrt{1-r^2}$. The system function for the k-th output is

$$H_k(z) = H_{o,k}(z) r \frac{1-g^2 z^{-2}}{1-r^2 g^2 z^{-2}} \left[\frac{t^2 g z^{-1}}{1-g^2 r^2 z^{-2}} \right]^k, \quad (1)$$

except for the last ($k=N$) output, for which

$$H_N(z) = H_{0,N}(z) \left[\frac{t^2 g z^{-1}}{1-g^2 r^2 z^{-2}} \right]^N, \quad (2)$$

where $H_{0,k}(z) = \left(-1\right)^k z^{-2(k+2)\frac{d}{L}}$. The frequency response is obtained by evaluating the system function on the unit circle $H_k(\Omega) = H_k\left(z^{-1} = e^{-j\Omega}\right)$, where $\Omega = \tau_0\omega$ is the normalized frequency. The z-dependent part in $H_{0,k}(z)$ is associated with the propagation delay in the waveguide sections of length d and will be neglected in the analysis.

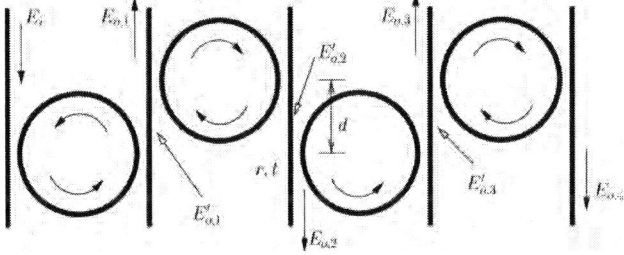

Fig. 2. Schematic of the zig-zag micro-ring resonator structure obtained as the cascade connection of $N=4$ stages of the type described in Figure 1.

The system function (1) has two multiple poles of order $k+1$ at $z=\pm rg$ for the intermediate outputs and two N-th order poles at the same positions for the last output. These are

978-1-4244-7043-3/10 $26.00 © 2010 IEEE   25

accompanied with a k-th order zero at the origin. But the main difference of the system response for $k{\neq}N$ is the presence of at zero at $z{=}{\pm}g$ due to the load effect of the next ring in the chain. The stability condition [6], given by $rg{<}1$, is equivalent to the criterion that the multiple pole must stay inside the unit circle.

Passive structures with a large N, recently demonstrated using SiN technology [7], provide sharp transmission peaks with high out-of-band rejection ratios. Figure 3 shows the transmittance in a $N{=}100$ chain for the final and several intermediate outputs. The propagation loss factor is $g{=}0.96$ and we have assumed $t{=}0.9$. When the loss is very low, $g{\rightarrow}1$, the zeroes of the transfer function of the $k{\neq}N$ outputs approach the unit circle and create dips in the response at resonances. In high-loss structures, where g is small, the introduction of gain in the straight connecting sections of the structure can be used for the compensation of the loss. Another effect that must be taken into account is the wavelength dependence of the propagation loss in the circular ring sections [8].

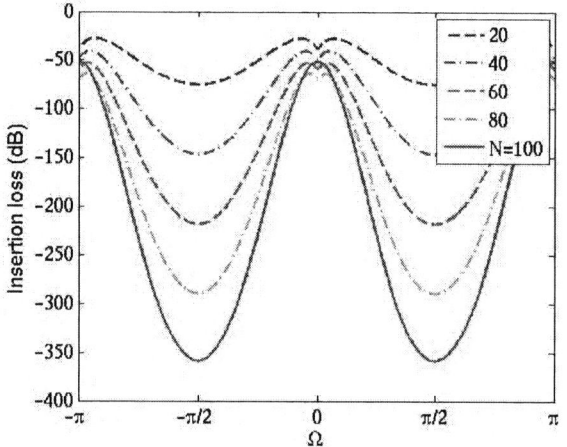

Fig. 3. Transmittance at the 20th, 40th, 60th, 80th and 100th output for a $N{=}100$ passive structure.

As regards the group delay in the structure, we find two regions of interest: at the system resonances, $\Omega{\rightarrow}0$, and right between resonances, $\Omega{=}\pi/2$. The normalised group delay at resonances can be obtained from the system transfer function (1) and is [6]

$$\frac{\tau_{g,k}}{\tau_0} = (k+1)\left[1+\frac{2g^2r^2}{1-g^2r^2}\right] \quad (3)$$

for $k{=}1{\ldots}N{-}1$. And, for the N-th output,

$$\frac{\tau_{g,N}}{\tau_0} = N\left[1+\frac{2g^2r^2}{1-g^2r^2}\right]. \quad (4)$$

Right between resonances, the group delay is

$$\frac{\tau_{g,k}}{\tau_0} = \frac{2g^2}{1+g^2}+k-(k+1)\frac{2g^2r^2}{1+g^2r^2} \quad (5)$$

for $k{=}1{\ldots}N{-}1$. And

$$\frac{\tau_{g,N}}{\tau_0} = N\left[1-\frac{2g^2r^2}{1+g^2r^2}\right] \quad (6)$$

for the Nth output. At resonances (Eqs. (3) and (4)) one obtains slow light. This can be achieved in passive structures with certain efficiency. One wants, ideally, $gr{\rightarrow}1$ for perfect stopping of light. This requires high couplings between the rings and waveguides, $r{\rightarrow}1$, and low loss, $g{\rightarrow}1$. But low loss means that the zeroes of the intermediate outputs approach the unit circle and these intermediate outputs will experience reduced amplitude transmission and, possibly, high distortion, depending on the signal bandwidth.

Active micro-ring structures, in general, hold very interesting properties [9]. In the case of our zig-zag structure, gain would permit to overcome the aforementioned limitations. For $g{>}1$ one can obtain nearly perfect signal stopping without the drawbacks found for $g{\rightarrow}1$. Moreover, new fast light effects can be found in this regime. When $\Omega{=}\pi/2$, one obtains equal transmission delays for all the intermediate outputs

$$\tau_{g,k} = \tau_0\frac{g^2-1}{g^2+1} \quad (7)$$

in the limit $rg{\rightarrow}1$. This means *any input signal will show-up at all intermediate outputs in a nearly simultaneous fashion* even though there is a finite separation among the outputs along the propagation direction. This novel fast light transmission effect has also been termed *ubiquitous light transmission*.

One drawback of fast light effects in this type of structures stems from the fact that we are working at the middle of the span between resonances. This means that we will possibly have a large deviation in the transmitted power at different outputs. This typical situation illustrated by the transmission spectra of Fig. 3. Nevertheless, for $g{>}1$, if the condition

$$r = \sqrt{\frac{g-1}{g(g+1)}} \quad (8)$$

is satisfied, one obtains equal transmission amplitudes at $\Omega{=}\pi/2$ for all intermediate outputs $k{=}1{\ldots}N{-}1$,

$$\left|H_k\left(\Omega=\frac{\pi}{2}\right)\right| = \frac{r(1+g^2)}{1+r^2g^2} \quad (9)$$

and

$$\left|H_N\left(\Omega=\frac{\pi}{2}\right)\right| = 1. \quad (10)$$

Figure 4 shows the transmission spectra for an active chain of $N{=}5$ rings with parameters $g{=}2.2260$ and $r{=}0.4132$ satisfying the condition (9) with $rg{\approx}0.92$. These results reveal an important limitation of active structures: At the operation point, the transmission gain is close to 0 dB, but, at resonances, the gain attains very large values which increase with the output order. With such high gain levels, it seems very difficult to keep the system from oscillating. This is not strange since we have decided to operate at the edge of the stability threshold set by the condition $gr{<}1$. Another limitation of this type of active structures is due to the spontaneous emission that significantly increases the power consumption [10].

CAOL*2010 International Conference on Advanced Optoelectronics & Lasers, 10-14 September, 2010, Sevastopol, Ukraine

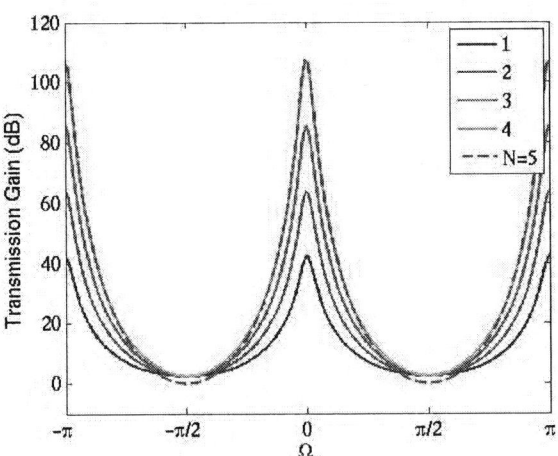

Fig. 4.Transmission spectra for an active chain of $N=5$ rings.

A solution to this intrinsic limitation has been proposed in [6]. The active structure is transformed into a laser array and the use of injection locking permits to set the oscillation in the whole chain at a wavelength with a given spectral separation with the transmitted signal. A series of vertically coupled lossy micro-rings with a round-trip time $\tau_a = q\tau_0$, with $q<1$ a rational number, permits to introduce the isolation between the lasing elements required to avoid saturation.

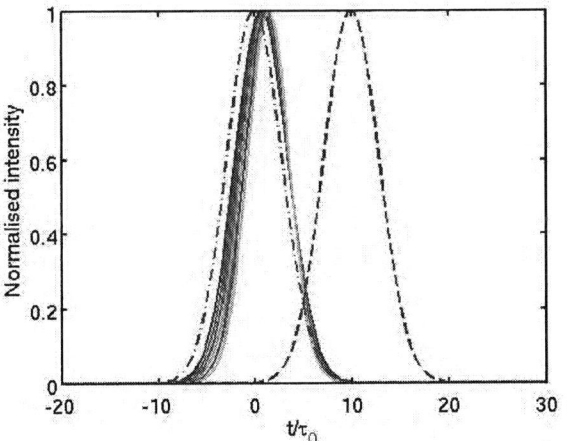

Fig. 5. Solid lines: normalised output intensities of a $N=10$ zig-zag microring laser array for k ranging from 1 to 6. Dotted-dashed line: normalised input intensity. Dashed line: reference output for the straight propagation along the equivalent length of N rings.

In this scheme, gain is naturally clamped close to the desired value $gr\sim1$ and the spontaneous emission noise is set under control [11]. Figure 5 shows the first six outputs obtained in the simulation of a chain of $N=10$ rings with unsaturated gain $g_0 = \sqrt{5}$, $r=0.5$ and $q=2/15$. Both the input (dashed-dotted line) and a reference output for the straight propagation along the equivalent length of N rings. Outputs $k=8$ to 10 (not

shown) have a center position and waveform similar to the ones shown in Figure 6, but they also develop a growing oscillatory tail after the pulse has passed [6], which is a manifestation of the limitations typically found in fast light phenomena.

In summary, new zig-zag micro-ring resonator structures have been analysed. These permit the realization of high-order optical filters and exhibit new fast- and slow- light effects. For the stable operation of active structures, a laser array configuration has been proposed and studied. This approach can be extended to operate other related active structures [9].

ACKNOWLEDGEMENTS

This work has been funded by the Spanish MICINN and FEDER, grant no. TEC2007-67429-C02-01 and 02, and JCyL VA001A08.

REFERENCES

[1] J. Capmany, P. Muñoz, J.D. Domenech and M.A. Muriel, "Apodized coupled resonator waveguides," Opt. Express, vol. 15, no. 16, pp. 10196-10206, 2007.

[2] G. Lenz, and C.K. Masden, "General optical all-pass filter structures for dispersion conrol WDM systems," J. Lightwave Technol., vol. 17, no. 7, pp. 1248-1254, 1999.

[3] C.Y. Chao, W. Fung and L.J. Guo, "Polymer microring resonators for biochemical sensing applications," IEEE J. Sel. Top. Quantum Electron., vol. 12, no. 1, pp. 134-142, 2006.

[4] J. S. Scheuer, G. Ploczi, J. Poon, and A. Yariv, Opt. Photonics News, vol. 16, pp. 16, 2005.

[5] B.E. Little, S.T. Chu, W. Pan, and Y. Kokubun, "Microring resonator arrays for VLSI photonics," IEEE Photon. Technol. Lett., vol. 12, no. 3, pp. 323-325, 2000.

[6] P. Chamorro-Posada, and F.J. Fraile-Pelaez, "Fast and slow light in zigzag microring resonator chains," Opt. Lett., vol. 34, no. 5, pp. 626-628, 2009.

[7] S.H. Tao, S.C. Mao, J.F. Song, Q. Fang, M.B. Yu, G.Q. Lo, and D.L. Kwong, "Ultra-high order ring resonator system with sharp transmission peaks," Opt. Express, vol. 18, no. 2, pp. 393-400, 2010.

[8] P. Chamorro-Posada and F.J. Fraile-Pelaez, "Superluminal propagation in resonant dissipative media," Opt. Commun, vol. 282, no. 6, pp. 1095-1098, 2009.

[9] F.J. Fraile-Peláez and P. Chamorro-Posada, "Active control and stability of micro-ring resonator chains," Opt. Expr., vol.15, no.6, pp. 3177-3189, 2007.

[10] J.B. Khurgin, "Dispersion and loss limitations on the performance of optical delay lines based on coupled resonant structures," Opt. Lett., vol. 32, no. 2, pp. 133-135, 2007.

[11] M. Zirngibl, "Gain control in Erbium-doped fibre amplifiers by an all-optical feedback loop," Electron. Lett., vol. 27, no.7, pp. 560-561, 2007.

Recent Progress on Deep-Ultraviolet Coherent Light Sources

Qinjun Peng, Zhimin Wang, Shenjin Zhang, Fengfeng Zhang, Feng Yang, Dafu Cui, Zuyan Xu

Research Center for Laser Physics and Technology,

Key Laboratory of Functional Crystal and Laser Technology,

Technical Institute of Physics and Chemistry, Chinese Academy of Sciences,

Beijing 100190, China

(Invited paper)

Abstract: The practical deep-ultraviolet (DUV, below 200nm) coherent light sources, which are considered as puzzles of laser field, can provide a kind of unique tool to explore the front fields of biology, physics, chemistry, material etc.. The methods to obtain the DUV coherent light sources are reviewed and compared, and all-solid-state DUV lasers are precise and practical compared with the others. Recent progresses on them and their engineering in our lab are reported, and some new phenomena such as Fermi pockets in superconductor are discovered with them.

I. INTRODUCTION

Coherent light sources in the DUV spectral range are essential tools for investigations of fundamental dynamics of atoms and molecules in physics, chemistry, biology, and material sciences. There are various techniques to produce coherent DUV laser:

1) High harmonic generation (HHG) in gases. In the HHG scheme, the pumping laser is focused into noble gases, when the laser electromagnetic fields is high enough to make the electron oscillate near its atomic orbit, the high harmonic generation can be obtained. For example, 118.2nm was obtained in Xe or the mixture of Xe and Ar with the maximum conversion efficiency of 2.8% [1,2,3], 72.5-83.5nm and 127-180nm were produced by sum-frequency generation (SFG) and difference frequency generation (DFG) in Kr, the corresponding conversion efficiency is 1.2×10^{-5} for SFG at 135 nm and 1.8×10^{-4} for DFG at 175nm, respectively [4]. Some molecular gases such as N_2, CO, H_2, CH_4 can also be used to generate harmonic wave. For example, the third harmonic generation 92-122nm laser with efficiency less than 1.0×10^{-7} is realized with N_2 [5]. HHG in gases can produce VUV laser as short as 100nm, however, the main limit for HHG is the low efficiency.

2) High harmonic generation (HHG) in plasma. The coherent X-ray can be generated in plasma with ultra-high intensity laser fields. DUV laser, even coherent soft-X ray at the water window, can be acquired with HHG in some plasma such as boron or noble gases plasma [6,7]. The HHG can only produce

the odd harmonic waves of the pumping laser, and there is a plateau in intensity after the steep decrease of low-order harmonics, and the cut-off of harmonic orders are not determined by the Ip + 3.17Up relation, where Up is the ponderomotive potential, Ip is the ionization potential of medium, but determined by the ionization-induced phase mismatch [7]. The pumping laser used for HHG in plasma is usually a fs laser because of the ultra-high laser intensity ($\sim10^{15}$W/cm^2) requirement to produce ionization. The efficiency of low order harmonic wave is about 1×10^{-4}, and it is several orders of magnitude lower in high order harmonic wave.

3) Four-wave frequency mixing. It can generate DUV laser using four-wave mixing (FWM) in metal vapors and gases. For example, the DUV radiation from 130nm to 200nm can be produced in noble gases with KrF (248nm), ArF (193nm) and Ti:sapphire lasers [8,9], and the 121.56nm lyman-α source was produced in Hg vapor with the SHGs of 1091nm fiber laser at 545.1nm, a Ti:sapphire at 816nm laser and a disc laser at 1015nm [10]. Obviously, the method is complicated and needs 3 lasers. In addition, the efficiency of this scheme is very lower, it is only about 1×10^{-7} or lower.

4) Excimer lasers. Some excimer lasers can produce DUV laser. ArF at 193nm. F_2 at 157nm. However, their wavelengths cannot be tuned, the pulse widths are limited in nanosecond, and it is complicated.

5) Anti-Stokes stimulated Raman scattering (SRS). Tunable DUV laser in the range of 138-190nm was produced by SRS in H_2 with maximum conversion efficiency of about 0.5×10^{-3} at 190nm [11]. Also tunable DUV laser in 116-189nm has been accomplished in H_2 by SRS with anti-stokes, and the efficiency is 0.33×10^{-3} at the third anti-stokes 190nm, and the efficiency is 5.5×10^{-6} for the 11th order anti-stokes 116nm [12]. Because of the pumping energy of the laser is distributed on different anti-stokes orders, the efficiency of each order is very

978-1-4244-7043-3/10 $26.00 © 2010 IEEE

low, especially for DUV lasers.

6) Free electron lasers (FELs). Free electron lasers (FELs) use the relativistic electron beams passing through a transverse periodic magnetic field to generate coherent electromagnetic radiation ranging from infrared to hard X—ray region. High-gain harmonic generation (HGHG) with seed laser can produce coherent DUV laser [13]. But it is very complicated and needs precise control of electrons and the volume of FEL is very huge, which prohibit its wide application.

7) Diode-pumped solid-state lasers (DPLs). The direct frequency doubling using nonlinear crystal to generate DUV-DPL is a practical method compared with other methods listed above. DUV-DPLs are compact, easy to attain high power high beam quality and narrow linewidth with a long life time and can work at ns, ps, fs pulse widths, and their repetition rates range from 1Hz to over 100 MHz, even to GHz. Up to now, DUV-DPLs are the only practical and precise coherent light sources at DUV wave band.

Ten years ago, it is very hard to generate DUV-DPL, and it was described as "200-nm wall in terms of deep-uv generation" [14]. The nonlinear crystal family of KBBF in DUV wave band are developed [15] and its practical prism-coupled technique was invented [16,17,18] to promote their wide application. The DUV-DPL is reported as new breakthrough in Laser Focus World in 2002 [19]. Up to now, we have developed several kinds of DUV-DPLs and produced their prototypes. With the DUV-DPL systems, some new phenomena such as Fermi pocket in superconductors are discovered [20].

II. HIGH POWER DUV-DPLS

High power DUV-DPLs have been developed in the recent years in our lab. The maximum average output powers 34.7 mW has been generated for nanosecond system a repetition rate of 10 kHz [21, 22], and the average output power can be as high as 41 mW for a mode-locked picoseconds operated at a repetition rate of 80 MHz[23] at 177.3 nm. With a femtosecond Ti:sapphire laser as the pump, the output is up to 102.1 mW at 193.5 nm [24], as is shown in Fig.1.

To acquire high output power ns DUV-DPL at 177.3nm, a high power and high beam quality ns 355 nm UV laser is developed in out lab. The 355 nm laser, with repetition rate of 10 kHz and pulse width of 49 ns, is used as the pumping laser of KBBF-SHG to produce 177.3 nm DUV laser. The maximum output power of 34.7 mW at 177.3 nm is obtained in 2.06 mm-thick KBBF with a pumping laser of 4.2 W at 355nm. With the same KBBF crystal, the maximum output power of 41 mW is obtained with a

7.55 W pumping laser at 355 nm, which is a frequency-tripled mode-locked Nd:YVO$_4$ laser with repetition of 80 MHz and pulse width of less than 15 ps. This is the highest average power at 177.3 nm to

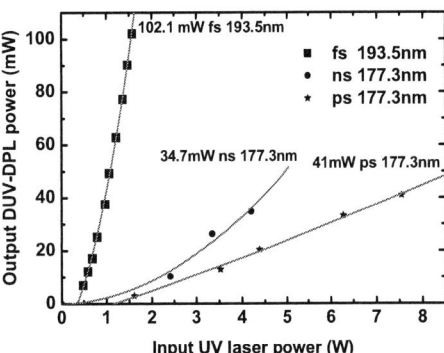

Fig.1. Output DUV-DPL power V.S. input UV laser power

our knowledge. The fs DUV-DPL is also developed in our lab, the maximum output power of 102.1 mW at 193.5 nm is achieved in 1.28 mm-thick KBBF with 1.57 W pumping laser at 387 nm produced by SHG from 774nm whose repetition rate is 80 MHz and pulse width is ~150 fs. Above experimental results were shown in Fig.1. With a thicker KBBF crystal, much higher DUV-DPL output power can be obtained. For example, 1.05 W 193.5 nm is generated in a 2.71 mm-thick KBBF using ~7 W 387 nm pumping laser with a repetition rate of 5 kHz and pulse width of 350 ps [25].

III. TUNABLE DUV-DPLS

Tunable DUV-DPLs are also developed in our lab. A high-power tunable Ti:sapphire laser with wavelengths from 680 to 1040 nm, repetition rate of 80 MHz and pulse duration width of ~150fs, is frequency-doubled in BBO to produce 340 - 520 nm laser, which is used as the pumping laser of KBBF-SHG , and the 170 - 232.5 nm widly tunable DUV-DPL is developed in our lab and the results are provided to Professor Chuangtian Chen's group who is responsible for growing KBBF crystals for publication[24]. More attention should be paid on that there is only one KBBF crystal is used to produce such a wide wavelength range, this is the widest tunable range in DUV-DPL ever produced, and has a maximum output of 119.8 mW at 202.5 nm. The shortest wavelength which has been obtained is at 170nm. It is an important promotion in the application of tunable DUV-DPLs compared with our previous results of ns 175-210nm tunable DUV-DPL with the repetition rate of 8 kHz and pulse width of 120 ns [26], in which it needs two KBBF crystals to cover such a wide wavelength range.

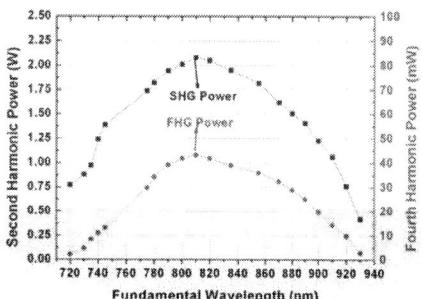

Fig.2 Output power of 4HG and 2HG V.S. fundamental wavelength

Another kind of tunable DUV-DPL with $RbBe_2BO_3F_2$ as the nonlinear crystal is also investigated in our lab, and the tunable wavelength range of 180 -232.5 nm is realized in one RBBF crystal with the size of $20\times6\times0.95mm^3$, the experimental data are provided to Professor Chuangtian Chen's group at Technical Institute of Physics and Chemistry [27], the tuning curve of the system is shown in Fig.2. The 4th harmonic generation (4HG) in the tunable range from 185 to 200 nm always exceeds 12 mW.

IV. DUV-DPL PROTOTYPES AND THEIR APPLICATION

Several kinds of DUV-DPL prototypes are made out in our lab and are provided to our copartners, the pictures of DUV-DPLs are shown below:

Fig.3 shows a 10 kHz ns 177.3 nm DUV-DPL used for deep-ultraviolet Raman spectrometer which is used to study the structure of molecules. This DUV-DPL can operate with stable output power of 4 mW, and the fluctuation of output power at 177.3 nm is ±3.4% in 2 hours. The spatial resolution for conventional Photoemission Electron Microscopy (PEEM) which is used to investigate the surface of solid samples, is about 20-50 nm. However, it will be reduced to less than 10 nm if a DUV-DPL is used, Fig.4 shows a 120 MHz ps DUV-DPL at 177.3 nm developed for deep-ultraviolet PEEM. Fig.5 shows a 10 Hz ps 177.3 nm DUV-DPL made to research the deep-ultraviolet photochemical reaction, which can probe high-energy exicitaion state of sample with a sensitivity of single deep-ultraviolet photon, and the time resolution will be improved obviously to picoseconds. An 80MHz fs 177.3nm DUV-DPL are developed for the research on the super-wide band

gap materials, such as AlN, BN, BeZnO, as is shown in Fig.6. Fig.7 shows the polarization-tunable 80 MHz ps 177.3 nm DUV-DPL developed for spin-resolved and angle-resolved spectroscopy, which can provide more information of the electronic structure in materials. A tunable 80 MHz fs DUV-DPL 175 -185 nm, as is shown in Fig.8, is

Fig.3 A 10 kHz nanoseond 177.3 nm DUV-DPL

Fig.4 A 120MHz picoseconds 177.3 nm DUV-DPL

Fig.5 A 10 Hz picosecond 177.3 nm DUV-DPL

Fig.6 An 80MHz femtosecond 177.3nm DUV-DPL

CAOL*2010 International Conference on Advanced Optoelectronics & Lasers, 10-14 September, 2010, Sevastopol, Ukraine

Fig.7 A polarization-tunable 80 MHz picosecond 177.3 nm DUV-DPL

Fig.8 An 80 MHz femtosecond DUV-DPL tunable between 175 nm to 185 nm

Fig.9 A photograph of 175 – 210 nm tunable 80MHz picosecond DUV-DPL

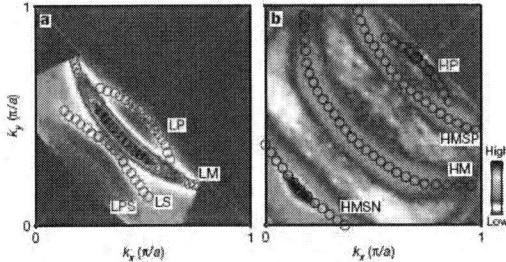

Fig.10 Fermi surface data measured using a DUV-DPL (left) and a helium discharge lamp (right)

developed for Photoassisted scanning tuned microscopy (PhotoSTM), which will make it possible to detect the excitation state of nanometer system. Fig.9 shows the 175 – 210 nm tunable 80MHz ps DUV-DPL being developed for photoemission system whicn can provide complete information of the electronic structure in material.

With the DUV-DPL as a unique tool, some new phenomena have been discovered. For example, with the angle-resolved photoemission spectroscopy (ARPES) based on the 80 MHz ps DUV-DPL developed by us before, whose energy resolution is better than 1 meV [28], Xingjiang Zhou's group compared the measured Fermi surface data in La-Bi2201 using a DUV-DPL and a helium discharge lamp, as is shown in Fig.10, the Fermi pocket is obvious using a DUV-DPL, while it cannot be seen using helium discharge lamp. The coexistence of Fermi arcs and Fermi pockets in a high-transition-temperature copper oxide superconductor that is not expected theoretically, was discovered [20]. Some important results were also published based on DUV-DPLs by Xingjiang Zhou's group [29, 30]. More new phenomena are expected to be discovered as the wide application of DUV-DPLs.

REFERENCES

[1] A.H. Kung, J.F. Young, and S.E. Harris, "Generation of 1182-A radiation in phase-matched mixtures of inert gases," Appl. Phys. Lett., Vol. 22, No.6, p301-302, 1973

[2] L. J. ZYCH AND JAMES F. YOUNG, "Limitation of 3547 to 1 I82 A Conversion Efficiency in Xe," IEEE JOURNAL OF QUANTUM ELECTRONICS, VOL. QE-14, NO. 3, p147-149, 1978

[3] R. A. Ganeev, I. A. Kulagin, T. Usmanov, and S. T. Khudalberganov, "Investigation of the generation of $\lambda = 118.2$ nm coherent radiation in rare gases," Sov. J. Quantum Electron. Vol.12, No.12, p1637-1640, 1982

[4] G. Hilber, A. Lago, and R. Wallenstein, "Broadly tunable vacuum-ultraviolet / extreme-ultraviolet radiation generated by resonant third-order frequency conversion in krypton," J. Opt. Soc. Am. B/Vol. 4, No. 11,p1753-1764, 1987

[5] Xie Xiaobo, wang Pengqian, Cao Chunshun, Sun Taoheng, "Generation of VUV/XUV coherent radiation in molecular gases," SCIENCE IN CHINA (series A), VOL43, NO.3, P280-290, 2000

[6] Rashid Ganeev, Masayuki Suzuki, Motoyoshi Baba, and Hiroto Kuroda, Tsuneyuki Ozaki, "High-order harmonic generation from boron plasma in the extreme-ultraviolet range, " OPTICS LETTERS , Vol. 30, No. 7, p768-770, 2005

[7] Emily A. Gibson,1 Ariel Paul,1 Nick Wagner, et al. "Coherent Soft X-ray Generation in the Water Window with Quasi–Phase Matching," SCIENCE VOL 302 P95-98, OCTOBER 2003

978-1-4244-7043-3/10 $26.00 © 2010 IEEE

[8] B. WELLEGEHAUSEN, H. WELLING, et al., "Generation of short-pulse VUV and XUV radiation", Optical and Quantum Electronics 28, p267-281,1996

[9] B. WELLEGEHAUSEN et al., "Short-pulse high-intensity excimer lasers – A powerful tool for the generation of coherent VUV and XUV radiation," Appl. Phys. B 63, 451-461, 10996

[10] Martin Scheid et al., "Continuous-wave Lyman-α generation with solid-state lasers," OPTICS EXPRESS, Vol.17, No.14, p11274-11280, 2009.

[11] Schomburg et al., "Generation of Tunable Narrow-Bandwidth VUV Radiation by Anti-Stokes SRS in H 2," H. Appl. Phys. B30, p131-134 ,1983

[12] Huo Yunsheng, et al., "Optimization of the gain coefficient in anti-stokes raman conversion, " Chinese Jouranl of Lasers, Vol.A21, No.8, p649-652, 1994.

[13] DENG Haixiao , DAI Zhimi, "On harmonic operation of Shanghai deep UV free electron laser," Nuclear Science and Techniques, 19, p7-12, 2008

[14] Hassaun Jones-Bey, "Deep-UV application await improved nonlinear optics," Laser Focus World, p127-129, 1998.

[15] L. Mei, C. He, and C. Chen, "Flux growth of KBe2BO3F2," J. Cryst. Growth 132(3-4), 609–610 (1993).

[16] Chinese patent ZL01115313.X

[17] US Patent 6859305 B2

[18] Japanese Patent 4074124

[19] Deep-ultraviolet coherent light shines from KBBF crystal, Laser Focus World 37 (7), 2002

[20] Jianqiao Meng et al., "Coexistence of Fermi arcs and Fermi pockets in a high-Tc copper oxide superconductor," p335-338, Nature, Vol 462, No19, November 2009

[21] Zhimin Wang, Jing-yuan Zhang et al. "Stable operation of 4 mW nanoseconds radiation at 177.3 nm by Second Harmonic Generation in $KBe_2BO_3F_2$ Crystals," Optics Express Vol. 17, No. 22, p. 20021–20032, 2009.

[22] F.Yang, Z,Wang, Q.Peng, et al. "Theoretical and Experimental investigations of nanosecond 177.3nm deep-ultraviolet light by SHG in KBBF, " Appl.Phys.B., 96, p415-422, 2009

[23] FengYang, Zhimin Wang, et al. "41 mW high average power picoseconds 1773 nm laser by second-harmonic generation in KBBF, " Opt.Commu., 2010(283),p142-145

[24] Xin Zhang, Zhimin Wang, "Widely tunable and high-average-power fourth-harmonic generation of a Ti:sapphire laser with a $KBe_2BO_3F_2$ prism-coupled device," OPTICS LETTERS, Vol. 34, No. 9, p1342-1344, 2009

[25] Teruto Kanai, Xiaoyan Wang, "Watt-level tunable deep ultraviolet light source by a KBBF prism-coupled device," OPTICS EXPRESS, Vol. 17, No. 10, p 8696-8703, 2009

[26] Hongbo Zhang, Guiling Wang, Lin Guo, Aicong Geng, Yong Bo, Dafu Cui et.al , "175-210nm tunable deep-ultraviolet light generation by the fourth harmonic of Ti:Sapphire laser with KBBF , " Applied Phys. B, p323-326, 2007

[27] Chuangtian Chen, Siyang Luo, Xiaoyang Wang, Guiling Wang, Xiaohong Wen, Huaxing Wu, Xin Zhang, and Zuyan Xu, "Deep UV nonlinear optical crystal:$RbBe_2BO_3F_2$," Vol. 26, No. 8, p1519-1525, J. Opt. Soc. Am. B, August 2009.

[28] Guodong Liu et al., "Development of a vacuum ultraviolet laser-based angle-resolved photoemission system with a superhigh energy resolution better than 1 meV," REVIEW OF SCIENTIFIC INSTRUMENTS 79, 023105, 2008.

[29] Wentao Zhang, Guodong Liu, Lin Zhao, Haiyun Liu, Jianqiao Meng, Xiaoli Dong,et al., "Identification of a New Form of Electron Coupling in the $Bi_2Sr_2CaCu_2O_8$ Superconductor by Laser-Based Angle-Resolved Photoemission Spectroscopy," Phys. Rev.Lett., 100, 107002 ,2008

[30] Wentao Zhang, Guodong Liu, Jianqiao Meng, Lin Zhao, Haiyun Liu, Xiaoli Dong, et al.,"High Energy Dispersion Relations for the High Temperature $Bi_2Sr_2CaCu_2O_8$ Superconductor from Laser-Based Angle-Resolved Photoemission Spectroscopy," Phys. Rev.Lett., 101, 017002, 2008.

Effective Mode Volume of a Natural Mode of an Open Dielectric Resonator with an Active Region

Alexander I. Nosich[1], *Fellow, IEEE*, Elena I. Smotrova[1], *Student Member, IEEE,*
Volodymyr O. Byelobrov[1], *Student Member, IEEE,* Trevor M. Benson[2], *Senior Member, IEEE,*
Ronan Sauleau[3], *Senior Member, IEEE*

[1] Institute of Radio-Physics and Electronics NASU, Kharkiv 61085, Ukraine
[2] G. Green Institute for Electromagnetics Research, University of Nottingham, Nottingham NG7 2RD, UK
[3] IETR, Universite de Rennes 1, Rennes Cedex 35042, France

(Invited *P*aper)

Abstract: We study the lasing eigenvalue problem for a generic open dielectric resonator with gain material. The gain is introduced within the active region via the "active" imaginary part of the refractive index. Each eigenvalue is constituted of two positive numbers, namely, the lasing wavenumber and the threshold value of material gain. This approach yields clear insight into the lasing thresholds of individual modes. The Optical Theorem, if applied to the lasing-mode field, puts familiar "gain=loss" condition on firm footing. It enables us to rigorously introduce the conception of the volume of an open resonator and then the effective-mode volume, both for the passive cavities and cavities with active regions.

The semiconductor, polymeric and crystalline microcavity lasers are among the most promising sources of waves from THz to UV. Their frequencies of lasing are determined mainly by the gain material system used and also the shape and size of the resonator, which is frequently a disk. Performance of such lasers as electron devices critically depends on the proper choice of the current-injection electrodes. As the density of carriers is the largest near the electrodes, their configuration, in fact, determines the location of the active region in the resonator, which can be viewed as dielectric cavity. Design and optimization of such devices relies heavily on the availability of computationally efficient and simultaneously accurate electromagnetic models. Here, two points are important. First, the typical size of the resonators is in the range of fractions to tens of the wavelength that makes the Geometrical Optics based modeling tools impractical. Therefore the use of the Maxwell equations is mandatory. However, FDTD codes popular today fail to characterize the lasing modes directly and suffer of a number of deficiencies. Second, account of the presence of active region is crucial. In the computations, it is necessary to have a simple and practical criterion for validation of numerical results. Our initial goal was deriving such a criterion from the Maxwell equations however obtained results have more general sense.

Consider a generic 3-D open dielectric resonator shown in Fig.1. Here V_d and V_a are passive-dielectric and active-dielectric regions with boundaries S_d and S_a, respectively, and $V_{\min}$ is the so-called minimum sphere, i.e. a sphere of the minimum radius $R_{\min}$ containing both V_d and V_a; note that it

may and may not contain a free space part, V_f, hence the whole passive part of the open cavity is $V_p = V_d + V_f$. Importance of the minimum sphere is in the fact that outside of it the field is superposition of solely outgoing waves while inside it also contains the incoming waves.

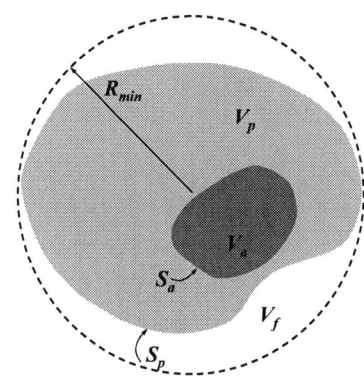

Fig. 1. Open dielectric resonator with a partial active region.

Following [1]-[5], we will be interested in the study of self-excitation threshold conditions of such a resonator. This means that we look for eigensolutions of electromagnetic-field problem characterized by the real-valued pairs of numbers, (k, γ). The first of them is normalized frequency $k = \omega/c$ and the second is material gain. They are the eigenvalues that generate non-zero time-harmonic modal fields $\{\vec{E}, \vec{H}\} e^{-i\omega t}$ solving, off S_d and S_a, the homogeneous Maxwell equations with piecewise-constant refractive index ν equal to 1 in V_f and out of $V_{\min}$, α_d in V_d ($\operatorname{Im} \alpha_d \geq 0$), and $\alpha_a - i\gamma$ ($\alpha_a, \gamma > 0$) in V_a (materials are non-magnetic). On S_d and S_a, the continuity of the tangential components is requested. Besides, the field energy must be locally integrable to prevent source-like singularities.

Further, a *condition at infinity*, at $R \to \infty$, must be added. If the domains V_d and V_a are finite and k is real-valued, this is

978-1-4244-7043-3/10 $26.00 © 2010 IEEE   33

the Silver-Muller condition of radiation [6].

The fundamental properties of the natural modes (eigenmodes) can be established for an arbitrary open cavity with an arbitrary active region. This is based on the analytical regularisation (see [1],[2]), i.e. equivalent reduction of the eigenvalue problem to a set of the Fredholm second-kind boundary IEs of Muller's type, and the use of the operator extensions of the Fredholm theorems. It is found that the eigenvalues form a discrete set on the plane (k, γ), so that they can be counted with the aid of some index, say s; each (k_s, γ_s) has finite multiplicity and depends on S_d, S_a and α_d, α_a in piece-continuous or piece-analytic manner, and this property can be lost only if eigenvalues coalesce. Note also that the gain per unit length, the traditional quantity in the descriptions of the Fabry-Perot cavities, is $g = k\gamma$.

A very instructive insight into the properties of natural modes can be obtained from the Complex Poynting Theorem (CPT) ([6], p. 98). The most general form of such expression, for the complex k, is

$$\Pi = -(1/2)\int_V (\vec{j}^{e*}\vec{E} + \vec{j}^m\vec{H}^*)dv$$
$$+(i/2)\int_V (k^*\varepsilon^* Z_0^{-1} |\vec{E}|^2 - k\mu Z_0 |\vec{H}|^2)dv, \tag{2}$$

where

$$\Pi = (1/2)\oint_S \vec{E} \times \vec{H}^* ds \tag{3}$$

is the total outward flux of the Poynting vector through the arbitrary boundary S enclosing a volume V containing all scatterers and sources, Z_0 is free-space impedance, $\varepsilon = v^2$ and μ are the relative permittivity and permeability, respectively, $\vec{j}^e$ and $\vec{j}^m$ are given electric and magnetic currents, respectively, and the asterisk means complex conjugation.

CPT (2) can be also applied to a *natural mode* number s (in this case, $\vec{j}^e = \vec{j}^m = 0$). At first, consider a *passive open cavity* (i.e., $V_a = 0$), having complex eigenfrequency k_s. On the extraction of the real part, we retrieve the formula,

$$-\frac{\operatorname{Re} k_s}{\operatorname{Im} k_s} = \frac{W_s}{W_{abs(s)} + W_{rad(s)}}, \tag{4}$$

$$W_s = (1/2)\int_{V_{min}} (Z_0^{-1} \operatorname{Re}\varepsilon |\vec{E}_s|^2 + Z_0 \operatorname{Re}\mu |\vec{H}_s|^2)dv, \tag{5}$$

$$W_{abs(s)} = (1/2)\int_{V_{min}} (Z_0^{-1} \operatorname{Im}\varepsilon |\vec{E}_s|^2 + Z_0 \operatorname{Im}\mu |\vec{H}_s|^2)dv, \tag{6}$$

$$W_{rad(s)} = \operatorname{Re}\Pi_s / \operatorname{Re} k_s, \tag{7}$$

where W_s, $W_{abs(s)}$ and $W_{rad(s)}$ are the powers stored in, absorbed in, and radiated from open cavity, and $\operatorname{Re}\Pi_s$ is the flux of the mode Poynting vector out of the minimum sphere.

Either side of (4) is simply twice the quality factor, $2Q_s$ and can be considered as its rigorous definition. It may take only discrete values linked to the mode.

Now, turn to a resonator with active region shown in Fig. 1 and apply (2) to the *lasing mode* field, $\{\vec{E}_s, \vec{H}_s\}$, taking into account that sources are absent, $\operatorname{Im} k_s = 0$, and $V_a \neq 0$. The result is the fundamental *"gain=loss"* expression whose simplified derivation is met in semi-classical theories [7],

$$\tilde{W}_{rad(s)} + \tilde{W}_{abs(s)} = \tilde{W}_{gain(s)} \tag{8}$$

$$\tilde{W}_{gain(s)} = Z_0^{-1}\gamma_s\alpha_a \int_{V_a} |\vec{E}_s(\vec{R}, k_s, \gamma_s)|^2 dv, \tag{9}$$

$$\tilde{W}_{abs(s)} = Z_0^{-1} \operatorname{Im}\alpha_d \operatorname{Re}\alpha_d \int_{V_d} |\vec{E}_s(\vec{R}, k_s, \gamma_s)|^2 dv, \tag{10}$$

where we use the mark $\sim$ to emphasize that the corresponding quantities are built on the eigensolutions and depend on γ.

Hence, for the s-th mode having the wavenumber k_s, the power lost for radiation is balanced by the "negative absorption" (i.e., modal gain as the power generated in the active region), provided that the material gain equals γ_s. So, this is the "gain = loss" condition derived in rigorous way.

However, besides of the real part, CPT expressed as (2) has also the imaginary part that leads to

$$2\operatorname{Im}\Pi_s = k_s \int_V [Z_0^{-1} \operatorname{Re}(v^2) |E_s|^2 - Z_0 |\vec{H}_s|^2]dv \tag{11}$$

where the domain V is arbitrary. In the limit of $S \to \infty$ as a circle of large radius, the left-hand part of (11) is zero due to the radiation condition, and the same is valid if $V = V_{min}$ because of the continuity of Π. Therefore, we obtain that

$$\int_{V_{min}} [Z_0^{-1} \operatorname{Re}(v^2) |\vec{E}_s|^2 - Z_0 |\vec{H}_s|^2]dv = 0, \tag{12}$$

which means that the fractions of the power contained in the electric and magnetic field of any mode inside the cavity volume V_{min} equal each other. The same is valid in the whole space. Note that this property holds true for any mode in both passive and active open cavities, on resonance.

The laser configurations where the active region does not coincide with the whole cavity (i.e. $V_p \neq 0$ in Fig. 1) are frequently met in practice. For instance, it is realized if one uses a sharply focused pump beam in optically pumped laser or, alternatively, the pump beam goes through an axicon. Besides, combination of separated active and passive regions is typical for the cavities with distributed Bragg reflectors and for the coupled-microcavity lasers using selective pumping. Moreover, this situation is common for all injection lasers, which are known as extremely vulnerable to the proper placing of electrodes.

The CPT for lasers sheds important light on the behavior of modal thresholds in the cavities with partial active regions. Indeed, can introduce the quantity $\Gamma_s^{(a)} \leq 1$ given by

$$\Gamma_s^{(a)} = \tilde{W}_s^{(a)} / \tilde{W}_s,$$

$$\tilde{W}_s^{(a)}(k_s, \gamma_s) = (1/2Z_0)(\alpha_a^2 - \gamma_s^2) \int_{V_a} |\vec{E}_s(\vec{R}, k_s, \gamma_s)|^2 dv, \quad (13)$$

$$\tilde{W}_s(k_s, \gamma_s) = (1/2Z_0) \int_{V_{\min}} \text{Re}(v^2) |\vec{E}_s(\vec{R}, k_s, \gamma_s)|^2 dv,$$

where $v = \alpha_d$ in V_d and 1 in V_f, and $V_{\min} = V_a + V_d + V_f$.

From this definition it is clear that $\Gamma_s^{(a)}$ is the fraction of E-field power contained in the active region. It is also the *overlap coefficient* between the active region and the modal *E*-field (a.k.a. *mode confinement factor*). This is a strictly discrete quantity having values linked to specific modes.

This enables us to re-write CPT (8) as follows:

$$\gamma_s = \alpha_a [\Gamma_s^{(a)}(k_s, \gamma_s) \, \tilde{Q}_s(k_s, \gamma_s)]^{-1}, \quad (14)$$

where now

$$\tilde{Q}_s = \tilde{W}_s / [\tilde{W}_{rad(s)} + \tilde{W}_{abs(s)}] \quad (15)$$

is the Q-factor of the *active cavity*. Further investigation of (14) assuming that the threshold is small, $\gamma_s \ll 1$, shows that the first-order approximation to γ_s is obtained if one takes the mode field components and the frequency as for a passive cavity ($\gamma_s = 0$),

$$\gamma_s = \alpha_s [\Gamma_s^{(a)}(k_s, 0) Q_s]^{-1} + O(\gamma_s^2) \quad (16)$$

Expression (16) tells that in order to achieve low threshold in the active (pump on) cavity, it is not enough to have high Q-factor of the same mode in the passive (pump off) cavity. The mode E-field overlap with active region is equally important and can dramatically counterbalance the Q-factor – this happens, for instance, with the quasi-WG modes in a stadium-cavity laser if the electrode is placed in the cavity centre.

If the modal electric field value in (5) or (13) is normalized by its maximum, then W_s or $\tilde{W}_s$ is the *effective mode volume* for a passive or an active cavity, respectively. Here, in view of (12) it is enough to take account of the first term of (5).

Effective mode volume plays very important role in the cavity quantum electrodynamics (QED) [8]. However, in cavity QED this quantity appears from heuristic considerations; besides, one and the same definition is used for active and passive cavities; the integration is usually taken only over V_p but sometimes is extended to a part of space outside the dielectrics and even outside of $V_{\min}$. In contrast, here we have introduced $\tilde{W}_s$ in rigorous and unambiguous way based only on mathematical manipulations with Maxwell equations, i.e. from first principles.

Note that in the cavity QED it is assumed that the smaller the effective mode volume, the lower the threshold of lasing. Our formula (14) convincingly shows that from the viewpoint of Maxwell equations this is not true. In fact, the role of the effective mode volume is just opposite as $\tilde{W}_s$ enters the denominator of r.h.p. of (14); however this role is balanced by the mode emission loss, $\tilde{W}_{rad(s)}$ in the numerator - these two quantities "breath" together. The real figure-of-merit of the mode in active cavity is its Q-factor, $\tilde{Q}_s$, which can be approximated by the passive-cavity counterpart, Q_s.

Thus, the Optical Theorem for the lasers considered in linear formulation has enabled us to propose a rigorous mathematical definitions of the open-resonator volume (as a minimum sphere), the effective mode volume, and the mode-active-region overlap coefficient – this, in fact, provides grounding to these quantities, widely used in semi-classical laser physics and QED as phenomenological ones.

This work has been partially supported by the European Science Foundation via the "Newfocus" network project.

REFERENCES

[1] A.I. Nosich, E.I. Smotrova, S.V. Boriskina, T. Benson, P. Sewell, "Trends in microdisk laser research and linear optical modeling," *Opt. Quant. Electronics*, 39, no 15, pp. 1993-1995, 2007.

[2] E.I. Smotrova, A.I. Nosich, T.M. Benson, P. Sewell, "Cold-cavity thresholds of microdisks with uniform and non-uniform gain: quasi-3D modelling with accurate 2D analysis," *IEEE J. Select. Topics Quant. Electron.*, vol. 11, no 5, pp. 1135-1142, 2005.

[3] V.O. Byelobrov, A.I. Nosich, "Mathematical analysis of the lasing eigenvalue problem for the optical modes in a layered dielectric cavity with a quantum well and distributed Bragg reflectors," *Opt. Quant. Electronics*, vol. 39, no 10-11, pp. 927-937, 2007.

[4] E.I. Smotrova, J. Ctyroky, T. Benson, P. Sewell, A.I. Nosich, "Lasing frequencies and thresholds of the dipole-type supermodes in an active microdisk concentrically coupled with a passive microring," *J. Opt. Soc. America A*, vol. 25, no 11, pp. 2884-2892, 2008.

[5] E.I. Smotrova, J. Ctyroky, T.M. Benson, P. Sewell, A.I. Nosich, "Optical fields of the lowest modes in a uniformly active thin sub-wavelength spiral microcavity," *Optics Letters*, 2009, vol. 34, no 24, pp. 3773-3775.

[6] N. Morita, N. Kumagai, J.R. Mautz, *Integral Equation Methods for Electromagnetics*, Boston, Artech House, 1990.

[7] A. Yariv, *Quantum Electronics*, New York, Wiley, 1989.

[8] H. Yokoyama, K. Ujihara (Eds.), *Spontaneous Emission and Laser Oscillation in Microcavities*, Roca Baton, CRC Publ., 1995.

CAOL*2010 International Conference on Advanced Optoelectronics & Lasers, 10-14 September, 2010, Sevastopol, Ukraine

Pneumatic photonic crystals: properties and multiscale indication of pressure

E. Ya. Glushko

Institute of Semiconductor Physics of NAS of Ukraine, Kiev Ukraine

(Invited Paper)

Abstract: A pneumatic optical medium (POM) sensitive to pressure and temperature due to gaseous voids inside is studied theoretically. The properties of glass, quartz and mica made POM in infrared and visual regions are analyzed. The Kerr type nonlinearity is predicted. Multiscale regimes of pressure and temperature indication are discussed.

Regarding the opto-mechanical properties the one-dimensional photonic crystals (PhCr) are mainly considered as controlled mirrors or flexible waveguides [1—2]. In our previous work, some artificial optical medium consisting of thin elastic layers divided by typed-in gaseous voids (opto-pneumatic medium (OPM)) was proposed. It was shown in [3] that an OPM sample should be optically sensitive to variations of the external pressure due to its elasticity. In this work, we study the novel pneumo-optical effects specific to pneumatic photonic crystals - a new kind of artificial periodic structures with impressed into them gaseous voids – that can vary the lattice period under the action of the external pressure, temperature and in dependence on the beam intensity.

Optical properties of elastic 1D PhCr with air voids were discussed in [3] for intrinsic and external incidence of the beam. OPM as a photoelastic media with parametric optical sensitivity to the external pressure can vary its geometry, and the reflectivity will be drastically changed depending on the external conditions.

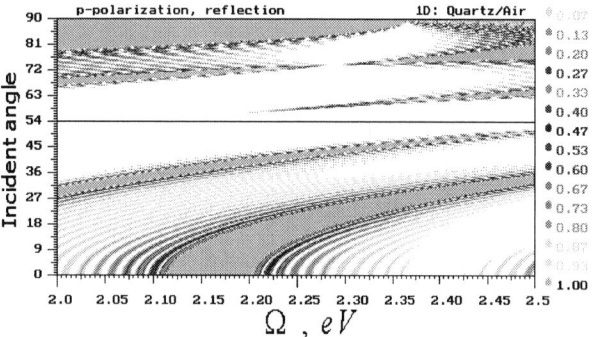

Fig. 1. 20 period quartz/air 1D PhCr. d_1=0.5 μm, d_2=1.0 μm. Frequency diagram for the reflection R of external incident light; the angles are (0 -- 90°). Photon energies are (2 -- 2.5) eV. Right column: color scale for reflection R. Horizontal line at 54.4° corresponds to the Brewster angle of the structure.

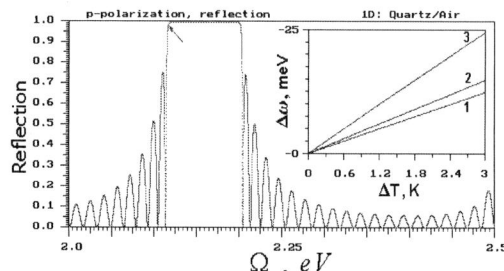

Fig. 2. 20 period quartz/air 1D PhCr. d_1=0.5 μm, d_2=1.0 μm. Normal incidence reflection vs photon energy. Frequency gap ($R≈1$) is distinguished by color; arrow shows beginning frequency. Inlet: gap beginning energy shift $\Delta\omega$ dependence on the temperature growth ΔT. 1, T_0=300 K, 2, T_0=250 K, 1, T_0=200 K.

In Fig. 1, we show the angular area representing the so-called external problem [3, 4]. The color diagram for reflection is presented for energy interval (2 -- 2.5) eV and angular interval (0° -- 90°). A transparency band near 54.4° is explained

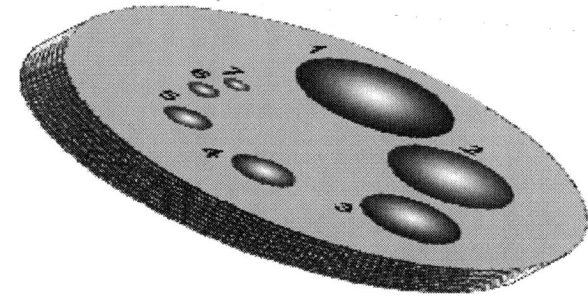

Fig. 3. A layered pneumosystem -- multiscale pressure indicator [3]. The stack of plates with circular OPM 1, 2, ...7 which show seven scales with sensibilities η_1=10^2, η_2=10^1 ,..., η_7=10^{-4} $mbar^{-1}$. The device covers seven orders of magnitude of measured pressure..

by the Brewster effect. In Fig. 2, the calculated reflection at normal incidence is plotted for energy interval (2 -- 2.5) eV as a section of Fig. 1 diagram at 0°. The reflection window (frequency gap) is observed in energy interval (2.13 -- 2.20) eV. A comprehensive analytical approach for a p-polarized wave propagating in a binary 1D PhCr structure consisting of

978-1-4244-7043-3/10 $26.00 © 2010 IEEE

Table 1. Isothermic optical sensibility $\eta = \left(\dfrac{\partial R}{\partial P}\right)_T$. ** -- negative magnitude.

Material/periods	Young modulus E, GPa	Sizes of material/air/R, μm; incident angle.	photon energy, eV	$|\eta|$, mbar^{-1}
Si/100	80.0 (amorphous)	1/4/550-800; 0°	0.350235	4.1-20.6
SiO$_2$/100	55.66	4/6/550-800; 0°	0.31660	12.1-53.9**
SiO$_2$/200		1/5/550-800; 0°	0.3501812	330-900
Mica/100	88.0	4/6/550-800; 0°	0.45254	54-238**
Mica/20		4/6/550-800; 36°	0.402	1.4-5.4**

N periods of two contacting materials and an end layer was considered in [3]. The sewing of two problems - internal and external and the bandgap structure transition near an external whispering incidence angle of $90°$ and an angle of total internal reflection was analyzed. It was shown there that on the boundary of the total internal reflection area and the area of transmitted waves the external whispering reflection at $90°$ as if feels the beginning of the bandgap structure inside the total internal reflection region. For the chosen frequency interval, the only reflection window is distinguished by color where the arrow marks off the initial frequency of the window.

A pneumatic PhCr may possess well expressed optical nonlinearity of Kerr type. Let us consider the effective coefficient of refraction for a binary periodic layered structure $n_{eff}=(n_1 d_1+ n_2 d_2)/(d_1+d_2)$, which changes if gaseous voids d_2 widens or narrows due to some reasons. viation of the refraction coefficient is proportional to the void width change

$$\Delta d_2 : \Delta n_{eff} = \frac{d_1 \Delta d_2 (n_2 - n_1)}{(d_1 + d_2)^2}.\qquad (1)$$

As a reason of this kind of optical nonlinearity, one can consider the irradiation absorption caused heating of walls and gaseous voids inside the structure. The absorption leads to temperature growth and, correspondingly, gaseous voids widen their widths. This phenomenon is proportional to light intensity and dominates in the total change of the effective refraction coefficient $n_{eff} = n_{eff}(0)(1+q_i I(\omega))$, where $I(\omega)$ is beam intensity, $n_{eff}(0)$ is effective refraction coefficient in the limit of weak intensity, q_i is the Kerr nonlinearity coefficient. The latter is proportional to the absorbed intensity of light

$$\Delta I(\omega) = I(\omega) a_\lambda \frac{d_1 + d_2}{2}, \qquad (2)$$

where a_λ is the mean absorption coefficient of considered medium, d_1 is the wall thickness. In a more detail approach the absorption of both media of binary structure should be considered separately. For a thin-walled OPM the temperature dependence of the width d_2 is determined by the expression $d_2=d_2(0)(1+\alpha\Delta T)$, where $\alpha \approx 1/T_0$. In our case $\alpha \approx 1/300$ K^{-1}. The temperature growth ΔT is the beam pulse caused temperature increasing inside the medium. In the inlet to Fig. 1b, the gap shift $\Delta\omega$ dependence on the temperature growth ΔT is shown for $T_0=200, 250$, and 300 K. A simple consideration gives for the nonlinearity coefficient q_i,

$$q_i = \frac{\alpha a_\lambda \Delta \tau d_2 (n_2 - n_1)}{2 C_w \rho_w (n_1 d_1 + n_2 d_2)}.\qquad (3)$$

Here, $\Delta\tau$ is the pulse duration, C_w and ρ_w are the material mean specific heat and density. For widespread parameters, an approximate formula for optical pneumatic nonlinearity arises $q_i \approx 10^{-4}\, a_\lambda \Delta\tau\ cm^2/W$, where a_λ is measured in cm^{-1} and $\Delta\tau$ is measured in seconds. Taking into account that in a wide frequency interval for glasses a_λ is ranged from 1 to 20 cm^{-1}. one can conclude from (3) that strong nonlinearity may be observed for $\Delta\tau > 10^{-6}$ s.

The high optical contrast and pressure and temperature depended elasticity are two main factors contributing to the high optical sensitivity of OPM. The results of calculations performed in [3] for pressure sensitivity of OPMs based on silicon, glass, and mica plates are presented in Table 1. One more novel optical characteristics -- the OPM system's isobaric optical sensitivity $\eta=(dR/dT)_P$, is measured in K^{-1} and means detected change of reflectivity when the system temperature is changed at constant pressure. For ideal gas inside a void, the temperature deviation of its width at a chosen standard temperature T_0 is proportional to the change of temperature. The gap edge shift $\Delta\omega$ dependence derivative in the inlet to Fig. 2 is proportional to isobaric optical sensitivity. Our estimation gives for 20 period OPM η ranges from 1 to 10 K^{-1} in temperature interval $100 - 300$ K.

The sensitivity strongly depends on the OPM radius r. It gives an opportunity to create an integrated multiscale indicators of pressure, temperature and intensity of light. The passage to lesser radii r strongly decreases the structure elasticity $\sim r^4$ and means a change in the scale of measured magnitudes.

REFERENCES

[1] A. Werber, and H. Zappe, "Tunable, membrane-based, pneumatic micro-mirrors", J. Opt. A: Pure Appl. Opt. vol. 8, pp. S313-S317, 2006.

[2] V. Pervak, I. Ahmad, M. K. Trubetskov, A. V Tikhonravov, and F. l. Krausz, "Double-angle multilayer mirrors with smooth dispersion characteristics", Opt. Express, vol. 17, pp.7943-7951, 2009.

[3] E. Ya. Glushko, "Pneumatic photonic crystals", Opt. Express, vol. 18, pp. 3071-3078, 2010.

[4] E. Ya. Glushko, A. E. Glushko, V. N. Evteev, and A. N. Stepanyuk, "All-optical signal processing in photonic crystals covered with fast nonlinear materials", Proceedings of SPIE vol. 6889, pp.19-30, 2008.

Non-stationary coherent population trapping in frequency-modulated fields

Vladimir L. Derbov and Alexey S. Lyubimov
Saratov State University, Saratov, Russia
(Invited *Paper*)

Abstract: Coherent population trapping (CPT) resonance formation is modeled numerically in a three-level Λ-system with one of the near-resonance fields being frequency-modulated. The model is based on density matrix equations in RW approximation with atomic relaxation properly taken into account. Slow modulation is shown to be equivalent to CW excitation with the frequency changed point by point. As the modulation period approaches the relaxation times, the delayed response of the system is shown to cause the CPT resonance shift and reshaping, as well as the appearance of transient oscillations of the energy level populations.

Coherent population trapping (CPT) in three-level double-resonance Λ-systems is known to produce narrow resonances under the condition of equal frequency detuning of two laser waves from two corresponding adjacent transitions. These resonances appear as sharp dips in the population of the upper state usually monitored via laser-induced fluorescence intensity measurements. Physical nature and numerous applications of CPT resonances, such as atomic clocks and high-resolution laser spectroscopy, are discussed in a number of papers (see, e.g., [1-4]).

To observe a CPT resonance it is common to fix the frequency of one laser wave and to vary that of the other. Here we consider this variation as a real-time process, i.e., the second laser is assumed to be frequency-modulated and the upper energy level population is plotted versus both the time and the instantaneous frequency. It is shown that when the period of modulation approaches the order of atomic relaxation times the delayed response of the atomic system manifests itself as transient features in both temporal and instantaneous frequency dependencies of the upper energy level population. These features may provide additional information about the relaxation properties of the system.

We consider a three-level system with energy levels $E_1 < E_2 < E_3$ excited with two laser waves at frequencies $\omega_p \approx \omega_{31}$ and $\omega(t) = \omega_s + \omega_1 \sin \Omega t \approx \omega_{32}$, where $\omega_{ij} = (E_i - E_j)/\hbar$ are the transition frequencies, ω_s is the carrier frequency of phase-modulated field, ω_1 and Ω are the amplitude and frequency of modulation, respectively. The full set of equations for the density matrix elements ρ_{ij} in the RW approximation was used to model numerically the interaction of the laser fields with the atom:

$$\dot{\tilde{\rho}}_{21} = \frac{1}{i}(\omega_{21} - (\omega_p - \omega_s) - i\gamma_{12})\tilde{\rho}_{21} + \frac{1}{\hbar i}((\tilde{\rho}_{31}(V_s^-)_{23} - \tilde{\rho}_{32}^*(V_p^+)_{31})$$

$$\dot{\tilde{\rho}}_{31} = \frac{1}{i}(\omega_{31} - \omega_p - i\gamma_{13})\tilde{\rho}_{31} + \frac{1}{\hbar i}((\rho_{11} - \rho_{33})(V_p^+)_{21} + \tilde{\rho}_{21}(V_s^+)_{32})$$

$$\dot{\tilde{\rho}}_{32} = \frac{1}{i}(\omega_{32} - \omega_s - i\gamma_{32})\tilde{\rho}_{32} + \frac{1}{\hbar i}((\rho_{22} - \rho_{33})(V_s^+)_{32} + \tilde{\rho}_{21}^*(V_p^+)_{31})$$

$$\dot{\rho}_{11} = \frac{1}{\hbar i}((V_p^-)_{13}\tilde{\rho}_{31} - c.c.) + W_{21}\rho_{22} + W_{31}\rho_{33}$$

$$\dot{\rho}_{22} = \frac{1}{\hbar i}((\tilde{\rho}_{32}(V_s^-)_{23} - c.c.) + W_{32}\rho_{33} - W_{21}\rho_{22}$$

$$\dot{\rho}_{33} = \frac{1}{\hbar i}((-\tilde{\rho}_{31}(V_p^-)_{13} - \tilde{\rho}_{32}(V_s^-)_{23} + c.c.) - W_{32}\rho_{33} - W_{31}\rho_{33}$$

Here the slow envelopes of the non-diagonal elements are defined as $\rho_{21} = \tilde{\rho}_{21}e^{-i(\omega_p - \omega_s)t}$, $\rho_{31} = \tilde{\rho}_{31}e^{-i\omega_p t}$, $\rho_{32} = \tilde{\rho}_{32}e^{-i\omega_s t}$, γ_{ij} are the transition line widths, W_{ij} are the relaxation transition rates, *c.c.* stands for complex conjugate. The slow envelopes of the interaction operators are expressed via the transition dipole matrix elements d_{ij} and the complex amplitudes of the fields $E_{p,s}^0$:

$$(V_{p,s}^+)_{ij} = -d_{ij}E_{p,s}^0, \quad (V_{p,s}^-)_{ij} = -d_{ij}E_{p,s}^{0*}$$

Initially the atom is assumed to be in the ground state $\rho_{11}(0) = 1$, $\rho_{ij}(0) = 0$, $i, j \neq 1$. The field E_s^0 is modulated in phase as $E_s^0 = A\exp[i(\omega_1/\Omega)\cos\Omega t]$, while E_p^0 is constant.

After switching the fields on, followed by transient behavior, we observed stable periodic variations of the level populations synchronized with the frequency modulation. For numerical simulation we scale the frequency differences to γ_{31}, so that $\gamma_{31} = 1$ and the unit of time is $2\pi/\gamma_{31}$. We put $\hbar = 1$, then it is easily shown that at $d_{13} = 1$ the amplitude $E_p^0 = 0.5$ corresponds to the saturation intensity, i.e., the intensity that reduces the population difference by the factor ½. Consider for comparison with [4] the following set of parameters: $\omega_{21} = 4$, $\omega_{31} = 6$, $\omega_{32} = 2$, for all transitions $d_{ij} = 1$, the relaxation parameters $\gamma_{31} = \gamma_{32} = 1$, $\gamma_{21} = 0.0001$, $W_{31} = W_{32} = 0.5$, $W_{21} = 0.0001$. Let the CPT resonance condition be valid for the carrier frequencies, i.e. $\omega_p = 6$ and $\omega_s = 2$, and $E_p = A = 2$ (strong saturation). First, let us consider slow modulation, namely, $\Omega = 0.005$, with the amplitude $\omega_1 = 3.5$ sufficient to span the excitation line.

978-1-4244-7043-3/10 $26.00 © 2010 IEEE

CAOL*2010 International Conference on Advanced Optoelectronics & Lasers, 10-14 September, 2010, Sevastopol, Ukraine

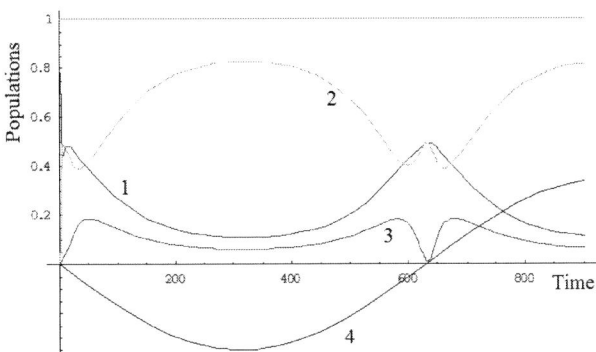

Fig. 1. Slow modulation of frequency produces dips in the level 3 population at the time of crossing the point where the two frequency detunings coincide. 1,2,3 – level populations, 4 – the frequency $\omega(t) - \omega_s = \omega_1 \sin \Omega t$

The results are presented in Fig.1 as plots of populations ρ_{ii} versus time. We can see that when the modulated part of the probing frequency passes zero the population ρ_{33} of level 3 (curve 3) goes sharply down. This is a manifestation of CPT, whose resonant nature becomes even more apparent if we plot ρ_{33} versus the instantaneous frequency (Fig. 2). Note, that this is rather a parametric plot $\rho_{33}(t)$ versus $\omega(t)$ than a Fourier spectrum $\rho_{33}(\omega)$. Slight shift to the blue from the exact resonance at $\omega = 2$ is due to small, but nonzero, delay in the system response. Increasing the field magnitudes under the same conditions is shown to decrease the residual population of level 3. However, this positive effect is accompanied by saturation broadening which is a disadvantage for spectroscopic goals.

Now increase the modulation frequency up to $\Omega = 0.05$, i.e., by ten. Similar plots in Fig. 3 and Fig. 4 reveal apparent manifestations of delayed response and transient behavior (damped oscillations) after each pass of the modulated frequency through the zero point.

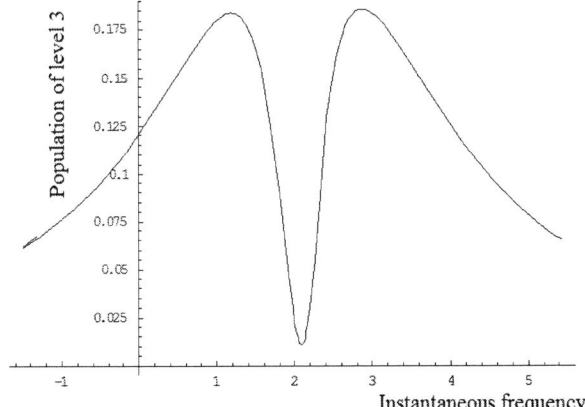

Fig. 2. Population of level 3 versus the instantaneous frequency at slow modulation. The narrow CPT dip is only slightly shifted to the blue due to the delayed response.

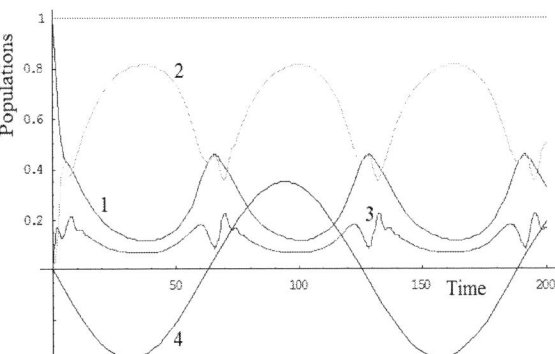

Fig. 3. Increasing the modulation frequency results in delayed CPT dips and transient oscillations.

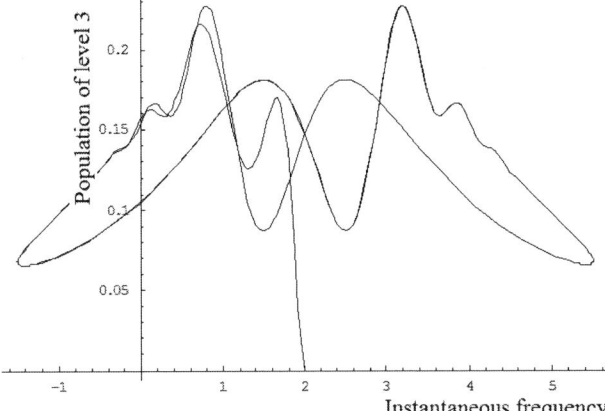

Fig. 4. Delayed CPT manifests itself as splitting of the dip and sideband oscillations in the $\rho_{33}(t)$ versus $\omega(t)$ plot.

Dips of $\rho_{33}(t)$ in Fig. 3 are obviously delayed with respect to frequency zero points. Each is followed by damped oscillations that remind free induction decay. In Fig. 4 these transient features are seen as symmetric splitting of the dip and sideband oscillations. We believe that plotting such curves on the base of frequency-modulation CPT experiment provides an additional tool for extracting relaxation rates from experimental data.

The work was supported by RFBR grant No. 08-01-00604-a.

REFERENCES

[1] J. Vanier, "Atomic clocks based on coherent population trapping: a review," *Appl. Phys. B*, vol. 81, no. 4, pp. 421-42, 2005

[2] Huss, R. Lammegger, L. Windholz, E. Alipieva, S. Gateva, L. Petrov, E. Taskova, and G. Todorov, "Polarization-dependent sensitivity of level-crossing, coherent-population-trapping resonances to stray magnetic fields," *J. Opt. Soc. Am. B*, vol. 23, no. 9, pp. 1729-1736, 2006.

[3] E. Arimondo, "Coherent population trapping in laser spectroscopy," *Prog. Opt.*, vol. 35, pp. 257-354, 1996.

[4] B D Agap'ev, M B Gornyĭ, B G Matisov, and Yu V Rozhdestvenskiĭ, "Coherent population trapping in quantum systems," *Phys.-Uspekhi*, vol. 36, no. 9, pp. 763-798, 1993.

978-1-4244-7043-3/10 $26.00 © 2010 IEEE

Mechanisms of holographic Recording in Photopolymers

(Invited Paper)

V.Obukhovsky[1], T.Smirnova[2],

[1]National Taras Shevchenko University of Kiev, 64, Volodymirska Str., Kiev 01601 Ukraine
[2]Institute of Physics, Prospect Nauki, 46, Kiev 03028 Ukraine

Abstract: Overview of physical mechanisms of photopolymers formations for laser and holographic applications.

The interest to photopolymers as the media for volume phase hologram recording is due to their practical advantages, such as relative cheapness, high diffraction efficiency of the recorded gratings and absence of "wet" development process [1-4].

Among variety of photopolymers, the materials exist where the hologram recording involves a spatial transfer of molecular components. These materials were called photoformers (PF). The distinguishing features of photoformers are: a) the mixture consists, at least, of two components (monomers) that photopolymerize independently, but with the substantially different rates; b) the refractive indices (or absorption coefficients) for the components are different; c) the spatial mass transfer takes place during the spatially inhomogeneous polymerization process when the mixture is exposed to light fringes.

Mutual diffusion of the components or local shrinkage of the materials during polymerization could cause the mass transfer. Both the refractive index and absorption coefficient can be modulated in that way. We develop the theoretical model of hologram formation in photoformers with the polymerization-diffusion mechanism of spatial separation of components. The importance of a nonlinear diffusion under inhomogeneous photopolymerization is underlined. The generalization of the Fick's law for diffusion in the systems with the phase transformations of the mixture components is discussed.

Generally, the theory of nonlinear diffusion [5-6] can be supplied to investigation of any multicomponent mixture of molecular liquids. If the condition of volume conservation takes place, the reverse diffusion flows taking into account automatically in the theory. It is shown, the interconnection between different molecular flows leads to nonlinearity of diffusion for the case of three (or more) components in mixture.

Process of diffusion mass-transfer are described by system of equations

$$\frac{\partial U_n}{\partial t} + div \vec{J}_n = S_n \quad , \tag{1}$$

$$\vec{J}_n = \sum_m d(n,m) \cdot [U_n \nabla U_m - U_m \nabla U_n] \quad , \tag{2}$$

$$\sum_n U_n = 1 \quad , \quad \sum_n \vec{J}_n = 0 \quad , \quad \sum_n S_n = 0 \quad . \tag{3}$$

Here U_n - relative volume of n-component; S_n - source functions (are determined by interactions between molecules in solution). Characteristics of diffusion in liquids are determined by matrix $d(n,m)$, that depends on physical properties of components only (but independent on their concentration).

It was proved, for two-components-mixtures the system of equations (1-3) can be converted into linear Fick's equations exactly. In more general case, a number of molecular complexes can be formed as a result of interaction between different components in process of its dilution.

The problem of diffusion coefficients measurements is discussed. In typical experiments, only «effective diffusivities» $D_j^{ef}(\{N_j\})$ are measured. Taking into consideration (1) and interrelation between "molecular complex" and "pure components" concentrations, functional dependence $D_j^{ef}(\{N_j\})$ can be found for standard models of liquid solutions.

Determination of the predominant mechanism of the polymer radical recombination in a specific photopolymer is an independent problem. It can be shown that the investigation of the recording kinetics of the holographic gratings in photopolymers can be used as a basis of the method for the determination of the recombination process order.

Suppose, for instance, that the binary recombination prevails. Then the polymerization kinetics dependence on the total light intensity I_0 can be found. Let's consider the phase (stage) of the polymerization process corresponding, for example, to 90% polymerization. In accordance with[5], the normalized (dimensionless) time of reaching this phase $t'_{0.9}$ is constant and does not depend on I_0. This means that the real time of reaching the 90%-polymerization stage depends on I_0 according with the rule

$$t_{0.9} = t'_{0.9} / I_0^2 \qquad (4)$$

(besides it has been taken into account that $I_0 \gg I_t$). In the case of linear recombination ($L_1 \neq 0$ $L_2 = 0$) it could be similarly obtained $t_{0.9} \sim 1/I_0$. Thus it can be determined whether the recombination is linear or quadratic by studying the dependence of $t_{0.9}$ on I_0.

The function $t_{0.9}(I_0)$ was experimentally found for both the ordinary photopolymer PP-A and photoformer PPC-488. The results are shown in Fig.1 (in double logarithmic scale).

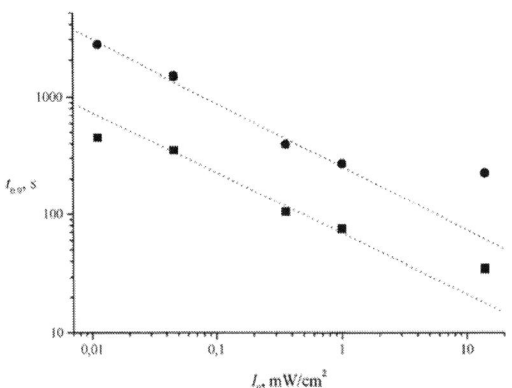

Fig. 1. Pump intensity dependence of characteristic polymerization time $t_{0.9}$ time of reaching of 90% level of n_1^{st} modulation.Dots are the experimental data for binary photoformer PPC-488, squares are that for the photopolymer PP-A, the lines are the best fit to linear arrroximation

As is obvious, a linear approximation can be applied over the medium intensity range. The experimentally determined average tangent of the inclination of $\ln t_{0.9}$ relative to $\ln I_0$ is $tg\alpha = -0.52\pm0.02$. On the other hand, it follows from Eq. (4) that in the case of binary recombination

$$tg\,\alpha = \frac{\partial(\ln t_{0.9})}{\partial \ln(I_0)} = -0.50 \qquad (5)$$

Therefore, we can arrive at the conclusion that bimolecular recombination prevails in the investigated materials PP-A, PPC-488. Then linear recombination may be neglected in our calculations, i.e. it can be supposed that $L_1=0$.

To check the validity of the approximation for the parameter of bimolecular chain termination suggested above, the experiments on the photopolymerization of the PP-A by homogeneous laser radiation beam ($\lambda = 488$ nm) have been performed. The polymerization degree of photopolymer composition was determined from measurement of the refractive index value $n(t)$ (using the interferometric method) for the plane samples investigated.

Let us suppose the initial refractive index to be n_1 and the final one to be n_2. Then, within a linear approximation, the polymerization degree P_Σ (the volume fraction of all the high-molecular components), at the moment of time t, can be found as $P_\Sigma(t) = \left[n_1 - n(t) \right] / \left[n_1 - n_2 \right]$. The data obtained allow plotting the graph of the auxiliary function $d\eta/dt'$, where $\eta(P_\Sigma) = 2(1 - P_\Sigma)^{-1/2}$ (see Fig. 2). $\qquad (22)$

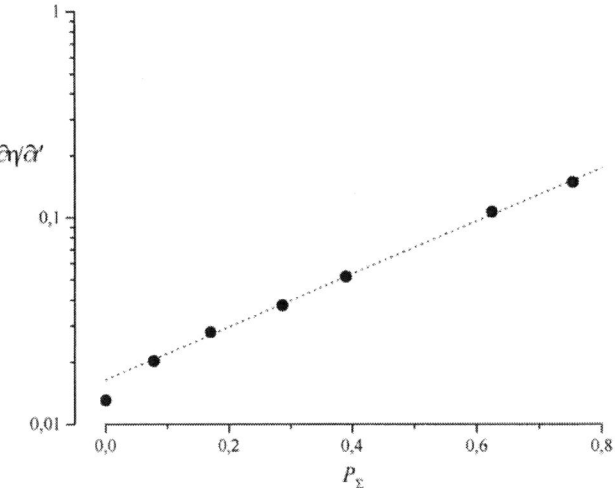

Fig. 2. Dependence of $d\eta/dt'$ on the polymer fraction under homogeneous illumination. Dots are the experimental results, the line is the best fit to linear dependence

In the experiment, the light intensity I_0 was chosen in such a way that the inequalities $\tau_r \ll \tau_p \ll \tau_c$ hold (this is not complicated if the initiator dissociates slowly, but the polymer radicals quickly decay). In this case

$$\frac{d\eta(P_\Sigma)}{dt'} = \exp(\gamma P_\Sigma / 2) \qquad (6)$$

It has been taken into account that $F(M) = M$, $C = C_0$ and, in absence of a neutral component, $M = 1$ (23). One can see from the results shown in Fig. 2 that the theoretical dependence (6) is well confirmed by the experimental data. The slope of the function $d\eta/dt'$ (taken on the logarithmic scale) allowed us to determine the parameter γ directly. We found $\gamma \approx 6.0$ for the photoformer PPC-488.

The modulation depth of the polymer component P_1^{st}, in a steady state, depends, certainly, on the presence of a neutral component. The theoretical dependence $n_1^{st}(N_0)$ is shown in Fig.3. In the same figure, the experimental data for the steady state modulation depth of the refractive index n_1^{st} are shown. It is clear that, in absence of an initial component ($N_0 = 0$), there is no stationary modulation of the refractive

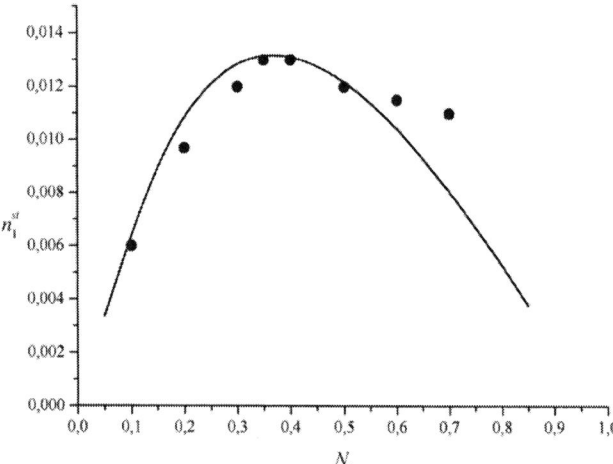

Fig. 3. Neutral component influence on refractive index modulation in PPC-488 under steady-state conditions. The solid line represents the theoretical results, the dots represent the experimental data

index $n_1^{st}(N_0 = 0) = 0$. The same results would be obtained as well in the case of $N_0 = 1$ (the monomer is completely absent). Thus we conclude that the optimum composition of mixtures should be about equal concentration $N_0 \approx M_0 \approx 0.5$. However the large concentration N decreases the average rate of polymerization and makes worse the condition of the spatial modulation $P(\mathbf{r},t)$. Therefore the optimum value N_0^{opt} (that ensures the maximum modulation depth) is less than 0.5. The calculations gave the optimum concentration of a neutral component is in a range $N_0^{opt} = 0.35\text{-}0.45$, depending on both the recording condition and mixture parameters. This corresponds fairly well to the results of our measurements.

Photopolymer gratings and gratings on the base of photopolymer-nanoparticles mixtures can be used as DFB cavities for microlaser production.

CONCLUSION

Photoformers are the special kind of photopolymerizing compositions with the essential differences as compared with the other photopolymers. The following features should be taken into account for their description
— the spatial inhomogeneity of photopolymerization;

— the regular (macroscopic) spatial transfer of the mixture components

For substantial phase transition due to photostimulated polymerization, the mass diffusion becomes highly nonlinear. In this case the mutual diffusion of both components of the mixture takes place and usual Fick's law have to be changed by the law of nonlinear diffusion.

As a consequence of the large difference of the refractive indexes of movable components ($|n_n^o - n_p^o| \approx 0.14$) the holograms recorded in PPC-488 are characterized by the large diffraction efficiency (up to 90). It is clear that the diffusion processes restrict the minimum values of the spatial frequencies K of holographic gratings. For example, at $\lambda = 488$ nm, $I_0 = 14$ mW/cm^2, the diffraction efficiency of recorded holograms begins to decrease for $K < 1.8 \cdot 10^3$ cm^{-1}. However, the maximum line resolution of PPC-488 is large enough and allows for the recording of information with the density up to $2 \cdot 10^5$ cm^{-1}.

To study both the polymerization kinetics and spatial transfer of substances in the photoformers, the optical (holographic) methods can be used. Based on these methods, it has been shown under the biradical mechanism of of reactionary chain termination takes place in the PPC-488. The dependence of the chain termination function L_2 on polymer concentration is well approximated by the exponential dependence.

The developed theory describes the main features of the holographic grating recording in photoformers and is in a good agreement with the experimental data. It allows for the evaluation of photoformer parameters as dark photopolymerization constant I_t, diffusivity D_0, and relaxation times for different processes.

It should be mentioned, our theory describes the binary photoformers with the main contribution to holographic recording caused by the polymerization-diffusion mechanism of mass transfer. The generalization of the present theory by taking into account the shrinkage in hologram formation is a subject for further investigations.

REFERENCES

[1]   W.J.Gambogi, W.K.Smouthers, K.W.Steijn, S.H.Stevenson and A.M.Weber, Proc. SPIE 2405 (1995) 62.
[2]   M.Kavabata, A.Sato, I.Sumiyoshi and T.Kubota, Appl. Opt. 33 (1994) 2152.
[3]   S.Piazzola and B.K.Jenkins, Opt. Letters 21 (1996) 1075.
[4]   C.Carre, D.J.Lougnot and J.P.Fouassier, Macromolecules 22 (1989) 791.
[5]   G. M. Karpov, V. V. Obukhovsky, T. N. Smirnova, V. V. Lemeshko. Optics communication, v. 174, 391-404, 2000 .
[6]   E. S. Gulnazarov, V. V. Obukhovsky, T. N. Smirnova, Holographic recording materials. Ed. by H. J. Bjelkhagen. SPIE Milestone Series, v. MS-130, 1997, pp. 473-475.

CAOL*2010 International Conference on Advanced Optoelectronics & Lasers, 10-14 September, 2010, Sevastopol, Ukraine

Superradiant Heterolasers

Vladimir V. Kocharovsky[1], Mikhail A. Garasoyv[1], Petr A. Kalinin[1], Ekaterina R. Kocharovskaya[1],
Alexey A. Belyanin[2], Vitaly V. Kocharovsky[1,2]

[1]Institute of Applied Physics RAS, Ulyanov str., 46, Nizhny Novgorod, 603950, Russia
[2]Department of Physics, Texas A&M University, College Station, Texas 77843-4242, USA

(Invited *Paper*)

Abstract: The phenomenon of collective recombination (superfluorescence or superradiance) of electrons and holes in semiconductor laser structures of various dimensions is reviewed. At the current level of technology, this phenomenon can be realized most easily in quantum-well and quantum-dot heterostructures. Under CW pumping, it could lead to the generation of powerful coherent radiation pulses which are shorter than incoherent relaxation times and have nothing to do with mode-locking technique. Particular attention is paid to the external magnetic field and distributed feedback of waves which favor suparradiant lasing.

Superfluorescence (SF) is one of very few observable QED phenomena in which a macroscopic ensemble of initially incoherent quantum oscillators demonstrates cooperative, coherent behavior due to efficient mutual self-phasing [1, 2]. For semiconductor samples, this means that incoherently excited and densely packed electrons and holes can emit coherent pulses of recombination radiation in a timescale shorter than the de-phasing time of their optical polarization. Similar pulses can also be generated in the general case of recombination superradiance (SR), when electron-hole oscillators are partially coherent from the very beginning. SF or SR places a fundamental upper limit on the rate of stimulated radiative transitions in the self-consistent electromagnetic field

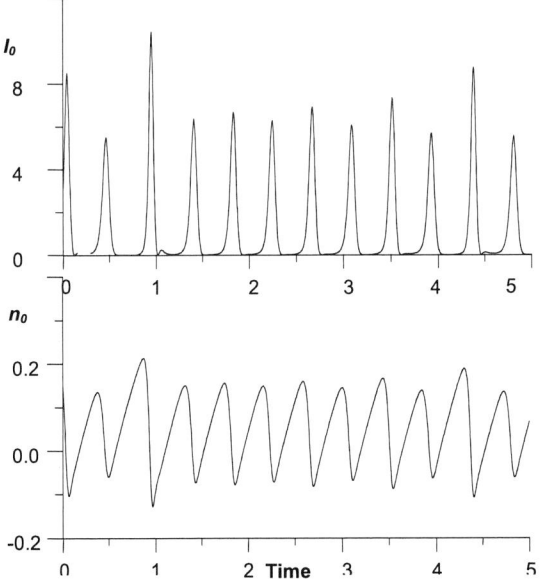

Fig. 1. Typical SR recombination of electrons and holes from the ground Landau levels in a MQW heterostructure [7-9] (in abr. units); I_0 is the radiation intensity, n_0 the inversion (difference of populations of these levels).

produced by electron-hole recombination in the presence of pulsed or continuous-wave (CW) incoherent pumping.

However, despite the fundamental physical beauty of the process, its applications have so far been limited mainly to refined spectroscopic experiments and superradiant lasing in semiconductor samples has not been achieved yet, because the conditions necessary for cooperative self-phasing are usually rather extreme. As a rule, the crucial problem is finding a way to overcome particle collisions destroying the phase of oscillating dipole moments of electron-hole pairs.

The aim of the present paper is to show that the modern semiconductor technology of heterostrustures makes it possible to overcome the threshold of SF/SR process. We will propose several methods of spectral selection of both electron-hole pairs and cavity modes in heterostructures which could result soon in experimental demonstration of suparradiant lasing. The timescale of expected SF/SR pulses falls into the picoseconds or sub-picosecond range. This could lead to creation of a compact semiconductor source of coherent ultrashort pulses which operates under CW pumping (injection or laser) at room temperature and does not require any mode-locking technique.

Fundamentals and simple schemes capable of the SF/SR process in heterostructures may be found in the papers [2-5] and references therein. In the present report we briefly discuss recent results of the theoretical and experimental investigations of the SF/SR phenomenon in semiconductors and consider more advanced heterolaser systems based on multiple quantum-well (MQW) or quantum-dot (MQD) heterostructures as well as heterostructures with quantum wires or quantized electron-hole plasma in a strong magnetic field. The advantages and disadvantages of superradiant vertical-cavity surface-emitting lasers (VCSELs) are also discussed. As promising prototypes of superradiant heterolasers, we analyze the magnetized MQW structures [6-9] and the sub-monolayer MQD structures with distributed feedback of the counter-propagating waves [10-12]. In the former case, the effects of inhomogeneous broadening of Landau levels, self-channeling of so-called hot polariton modes, possible multi-color SR lasing, and nonlinear frequency mixing of resulted SR pulses are pointed out. In the latter case, we focus on the interplay of Bragg mode selection and inhomogeneous broadening of MQD spectral line, namely, we indicate the domain of heterostructure parameters and the range of incoherent pumping where the conditions of superradiant lasing are less restrictive. Various superradiance regimes of expected laser operation are outlined.

The resonant interaction of electron-hole plasma with radiation in heterostructures is adequately described by the

978-1-4244-7043-3/10 $26.00 © 2010 IEEE

semiconductor Maxwell–Bloch equations in the mean-field approximation (they are discussed, e.g., in [2]). We expect that the superradiant lasing in heterostructures will proceed in the low-quality mode rather than unidirectional SF/SR regime [2-5]. Typical illustrations are given on Figs. 1 – 3 for two above-mentioned cases of MQW and MQD heterolasers.

In the first case, the combination of the reduced MQW dimensionality and the magnetic field fully quantizes the electron-hole pairs into an atomic-like system with a series of Landau energy levels, thus strongly increasing the spatial and energy densities of states and making the superradiant lasing possible in the magnetic field of the order of 10 – 50 T. (An example of SR pulsed lasing between the ground Landau levels is given on Fig. 1.) Typical injection or optical pumping of a heterostructure with magnetized MQWs creates carriers with energies considerably higher than the energies of the lower levels of dimensional quantization and Landau sublevels of the lower dimensional quantization level. Then, the intraband scattering of carriers provides cascade pumping of neighboring recombination transitions, which can lead, in principle, to the simultaneous generation of corresponding resonant modes. (This was already observed in the experiment [13] with ultrafast pulsed pumping which resulted in high density and energy confinement of electro-hole pairs sufficient for the generation of their spontaneous macroscopic polarization that decayed through the coherent emission of SF pulses.) We present the analysis of the SR generation of two modes that are resonant with the two transitions between pairs

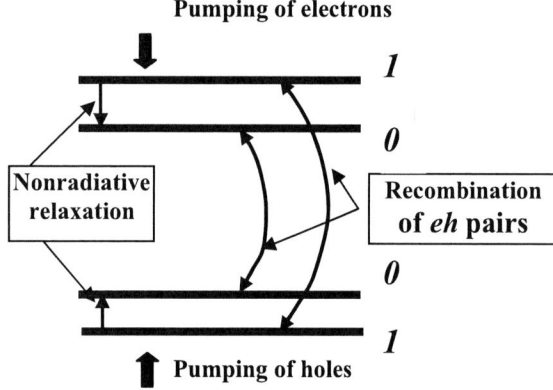

Pumping of electrons

Fig. 2. Cascade pumping of electrons and holes at the ground and first Landau levels in a MQW heterostructure.

of neighboring Landau levels of electrons and holes (Fig. 2). Numerical solution to the appropriate Maxwell–Bloch equations shows that there is the two-color SR lasing under CW pumping. In this regime, the pulse sequences generated on the adjacent recombination transitions can be essentially different (with respect to the repetition rate, power and duration). However, each pulse is accompanied by a changing of the inversion sign on the corresponding transition, which is typical for the SR lasing and unusual for the standard lasing.

In the second case, we analyze the pulsed regime of SR generation in a low-quality cavity of sub-monolayer MQD

heterolaser with distributed feedback of the counter-propagating waves. Superradient lasing would be suppressed by strong inhomogeneous broadening of MQD spectral line and becomes possible only due to proper Bragg selection of the cavity modes owning to this distributed feedback. The proposed heterolaser generates a chaotic or quasiperiodic sequence of SR pulses under cw pumping, as shown on Fig. 3.

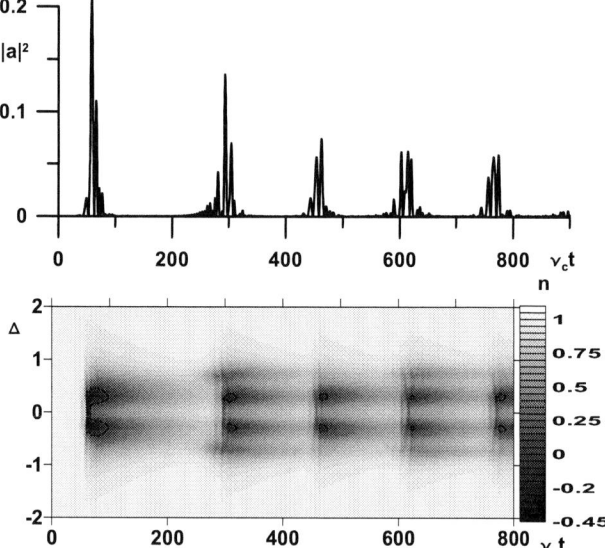

Fig. 3. Typical pulsed oscillogram of the radiation intensity, $|a|^2$, and dynamical spectrum of the quantum-dot inversion at a heterostructure facet, n, in the MQD heterolaser with distributed feedback [12] (in abr. units).

We use the dynamical spectra approach as a powerful representation of the inversion and field evalution and give clear physical interpratation of various features of the superradiant lasing on time-scales from fractions of a MQD polarization relaxation times to several MQD inversion relaxation times. In particular, we resolve and explain the formation of pulsed spectral hole burning and quasiperiodic evolution of mode-related spectral channels in an inhomogeneously broaden line of a MQD heterostructure. Again, we emphasize that population inversion vanishes after each SR pulse and the field intensity in the cavity is attenuated by several orders of magnitude during recovery of population inversion by powerful pumping. This is exactly why each separate pulse of a given spiking regime is qualitatively similar to a standard pulse of mode SR or mode SF [2-5].

In both cases we point out that there is the nonlinear difference-frequency mixing (due to either electron or lattice nonlinearities) of two optical SR pulses or an optical SR pulse and a CW laser field, generated in one and the same heterostructure. This mixing could results in the formation of the THz or far-infrared pulses which contain only one or few periods of the field oscillations (cf. the talk "THz Emission Efficiency of Grating-Outcoupled Nonlinear-Mixing Heterolasers" by A.A. Andrianov, A.A. Belyanin, V.V. Kocharovsky, and Vl.V. Kocharovsky at the parallel workshop TERA 2010).

For more details on possible schemes and generation regimes of the superradiant heterolasers, see below the list of references to some our works related to the problem. Note also the talk "Polariton mode lasing in quantum-well traps for Bose-condensation of dipolar excitons" by P.A. Kalinin, V.V. Kocharovsky, and Vl.V. Kocharovsky at the present conference CAOL 2010 where coherent superradiant emission from the semiconductor traps for Bose-Einstein exciton condensation under CW incoherent laser pumping is discussed.

If achieved experimentally, the superradiant heterolasing would offer a wide range of potential applications in optoelectronics, communications, and spectroscopy. In particular, the realization of superradiant lasers based on indirect band-gap GeSi/Ge low-dimensional heterostructures may have the strongest impact on semiconductor optoelectronics, since it would open the door for highly advanced Ge and Si technology. Another important application may be related to the development of superradiant lasing in the blue and UV ranges using wide-gap semiconductors, in which conventional lasing schemes encounter difficulties due to very low quantum efficiency. Possible candidates are group III–V nitrides and II–VI compounds.

In the case of traditional semiconductor lasing materials, the SF/SR process opens up new opportunities for ultrashort pulse generation. For example, the coherent SF/SR pulses generated in AlGaAs/GaAs heterolasers of ~10-100 μm length should have a duration of the order of 1-0.1 ps, a repetition rate in the THz range, and a peak intensity exceeding 100 MW/cm^2 (i.e. peak power ~ 1 W). The limiting duration and intensity of these pulses are defined by the so-called cooperative frequency of a given active medium and do not depend on any special technique. Moreover, since the SF/SR recombination is so fast, these sources can be modulated at an extremely high rate — up to several THz. Therefore, potential applications of superradiant heterolasers may include ultrafast switching and data processing.

The work was partially supported by the program of fundamental research "Fundamentals of basic research in nanotechnologies and nanomaterials" of the Presidium of the Russian Academy of Science.

REFERENCES

[1] V.V. Zheleznyakov, V.V. Kocharovsky, and Vl.V. Kocharovsky, "Polatization waves and superradiance in active media", *Sov. Phys. - Usp.* **32**, p. 835, 1989 (*Usp. Fiz. Nauk* **159**, p. 193-260, 1989, in Russian).

[2] A.A. Belyanin, V.V. Kocharovsky, and Vl.V. Kocharovsky, "Collective QED processes of electron-hole recombination and electron-positron annihilation in a strong magnetic field", *Quantum and Semiclassic. Optics* **9**, p. 1-44, 1997.

[3] E.R. Golubyatnikova, V.V. Kocharovsky, and Vl.V. Kocharovsky, "Nonlinear theory of mode superradiance in a low-Q resonator", *Laser Physics* 5, p. 801-813, 1995.

[4] E.R. Golubyatnikova, V.V. Kocharovsky, and Vl.V. Kocharovsky, "Mode Instability and Nonlinear Superradiance Phenomena in Open Fabry-Perot Cavity", *International Journal of Computers and Mathematics with Applications* 34, n. 7/8, p. 773-793, 1997.

[5] A.A. Belyanin, V.V. Kocharovsky, and Vl.V. Kocharovsky, "Superradiant generation of femtosecond pulses in quantum-well heterostructures", *Quantum and Semiclass. Optics* **10**, p. L13-L19, 1998.

[6] A.A. Belyanin, V.V. Kocharovsky, and Vl.V. Kocharovsky, "Superradiant generation in magnetized quantum wells", *Bull. of the Russian Academy of Sciences. Physics* **64**, n. 2, p. 205-208, 2000.

[7] A.A. Belyanin, V.V. Kocharovsky, Vl.V. Kocharovsky, and D.S. Pestov, "Features of superradiance of cyclotron quantum-dot heterolaser under continuous pumping", *Radiophysics and Quantum Electronics* 44, n.1-2, p. 184-195, 2001.

[8] A.A. Belyanin, V.V. Kocharovsky, Vl.V. Kocharovsky, and D.S. Pestov, "One- and two-colour superradiant lasing in magnetized quantum-well heterostructures", *Nanotechnology* **12**, p. 581-584, 2001.

[9] A.A. Belyanin, V.V. Kocharovsky, Vl.V. Kocharovsky, and D.S. Pestov, "Novel Schemes and Prospects of Superradiant Lasing in Heterostructures", *Laser Phys*ics **13**, n. 2, p. 1-7, 2003.

[10] N.S. Ginzburg, E.R. Kocharovskaya, and A.S. Sergeev, "Photonic-crystal superradiant lasers: Effects of homogeneous and inhomogeneous broadening of an active medium", *in Photonics, Devices and Systems IV*, Proc. SPIE **7138**, 71381C (SPIE, Bellingham, WA, 2008).

[11] N.S. Ginzburg, E.R. Kocharovskaya, and A.S. Sergeev, "Collective spontaneous emission in a laser with distributed feedback under the condition of inhomogeneous broadening of an active medium", *Bull. of the Russian Academy of Sciences. Physics* **74**, n. 7, 2010, in press (Izv. RAS, ser. Fizika **74**, n. 7, p. 946-949, 2010, in Russian).

[12] Vl.V.Kocharovsky, M.A.Garasyov, P.A.Kalinin, E.R.Kocharovskaya, „Perspectives of superradiant heterolasering", *in Proceeding of the Symposium on coherent optical emission of semiconductor materials and structures (Zvenigorod, 2009)*, Moscow, Lebedev Institute, 2010.

[13] Y.D. Jho, X. Wang, J. Kono, D.H. Reitze, X. Wei, A.A. Belyanin, V.V. Kocharovsky, Vl.V. Kocharovsky, and G.S. Solomon, "Cooperative Recombination of a Quantized High-Density Electron-Hole Plasma in Semiconductor Quantum Wells", *Phys. Rev. Lett.* **96**, 237401, 2006.

CAOL*2010 International Conference on Advanced Optoelectronics & Lasers, 10-14 September, 2010, Sevastopol, Ukraine

Selected fundamental and applied problems of laser mode formation

V.G. Niziev

Institute on Laser and Information Technologies, Shatura, RUSSIA

(Invited Paper)

Abstract: Several problems are considered in this review. Using our dipole wave theory of diffraction the analytical formulas for diffraction from a narrow ring for different polarization were obtained. They were used in the new wave numerical model of an open resonator. A number of parameters of mode formation were investigated. It was shown that multipass geometrical "modes" win in competition with single pass Laguerre-Gaussian (LG) non-main modes during the relaxation oscillations. A method has been suggested for description of the azimuthally (A) and radially (R) polarized modes that excludes any inherent contradictions and unjustified approximations. It allows the analytical calculation of the field components for these modes, including the longitudinal one. Generation of high power beams with axial symmetric polarization in stable resonator of industrial CO_2-laser with diffraction gratings has been realized. The transformation of azimuthally polarized mode to R-polarized mode was demonstrated. R-polarized beam was successfully used for cutting metal. The principal scheme based on a modified Sagnac interferometer was proposed and investigated experimentally. This external cavity technique was used for generation of R- and A-polarized laser beam.

In many cases of modern research the classical approach of describing the transversal laser beam structure of the open resonators [1] is not enough. LG modes TEM_{pq} belong to the self-reproducing solutions of the scalar wave equation. These solutions are in direct contradiction to Maxwell equation $\nabla \cdot \mathbf{E} = 0$. The longitudinal field is neglected at the describing laser beams in paraxial approximation. The solution of scalar wave equation relates to the homogenously polarized beams. The direction of the electric field in every point of the beam cross section remains the same for such beams. There are many other beams with inhomogeneous polarization. There also exists the problem of correlation between the wave and geometric optics in describing stable resonators. The report is devoted to these very problems. It is a review of our results in this field for the last years.

1. Dipole-wave theory of diffraction.

The Kirchhoff integral is scalar so it can be applied for strictly limited number of problems.

$$E(\mathbf{r}) = \int_{S'} [G(\mathbf{n} \cdot \nabla)E_0 - E_0(\mathbf{n} \cdot \nabla)G]\, dS'$$

If we have the initial field $E_0(\mathbf{r}')$ on the given surface S', the field $E(\mathbf{r})$ at the observation point can be calculated. Here $\mathbf{n}$ is a unit normal to the surface S', and $G(\mathbf{r}, \mathbf{r}')$ is the Green's function of the scalar wave equation. Strictly saying the used

form of $G(\mathbf{r}, \mathbf{r}') = \dfrac{e^{ik|\mathbf{r}-\mathbf{r}'|}}{|\mathbf{r}-\mathbf{r}'|}$ is incorrect for the point source of vector electromagnetic wave.

This integral can be used for obtaining vector solutions of diffractive problems by submission Hertz vector instead of electric field E.

$$\mathbf{Z}(\mathbf{r}) = \int_{S'} [G(\mathbf{n}\,\nabla)\mathbf{Z}_0 - \mathbf{Z}_0(\mathbf{n}\,\nabla G)]\, dS'$$

Then the fields can be found by the formulas: $\mathbf{E} = \nabla \times \nabla \times \mathbf{Z}$, $\mathbf{H} = ik\nabla \times \mathbf{Z}$. The main problem here is finding $\mathbf{Z}_0$ if we know $\mathbf{E}_0$. But for a widespread case of a plane polarized wave $\mathbf{E}_0(\mathbf{r}')$ and $\mathbf{Z}_0(\mathbf{r}')$ on S' surface are related by a simple expression $\mathbf{E}_0(\mathbf{r}') = -k^2 \mathbf{Z}_0(\mathbf{r}')$. Solutions obtained by this method automatically satisfy the Maxwell equation $\nabla \cdot \mathbf{E} = 0$. Several vector problems were solved by this method analytically in [2, 3]. In particular the diffraction field from narrow ring slit was calculated for different polarization of initial field E_0. For the plane and azimuthal polarization respectively:

$$ERL(L, \theta, r_0) \approx 2\pi r_0 ik \frac{e^{ikL}}{L} e^{ik\frac{r_0^2}{2L}} J_0(kr_0\theta) \tag{1}$$

$$ERA(L, \theta, r_0) \approx 2\pi r_0 ik \frac{e^{ikL}}{L} e^{ik\frac{r_0^2}{2L}} J_1(kr_0\theta) \tag{2}$$

These formulas are used for numerical simulation of resonator in the next section.

2. Dynamics of mode formation in open resonator

Despite the great number of works in the field of open resonators, this problem cannot be regarded as a finally solved one. This work is aimed at the investigation of the factors affecting the time of attaining the field stationary distribution, the physical phenomena accompanying this process for various configurations of empty resonators, as well as at the elucidation of the relationship between the conclusions resulting from the geometric and wave approaches to the analysis of stable resonators. Results can be important for high power lasers with low quality resonators and for the lasers having short pulses of generation.

A numerical wave model has been developed for the description of an axially symmetric empty resonator. The model employs an analytical description of radiation diffraction from a narrow ring (1), (2). Reflection of an incident wave with a specified amplitude-phase distribution from a mirror is regarded as Green problem.

978-1-4244-7043-3/10 $26.00 © 2010 IEEE

$$E(L,\theta) \approx 2\pi \frac{e^{ikL}}{L} \int_0^{r_m} r_0\, E_0(r_0) \exp\left[\frac{ik}{2}\left(\frac{r_0^2}{L} - 2\frac{r_0^2}{R}\right)\right] J_0(kr_0\theta)\, dr_0 \quad (3)$$

The "phase exponent" in the integral depends on the form of surfaces where incident field is specified and where diffraction field is calculated. Presented integral is for flat initial wave front of $E_0(r_0)$. Results are calculated at the spherical surface of the radius L (resonator length).

The resonator simulation based on expression (3) is rational and versatile. This allows the comparative research of the dynamics of mode formation in the resonator to be carried out over a wide range of parameters.

The process of mode formation has the character of relaxation oscillations of various frequencies, depending on the resonator parameters. The time of attaining a field distribution in the resonator, corresponding to the main mode, is dependent on losses at the mirror edges, i.e. on the ratio between the caustic radius on the mirror and the radius of the mirror itself. Fig.1, in particular, is indicative of the advantage of a convex-concave stable resonator: the time of mode formation here is minimal because this resonator displays a maximum caustic size of the main mode.

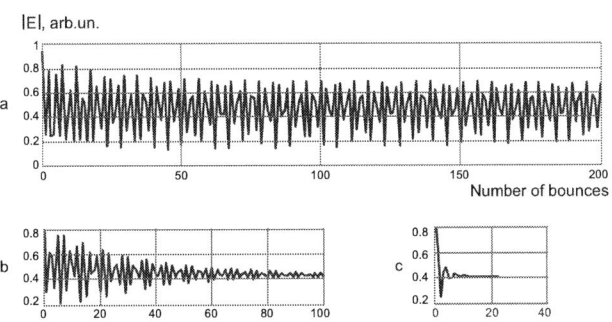

Fig.1. The field amplitude on the second mirror at r=0.14·r_m. Parameters: r_m·k=2900, L=195·r_m. The initial field amplitude is uniform, the front is flat. (a) R1= R2=2.5·L, (b) R1= 2.5·L, R2 = ∞, (c) R1=2.5·L, R2= -2.5·L.

The amplitude of the field oscillation depends on the kind of initial distribution of the field. From the variants studied the peak amplitudes of field fluctuations are observed in the initial field as plane waves. If we specify the front of initial wave coinciding with the mirror surface and the distribution of field amplitude corresponding to TEM$_{00}$ mode for the given resonator configuration, the field fluctuations are absent.

The period of field oscillations and the character of mode formation in the resonator differ sharply depending on the resonator parameters. The field amplitude evolution can be aperiodic, or manifest itself as beats of various frequencies. The parameters of resonators where beats are lacking (Fig.2a) are in conformity with the condition of paraxial resonance derived in the context of resonator describing by the methods of geometric optics [4]:

$$g_1 \cdot g_2 = \frac{1+\cos\theta}{2}; \quad \theta = 2\pi\frac{K}{N}; \quad 0 \le K \le N/2 \quad (4)$$

N is the number of resonator round-trips required to form a closed ray trajectory. $g_i = 1 - \dfrac{L}{R_i}$; i = 1, 2 are the parameters of stability diagram of open resonators.

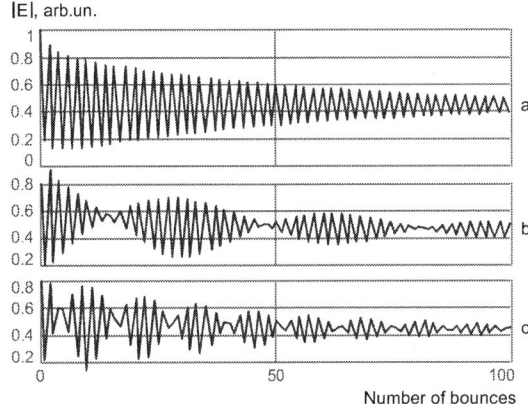

Fig.2. Evolution of the field amplitude on the second mirror at r=0.14·r_m. Parameters: r_m·k=2900, L=195·r_m, the second mirror is flat. The initial field amplitude is uniform, the front is flat. (a) R1= 2.0·L, (b) R1=2.1·L, (c) R1= 2.3·L.

The results of numerical calculations done in the context of wave approach (Fig.3) are uniquely indicative of the presence of paraxial resonance effects predicted in the geometric optics. The cases are illustrated when the oscillation period of the field corresponds to 3, 5 and 7 round-trips of the resonator. This phenomenon can be explained in the following way. The conclusions concerning paraxial resonances are applicable to the rays located at any distance from the axis; this is seen from formula (4) of paraxial resonance that involves no coordinates. However, there exist the boundaries of applicability of geometric consideration. A relatively large value of r, $F = r^2/\lambda L > 1$, permits paraxial resonances to be observed, but at small radius (F≲1), where the principal TEM$_{00}$ mode is localized, the laws of geometric optics are inapplicable.

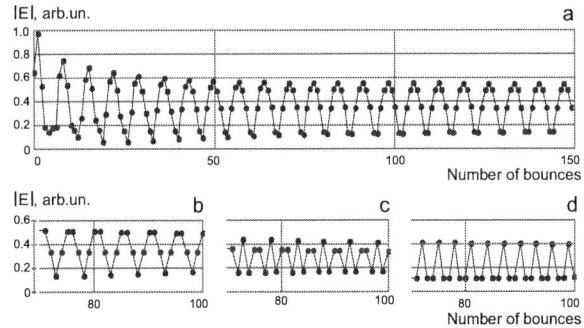

Fig.3. Examples of paraxial resonances. The field amplitude at the mirror center as a function of the number of bounces. The points on the curve correspond to consequent reflections from the mirror. (a) Seven-pass resonance at g_1=g_2= 0.222. (b) Five-pass resonance at g_1=g_2= 0.309; (c) Five-pass resonance at g_1=g_2= 0.809; (d) Three-pass resonance at g_1=g_2= 0.5.

978-1-4244-7043-3/10 $26.00 © 2010 IEEE

The resonances mentioned above can be observed independently on mirror radius r_m and initial field. These two factors can only influence to the amplitude of oscillations or the time relaxation, but not to the oscillation period. With "spontaneous noise" adding in each of the bounces the system arrives at a quasi-stationary state with stationary oscillations of the field distribution on the mirrors.

If we take initial field strictly corresponding to TEM_{10} (wave front form and amplitude distribution) for chosen resonator any fluctuations are not observed. But, any deviations of the initial field from the ideal one cause a cardinal change in the situation. In Fig.4 the front of the initial field is only changed to flat. The initial distribution appropriate for the TEM_{10} mode is "decomposed" and evolves to the main TEM_{00} mode via relaxation oscillations.

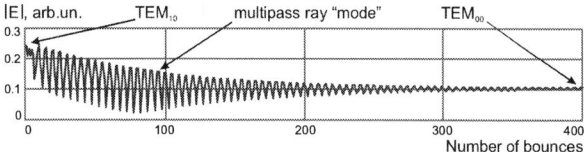

Fig.4. The field temporal evolution from the initial state (TEM_{10} with plane front) to the main mode (TEM_{00}) at the centre of the mirror. Resonator parameters are $g_1=g_2=0.309$.

Entering an absorbing annular zone on the mirrors in our numerical experiments resulted in suppression of MP modes and in generation of a TEM_{10} mode. The radius of the absorbing ring equaled that of the zero field of the TEM_{10} mode, and the area of the ring comprised less than 1% of the mirror area. In the absence of the absorbing ring relaxation oscillations gave rise to formation of the main TEM_{00} mode, the time of achieving the stationary filed distribution being considerably longer.

3. Description of R- and A-polarized laser modes.

Except of "usual" modes with homogeneous polarization there is a big class of solutions of the vector wave equation presented inhomogeneously polarized modes (IPM) with unique physical properties. The modes with R and A polarization are the most interesting from the practical point of view because they have full axial symmetry of all the beam parameters including polarization. This property is often useful for applications. R-polarized beams have been proposed for the laser cutting of metals and A polarization for hole punching. The R-polarized beam appears more efficient than a linearly polarized beam in experiments on laser heating plasma, due to its higher resonance absorption.

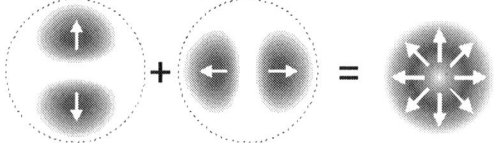

Fig.5. The scheme of formation of R-polarized beam as coherent superposition of "usual" modes.

It was proved that a R-polarized beam can be focused more sharply than a linearly polarized beam. These modes can be applied for trapping cold atoms, and the helical modes are

proposed in diagnostic and metrological systems. The longitudinal component of an electric field for a sharply focused R-polarized beam can be used for the acceleration of relativistic electrons [12, 13].

The scheme presented in Fig.5 is widely known. It schematically explains the principle of formation of IPM, R-polarized beam. However the description of IPM following this scheme by using the classic solutions for LG modes is not resulting. The principal difficulties on this way are connected with serious inner contradictions of classic solutions not acceptable for analysis of IPM. A method of theoretical description of the R or A polarized beams was suggested that excludes any inherent contradictions and unjustified approximations [5-7]. A solution will be sought in the class of azimuthally polarized modes. Represent the desired function in the form $\mathbf{H}=H_\varphi(r,z)\cdot\mathbf{e}_\varphi(\varphi)$. The equation $\nabla\mathbf{H}=0$ in this case is satisfied, and the vector wave equation is reduced to the scalar type:

$$\frac{1}{r}\frac{\partial}{\partial r}r\frac{\partial H_\varphi}{\partial r}+\frac{\partial^2 H_\varphi}{\partial z^2}+\left(k^2-\frac{1}{r^2}\right)H_\varphi=0$$

The solution of the last equation in paraxial approximation is r-z part of expression for LG modes TEM_{pq} at $q=1$.

$$H_\varphi=\sqrt{\frac{2p!}{\pi(p+1)!}}\cdot\frac{1}{w}\cdot\left(\sqrt{2}\cdot R\right)\cdot L_p^1\left(2\cdot R^2\right)\cdot\exp\left(-R^2\right)\cdot\exp(i\theta)$$

$$\theta=2arctgZ-2Z\frac{z_0^2}{w_0^2}-ZR^2;\quad R=r/w;\quad R_0=r/w0;\quad Z=$$

$$z/z_0;\quad z_0=\frac{\pi w_0^2}{\lambda};\quad w^2=w_0^2\cdot\left(1+Z^2\right)$$

$$L_p^1(x)=\sum_{m=0}^{p}(-1)^m\frac{(p+1)!}{(p-m)!(m+1)!m!}x^m$$

E_z, E_r, arb.un.

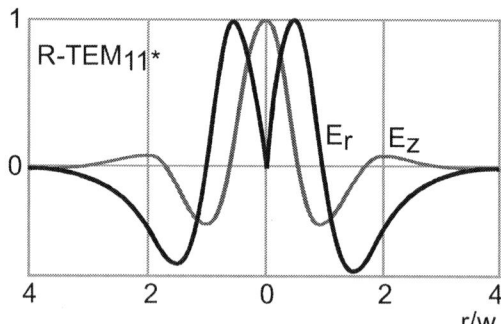

Fig.6. Calculated distributions of longitudinal and radial components of electric field E_z and E_r in the waist of R-polarized modes TEM_{11}.

An approximate equality of E_r and H_φ takes place under the condition $\frac{\lambda^2}{\pi^2 w_0^2}\ll 1$. The presented method permits the field components to be calculated for two classes of modes:

$$\mathbf{H}=H_\varphi(r,z)\cdot\mathbf{e}_\varphi(\varphi),\ \mathbf{E}=E_r(r,z)\cdot\mathbf{e}_r(\varphi)+E_z(r,z)\cdot\mathbf{e}_z.$$
$$\mathbf{E}=E_\varphi(r,z)\cdot\mathbf{e}_\varphi(\varphi),\ \mathbf{H}=H_r(r,z)\cdot\mathbf{e}_r(\varphi)+H_z(r,z)\cdot\mathbf{e}_z$$

4. Generation of R- and A-polarized modes in high power lasers.

There are two main conceptions of obtaining IPM: internal-cavity and external-cavity techniques. The most popular internal-cavity technique which is using a diffractive mirror as one of resonator mirrors. It is preferable for lasers with high gain of the active media and low resonator quality. The long wavelength of radiation (CO_2-lasers) is an additional positive factor in this case, because the period of diffractive structure is directly proportional to the wavelength. An external cavity technique (considered in the next section) can be effectively used for lasers with a short wavelength, low gain and high resonator quality. The mode quality in such lasers is much higher. These "high quality modes" can therefore interact coherently outside the resonator.

 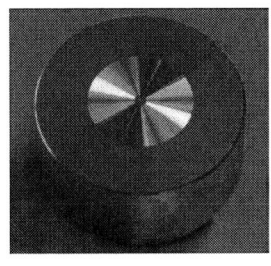

a   b

Fig.7. Diffractive mirrors. (a) Photolithography and wet etching. Reflectivity 94% - 72% for $E_\perp$-$E_\parallel$ components. ILIT RAS, Russia, 2000. (b) Diamond turning, gold coating. Reflectivity 94% - 20%. II-VI Inc. USA, 2006.

We proposed [8, 9] to use a diffractive mirror as a rear resonator mirror for obtaining IPM in high power lasers. The property of polarization selectivity for diffraction grating was used for generation of R- or A-polarized laser beam. The couple of diffractive mirrors are compared on the Fig.7.

All experiments for generation of R- and A-polarized radiation were performed on an industrial transverse gas-flow, transverse-discharge CO_2 laser was generated in. The five-pass laser resonator of length 8 m had a plane mirror and the output mirror with the radius of curvature 30m. Fig.8, 9 demonstrate the possibility of generation as a TEM_{01*} mode so a TEM_{11*} mode too.

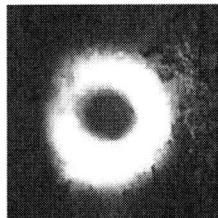 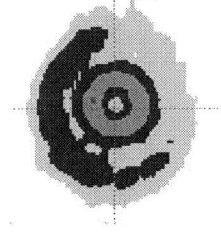

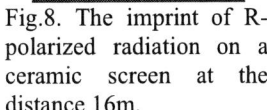

Fig.8. The imprint of R-polarized radiation on a ceramic screen at the distance 16m.

Fig.9. Far-feld intensity distribution for the R-polarised TEM_{11*} mode

The transformation of an A-polarized beam to a R-polarized beam (and vice versa) was realized by using a pair of $\lambda/2$

reflection phase shifters following to the scheme Fig.10. One of the most interesting applications of R-polarized radiation is laser cutting metals. As it was shown theoretically in [10] absorption of laser radiation on the cut walls should be by two times higher for R-polarized beam in comparison with widely used circularly polarized beam.

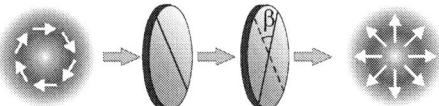

Fig.10. Scheme of transformation of A polarisation to R by two $\lambda/2$ phase shifters; $\beta=45°$.

It will be realized in higher cutting velocity or in thicker metal sheet. This prediction has been confirmed experimentally recently [11]. Experiments were performed with participation and support by biggest in the world producer of laser cutting machines the company Trumpf, Germany. It was claimed a 50% increasing cutting velocity and 2 times higher quality were obtained with using R-polarized beam in comparison with circularly polarized one.

5. Generation of R- and A-polarized modes in low power lasers [12].

Here attention is focused on the interferometric methods of reconstruction of a mode with inhomogeneous polarization following to the scheme Fig.5. The main advantage of this method is its universality. It can be applied for any wavelength and for any type of IPM. The most serious problem is the alignment of the interferometer and its temporal stability. A Melles Griot He–Ne laser 05-LPB-670 was used with an output power of 6.0 mW. This laser has a tube with a sealed Brewster window at one end and a mirror at the other. This laser can generate many transversal modes. The mode TEM_{01} with a controlled direction of plane polarization can be selected through the aperture. We used a Sagnac interferometer. An attractive feature of the Sagnac interferometer is that both interfering beams sample the same optical path with the same elements, so distortions of the optics have a minimal effect on the sensitivity of the differential signal. A simplified scheme of a Sagnac interferometer, which is based on standard optical components, is presented in Fig.11a. The intensity distribution of two split beams can be mutually rotated by inserting a Dove prism (DP), as seen in Fig.11b. The spatial orientation of the intensity distributions of the beams in plane F just after the lateral displacement polarizing beam splitter is absolutely the same. Then each beam passes the DP and an angle reflector (AR) in different sequence and comes to the lateral displacement beam splitter, which works now as a combiner. Both the DP and AR rotate the intensity distribution of a beam like a half-wavelength phase shifter (PS) rotates the electric field vector. If both elements of the DP and AR have the same orientation in space, their common effect of rotating the beam intensity distribution will be zero. If the DP and AR possess a mutual orientation, as shown in Fig.11c, $\theta=22.5°$, the intensity distribution of each two beams will be rotated around the beam axis in opposite directions at an angle of $2\theta=45°$ so that the total mutual rotation angle is $4\theta=90°$.

Four $\lambda/2$ phase shifters were installed in this scheme in the places indicated by arrows. PS1 was used for correction of amplitudes of two beams after splitting. Correction was performed by rotating PS1 around the beam axis. The axes of the second and third PS are parallel each other and oriented at an angle $\beta=10.25°$ along the bisector of the angle θ between the DP and AR. These phase shifters are necessary to avoid changes of polarization made by the Dove prism. The axis of this PS must be parallel ($\gamma=0°$) or perpendicular ($\gamma=90°$) to the AR edge. The correction of phase shift is performed by turning the PS around the line, which is perpendicular to the plane of drawing in Fig.11. Fig.12 shows the corresponding setup mounting on the optical table.

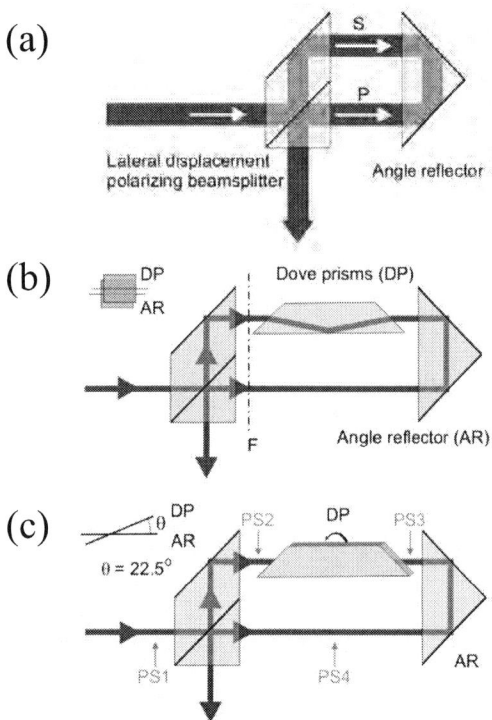

Fig.11. The scheme of a Sagnac interferometer. The simplified scheme (a); P and S are the corresponding polarization of two beams. The scheme with a Dove prism (b). The modified Sagnac interferometer configuration to produce laser beams with inhomogeneous polarization (c).

The modes with radial and azimuthal directions of the electric field are only two representatives of the large family of inhomogeneously polarized modes that are solutions of the vector wave equation. Some of them are presented in Fig.13. The resulting distribution of parameters characterizing the state of polarization over the cross section of a laser beam depends on the mutual positions of these two modes, the amplitude and the orientation of the electric field and a phase shift of their oscillations. The couple of LG modes TEM_{pl} (with $p\geq0$ and the same indexes) can coherently interact with each other creating,

for example, a radially or azimuthally polarized mode of higher order.

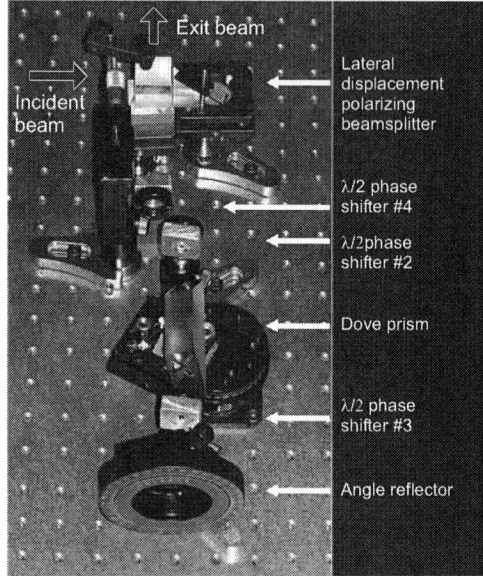

Fig.12. Modernized Sagnac interferometer mounting on an optical table.

The suggested scheme based on a Sagnac interferometer is simple and inherently stable. The amplitudes of two interfering beams and their mutual phase shift are controlled and variable. The electric field vector of the linearly polarized incident beam must be at $45°$ to the plane of the interferometer, giving the same amplitude of the two beams after being split.

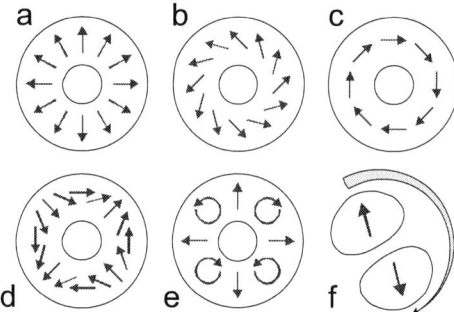

Fig.13. Examples of inhomogeneously polarized resonator modes. a) The radially polarized mode $R\text{-}TEM_{01*}$. b) The angle between $\mathbf{E}$ and radius is $45°$. c) The azimuthally polarized mode $A\text{-}TEM_{01*}$. d) The mode with different direction of electric field. e) The mode with different type of polarization. f) Helical mode with plane polarization.

Experimental results of the typical diagnostics of radially and azimuthally polarized beams are presented in Fig.14. The reconstructed beam from a Sagnac interferometer has a ring-type distribution of intensity, Fig.14a. This beam passes through a polarizer-analyzer and the resulting picture is fixed on the screen. The Fig.14b shows the experimental pictures for a mode indicated in Fig.13e. The single difference between the mode in Fig.13e and a radially polarized mode in Fig.13a consists of a phase shift between two mode patterns combined at the exit of the interferometer. The diagnostically obtained pictures for a radially polarized beam are presented in Fig. 6c. The spots on the screen look like a pure TEM_{01} mode and rotate synchronically with rotation of the polarizer axis. In the case of an azimuthally polarized beam, the polarizer axis and the mode pattern are mutually turned at 90° compared to the radially polarized beam. At any orientation of the polarizer axis, two spots will have the same position relative to this axis.

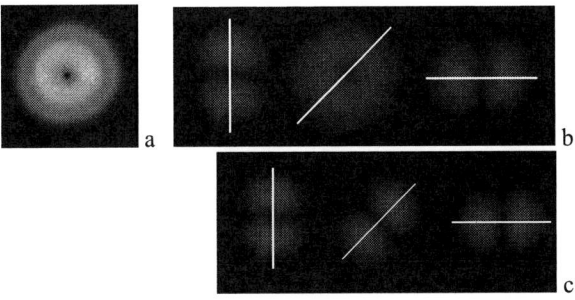

Fig.14. Experimentally obtained pictures of the modes with axially symmetric polarization. a) Intensity distribution in the cross section of the laser beam. b) The diagnostics of the mode with inhomogeneous polarization indicated in Fig.13e. The intensity distribution is just after the polarizer-analyzer. The white line is the axis of the polarizer. c) The diagnostics the radially polarized beam indicated in Fig.13a. The white line is the axis of the polarizer. The mode pattern rotates with rotation of a polarizer around the beam axis.

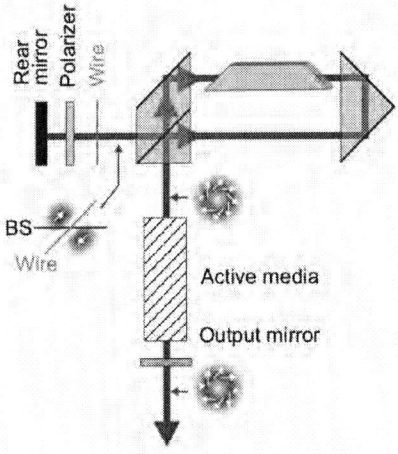

Fig.15. The scheme of using a Sagnac interferometer in a laser resonator as a rear mirror to generate inhomogeneously polarized modes.of a polarizer around the beam axis.

An interesting additional possibility of such a scheme should also be mentioned. Because this interferometer has one path exit, it can be used as "an interferometric rear mirror" of the laser resonator, Fig.15 with a controlled state of polarization of an inhomogeneously polarized beam (radial, azimuthal and others). The orientation of the electric vector determined by the polarizer (in this part of the scheme) should always be at 45° to the polarizing beam splitter (BS). The

orientation of the mode here can be changed by rotating a wire around the beam axis. It leads to the different states of polarization of the exit beam, from radial to azimuthal. This idea has not be tested yet. In order to do this, one would need a laser with a higher gain than a He-Ne laser has.

References

[1] S. Solimeno, B. Crosignani, P. DiPorto, "Guiding, Diffraction and Confinement of Optical Radiation" Academic Press, New York, 1986.

[2] V.G. Niziev "Dipole-wave theory of electromagnetic diffraction" *Physics-Uspekhi*, 45(5), pp. 553-559, 2002.

[3] A.V. Nesterov and V.G. Niziev "Vector solution of the diffraction task using the Hertz vector" *Physical Review, E* **71**, 4, 046608, 2005.

[4] I.A. Ramsay and J.J. Degnan, A Ray "Analysis of Optical Resonators Formed by Two Spherical Mirrors", *Applied Optics*, vol. 9, no. 2, pp. 385-398, 1970.

[5] V.G.Niziev, A.V.Nesterov. "Influence of Beam Polarization on Laser Cutting Efficiency" *J. of Physics D Appl. Phys.*, vol. 32, p. 1455-1461, 1999.

[6] A.V. Nesterov, V.G. Niziev "Propagation Features of Beams with Axially Symmetric Polarization", *J. Opt. B: Quantum and Semiclassical Optics*, vol.3, no.2, pp. 215-219, 2001.

[7] V.G. Niziev, A.V. Nesterov "Longitudinal fields in cylindrical and spherical modes" *J. of Optics A: Pure and Applied Optics*, 10, 085005 (7pp), 2008.

[8] A.V. Nesterov, V.G. Niziev, V.P. Yakunin "Generation of high-power radially polarized beam" *J. of Physics D Appl. Phys.*, vol.32, pp. 2871-2875, 1999.

[9] V.G. Niziev, V.P. Yakunin, N.G. Turkin, "Generation of polarisation-nonuniform modes in a high-power CO_2 laser", *Quantum Electronics* 39 (6), pp. 505-514, 2009.

[10] V.G.Niziev, A.V.Nesterov "Influence of Beam Polarization on Laser Cutting Efficiency" *J of Physics D Appl. Phys.*, vol.32, pp. 1455-1461, 1999.

[11] M.A. Ahmed, A. Voss, M.M. Vogel, A. Austerschulte, J. Schulz, V. Metsch, T. Moser, T. Graf, "Radially Polarized High-Power Lasers" XVII Int. Symp. on Gas Flow and Chemical Lasers & High Power Lasers, 15-19 Sept. 2008 Lisbon, Portugal *Proc. SPIE* Vol. 7131, 71311I, 2009.

[12] V. G. Niziev, R. S. Chang, A. V. Nesterov "Generation of inhomogeneously polarized laser beams by use of a Sagnac interferometer", *Applied Optics*, vol. 45, no. 33, pp. 8393-8399, 2006.

CAOL*2010 International Conference on Advanced Optoelectronics & Lasers, 10-14 September, 2010, Sevastopol, Ukraine

Properties of the Interaction of Laser Radiation with a Gaseous Dust Medium

A.F. Glova[1], A.Yu. Lysikov[1], M.M. Zverev[2]

[1]State Research Center of the Russian Federation "Troitsk Institute for Innovation and Fusion Research",
142190, Troitsk, Moscow region, Russia, e-mail: afglova@triniti.ru, lysikov@triniti.ru;
[2]Moscow State Institute of Radio-Engineering, Electronics and Automation (Technical University),
119454, prosp. Vernadskogo 78, Moscow, Russia, e-mail: mzverev@mail.ru

(Invited *Paper*)

Abstract-**Radiation power for the combustion of carbon particles is measured at particles irradiation in air by a cw CO_2 laser. For the medium in the form of a free vertical jet the dependences of the threshold radiation intensity for the evaporation of aluminium and carbon particles in nitrogen on their radius are obtained.**

I. INTRODUCTION

One of the methods for producing films on surfaces and applying coatings is the laser method based on the evaporation of a material during irradiation of solid targets by high-power pulsed lasers [1]. This method is relatively simple and can be use to evaporate almost any materials. Its main disadvantage is the possibility of the deposition of microfragments of the evaporated material on the surface in the form of solidified drops or particles of arbitrary shape, which impairs the film quality. The film quality can be improved by evaporating additionally microfragments escaping from the surface in crossed laser beams [1,2]. Note that the efficiency of the method of crossed beams depends on the interaction time of particles with the additional radiation field and can be insufficient in the case of a large initial dispersion velocity of particles initiated by main high-power laser radiation.

The aim of this paper is to study the possibility of efficient evaporation of a material under conditions when the initial object irradiated by a laser is a gaseous dust medium containing micron particles with a controllable translation motion velocity. Under these conditions, due to weak heat exchange between particles and surrounding gas, the particles can be completely evaporated in the radiation field of even a relatively low-power laser.

II. EXPERIMENTAL

We used in our experiments a CO_2 laser operating either in the cw or quasi-cw regime with controllable power and pulse duration. The horizontally oriented laser beam was focused into a region containing a mixture of particles with the atmospheric air at the normal pressure or with argon or nitrogen at controllable pressure. Carbon particles of size 30-300 μm and CdS and Al particles of size ~ 10 μm were used. A gaseous dust medium was produced mechanically. In one case, it was a quasi-stationary gaseous dust cloud, while in another – a vertical jet of particles in gas directed downward. Experiments were mainly performed with the jet of particles allowing the control of the initial parameters of the medium.

III. MEASUREMENTS AND CALCULATION RESULTS AND DISCUSSION

A. Irradiation of a carbon microparticle jet in the atmospheric air by a cw laser

Fig. 1 presents the power dependences of the parameter $\eta = (W - W_1)/W$, characterizing the decrease of the power W due to the adsorption and scattering of radiation by particles in the medium (W_1 is the laser power after the propagation of radiation through the jet). These dependences were obtained for distances h from the jet initiation site to the lens focus equal to 2 and 4 cm. The particle flow rate was 21.7 mg/s and the jet diameter was 14 and 16 mm for $h = 2$ and 4 cm, respectively. The divergence of laser radiation was 2 mrad, the laser beam diameter was 6 cm, and the focal distance of the lens was 30 cm. The value of η begins to increase when the threshold power W_{th} is achieved at which particles begin to burn in the region of their interaction with laser radiation. The threshold power depends on h, being ~ 10 W for $h = 2$ cm, which, taking into account the attenuation of radiation, corresponds to the threshold radiation intensity in the focal spot ~ $8 \cdot 10^2$ W/cm². For $W > W_{th}$, the bright emission in the interaction region acquired nearly symmetric shape with respect to the beam axis and focal plane. The maximal longitudinal size of the emission region was approximately equal to the jet diameter. For brevity, this emission will be referred as the discharge. Note that the value W_{th} ~ 10 W is consistent with the average threshold power ~ 7 W for the stable ignition of dispersed hard coal particles of size ~ 100 μm in the nitrogen-oxygen mixture jet ($h = 2$ cm, and $[O_2] = [N_2]$) irradiated by a repetitively pulsed Nd:YAG laser [3].

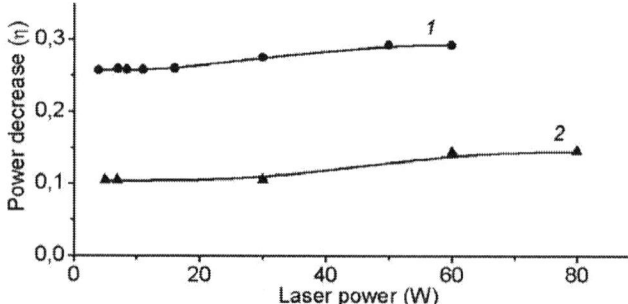

Fig. 1. Dependences of the parameter η on the power of a cw CO_2 laser irradiating a jet of carbon microparticles of size $30 - 300$ μm in the atmospheric air for $h = 2$ (*1*) and 4 cm (*2*).

978-1-4244-7043-3/10 $26.00 © 2010 IEEE

The decrease in η with increasing h (Fig. 1) is explained by the increase in the jet transparency due to the decrease in the concentration of particles in it. The average concentration of particles in the given section of the jet in the case of the homogeneous initial size distribution of particles can be estimated from the expression $<N> = G/(<m><v>s)$, where G is the total flow rate of all particles; $<m>$ is the mass of particles averaged over their sizes; $<v>$ is the average velocity of particles depending on $<m>$ and h; and $s = s(h)$ is the cross-sectional area of the jet. By neglecting collisions between spherical particles during their motion and averaging over three radii of particles $r = 30, 70,$ and 120 μm for $h = 2$ and 4 cm and $G = 21.7$ mg/s, we obtain $<N> \approx 50$ and ~ 30 cm⁻³, respectively. Note that $<N>$ decreases with increasing h mainly due to the increase in the average velocity $<v>$, because the cross-sectional area s of the jet weakly depends on the distance h. As for the increase in η for $W > W_{th}$ compared to η for $W < W_{th}$, it probably occurs due to the additional absorption and scattering of radiation by the combustion products of particles in the air oxygen, which compensate the possible decrease in the initial concentration of particles during their interaction with radiation.

It is obvious that heated carbon particles can acquire a noticeable positive charge due to thermionic emission. Let us estimate this charge under our conditions. Note preliminarily that for radiation intensities $\sim 10^3$ W/cm², the avalanche multiplication of electrons can be neglected [4]. At the plasma temperature of the order of the gas temperature ($\sim 10^3$ K) and ionization potential ~ 10 eV, the equilibrium electron concentration can be also neglected. Therefore, we will assume that electrons in the discharge are thermoelectrons at the concentration $n_e \sim q<N>$, where q is a particle charge in elementary charge units. To determine the electron concentration in the discharge by the thermionic emission current density [5], we assume that the current transfer by electrons emitted from the surface of particles occurs at the electron velocity corresponds to the gas temperature $\sim 10^3$ K. By using then experimental results from Ref. [3], according to which the maximum temperature of the particle surface achieves 2000 K at the average temperature 1700 K, taking into account that $<N> \sim 10^2$ cm⁻³, we obtain the maximum and average charges $q \sim 10^6$ and $\sim 10^4$, respectively. By considering the discharge under study as the approximate analogue of a dust plasma, the estimates of q presented above can be compared with experimental data in Ref. [6] obtained for a dust plasma containing positively charged particles, according to which the charge of particles in the plasma can achieve $\sim 10^4$.

Based on the results obtained in Ref. [6] and estimates presented above, it is reasonable to assume that the presence of macroscopic particles in the discharge, which have such a large charge and concentration $\sim 10^2$ cm⁻³, should affect the flow of current through the discharge. The current was measured with metal electrodes of diameter 2 mm with flat ends. The interelectrode gap of length 3 mm was oriented perpendicular to the jet axis and the laser beam, the lens focus being located in the central part of the gap. When a dc voltage ~ 100 V was applied to the electrodes in the presence of the discharge, the average quasi-stationary current flowing through

the gap was 50 mA, corresponding to the average conducting of the gap equal to $5 \cdot 10^{-3}$ Ω⁻¹ cm⁻¹. This conduction noticeably exceeds the electron conduction of the plasma in a constant field calculated by expressions from [4]. For example, even for the maximal estimated electron concentration $n_e \sim q<N> \sim 10^8$ cm⁻³, the electron conduction does not exceed 10^{-7} Ω⁻¹ cm⁻¹. We assume that the obtained conduction of $5 \cdot 10^{-3}$ Ω⁻¹ cm⁻¹ is related to one or several current filaments formed by carbon microparticles of diameter $\sim 0.1 - 1$ mm conducting current, which we observed in our experiments. Note that the quasi-stationary type of the current is probably explained by the periodic destruction of filaments by particles entering the gap. Note also that, when the voltage ~ 100 V is applied across the gap in the absence of the discharge, no filaments are for-med and the electric breakdown of the gap filled with particles of this size and concentration does not occur.

Consider a tentative mechanism of formation of an individual filament, by neglecting the role of induced polarization during the interaction of particles. A positively charged carbon microparticle located near the cathode settles down on it under the action of the attraction force, is retained on it due to sintering, and becomes a component of negatively charged cathode, forming an asperity on its surface. On this asperity the electric field of the interelectrode gap is concentrated. As a result, a charged microparticle located most closely to the asperity settles down on it rather than in its plane vicinity. The deposition process repeats, the attraction force from the anode side orienting this process by forming a filament cathode "growing" to the anode side. The locking of the interelectrode gap and formation of the filament can be treated as the result of the action of the same force from the anode side, resulting in the expansion of the filament cathode. It should be emphasized that thus mechanism is only hypothetical, and its refinement or the establishment of another mechanism requires special investigations.

B. Threshold radiation intensity for evaporating particles in a jet

Let us estimate the threshold laser radiation intensity required to evaporate particles in a jet. Preliminarily estimates show that heat conduction loss can be neglected compared to the radiative loss. By neglecting also the energy consumption for heating and melting of particles compared to the energy spent for evaporation, we can write the energy balance equation for a particle in the form

$$mQ = SA\delta t(\alpha I - \sigma T^4), \qquad (1)$$

where S is the particle surface area; A is the absorption capacity of the particle material; Q is the specific evaporation energy; I is the threshold radiation intensity required to evaporate the particle; T is the evaporation temperature; σ is the Stefan-Boltzmann constant; and $\delta t = 2<R>/v$ is the flight time of the particle in the interaction region with the average radius $<R>$. The coefficient $\alpha = 1/2$ in the right hand side of expression (1) approximately takes into account the absorption of radiation only by half the particle surface. Let us represent the interaction region in the form of a straight cylinder with the focal plane in its centre. The cylinder generatrix is oriented along the laser beam and its length is equal to the jet diameter. Such a choice of the interaction region shape is based on the visible longitudinal size of the discharge. Let us define the average radius of the interaction region as $<R> = (R + R_f)/2$, where $R = Dd/(4f) + R_f$; D is the laser beam diameter; d is the jet diameter; $R_f = \gamma f$ is the focal spot radius; γ is the radiation divergence; and f is the focal distance of the lens.

Fig. 2 presents the calculated dependences of the threshold evaporation intensity I on the radius r of aluminium and carbon particles forming a vertical jet in nitrogen at the atmospheric pressure. The calculations were performed for $D = 9$ cm, $d = 1$ cm, $f = 40$ cm, and $\gamma = 2$ mrad, which gives $<R> \approx 1$ mm, for $h = 2$ cm, corresponding to the change in δt from ~ 300 to 4 ms when r is increased from 5 to 100 μm. We used in calculations the parameters $A = 0.05$ and 1, $Q = 10.9$ and 60 kJ/g, $T = 2450$ and 3600 °C for aluminium and carbon, respectively. The increase of A for aluminium upon heating was neglected. Note that due to a weak pressure dependence of the dynamic viscosity of nitrogen [5], the results of calculations of the dependences of I on r at other pressures weakly differ from the results obtained at the atmospheric pressure.

One can see from the curves presented in Fig. 2 that, depending on r, the ratio I_{Al}/I_C can be either greater or smaller than unity (hereafter, the quantities with subscripts Al and C are related to aluminium and carbon, respectively). This is related to the different contributions of the radiative loss to the energy balance of particles depending on their radius. It is obvious that the radiative losses can be neglected beginning from certain values of r due to their weak dependence on r compared to the evaporation laser radiation energy loss. Then, taking into account that masses and velocities of aluminium and carbon particles are approximately equal, we obtain from (1) that $I_{Al}/I_C \approx (Q_{Al}/Q_C)(A_C/A_{Al}) > 1$. As the particle radius r decreases, when the contribution of radiative loss increases, the ratio I_{Al}/I_C can be most simply found when both types of losses are equal for each of the materials: $I_{Al}/I_C \approx (T_{Al}/T_C)^4 < 1$. According to calculations, the radiative losses are negligible beginning from $r > 10 - 20$ μm, and for $r \leq 10$ μm, they become comparable with the evaporation loss, which is reflected in Fog. 2 in accordance with estimated presented above.

Despite the large enough average radius ($<R> \approx 1$ mm) of the interaction region used in calculations, the laser radiation power $\sim I\pi<R>^2$ required for evaporation of small particles proves to be low. Thus, for $r = 10$ μm, this power for aluminium and carbon is ~ 170 and ~ 100 W, respectively. We can say that the efficient evaporation of such particles can occur in the radiation field of a comparatively low-power cw laser. As the particle radius increases, the laser power at the fixed $<R>$ should be increased, and the choice of a proper laser is determined by the particle size. This can be either cw or repetitively pulsed laser with a moderate average output power.

Note that the authors of Ref. [7] studied the change in the concentration of Al particles of size ~ 1 μm due to evaporation in argon at the atmospheric pressure upon irradiation by several 10-J, 0.5-μs pulses from a CO_2 laser. The radiation intensi-

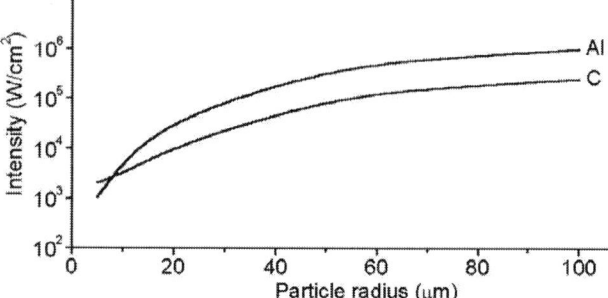

Fig. 2. Calculated dependences of the evaporation threshold radiation intensity on the radius of aluminium and carbon particles.

ty depended on the beam formation method and was varied from the maximum value $6 \cdot 10^7$ W/cm^2 down to $\sim 10^6$ and $3 \cdot 10^5$ W/cm^2. Calculations by using (1) for $\delta t = 0.5$ μs show that the radiation intensity ($3.5 \cdot 10^7$ W/cm^2) required for evaporation of these particles corresponds to the maximum intensity. It is possible that particles were evaporated at lower intensities due to heating in the optical discharge plasma produced in [7].

The possibility of the efficient evaporation of particles of radius $5 - 10$ μm upon irradiation of a gaseous dust medium by a comparatively low-power laser for producing films on a surface was verified experimentally. Substrates were placed above the geometrical focus of the lens at a distance from it precluding the contact of the substrate with the visible boundary of the discharge. A jet containing CdS or Al particles of this size was irradiated by a ~ 500-W quasi-cw CO_2 laser ($f = 40$ cm, $D = 9$ cm) in the argon or nitrogen atmosphere at a pressure of $100 - 300$ Torr. CdS and AlN films of area ~ 1 cm^2 were prepared after irradiation by $5 - 10$ 50-ms pulses from a CO_2 laser. The preliminarily visualization of the surfaces of films demonstrated that they did not contain microfragments of size exceeding 1 μm.

IV. CONCLUSIONS

Laser evaporation of particles from a gaseous dust medium is promising for preparing films and coatings from various materials without using the expensive equipment. We have shown by the example of an air-carbon medium irradiated by a cw CO_2 laser that the medium can be ignited at comparatively low threshold radiation intensity. The investigation of the conditions for initiating combustion in various gaseous dust media and of the energy balance in the medium can be of interest for laser nanotechnologies, the physics of dust plasmas, and a number of other applications.

REFERENCES

[1] A.A. Voevodin, M.S. Donley, *Surf. Coating. Thechn.*, vol.82, p.199, 1996.
[2] I.V. Grekhov, I.A. Liniichuk, I.E. Titkov, *Pis'ma Zh. Tekh. Fiz.*, vol. 32, p. 24, 2006.
[3] J.C. Chen, M. Taniguchi, K. Ito, *Fuel*, vol. 74, p. 323, 1995.
[4] Yu. P. Raizer, *Physics of a Gas Discharge,* Moscow, Nauka, 1987.
[5] I.K. Kikoin (Ed.), *Handbook of Physical Quantities,* Moscow, Atomizdat, 1976.
[6] V.E. Fortov, A.P. Nefedov, O.F. Petrov, *Zh. Eksp. Teor. Fiz.*, vol. 111, p. 467, 1997.
[7] I.A. Bakulin, V.S. Kazakevich, S.Yu. Pichugin, *Zh. Tekh. Fiz.*, vol. 76, p. 96, 2006.

DIFFRACTION FEATURE AND APPLICATIONS OF
HOLOGRAPHIC BRAGG GRATINGS
(Invited Paper)

Eugene A.Tikhonov, SPIE member

Institute of Physics, National Academy Science, Kiev-28, prospect of Sciences 46, 03680, Ukraine, e-mail: etikh@iop.kiev.ua, fax:38-044-5251589

Abstract - Bragg character of a light diffraction on volume phase holographic gratings is considered as the basic of their applications in a number of classical and new optical instruments. Owing to high angular/spectral selectivity of the Bragg grating monochromators can be realised without application of input/output slits. Strengthening of an spectral dispersion due to the 3D-diffraction provides a single-mode oscillation in dye lasers even with nanosecond pumping. The polychromator with a such grating is attained in divergent "white" light beam in a plane of a grating dispersion. Two diffracted beams controlled power on a grating exit in a combination with the simple lens optics for beam crossing implement a 2-beam interferometer with controlled spatial resolution. Calibrated on a spatial frequency the grating on a goniometric knot forms the high-precision wavelengthmeter. High angular selectivity of such grating allows to implement optical vibration sensor with sensitivity comparable with that at modern seismographer.

Keywords: transmission phase grating, Bragg diffraction, holographic photopolymer, frequency dispersion, diffraction efficiency, angular/spectral selectivity, spectral resolution, 2-beam interferometer, polychromator, monochromator, λ-wave measurer, vibration sensor

I. Applications of volume diffraction gratings in a optical instrumentation leans upon their Bragg diffraction parameters and for the first time-high diffaction efficiency η and angular/spectral selectivity: Λ/T and $\lambda\Lambda/T$, where λ, Λ, T – wavelength, spatial period and grating thickness. If T<5mkm selectivity is small and grating parameters approach to relief grating, for T>50-100mkm angular selectivity attains 1 grade and less that important to take into account under device designing. In the report application of a such volume phase gratings in optical instrumentation in transparency windows 300-1200nm are presented. At theoretical description of similar holographic gratings the 2-wave interaction model under conditions of the harmonic isotropic recording response has been accepted [1,2,3]. Because grating efficiency is most important parameter it is necessary to analyze its spectral behavior first of all.

$$\eta_{max}(s)=\sin^2(\pi n_1(\lambda)T/\lambda\cos(\theta_{in})= \quad (1)$$
$$\sin^2(\pi n_1(\lambda)T/\lambda(1-\lambda^2/4\Lambda^2)^{0,5})$$

The expression present dependence η=f(λ) for s-polarized emission without taking into account less weighty natural dispersion dependence n_1=f(λ). From (1) it becomes clear that is impossible to conserve argument $\sin^2(..)$ equal π/2 in all spectral window of phase grating.

As results the angular/spectral efficiency distribution of grating that determines the spread function of grating intolerably changes with λ fig.1. But if to record grating efficiency in red band of spectra equal 50-60% instead 90% as on fig.1. formfactor of angular/spectral dependence will conserve in working spectral band.

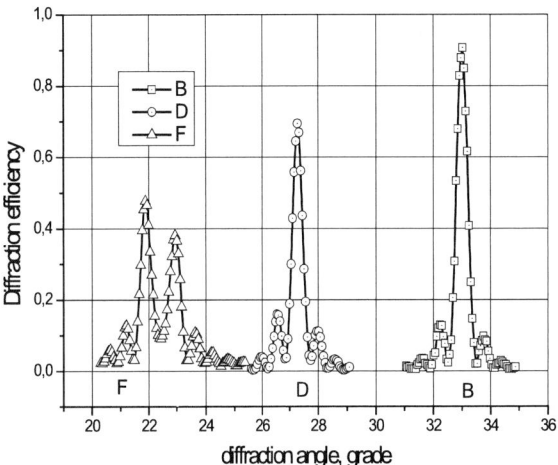

II. Next steps are choice of light illumination geometry for transmission Bragg for the case of monochromator and polychromator usages. In the first. case it is necassery to illuminate grating by collimated paraxial "white" beam with devergency $\delta\varphi \leq \Lambda/T$ under Bragg angle in order to achieve highest diffraction efficiency for given wavelength. Under this condition the output beam will be chracterized by linewidth and devergency equal $\lambda\Lambda/T$ and Λ/T correspondentely. No input/output slits is needed to obtain monochromatic light so that a cross-section of beam will limited by grating aperture only. It provides the highes efficiency of monochromization that is possible.Two identical gratings can be set up in series to deduct dispersion or to add it. The last case wavelength tuning under rotation of 2-grating knot does not follow by beam spatial displacement.

III. To realise polychromatic functioning with Bragg grating one needs to illuminate it by broadband "white" light beam divergent (or convergent) in dispersion plane used grating fig.2.

978-1-4244-7043-3/10 $26.00 © 2010 IEEE   55

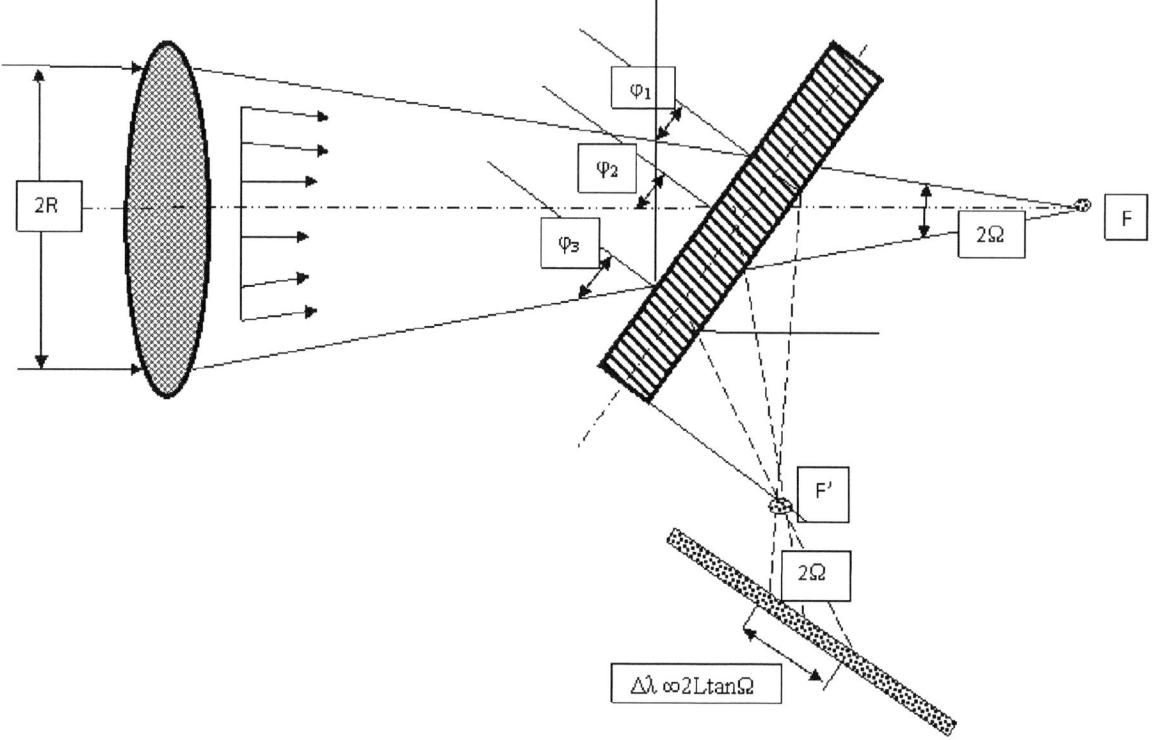

fig.2.

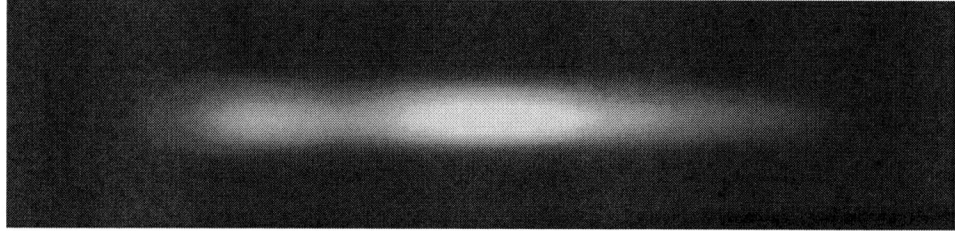

The spectral range, which satisfies Bragg's condition on this grating, is determined by (2) and is showns on fig.3.:

$$\Delta\lambda = 2\Lambda(\sin\varphi_3 - \sin\varphi_1) = 4\Lambda\sin[(\varphi_3+\varphi_1)/2]\cos[(\varphi_3-\varphi_1)/2] = 4\Lambda\cos(\varphi_2)\sin(\text{arctg}(R/F)) \quad (2)$$

fig.3.

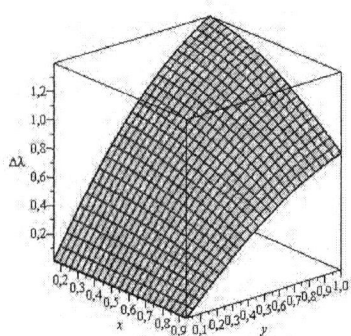

where $\Delta\lambda$ in mkm, $x=\varphi_2$, $y=R/F$ in accordance with fig.2. designation.

As the result Bragg diffraction on the different λ the addition of diffracted beams leads to formation in the space F' the focused region of initial "white" light. The size of the necking in of caustic curve in the plane of dispersion for the divergent light beam is determined by the angular selectivity of this grating Λ/T, and height is determined by the size of beam into the plane, $\perp$ the plane of dispersion. For the convergent light beam (fig.2.) this size is determined by that and additionally by focusing ability of lens. Color picture fig.2. shows "white" light spectra of a light diode outside point F'.

The spectral selectivity of Bragg gratings determines the width of the contour of the diffraction efficiency as function of wavelength for the preset Bragg angle. This value does not depend on the width of the illuminated region of grating. The component of the grating resolution is equal:

$$(\lambda/\Delta\lambda)_1 = T/\Lambda \quad (3)$$

The second component of grating resolution, similar to the relief grating, can be defined as:

$$(\lambda/\Delta\lambda)_2 = N = W/\Lambda \qquad (4)$$

where N, W are total number of illuminated phase planes and width of illuminated grating region. Complete spectral resolution of this system can be calculated as product (3) and (4):

$$(\lambda/\Delta\lambda) = TW/\Lambda^2 \qquad (5)$$

For the typical values of the phase grating parameters, recorded on the photopolymer material PhPC-488 $\Lambda=2000mm^{-1}$, T=0,5mm, W=20mm, one obtains $R=4*10^7$, which is compared with the resolution of Fabri-Perot interferometer /4/.

IV. Dependences of the diffraction efficiency of Bragg gratings from the diffraction angle on fig.1. show that if a some scanning of laser beam or grating takes place (for the beam divergence the smaller angular selectivity of grating) in the diffraction plane the antiphase amplitude modulation of the power of transmited and diffracted radiations occurs. The magnitude of the modulation under the designated conditions can be calculated from Taylor line with the values of derivatives on the diffraction angle θ at the operating points θ'.

$$P_{out}(t) = P_0\eta(\theta' \pm \delta_0 \sin(\Omega t)) \simeq P_0[\eta(\theta') \pm \delta_0 \sin(\Omega t)(\frac{d\eta(\theta)}{d\theta})_{\theta'} \pm$$
$$\frac{1}{2!}(\delta_0 \sin(\Omega t))^2 (\frac{d^2\eta(\theta)}{d\theta^2})_{\theta'} \pm ...) \qquad (6)$$

A similar system can be used as the sensitive vibration pickup, when the vibrations cause angular scanning of one of the elements of sensor and attain several minutes. Depending on the selection of operating point on the angle θ', η (θ) and its derivatives will have finite or zero values, which will determine the relationship between the vibration frequencies at the grating entrance and at the output beam power after diffraction. Measurement of vibrations when laser emission had the divergence of ≈1mrad showed that when the scanning amplitude comes near the divergence of the laser beam the signal from the sensor in the form modulated radiation approachs to zero.

V. The property of the transmission Bragg gratings to form 2 adjusted according to the power of exit beam with the angle of devergence the equal to the doubled angle of Bragg allowes to realize the new type of double-beam interferometer. The subsequent crossing of beams at different angles by lens makes it possible to change the resolution of interferometer via the simple translation of the lens optics relative to the conditional focus of the appearance of exit beams on the grating. A change of the power relationship in the diffracted beams

with the smooth turn of grating is necessary for the interferometry of the absorbing objects. There are possibility to rotate the interference plane by simple rotation of beam splitting grating around axes that is ⊥ to its plane. In order to take into account a variation of special frequency of interference pattern in the limit of interference pattern due to beam divergence the simple approximation by Macloren line is proposed:

$$N(\alpha \mp \Delta\varphi) = 2\sin(\alpha)/\lambda \mp 2\cos(\alpha)\Delta\varphi/2\lambda + \sin^2(\alpha)(\Delta\varphi)^2/12\lambda \mp ...) \qquad (7)$$

For the case λ=0,6328мкм, $\alpha=2^0$, $\Delta\varphi=0..1^0$ calculation with taking into account third member of series is presented on fig.4. Precision of the used approximation is getting better as higher is the averaged angle of beam crossing α.

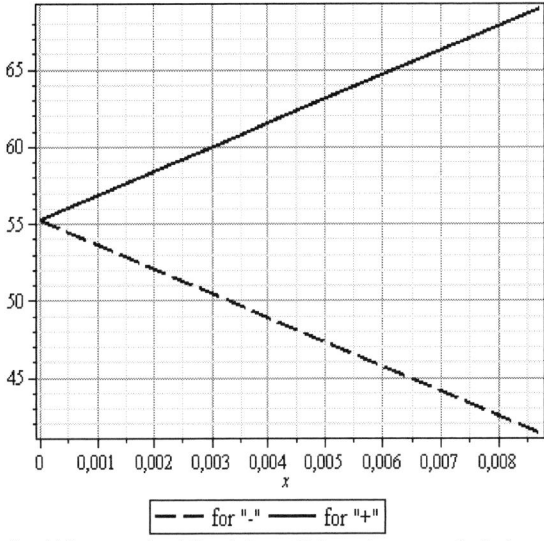

Spatial frequency (mm^-1) variations with beam devergence (radian)

Fig.4.

References

1. H. Kogelnik, «Coupled wave theory for thick hologram grating », Bell Sys.Techn. J.,v.48, #9, p.2909, 1969
2. E.A.Tikhonov, "Analysis and measuring of parametres of holographic Bragg gratings, Vestnik S-Peterburg Akademija of engineering sciences, №4, pp.57-78, 2008
3. E.A.Tikhonov " Study of holographic Bragg gratings: recording, measurement and application, Advanced Optoelectronics and Lasers, 2008. CAOL 2008. 4th International Conference on Publication Date: Sept. 29 2008-Oct. 4. 2008, On page(s): 155 - 158 Print ISBN: 978-1-4244-1973-9 INSPEC Acces Numr: 10388938, Digital Object Identifier: 10.1109/CAOL.2008.4671980
4. E.A.Tikhonov, A. A.Tyutyunnik, «Measuring of light wavelength with Bragg diffraction grating», Journal of Optical Technology, vol. 74, #8, pp. 521-525,2007

(110) Quantum Well Based Spin VCSELs

(Invited Paper)

Hitoshi Kawaguchi, *Member, IEEE,* Shinji Koh

Graduate School of Materials Science, Nara Institute of Science and Technology

8916-5, Takayama-cho, Ikoma, Nara 630-0192, Japan

Fax: +81-743-72-6183 e-mail: koh@ms.naist.jp

Abstract: A vertical-cavity surface-emitting laser (VCSEL) based on (110) InGaAs/GaAs multiple quantum wells (MQWs) was fabricated and characterized. Circularly polarized lasing by optical pumping was demonstrated. A high degree of circular polarization, 0.94, was observed at 77 K, reflecting the long electron spin relaxation time in the (110) MQWs.

I. INTRODUCTION

Spin vertical-cavity surface-emitting lasers (VCSELs) have been attracting much attention because they allow control of lasing polarization via the optical selection rules (Fig. 1), where the total spin of electrons and holes must equal to the angular momentum of the photon. Spin VCSELs are thereby expected to enhance the bandwidth and security performance in optical communication systems. Furthermore, spin-controlled operations can lead to a reduction of the threshold current [1]-[4]. A spin-controlled VCSEL was first demonstrated by Hallstein *et al.*, where modulation of circular polarization synchronized to the electron spin precession was observed [5]. Optically pumped, circularly polarized lasing in a (100) VCSEL was demonstrated with an active layer of bulk GaAs [6]. Although circularly polarized lasing was observed in this case, quantum wells (QWs) are better suited for active layers. This is because the degeneracy between heavy hole (HH) and light hole (LH) bands is lifted in QWs and cavity resonance can only be tuned to the HH band-relevant transitions, resulting in a higher degree of circular polarization. More recently, circularly polarized lasing and reduced threshold current in (100)-QW VCSELs were demonstrated via optical and electrical spin injection [1]-[4], [7]. However, the performance of the spin-VCSELs was limited by short electron spin relaxation times τ_s in (100) QWs (<100 ps at RT). (110) QWs have attracted significant interest because τ_s is much longer than that found in (100) QWs [8]-[10]. The long τ_s has major advantages for realizing practical spin-controlled VCSELs. The main challenge in fabricating (110)-QW VCSELs is crystal growth since the growth condition window for GaAs (110) is

much narrower than that of GaAs (100). Optically pumped lasing in a (110) InGaAs/GaAs QW VCSEL was reported [11]. However, the scope of that study did not include demonstrating spin-controlled VCSELs, but rather focused on stabilizing polarization by utilizing the optical gain anisotropy on (110) surfaces. (110)-QW spin-controlled VCSELs have not been demonstrated yet although their potential has been recognized.

In this study, we fabricated a (110)-oriented InGaAs/GaAs QW VCSEL [12], [13]. Circularly polarized lasing in the VCSEL by optical spin injection was demonstrated. Under excitation by circularly polarized pump pulses, the polarization characteristics of photoluminescence (PL) and laser dynamics were measured.

II. SPIN DYNAMICS IN (110) INGAAS/GAAS MQWS

In our previous study [14], molecular beam epitaxy (MBE) growth conditions for GaAs (110) substrates were systematically investigated. Suitable conditions were found as a growth rate of 0.5 μm/h, an As/Ga flux ratio of 80 and a growth temperature of 480 °C, and high quality (110) GaAs/AlGaAs multiple quantum wells (MQWs) were grown. In this study, we fabricated *strained* InGaAs/GaAs MQWs for the active layer of the spin VCSEL. In the five $In_{0.1}Ga_{0.9}As$/GaAs MQWs, 6-nm-thick InGaAs wells were separated by 10-nm-thick GaAs barrier layers. These MQWs were grown on semi-insulating (SI) GaAs (110) substrates by MBE at a substrate temperature of 440 °C under an As/Ga flux ratio of 80 and a GaAs growth rate of 0.5 μm/h. To increase luminescence efficiency of the QWs, we performed a 30-second annealing at 600 °C before the growth of each InGaAs layer as well as a 10-min post-growth annealing at 650 °C.

τ_s and carrier lifetimes τ_c in the MQWs were evaluated at 77 K and room temperature (RT) using polarization- and time-resolved photoluminescence (TRPL) measurements. A mode-locked Ti:sapphire laser with a 70 fs duration and a 80 MHz repetition rate was used as the pump laser. The wavelengths and the power of the pump laser were 750 nm and 1 mW at 77 K, and 790 nm and 20 mW at RT, respectively. Spin-polarized electrons were generated by circularly polarized pump pulses via the selection rules of Faraday geometry (Fig. 1). The time evolution of the PL intensities for right- (σ+) and left- (σ-) circularly polarized lights, which reflect a decrease in the population of conduction band down- and up-spin electrons, were measured separately using a λ/4 Babinet-Soleil compensator and a streak camera. The peak wavelength and full-width-at-half-maximum (FWHM) of the PL spectra (not shown) were 865 nm and 23 meV at 77 K, and 917 nm and 36

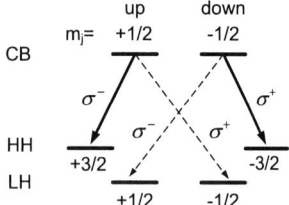

Fig. 1. Optical selection rules in zinc-blende-type direct-transition semiconductor quantum wells for Faraday geometry.

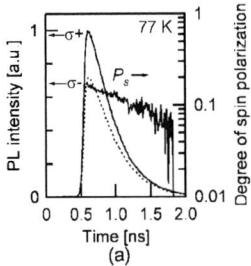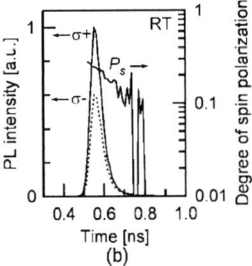

Fig. 2. PL intensities of the σ+ (solid line) and σ- (dashed line) components, and the degree of spin polarization plotted as a function of time at (a) 77 K and (b) RT.

meV at RT, respectively. The PL peak wavelengths corresponded to the transition energies between the ground states of the conduction band and the HH band. 10-nm-wide spectral regions around the PL peak were selected by a band pass filter. PL intensities of the σ+ and σ- components are plotted as a function of time in Fig. 2 at (a) 77 K and (b) RT. The polarization of the pump pulses was σ+. The time evolution of the degree of spin polarization P_s, defined as $P_s = (I_{\sigma+} - I_{\sigma-})/(I_{\sigma+} + I_{\sigma-})$, is also plotted. The time dependence of P_s is given by $P_s(t) \sim \exp(-2t/\tau_s)$, and hence τ_s can be derived from the slope of the P_s decay curve. τ_s in the (110) MQWs was measured to be 2.8 ± 0.3 ns at 77 K. τ_c was measured to be 420 ps from the decay curve of the sum of the two components. At RT, τ_s and τ_c were 440 ± 50 ps and 40 ps, respectively.

III. FABRICATION OF (110) QW BASED SPIN VCSEL

We fabricated a VCSEL on a SI GaAs (110) substrate by MBE. A schematic structure is shown in Fig. 3(a). A λ-cavity was formed with a top 22-pair and a bottom 22.5-pair AlAs/GaAs distributed Bragg reflector (DBR). Active layers of five InGaAs/GaAs MQWs were inserted into the center of the cavity. The VCSEL was grown under an As/Ga flux ratio of 80, and growth rates of GaAs and AlAs of 0.5 μm/h. A 300-nm thick GaAs buffer layer and the bottom DBR were first grown at 480 °C. The growth was interrupted before growing the active layer, and the substrate temperature was cooled to 440 °C during the interruption. A 30-second annealing at 600 °C before the growth of each InGaAs well layer was carried out. The process was again interrupted for 1 min

following the growth of active layers, and the substrate was then heated to 480 °C. A 10-min post-growth annealing at 650 °C was performed after the growth of the whole structure. Figure 3(b) shows a scanning electron microscope (SEM) image of the cross-section of the VCSEL.

IV. LASING CHARACTERISTICS OF SPIN VCSEL

The (110) VCSEL was optically pumped at 77 K and RT. Thickness variations of the MBE-grown layers in the wafer allowed wavelength matching between the optical gain and the cavity mode at different temperatures in a single wafer. The pump condition was the same as that in the TRPL measurement, except that the wavelengths were tuned to 780 nm and 820 nm at 77 K and RT, respectively. These wavelengths correspond to the reflection minima of the cavity at wavelengths shorter than the stop bands. To evaluate polarization characteristics, the σ+ and σ- components of the VCSEL output were detected separately using a CCD spectrometer. In Fig. 4, the σ+ and σ- components of the output intensities measured at (a) 77 K and (b) RT are plotted versus the power of σ+ pump pulses. At 77 K, the VCSEL output showed apparent threshold characteristics for the σ+ component at a pump power of about 2.6 mW. Above the threshold, the output was dominated by the σ+ component, and the σ- component did not show a clear laser threshold. The lasing spectra at a pump power of 3 mW measured at 77 K are shown in Fig. 4(c). Circularly polarized and single-longitudinal-mode lasing at 865 nm was observed. The intensity of the σ+ component was approximately 30 times larger than that of the σ- component. On the other hand, at RT, the lasing threshold of the VCSEL was around 15 mW both for the σ+ and σ- components (Fig. 4(b)). However, with increasing pump power, the σ+ component increased more steeply than the σ- component. The lasing spectra at a pump power of 21 mW measured at RT are shown in Fig. 4(d). The VCSEL showed single-longitudinal-mode lasing at 917 nm and intensity of the σ+ component was about 2.5 times larger than that of the σ- component.

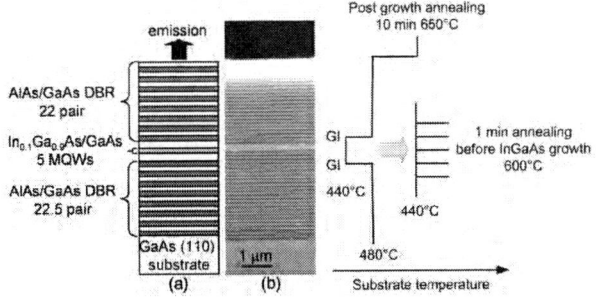

Fig. 3. (a) Schematic structure and (b) a cross-sectional SEM image of the (110) VCSEL. The substrate temperature in the MBE growth is also shown, in which GI denotes growth interruption.

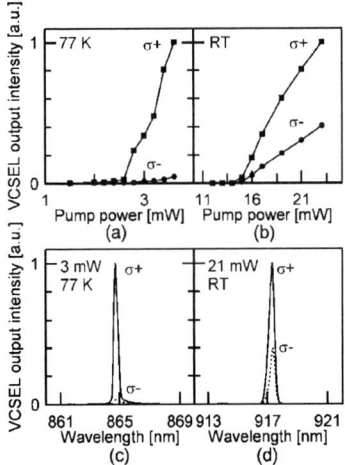

Fig. 4. VCSEL output intensities of the σ+ and σ- components plotted as a function of the time-averaged pump power at (a) 77 K and (b) RT. The VCSEL output spectra of σ+ and σ- at (c) 77 K and (d) RT.

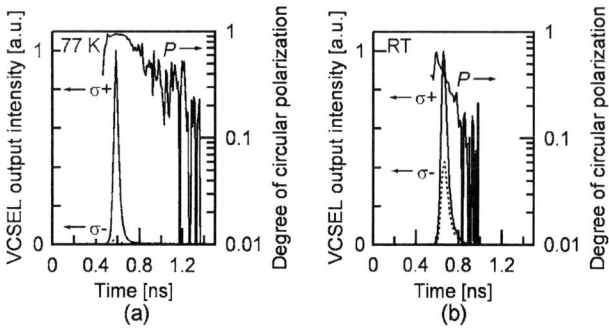

Fig. 5. VCSEL output intensities of the σ+ (solid line) and σ- (dashed line) components and the degree of circular polarization P (right axis) plotted as a function of time at (a) 77 K and (b) RT.

The threshold characteristics observed at 77 K are due to the electron spin polarization in the active layers. If we consider only radiative recombination transitions relevant to the HH bands, down-spin electrons selectively recombine with down-spin heavy holes according to the selection rules, which results in the emission of σ+ photons. When down-spin electrons are supplied preferentially, the optical gain for the σ+ mode becomes larger than that of the σ- mode. Therefore, it is possible to achieve a situation where the gain for the σ+ mode exceeds the threshold while that for the σ- mode is still below the threshold. The optical gain difference between the two modes is preserved during lasing when τ_s is long enough to maintain the spin-polarized states. Consequently, a large difference between the threshold characteristics of the two lasing modes was observed at 77 K due to the long τ_s, while at RT, the two modes showed almost the same lasing threshold because the short τ_s leads to only a small difference in the optical gains of the two modes.

The lasing dynamics were evaluated by polarization- and time-resolved measurements. In Fig. 5, output intensities of the σ+ and σ- components and the degree of circular polarization P_l are plotted as a function of time at (a) 77 K and (b) RT. P_l is defined as having the same equation as P_s. The power of the pump pulses with σ+ polarization was 3 mW and 21 mW at 77 K and RT, respectively. Laser oscillation built up within a duration of about 200 ps, where the output was dominated by the σ+ component. At 77 K, the maximum P_l was obtained at the peak intensity of the σ+ component and was found to be 0.94. High P_l was maintained during the lasing, because τ_s is much longer than τ_c and the lasing duration (200 ps) and hence the electron spin polarization was almost completely maintained until the electrons recombined with the holes. The result at 77 K indicated that the VCSEL emitted nearly pure circularly polarized lasing light, and the lasing polarization was determined not by the gain anisotropy on (110) surfaces [11], but by the optical selection rules. On the other hand, at RT, P_l at the peak intensity of the σ+ component was relatively short τ_s (440 ps). Although P_l at RT was smaller than that at 77 K, τ_s was still longer than that in (100) QWs and spin-controlled VCSEL emission, though small, was observed even at RT.

V. Summary

In summary, a (110) InGaAs/GaAs QW based VCSEL structure was fabricated. Circularly polarized lasing in the (110) VCSEL by optical spin injection was demonstrated at 77 K and RT. A high degree of circular polarization, 0.94 was observed at 77 K, which reflected the electron long spin relaxation time in the (110) InGaAs/GaAs MQWs.

Acknowledgment

This work was partly supported by Research Foundation for Opto-Science and Technology.

References

[1] J. Rudolph, D. Hägele, H. M. Gibbs, G. Khitrova, and M. Oestreich: Laser threshold reduction in a spintronic device, *Appl. Phys. Lett*, vol. 82, no. 25, pp. 4516-4518, Jun. 2003.

[2] J. Rudolph, S. Döhrmann, D. Hägele, M. Oestreich, and W. Stolz: Room-temperature threshold reduction in vertical-cavity surface-emitting lasers by injection of spin-polarized electrons, *Appl. Phys. Lett.*, vol. 87, no. 24, p. 241117, 2005.

[3] M. Holub, J. Shin, S. Chakrabarti, and P. Bhattacharya: Electrically injected spin-polarized vertical-cavity surface-emitting lasers, *Appl. Phys. Lett.*, vol. 87, no. 9, p. 091108 Aug. 2005; vol. 88, no. 5, p. 056102, Jan. 2005.

[4] M. Holub, J. Shin, D. Saha, and P. Bhattacharya: Electrical spin injection and threshold reduction in a semiconductor laser, *Phys. Rev. Lett.*, vol. 98, no. 14, p. 146603, Apr. 2007.

[5] S. Hallstein, J. D. Berger, M. Hilpert, H. C. Schneider, W. W. Rühle, F. Jahnke, S. W. Koch, H. M. Gibbs, G. Khitrova, and M. Oestreich: Manifestation of coherent spin precession in stimulated semiconductor emission dynamics, *Phys. Rev. B.*, vol. 56, no. 12, pp. R7076-R7079, Sep. 1997.

[6] H. Ando, T. Sogawa, and H. Gotoh: Photon-spin controlled lasing oscillation in surface-emitting lasers, *Appl. Phys. Lett.*, vol. 73, no. 5, pp. 566-568, Aug. 1998.

[7] S. Hövel, A. Bischoff, N. C. Gerhardt, M. R. Hofmann, T. Ackemann, A. Kroner, and R. Michalzik: Optical spin manipulation of electrically pumped vertical-cavity surface-emitting lasers, *Appl. Phys. Lett.*, vol. 92, no. 4, p. 041118, Jan. 2008.

[8] Y. Ohno, R. Terauchi, T. Adachi, F. Matsukura, and H. Ohno: Spin relaxation in GaAs(110) quantum wells, *Phys. Rev. Lett.*, vol. 83, no. 20, pp. 4196-4199, Nov. 1999.

[9] K. Morita, H. Sanada, S. Matsuzaka, C. Y. Hu, Y. Ohno, and H. Ohno: Strong anisotropic spin dynamics in narrow *n*-InGaAs/AlGaAs (110) quantum wells, *Appl. Phys. Lett.*, vol. 87, no. 17, p. 171905, Oct. 2005.

[10] L. Schreiber, D. Duda, B. Beschoten, G. Güntherodt, H.-P. Schönherr, and J. Herfort: Spin lifetime of (In,Ga)As/GaAs (110) quantum wells, *Phys. Stat. Sol. (b)*, vol. 244, no. 8, pp. 2960-2970, Aug. 2007.

[11] D. Sun, E. Towe, P. H. Ostdiek, J. W. Grantham, and G. J. Vansuch: Polarization control of vertical-cavity surface-emitting lasers through use of an anisotropic gain distribution in [110]-oriented strained quantumwell structures, *IEEE J. Sel. Top. Quantum Electron.*, vol. 1, no. 2, pp. 674-680, Jun. 1995.

[12] H. Fujino, S. Koh, S. Iba, T. Fujimoto, and H. Kawaguchi: Circularly polarized lasing in a (110)-oriented quantum well vertical-cavity surface-emitting laser under optical spin injection, *Appl. Phys. Lett.*, vol. 94, no. 13, p. 131108, Apr. 2009.

[13] H. Fujino, S. Iba, T. Fujimoto, S. Koh, and H. Kawaguchi: Optically-pumped circularly-polarized lasing in a (110)-oriented VCSEL based on InGaAs/GaAs QWs, *in Proc. CLEO/IQEC 2009*, Baltimore, MD, Jun. 2009, paper CMSS5.

[14] S. Iba, H. Fujino, T. Fujimoto, S. Koh, and H. Kawaguchi: Correlation between electron spin relaxation time and hetero-interface roughness in (110)-oriented GaAs/AlGaAs multiple-quantum wells, *Physica E*, vol. 41, pp. 870-875, Jan. 2009.

CAOL*2010 International Conference on Advanced Optoelectronics & Lasers, 10-14 September, 2010, Sevastopol, Ukraine

Dark Resonances in Rubidium Atoms Excited by Optical Comb

A. M. Negriyko, R. A. Malitskiy, V. M. Khodakovskiy, V. I. Romanenko,
I. V. Matsnev, L. P. Yatsenko

Institute of Physics, National Academy of Sciences of Ukraine, Nauki avenue 46, Kiev-28, Ukraine
negriyko@iop.kiev.ua

(Invited Paper)

Abstract— In a three-level lambda system, a destructive interference between excitation pathways at laser excitation can provide the resonant reduction of fluorescence and absorption ("dark resonance"). The dark resonances in rubidium vapor excited by a femtosecond laser or laser with frequency shifted feedback have been studied theoretically and experimentally.

Index Terms— Coherent population trapping, Rubidium, Magnetic field

I. INTRODUCTION

THE study of phenomena based on the coherence of atomic and molecular quantum states attracts the attention of researchers for the last three decades. One of the most interesting among them is the coherent population trapping. Population trapping manifests itself experimentally as a reduction of the atomic fluorescence intensity under the influence of two fields, which couple two long-lived atomic states (one of which can be the ground state) with a short-lived excited state in a narrow frequency interval, when the frequency of either field changes. This dip in the fluorescence intensity is observed, if the difference between the energy quanta of the light fields acting upon the atom is equal to the difference between the energies of the long-lived states (the condition of two-photon resonance). This phenomenon is referred to as a coherent population trapping (CPT) or "dark resonance" [1]. Its origin comes from formation of a "dark state"t, which is a superposition of the long-lived states. The atom in such a state does not emit radiation. The dark resonance width is governed by the duration of the atom-field interaction and the relaxation times of the density matrix elements which describe the long-lived state. It can be very narrow (in a rubidium cell filled with a buffer gas, there were observed resonances with width 30 Hz [2]). Such resonances are of great interest in the domains of frequency standards [3] and magnetometry [4], [5].

It is well known that a "frequency comb" which overlaps a wide frequency range by equidistant frequency components can be generated by femtosecond lasers. The fixed time interval between laser pulses determines the frequency shift between the "comb" components, and the wide spectral range is provided by a small duration of laser pulses. The frequency comb is widely used today in metrology, mainly for optical frequency measurements [6]. Besides its application in metrology, the "frequency comb" can be used in experiments, where the

coherent interaction between laser light and atoms is studied. Let the width of the frequency comb exceed the frequency ω_{12} that corresponds to the difference between the energies of long-lived states. If the difference between the "comb" frequency components Ω_m is so selected that $N\Omega_m \approx \omega_{12}$, where N is an integer, then, provided $N\Omega_m = \omega_{12}$, the coherent trapping of population and, respectively, a dip in the fluorescence spectrum can be expected, if the interval between frequency components is varied in the vicinity of $\Omega_m = \omega_{12}/N$ [7].

II. EXPERIMENT

A. Setup

Fluorescence of rubidium was registered by means of a photoelectronic multiplier. The cell was located at the center of a system of Helmholtz rings, two pairs of which were used to compensate the Earths magnetic field, and the third pair created the working magnetic field along the direction of beam propagation. The block of the current control by means of Helmholtz rings allowed us to scan the magnetic field in accordance with a triangular signal given by an external generator and also to modulate the field with a frequency of 900 Hz. The deviation amplitude was 0.2 G. The signal from a photoelectronic multiplier was synchronously detected at the same frequency of 900 Hz and recorded into the computer memory with the help of an ADC. To register the current in

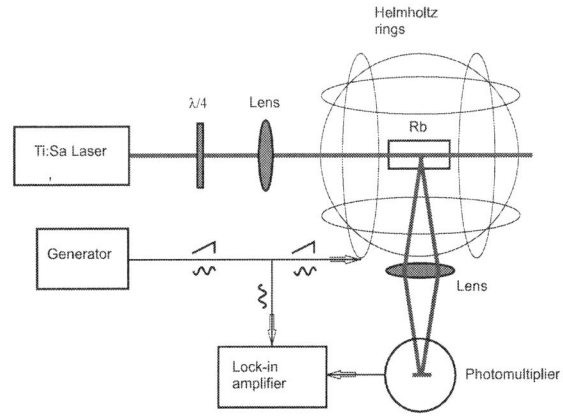

Fig. 1. Scheme of the experimental setup.

978-1-4244-7043-3/10 $26.00 © 2010 IEEE

the coils, and, respectively, the magnetic field magnitudes in rubidium vapor, the voltage drop across a resistor connected in series with the coils was measured.

For the observation of a dark resonance, a femtosecond laser with the wavelength $\lambda = 780.24$ nm was used. By filtering the laser radiation, a spectral interval of about $\Delta\lambda = 0.2$ nm was selected. The radiation power in this spectral interval was 3 mW. The pulse repetition frequency ν was different in every experiment, by varying around 75.65 MHz. This means that the radiation that acted upon the atom consisted of $N = c\Delta\lambda/\lambda^2\Delta\nu = 1300$ modes, and the power of a mode was $P = 3000\ \mu W/1300 = 2.3\ \mu W$.

The scheme of the experimental installation is shown in Fig. 1. A beam of about 6 mm in diameter generated by a femtosecond laser passed through a quarter-wave plate to be focused by a lens with a focal length of 40 cm in a cell with rubidium vapor (the natural mixture of isotopes ^{87}Rb:^{85}Rb=27.835%:72.165% at a temperature of 25°C. The laser beam diameter at the fluorescence registration site was about 2 mm, so that the intensity of one mode was $I = P/(\pi R^2) = 73 \cdot 10^{-6} W/cm^2$. This intensity corresponds to Rabi frequency $\Omega_0/(2\pi) = 2.6$ MHz.

B. Results

Before investigation of magneto-optic resonances corresponding D_2 line in rubidium, we recorded the fluorescence spectrum in the vicinity of the line excited by radiation of femtosecond laser. As it was expected, fluorescence spectrum with the maximum near 780 nm.

Figure 2 shows magneto-optical resonances in a cell with rubidium vapor irradiated by σ^+ polarized radiation of the femtosecond laser. The metastable levels are characterized by $F = 1$, $F = 2$ (^{87}Rb), $F = 2$, $F = 3$ (^{85}Rb) and magnetic quantum number m. Small (not marked) resonance nearly zero field is produced by noncompensated Earth magnetic field. Depending on magnetic field, resonances with $N = 39, 40, 41$

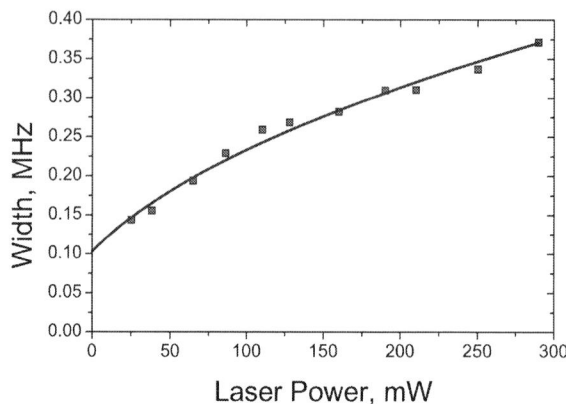

Fig. 3. Resonance width vs the laser power.

for ^{85}Rb and $N = 90, 91$ for ^{87}Rb in the interval $0 < B < 50$ G are observed.

Figure 3 shows that the power broadening of the CPT resonance is rather well described by the formula

$$\Gamma = \sqrt{\frac{1}{\tau_0^2} + \beta I},\tag{1}$$

where τ_0 is a time of flight of an atom through laser beam, I is the intensity of laser radiation. The line shape of the first derivative of the signal on the magnetic field in the vicinity of resonance is described by the formula

$$S = S_0 \left[\frac{2k(\nu - \nu_0)}{1 + \frac{(\nu - \nu_0)^2}{\Gamma^2}} - \alpha \right].\tag{2}$$

Light shifts of the CPT resonances were investigated for different maxima of radiation spectrum of femtosecond Ti:Sa-laser. When the position of the maximum varied, the CPT resonances for small laser intensities where observed for different magnetic field. Figure 4 shows the power dependences of the light shifts for different maxima of radiation spectrum. The depicted results shows that the position of the laser

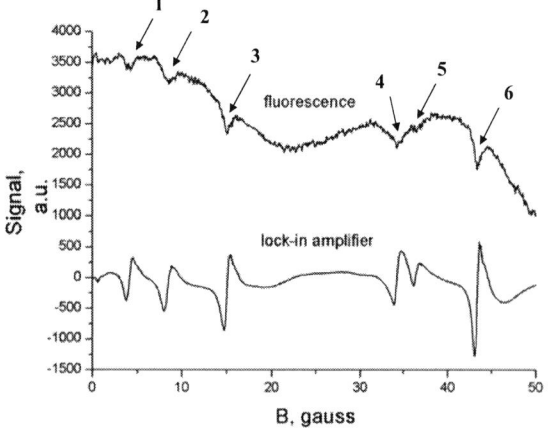

Fig. 2. Dark resonances in the rubidium vapor in the σ^+ polarized radiation field of femtosecond laser. 1: $m = -2$, $N = 40$, ^{85}Rb; 2: $m = -1$, $N = 40$, ^{85}Rb; 3: $m = -1$, $N = 90$, ^{87}Rb; 4: $m = 2$, $N = 41$, ^{85}Rb; 5: $m = 1$, $N = 91$, ^{87}Rb; 6: $m = -2$, $N = 39$, ^{85}Rb.

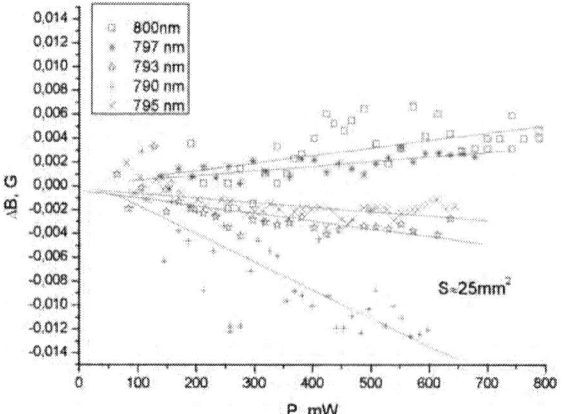

Fig. 4. Dependence of the resonance width on the laser power.

radiation spectral maximum correlates with the sign of the light shift. This result is in a qualitative agreement with a theoretical analysis of the simple model of CPT resonances in a polychromatic field [8].

III. CONCLUSION

We studied the magneto-optical CPT resonances in [87]Rb and [85]Rb fluorescence, excited by the femtosecond Ti:Sa-laser radiation or FSF laser radiation resonant with D_2 line. Power broadening is well described by simple theoretical formula. Light shifts of the resonances strongly depend on the position of the maximum laser spectrum and change the sign in qualitative agreement with the theory describing CPT resonances in three-level atoms excited with polychromatic field.

ACKNOWLEDGMENT

Authors acknowledge support from NASU projects V136, V137, VC139 and joint Russian-Ukrainian grant RFFR/1-09-25.

REFERENCES

[1] Shore, B. W., "Coherent manipulations of atoms using laser light," *Acta Phys. Slov.* **58**, 243–486 (2008).

[2] Erhard, M. and Helm, H., "Buffer-gas effects on dark resonances: Theory and experiment," *Phys. Rev. A* **63**, 043814 (2001).

[3] Knappe, S., Wynands, R., Kitching, Robinson, and Hollberg, L., "Characterization of coherent population-trapping resonances as atomic frequency references," *J. Opt. Soc. Am. B* **18**, 1545–1553 (2001).

[4] Nagel, A., Graf, L., Naumov, F., Mariotti, E., Biancalana, V., Meschede, D., and Wynands, R., "Experimental realization of coherent dark-state magnetometers," *Europhys. Lett.* **44**, 31–36 (1998).

[5] Schwindt, P. D. D., Knappe, S., Shah, V., Hollberg, L., Kitching, J., Liew, L.-A., and Moreland, J., "Chip-scale atomic magnetometer," *Appl. Phys. Lett.* **85**, 6409–6411 (2004).

[6] Cundiff, S. T. and Ye, F. J., "Colloquium: Femtosecond optical frequency combs," *Rev. Mod. Phys.* **75**, 325–342 (2003).

[7] Arissian, L. and Diels, J.-C., "Repetition rate spectroscopy of the dark line resonance in rubidium," *Opt. Commun.* **264**, 169–173 (2006).

[8] Romanenko, V. I., Romanenko, A. V., Yatsenko, L. P., Kazakov, G. A., Litvinov, A. N., Matisov, B. G., and Rozhdestvensky, Y. V., "Dark resonances in the field of frequency shifted feedback laser radiation." arXiv:1006.5595v1 [quant-phys] (June 2010). http://arxiv.org/abs/1006.5595v1.

CAOL*2010 International Conference on Advanced Optoelectronics & Lasers, 10-14 September, 2010, Sevastopol, Ukraine

589nm High Power Sum-Frequency Generation with Quasi-Continuous-Wave Diode-Pumped Nd:YAG Lasers

Bo Yong, Xie Shiyong, Wang Zhichao, Shen Yu, Xu Jialin, Cui Dafu, Peng Qinjun, Xu Zuyan

Technical Institute of Physics and Chemistry, Chinese Academy of Sciences, Beijing, 100190 China

(Invited *Paper*)

Abstract: A 12.7 W 589 nm coherent light source with the linewidth of less than 1 GHz and the beam quality of $M^2 = 1.2$ is achieved by extra-cavity sum-frequency mixing 1064 nm and 1319 nm quasi-continuous-wave diode-pumped Nd:YAG lasers with a LBO crystal.

Keywords: sum–frequency generation, yellow laser, solid-stated laser, diode-pumped.

Diode-pumped solid-state yellow lasers have become interesting for many applications in medicine, biotechnology, display, communications, and remote sensing. Especially, high power and high beam quality and narrow linewidth yellow laser at the sodium D2 line of 589nm can be used to create an artificial guide star for adaptive optical telescope systems by exciting atomic sodium in the upper atmosphere (85 to 105km) [1]. It is very difficult to achieve high power yellow solid-state laser with directly frequency doubling in general because the fluorescence spectrum intensity of corresponding fundamental frequency for Nd or Yb laser is too weak. At present, there are two main methods to produce high average power and narrow linewidth solid-stated yellow laser. One is sum-frequency mixing the two lines of Nd:YAG lasers operating near 1064 and 1319 nm [2]. For example, output power of 50W is demonstrated with sum–frequency mixing of injection-locked Nd:YAG continuous-wave (CW) lasers [3]. The other is frequency doubling of 1178nm CW Raman fiber amplifiers and output power of 50W is achieved recently [4].

In this letter, we have developed a high beam quality and narrow linewidth yellow beam at 589nm by extra-cavity sum-frequency mixing 1064 and 1319 nm radiations of quasi-continuous-wave (QCW) diode-pumped Nd:YAG lasers in a LBO crystal. The 1064 and 1319nm lasers are generated by means of a diode-pumped Nd:YAG rod master-oscillator power-amplifier (MOPA) system. The thermally near-unstable resonators [5] with two-rod birefringence compensation [6] are designed and employed in the 1064 and 1319 nm ring oscillators. Both 1064 and 1319 nm oscillators operate at the border of the stable region near the unstable region where they have large fundamental mode size at the gain media of Nd:YAG to achieve high beam quality output. A solid etalon is placed inside the ring cavity to compress the linewidth and tune the wavelength of the ring oscillator. Both 1064 and 1319 nm oscillators are amplified by two laser heads. Moreover the beam width of the 1064 and 1319 nm lasers inside the

nonlinear optical crystal of LBO is optimized. As a result, QCW average output power of 12.7 W with the linewidth of less than 1 GHz and the beam quality factor of $M^2 = 1.2$ at 589nm is achieved by extra-cavity sum-frequency mixing of the diode-pumped Nd:YAG MOPA systems.

The schematic of the overall yellow laser system for extra-cavity sum-frequency generation is shown in Fig. 1. There are three subsystems: an 1064nm QCW diode-pumped Nd:YAG MOPA system, an 1319nm QCW diode-pumped Nd:YAG MOPA system and a sum-frequency generator. The two MOPA laser systems are similar in layout and function. For example, the 1064nm MOPA system is made up of a master oscillator, an isolator, a beam forming and the 1064nm amplifier.

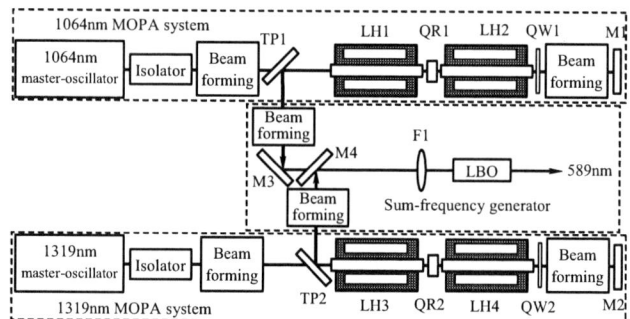

Fig. 1. Schematic of the laser system showing the three subsystem: the 1064nm MOPA system, the 1319nm MOPA system and the sum-frequency generator. M1: high reflection at 1064nm; M2: high reflection at 1319nm; M3: 45° high reflection at 1064 nm; M4: 45° high reflection at 1319 nm and antireflection at 1064 nm; LH1 and LH2: 1064nm laser head; LH3 and LH4: 1319nm laser head; QR1, 1064nm quartz rotator; QR2, 1319nm quartz rotator; TP1, 1064nm thin film polarizer; TP2, 1319nm thin film polarizer; QW1, 1064nm quarter-wave plate; QW2, 1319nm quarter-wave plate.

The master oscillator is described in detail elsewhere [7]. In brief, it is a ring laser with three mirrors folded cavity. Each ring laser contains two identical diode-side-pumped laser heads. Each laser head consists of a Nd:YAG rod (diameter, 3mm; length, 65mm;) and three diode arrays. Each array contains three QCW diode bars with average power of 20W and peak power of 100W at wavelength of 808nm. A quartz 90° polarization rotator is placed between two Nd:YAG rods

978-1-4244-7043-3/10 $26.00 © 2010 IEEE

for polarization dependent birefringence compensation. Unidirectional operation of the ring lasers is obtained by the combination of the Faraday rotator, the thin film polarizer and the half-wavelength waveplate. For the homogeneously broadened linewidth of the Nd:YAG laser transitions at 1064 and 1319 nm, the narrow linewidth or single-frequency laser can be obtained by using the ring cavity configuration to avoid spatial hole burning. To eliminate the mode-hopping and tune the wavelength of the ring laser, a solid F-P etalon is inserted in the cavity. The wavelength of the laser is controlled by the temperature and angle control of etalon. The QCW 1064nm master oscillator provided average output power of 7.5W with the beam quality of $M^2=1.19$ and the linewidth of 0.4GHz, the QCW 1319nm master oscillator provided average output power of 3.2W with the beam quality of $M^2=1.14$ and the linewidth of 0.4GHz.

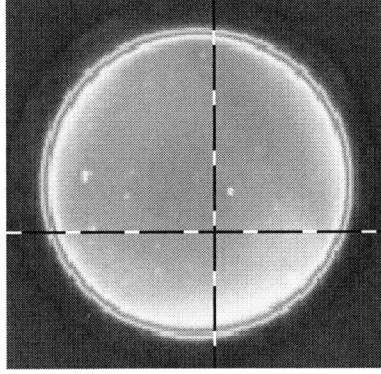

Fig.2. Spatial profile of fluorescence from the Nd:YAG rod.

The output beams of the 1064 and 1319 nm master oscillator are expanded and collimated by the beam forming to match the Nd:YAG rod of the amplifiers. Each amplifier contained two QCW diode-side-pumped laser heads. The laser head also consists of a Nd:YAG rod (diameter, 3mm; length, 65mm;) and three diode arrays with three 20W QCW diode bars. So the total average pump power of each laser head is 180W. In order to obtain a uniform pump distribution in the rod of the laser head for minimizing the influence of thermal aberration for the laser beam, the parameters of the laser head are optimized, including the wavelength of the diode bars, the distance between the diode arrays and the Nd:YAG rod, the diameter and Nd concentration of the rod. This design leads to a uniform pump distribution in the rod as shown in the Fig. 2. The thermal focal lengths of the rods for both 1064 and 1319 nm laser operations are measured by using the unstable-resonator method,[8] and the results are shown in Fig. 3. It can be seen that the thermal focal length of the rods are long for both 1064 and 1319nm operations. For example, when the diode pump power is 157.5W, the thermal focal length of the former is 528mm and the later 450mm. The reason for the former is longer than the later is because the quantum defect of Nd:YAG for 1064nm operation is less than for 1319nm operation.

A quartz 90° polarization rotator is inserted between two Nd:YAG rods for polarization dependent birefringence compensation. After first-pass amplification, the beam is reflected back through the amplifier chain by the high reflectivity mirror of M1 or M2. To maximize the extraction energy from two rods in double-pass propagation and reduce the Fresnel diffraction and the alignment sensitivity of the system, an imaging relay optics is placed in the front of the high reflectivity mirror, and suitable propagation in second-pass amplification can be obtained. The quarter-wave plate and the polarizer serve as power coupling from the amplifier.

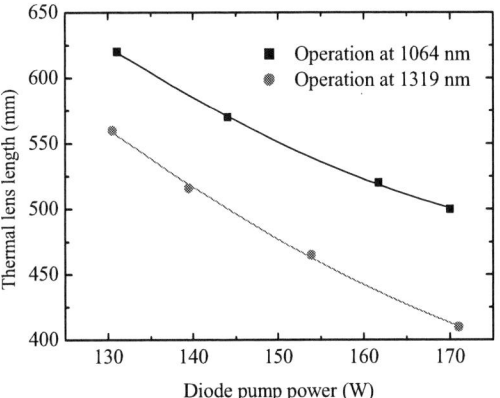

Fig.3. Thermal lens length of the Nd:YAG rod.

After double-pass amplification, the QCW 1064nm output power is up to the highest value of 45W with the beam quality of $M^2=1.37$ ($Mx^2=1.35$, $My^2=1.40$), and the QCW 1319nm output power is up to the highest value of 31W with the beam quality of $M^2=1.34$ ($Mx^2=1.36$, $My^2=1.33$). The indicated M^2 values are the average values between Mx^2 (M^2 at x axis) and My^2 (M^2 at y axis), which are measured by using a beam quality analyzer (M2-200, Spiricon Inc.).

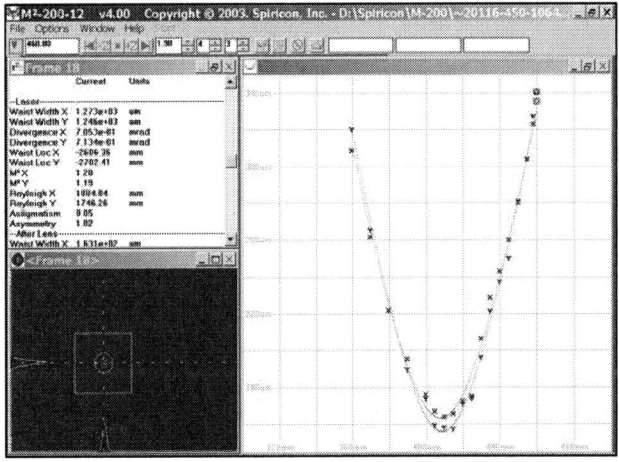

Fig.4. Laser beam quality measured at 12.7 W output power.

The 1064nm beam and the 1319nm beam are expanded and collimated, then synthesized to be one beam by the dichromatic mirror M4. The 1064 and 1319 beam is focused inside the LBO crystal to achieve sum-frequency mixing by the lens F1.

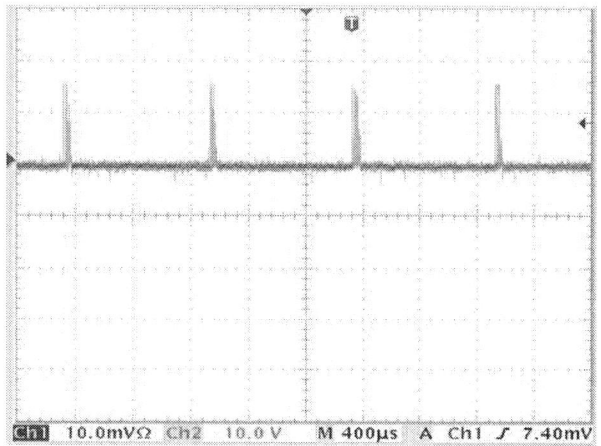

Fig.5. Oscilloscope traces for laser pulse.

The LBO crystal is noncritical type I phase-matched, so it is with no walkoff effect. The aperture of the LBO crystal is 4 mm. The crystal has a triple band anti-reflection coating on both faces. The sum frequency process requires optimization of the co-linearity, spatial overlap, spot size and polarization orientation of the input 1064 and 1319 nm laser. The expanded 1064 and 1319 nm beam is focused to an optimized spot size by a suitable lens and pass through the LBO crystal for sun-frequency generation. When input powers at 1064 and 1319 nm are 45W and 31W respectively, the maximum average output power of 12.7W is obtained at 589nm with sum frequency efficiency of 16.7%. The beam quality of the yellow laser is measured to be $M^2 = 1.2$ ($M_x^2=1.20$, $M_y^2=1.19$) as shown in the Fig.4. The repetition rate is 1kHz with the pulse width of 80µs as shown in the Fig.5. The sum frequency of 589nm can be precisely tuned with temperature and angle control of the solid etalon in the 1064nm laser, and the solid etalon in the 1319nm laser is fixed at given angle and temperature. The linewidth and the wavelength of the sum frequency laser are measured by using the wavelength meter (WS-7, HighFinesse GmbH). As shown in the Fig.6, the laser linewidth is approximately 0.4GHz.

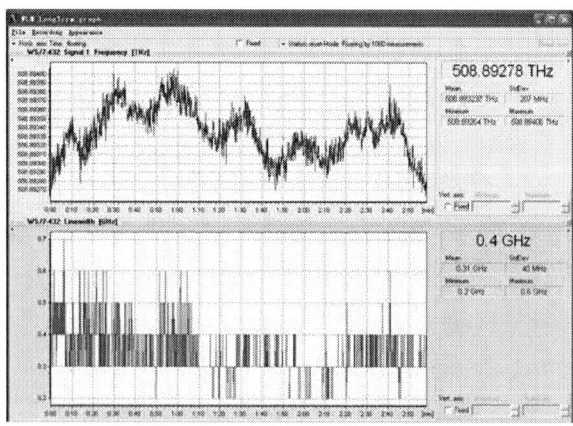

Fig.6. Laser wavelength and linewidth measured byWS-7

In conclusion, 12.7W QCW solid-state laser source with the linewidth of less than 1GHz and the beam quality of $M^2 = 1.2$ at the wavelength of 589nm is achieved by external-cavity sum-frequency mixing of the 1064 and 1319 nm MOPA systems in a LBO crystal.

REFERENCES

[1] Allen K. Hankla, Jarett Bartholomew, Ken Groff, and Ian Lee, et al., "20 W and 50 W Solid-State Sodium Beacon Guidestar Laser", Proc. of SPIE, Vol. 6272, P. 62721G, 2006.

[2] A. C. Geng, Y. Bo, X. D. Yang, H. Q. Li, Z. P. Sun, Q. J. Peng, X. J. Wang, G. L. Wang, D.F.Cui and Z. Y. Xu, "Intracavity sum-frequency generation of 3.23 W continuous-wave yellow light in an Nd:YAG laser", Optics Communications, Vol. 255, P. 248, 2005.

[3] J. C. Bienfang, C. A. Denman, B. W. Grime, P. D. Hillman, G. T. Moore and J. M. Telle, "20 W of continuous-wave sodium D2 resonance radiation from sum-frequency generation with injection-locked lasers", Optics Letters, Vol. 28, P. 2219, 2003.

[4] Luke R. Taylor, Yan Feng, Domenico Bonaccini Calia, "50W CW visible laser source at 589nm obtained via frequency doubling of three coherently combined narrow-band Raman fibre amplifiers", Optics Express, Vol. 18 No. 8, P. 8541, 2010.

[5] Y. Feng, Y. Bi, Z. Y. Xu, and G. Y.Zhang, "Thermally near-unstable cavity design for solid state lasers," in Laser Resonators and Beam Control, Proc. of SPIE, Vol. 4969, P. 227, 2003.

[6] S. Konno, S. Fujikawa, and K. Yasui, "80 W cw TEM00 1064 nm beam generation by use of a laser-diode-side-pumped Nd:YAG rod laser", Appl. Phys. Lett. Vol. 70, P. 2650, 1997.

[7] Peng Qinjun, Bo Yong, Lu Yuanfu, Xie Shiyong, Zong Nan, Jialin Xu, Cui Dafu, Xu Zuyan, "Tunable and narrow linewidth yellow beam generation by sum–frequency mixing of diode-pumped Nd:YAG ring lasers", CAOL 2008, September 29 - October 4, Alushta, Crimea, Ukraine.

[8] D. G. Lancaster and J. M. Dawes, "Thermal-lens measurement of a quasi steady-state repetitively flashlamp-pumped Cr, Tm, Ho:YAG laser", Optics & Laser Technology, Vol. 30, P. 103, 1998.

High Power Wavegude CO_2 Lasers for Technologies and Medicine

(Invited Paper)

V. Ya. Panchenko, V. S. Golubev, V. V. Vasiltsov[*], E.N.Egorov

Institute on Laser and Information Technologies, Russian Academy of Sciences

140700, Shatura, Russia,

http://www.laser.ru [*]v.vasiltsov@mail.ru

1. Introduction

The ongoing development of the laser technologies of material processing places heavy demands on the sources of laser radiation, such as reliability, high beam quality, low operating costs, long service life, and low price.

The CO_2 lasers keep on heading the list of lasers applied in various technologies [1]. ILIT RAS has been for years developing the multichannel waveguide diffusion cooled CO_2 lasers excited by sound-frequency ac discharge [2-4].

The paper considers the recent advances in the development of the high-power (kilowatt-level) waveguide CO_2 lasers meant for application in processing of various materials, specifically in precision cutting.

The technical characteristics of the new three models of diffusion-cooled multichannel waveguide industrial CO_2 lasers excited with acoustic-frequency ac discharge are presented.

The industrial lasers of this type, being developed for years at ILIT RAS, offer a number of merits, such as

- high quality (single waveguide mode EH11) and stability of radiation at the expense of waveguide generation;
- unprecedented weight and dimensions;
- possibility to use CO_2:He:air mixture;
- low operating costs.

The above lasers can be used to advantage in the laser processing systems intended for precision cutting of metallic (thickness to 10 mm) and non-metallic (thickness to 40 mm) materials; welding; surfacing and fabrication of parts from composite and metallic powder materials.

As well as the paper provides the description and the technical characteristics of medical cardio-surgery laser systems of "Perfocor" series, based on high-power waveguide CO_2 lasers of 1 kW average power with pulse energy to 60 J, developed at the IPLIT RAS for the transmyocardial laser revascularization (TMLR) presents a promising method to cure the ischemic disease of heart [5].

The waveguide lasers covers the gaseous CO_2 and CO lasers, and the excimer Xe and KrF lasers. In these lasers the propagation of radiation through the active medium occurs in the waveguide. The idea of employing hollow waveguides in the gas discharge lasers was first advanced by Markatilly and Schmeltzer in 1964 [6]. The gas discharge waveguide lasers are distinguished from the conventional gaseous lasers by the presence of a hollow (e.g. dielectric) waveguide between the mirrors, which performs several functions. On the one hand, its inner surface limits the volume filled with the active medium; on the other hand, it serves as a channel for laser radiation propagation, presents a component part of the waveguide resonator, and specifies the mode structure of radiation.

The main merit of the waveguide CO_2 laser as a technical system (the design compactness permitting the laser weight and dimensions to be reduced) is easily realized.

The waveguide in the CO_2 lasers presents, as a rule, a gas discharge tube that has the length much more of the inner diameter which, in its turn, can be hundreds of times longer than the radiation wave length.

Depending on the optical scheme of tube connection, parallel or series, the multibeam or single generation can be obtained. The multichannel waveguide arrays containing to 60 tubes generated up to 6 kW average power, and the single-beam oscillator produced the power of about 1.5 kW, with the specific power of radiation reaching 50 W/m. The multibeam waveguide oscillators are advantageously used in the laser based industrial systems for treatment of various materials. The cardio-surgery laser plant to carry out the operations on heart is built around a single-beam single-mode waveguide oscillator of 1 kW average power.

2. Short description of the TL-300, TL-1200, and TL-1500 models

These three models of the multichannel single-beam lasers have been designed as the monobloc units.

Fig. 1 shows the optical scheme of oscillators. The corner reflectors ensure the passage of radiation from the 9 to 27 parallel tubes through all the tubes, thus forming a single extended resonator. The radiation is extracted with the use of a half-transmitting plane mirror of ZnSe.

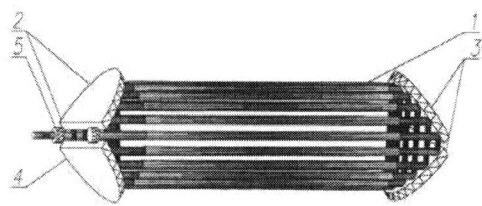

Fig. 1. Optical scheme of oscillators multichannel waveguide diffusion cooled CO_2 lasers:
1 – gas discharge tubes, 2,3 – foiling mirrors, 4 – rear mirror of resonator, 5 – exit mirror of resonator

- Generation of low (to 400 W) average power proved to be technically realizable through air cooling of the oscillator, which makes the laser performance even more attractive. This scheme was embodied in the TL-300 model; Fig. 2 illustrates its external view.

- In the models of high average power the liquid cooling of oscillator is used as in the previous models of the MTL family. Fig. 3 illustrates the external view of the TL-1200 model.

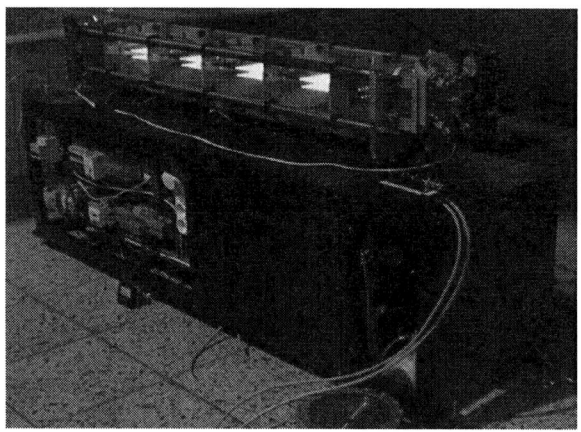

Fig. 2. External view of the TL-300 model (without casing).

Fig. 3. External view of the TL-1200 model.

The lasers of this family feature small dimensions and operating costs, since they offer rather high technical efficiency (more than 10%), and low consumption of the working mixture.

Besides, the waveguide mode permits the single-mode generation to be easily realized and maintained at high long-term stability. The distribution of radiation produced by the TL-1200 model is shown in Figs. 4. The generation of a nearly pure first waveguide mode is obvious.

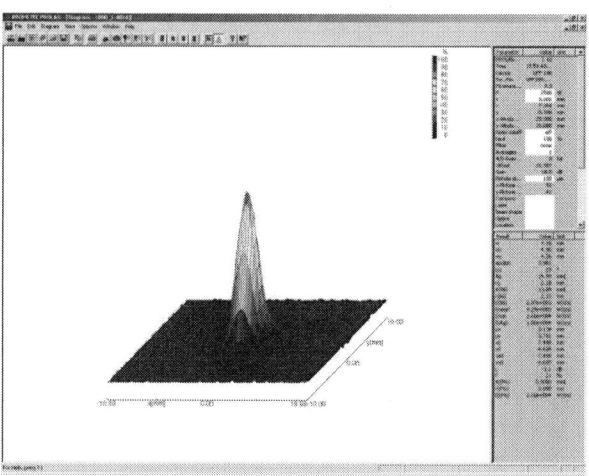

Fig. 4. Distribution of power density (TL-1200) divergence at the power level 0.86 = 1.57 mrad divergence at the power level 0.9 = 1.67 mrad P=1250 W, level > 17%, P=1380 W, level > 4%

Table 1. Main technical characteristics

Parameter/model	TL-300	TL-1200	TL-1500
Beam wave length, μm	10.6	10.6	10.6
Rated beam power, W	320	1200	1500
Pulsed-periodic mode, kHz	0.2-2.5	the same	the same
Beam power instability, %	<2	the same	the same
Aperture, mm	20 (telescope)	the same	the same
Beam divergence, mrad	0.8 (telescope)	the same	the same
Gas mixture consumption, nl/h	0.5	1.5	1.7
Cooling water consumption, m^3/h	none	0.7	0.8
Technical efficiency, %	10	11	11
Dimensions, m	0.6x0.6x1.5	0.6x1.2x2.2	0.6x1.2x2.5
Weight, kg	220	380	400

Feature: cooling	air	liquid	liquid

3. Technological applications

The above lasers can be used to advantage in the laser processing systems intended for precision cutting of metallic (thickness to 10 mm) and non-metallic (thickness to 40 mm) materials; welding; surfacing, and fabrication of parts from composite and metallic powder mater

4. Laser systems "Perfocor" for the transmyocardial laser revascularization (TMLR).

The paper provides the description and the technical characteristics of medical cardiosurgery laser systems of "Perfocor" series, based on high-power waveguide CO_2 lasers of 1 kW average power with pulse energy to 60 J, developed at the IPLIT RAS for the transmyocardial laser revascularization (TMLR) presents a promising method to cure the ischemic disease of heart.

The ischemic disease of heart is one of the main causes of human mortality in the industrially developed countries. According to the data available from the World Health Organization, the contribution of ischemia to the mortality from different cardio-vascular diseases is about 60%.

In the middle 90s, a fundamentally new method for treatment of the ischemic disease gained acceptance in the clinical medicine. It was the so called Transmyocardial Laser Revascularization (TMLR). During the operation, powerful laser beam is used to make channels in the myocardium left ventricle, which are opened to the heart cavity. These channels favor the restoration of blood stream in the myocardium zones invaded by ischemia, and prevent the progression of myocardium infarction. The method was naturally based on the fact that the heart of reptiles has a poorly developed system of coronary vessels. Nutrition of myocardium (oxygen delivery and metabolism) is affected by blood through the net of radically located intra-ventricle channels (trabeculae).

The TMLR method involves the formation of several tens of channels with 0.3 to 1.5 mm diameter in the myocardium (typical thickness about 20 mm).

Among the medical laser systems under discussion (solid-state, Nd:YAG, Er:YAG, Ho:YAG, excimer lasers, etc.) the high-power CO_2 laser with pulse energy to 40 J is most suited to produce channels in myocardium.

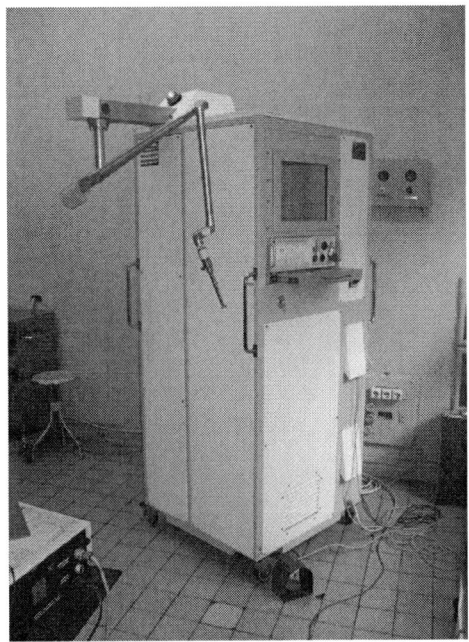

Fig. 5. The "PERFOCOR" laser system for Transmyocardial Laser Revascularization

Table 2. Main technical characteristics

Behaviors	Perfokor ILIT RAS	Heart Laser Eng. Inc. Milford M.A. USA
Wavelength, μm	10.6	10.6
Power, W	1000	800
Oscillating mode	Pulse-periodical	Pulse-periodical
T pulse, ms	10-100	1-99
E pulse, J	5-150	–
Pulse delay, ms	10-1000	1-999
Radiation structure	single-mode	single-mode
Consumption of mixture, l/h	1	60
Dimension of array		
Propulsion module, cm	70x80x200	99x167x114
Mass, kg	260	952
Price, $	250000	1000000
Manufacture	1996-2005	1995-2005

A high-power (about 1 kW) waveguide CO_2 laser served as the basis for production of the intellectual laser medical system "Perfocor" and its application in clinical practice. The system is intended to perform operations on heart with using TMLR procedure.

The system has no analogs in Russia and in Europe. It has been granted the certificate by the Ministry of Health RF.

More than 650 operations have been successfully performed from 1997 to 2006.

REFERENCES

[1] Holger Schlueter, Advances in industrial power lasers, Proc. SPIE, Vol. 5777, pp. 8-15, 2005.

[2] V.V. Vasiltsov, M.G. Galushkin, V.S. Golubev, V.Ya. Panchenko, High power waveguide industrial CO_2 lasers, Proc. SPIE, Vol. 2713, pp. 76-84, 1995.

[3] A.I. Bondarenko, V.V. Vasiltsov, M.G. Galushkin, V.S. Golubev, V.G. Niziev, V.Ya. Panchenko et al., High power high-beam-quality industrial CO_2 lasers, Perspectivnye Materialy, № 2, pp. 60-67, 1999.

[4] V.V. Vasiltsov, M.G. Galushkin, A.P. Roshin, A.V. Solovyev, Waveguide high power industrial CO_2 lasers, Proc. SPIE, Vol. 4165, pp. 169-177, 2000.

[5] I.I. Berishvili, L.A. Bokeria, V.V. Vasiltsov et al., High power single-mode diffusion-cooled waveguide CO_2 laser for transmyocardial revascularization, Izvestiya RAN, ser. Fizicheskaya, Vol. 63, № 10, pp. 2059-2065, 1999.

[6] J.J. Degnan, The waveguide laser: A review, Appl. Phys., Vol. 11, № 1, pp. 1-33, 1976.

CAOL*2010 International Conference on Advanced Optoelectronics & Lasers, 10-14 September, 2010, Sevastopol, Ukraine

3D Model of Vertical Cavity Surface Emitting Laser with Resonant Array of Quantum Wells and External Mirror

N. N. Elkin, A. P. Napartovich, *Member, IEEE*, D. V. Vysotsky

SRC RF Troitsk Institute for Innovation and Thermonuclear Research

142190 Troitsk, Moscow province, Russia, elkin@triniti.ru

(Invited *P*aper)

Abstract: Bidirectional beam propagation method is modified to describe wave field interference in laser cavity with multiple quantum wells. Properties of optical modes are found numerically for cylindrical vertical cavity surface emitting laser with above threshold e-beam pumping and external mirror. The stability limits of single-mode operation are determined.

I. INTRODUCTION

A resonant heterostructure of an array of quantum wells (QW) is of practical interest for application in vertical external cavity surface emitting lasers (VECSELs). The steady-state oscillating modes of a laser are described by non-linear partial differential equations containing eigenvalues. For the first time, the applications of bidirectional beam propagation method (BiBPM) for laser devices [1] were restricted by linear eigenvalue problem neglecting influence of the light beam on gain and index of the active medium. The eigenvalue problems for a non-hermitian matrix of high dimension were solved numerically in [1]. Next, the BiBPM combined with the round-trip operator technique was developed for self-consistent solution of wave field and material equations [2]. The Krylov subspace methods applied in [2] to calculate the eigenfunctions of the linear wave equation is considerably more effective in comparison with the matrix method [1]. The presented numerical algorithm evolves the method elaborated previously for the vertical cavity surface emitting laser (VCSEL) [2].

II. DESCRIPTION OF THE VECSEL AND BASIC EQUATIONS

The scheme of the VECSEL containing a resonant heterostructure is presented in Fig. 1. Assuming a vertical *z*-axis we represent the VECSEL as a pile of layers, separated by planes $\{z=z_k, k=0,...,M\}$ where M is the total number of layers. The index and absorption are constant in each layer except the active layers (QWs) where non-uniform distributions are controlled by electrical current and light intensity. To distinguish QWs from other layers we define the index array $\{v(l), l=1,...,q\}$, where $q=25$ is the number of QWs. If $k=v(l)$ then the layer $[z_{k-1}, z_k]$ is the l-th QW. The external spherical mirror is a top member of the VECSEL.

This work is supported by Russian Foundation for Basic Research, project #08-02-00796-a.

We assume that the scalar diffraction theory is applicable. The pump profile has circular symmetry according to assumption. Therefore, we use cylindrical coordinates. Laser modes have a time dependence of the form $E(r,\varphi,z,t)=U(r,\varphi,z)\exp(-i\Omega t)$, $\Omega=\omega_0+\Delta\omega-i\delta$, where ω_0 is the reference frequency, $\Delta\omega=\omega-\omega_0$ is the frequency shift and δ is the attenuation factor. The reference wavenumber and reference wavelength are defined by standard relations: $\omega_0=k_0 c$, $k_0=2\pi/\lambda_0$. The solutions of a form $U(r,\varphi,z)=U_m(r,z)\exp(im\varphi)$ are subjects of investigation. Introducing new variables $g_r=2\delta/c$, $\Delta k=\Delta\omega/c$, $\beta=g_r+i2\Delta k$, we have obtained the Helmholtz equation for m-th angular harmonic:

$$\frac{\partial^2 U_m}{\partial z^2}+Q^2 U_m=0, \quad Q=\left(k_0^2 n^2-ik_0 g-ik_0 n^2\beta+\frac{\partial^2}{\partial r^2}+\frac{1}{r}\frac{\partial}{\partial r}-\frac{m^2}{r^2}\right)^{1/2}. \quad (1)$$

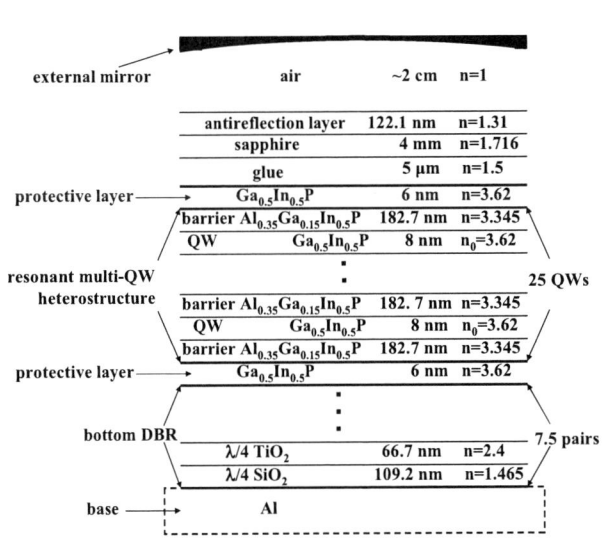

Fig. 1. Scheme of the VECSEL (Cross-Section View).

The equation (1) contains a complex eigenvalue β. The real part of β is the decay rate of wave field expressed in units of inverse length, the imaginary part is the twice wavenumber shift relative to the reference value. Here n and g are index and gain respectively, Q is the operator of longitudinal wavenumber.

978-1-4244-7043-3/10 $26.00 © 2010 IEEE   71

The boundary conditions at the interfaces between adjoining layers and at the lateral boundary were determined. We use condition of continuity for the wave field U_m and its normal derivative at the interfaces. The zero boundary conditions at the lateral boundary were set because the active layers have strong attenuation in the absence of pump. The boundary condition at the mirror suits a good reflecting surface. The set of quantum wells forms a finite periodical structure so as the optical length of one period is equal to $\lambda_s = 640$ nm.

The main problem consists of self-consistent solving of the wave field equation and material equations in order to find the spatial profile of a laser electromagnetic field and its frequency in steady-state mode of operation. We restrict our consideration with axisymmetric laser modes. According to this condition, we are to solve the axisymmetric ($m = 0$) equation (1) jointly with the set of non-linear diffusion equations [3]:

$$\frac{1}{r}\frac{\partial}{\partial r}\left(r\frac{\partial Y_l}{\partial r}\right) - \frac{Y_l}{D\tau_{nr}} - \frac{B}{D}N_{tr}Y_l^2 - \frac{|U_0|^2\ln(\chi(Y_l))}{D\tau_{nr}} =$$
$$= -\frac{kE_e j}{DN_{tr}3E_g qed}, \quad l = 1,...,q \qquad (2)$$

for normalized carrier density $Y_l = N_l/N_{tr}$ at the l-th active layers. Here N_l is the carrier density, D is the diffusion coefficient, τ_{nr} is a recombination time, B is a coefficient of nonlinearity, d is thickness of the QW, e is the elementary charge, $N_{tr} = \left(-1/\tau_{nr} + \sqrt{1/\tau_{nr}^2 + (4BkE_e j_{tr})/(3E_g qed)}\right)/(2B)$ is the carrier density for conditions of transparency, j_{tr} is the injection current density for conditions of transparency, $|U_0|^2$ is the normalized light intensity, E_e is the energy of electrons, k is the part of the energy of electrons inputted into QWs, E_g is band gap of the barrier layers, $j = If(r/r_0)/2\pi\int f(r/r_0)rdr$ is the current density of the electron beam (e-beam), I is the total current of the beam, $f(\rho)$ is the pump profile function, r_0 is the pump region radius, j_{tr} is the current density for conditions of transparency. Zero boundary conditions for $Y_l(r)$ are set at the lateral boundary of the active layer. The function $\chi(Y)$, gain and index at the active layers are approximated by the formulas: $\chi(Y) = \alpha + (1-\alpha)Y^{1/(1-\alpha)}$ if $Y < 1$ and $\chi(Y) = Y$ if $Y \geq 1$; $g_l = g_0\ln(\chi(Y_l))$, $n_l = n_0 - R(g_l - g_{min})/(2k_0)$, where $\alpha = \exp(g_{min}/g_0)$, g_0 and g_{min} are gain parameter, n_0 is the refractive index in the absence of carriers, R is the line enhancement factor.

The equation (1) at $m = 0$ jointly with the equations (2) supplemented with corresponding boundary conditions form the eigenvalue problem for a non-linear operator. The supplementary condition $\delta = 0$ (Re(β) = 0) is required for steady-state operation.

We consider also a subsidiary problem when we neglect dependence of material characteristics on electromagnetic field intensity. It is so called case of "frozen" active medium. The equation (1) with the boundary conditions described must be solved in order to find the spatial profile of an eigenfunction and the complex eigenvalue β. The angular-dependent solutions ($m \neq 0$) are considered in this case also.

III. NUMERICAL METHOD

According to BiBPM we represent a wave field in each horizontal plane as a vector $(V^+ \; V^-)^T$ of the upward and downward propagating waves, V^+ and V^-. The wave fields in two arbitrary planes, marked by symbols t and b are bounded by a transfer equation: $(V_t^+ \; V_t^-)^T = \mathbf{M}(V_b^+ \; V_b^-)^T$, where $\mathbf{M}$ is a transfer matrix. Transfer matrix for set of layers can be calculated as a product of the elementary interface and propagation matrices [1]:

$$\mathbf{T}_k = \frac{1}{2}\begin{pmatrix} 1+Q_{k+1}^{-1}Q_k & 1-Q_{k+1}^{-1}Q_k \\ 1-Q_{k+1}^{-1}Q_k & 1+Q_{k+1}^{-1}Q_k \end{pmatrix}, \quad \mathbf{P}_k = \begin{pmatrix} \exp(iQ_k h_k) & 0 \\ 0 & \exp(-iQ_k h_k) \end{pmatrix},$$

where $h_k = z_k - z_{k-1}$, Q_k is the operator of longitudinal wavenumber in the k-th layer. For example, $\mathbf{M} = \mathbf{T}_M\mathbf{P}_M\cdot...\cdot\mathbf{T}_{\nu(q)+1}\mathbf{P}_{\nu(q)+1}$ is the transfer matrix for region above the top QW, $\mathbf{M} = \mathbf{T}_{\nu(l)}\mathbf{P}_{\nu(l)}\mathbf{T}_{\nu(l)-1}$ ($l = 1,...,q$) are the transfer matrices for QWs, $\mathbf{M} = \mathbf{P}_{\nu(l)-1}$ ($l = 2,...,q$) are the transfer matrices for the barrier layers and $\mathbf{M} = \mathbf{P}_{\nu(1)-1}\mathbf{T}_{\nu(1)-2}\cdot...\cdot\mathbf{P}_1\mathbf{T}_0$ is the transfer matrix for bottom DBR region. The fast Hankel transform algorithm was used for effective calculations with the transfer matrices. In the wavenumber space the operator Q_k is replaced by the number q_k, the operational matrices $\mathbf{T}_k$ and $\mathbf{P}_k$ are turned into numerical matrices. The calculations in the QW regions were performed in the physical space because of non-uniform transverse gain and index distributions. The approach of locally uniform wave field was used. This approach is admissible since thickness of the QW is far less then the wavelength.

Joining the set of every possible transfer equations with the condition of absence of externally injected electromagnetic fields we can obtain the closed system of equations and represent it as an eigenvalue problem:

$$\mathbf{P}(g, n, \beta)u = \gamma u, \qquad (3)$$

where $\mathbf{P}$ is the round-trip operator, u is the upward propagating wave at the preselected plane, $\gamma = 1$.

Our approach to problem (3) consists in solution of the auxiliary problem for a function u and eigenvalue γ to be found provided the value of β is specified. The value β is adjusted until $\gamma = 1$ within a certain tolerance. Generally, calculations were organized as follows: an inner iteration procedure solves the equation (3) at fixed value of β to find one or several eigenpairs (u, γ); the external iterative cycle encloses inner cycle and serves to find the value β where $\gamma = 1$.

In case of self-consistent solution the eigenvalue β is an imaginary number, $\beta = i2\Delta k$. The problem for fixed β is the eigenvalue problem for a non-linear operator because g and n are determined by equation (2) and depend on u. This problem is solved by the Fox-Li iteration method [4]. The value Δk is adjusted in an external cycle using the secant method.

For case of "frozen" active medium we have the linear non-hermitian eigen-value problem if β is fixed. Only several eigenpairs (u, γ) are required. The standard Arnoldi method is efficient in this case. Calculation and storing of matrix of $\mathbf{P}(g, n, \beta)$ is not required. It is necessary to calculate elements of vector $\mathbf{P}(g, n, \beta)\,u$ only. The complex eigenvalue β is adjusted in an external cycle using the Broyden method [5].

IV. RESULTS AND DISCUSSION

The parameters for calculations were given as follows:
$D = 0.5$ cm^2 s^{-1}, $\tau_{nr} = 10^{-9}$ s, $B = 3.5 \times 10^{-10}$ cm^{3}s^{-1}, $k = 0.75$, $E_e = 4 \times 10^4$ eV, $E_g = 2.36$ eV, $j_{tr} = 2.35$ A cm^{-2}, $r_0 = 25\ \mu m$, $g_0 = 3400$ cm^{-1}, $g_{min} = -1000$ cm^{-1}, $R = 2.5$, $I = 2.35$ mA, $f(\rho) = (1 + \rho^4)^{-1}$. The external spherical mirror has radius of curvature 3 cm, transverse size $400\ \mu m$ and reflection coefficient 0.985. Optical length L_{opt} of the space between mirror and heterostructure is a variable parameter. Test calculations reveal that for practical purposes 256 mesh nodes over polar radius r is good choice. The lasing mode intensity in physical units is calculated by the formula $J = J_s |U_0|^2$, where $J_s = (hcN_{tr})/(\lambda_0 g_0 \tau_{nr})$ is the intensity of saturation. We use standard notation TEM$_{nm}$ for optical modes in a VECSEL where m is the angular quantum number responding to dependence $\sim \exp(im\varphi)$ and n is the radial quantum number.

We have performed calculations with the framework of self-consistent problem defined by equation (1)-(2) and boundary and supplementary conditions. The output power $P_{out} = 2\pi \int J\, r dr$ depends smoothly on L_{opt} except for small neighbourhood of the value $L_{opt} = 2.3$ cm where it jumps to zero (Fig. 2).

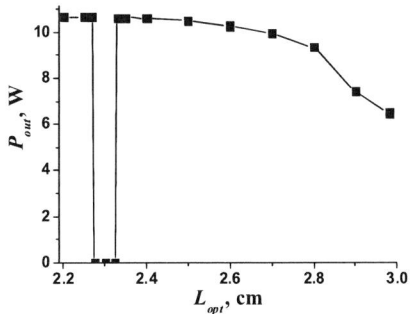

Fig. 2. Output power vs mirror position.

To understand this effect we have performed calculations for "frozen" active medium formed by e-beam pump ignoring saturation of the active media by the light intensity. One can see in Fig. 3 that in interval 2.5 cm $< L_{opt} <$ 2.98 cm modes are strongly discriminated on losses. It follows from fact that loss caused by diffraction on the mirror edge increase at near-concentric configuration. If $L_{opt} <$ 2.5 cm then transverse sizes of modes at the external mirror become lesser then the size of mirror and diffraction losses on the mirror edge become negligible. On the contrary, sizes of modes at the milti-QW structure increase with decreasing L_{opt} and become approximately equal to size of the pump spot. The non-uniform gain-index profile formed by e-beam pumping plays a key role

in mode profile formation. As a result, mode patterns may remarkably deviate from patterns of classic Laguerre-Gaussian beams. In the interval 2.2 cm $< L_{opt} <$ 2.34 cm all the modes have positive decay rate g_t, and so they are subthreshold. This circumstance explains why we have not obtained laser generation at $L_{opt} \approx 2.3$ cm.

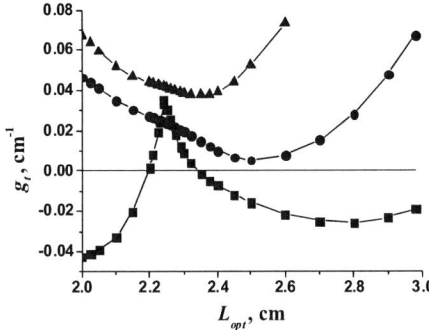

Fig. 3. Decay rate of modes with highest Q−factor: TEM$_{00}$ (squares), TEM$_{01}$ (circles) and TEM$_{10}$ (triangles). Unsaturated medium.

V. CONCLUSION

The BiBPM developed for multilayer media can be successfully joined up with the well-known round-trip operator technique for optical resonators including Fox-Li iterations and Krylov subspace methods. As a result we have developed the efficient numerical method for modeling lasers with multilayer structure including linear and non-linear regimes of operation.

REFERENCES

[1] H. Rao, M. J. Steel, R. Scarmozzino and R. M. Osgood, Jr., "High-power single-mode antiresonant reflecting optical waveguide-type vertical-cavity surface-emitting lasers," IEEE J. Quantum Electron., vol. 37, pp. 1435-1440, 2001.

[2] N. N. Elkin, A. P. Napartovich, D. V. Vysotsky, B. M. Lavrushin, V. I. Kozlovsky, "Modeling of a Vertical Cavity Surface Emitting Laser with a Resonant Array of Quantum Wells", AIP Conference Proc., vol. 1168, pp. 436-439, 2009.

[3] G. R. Hadley, "Modeling of diode laser arrays", Chapter 4 in Diode Laser Arrays, D. Botez and D.R. Scifres, Eds, Cambridge, U.K., Cambridge Univ. Press, 1-72, 1994.

[4] A.G.Fox, T.Li, "Effect of gain saturation on the oscillating modes of optical masers," IEEE Journal of Quantum Electronics, vol. QE-2, pp. 774-783, 1966.

[5] C. G. Broyden, "A Class of Methods for Solving Nonlinear Simultaneous Equations," Mathematics of Computation, vol. 19 (92), pp. 577-593, 1965.

CAOL*2010 International Conference on Advanced Optoelectronics & Lasers, 10-14 September, 2010, Sevastopol, Ukraine

Light fields of complex polarization structure

V.G. Volostnikov[1], S.P. Kotova[1], O. Yu. Moiseev[2], A.V. Volkov[2], E.N. Vorontsov[1], D.M. Yakunenkova[2]

[1]Lebedev Physical Institute of Russian Academy of Sciences, Samara Branch, Samara, Russia

[2]Image Processing Systems Institute of the RAS, Samara, Russia

(Invited *Paper*)

Abstract: Abstract: Of a high interest now are the light fields with spatially inhomogeneous polarization state. These fields give new possibilities for many aspects of optical science and engineering. We consider a new effective method for the synthesis of light fields with predetermined polarization structure by means of specific diffractive optical element (DOE).

Light fields of a complex polarization structure or vector beams have recently become of interest for a wide range of applications. Laser beams with radial polarization for instance, have a better efficiency for cutting metals [1].

The "doughnut" structure of the light field is preferable for the emission of fluorescence owing to the comparatively weak background [2]. These fields are also effective for a sharp focusing needed for the laser micromanipulation [3].

The complex structure of polarization provides specific properties of these fields in the focus, in particular, radial polarization beams have nonzero axial electric field and the fields with azimuthal polarization have nonzero magnetic fields in the beam center. These properties are useful for the polarization spectroscopy and for the microparticles acceleration [4].

Naturally, the attempts to effectively generate such vector beams are numerous. The experimental setup of the last investigation is shown in Fig. 1 [5]. It should be noted that the energy efficiency is less than 10 % in this approach. Besides the resulting wave fields are not genuine Laguerre-Gaussian modes and therefore are structurally unstable under propagation and focusing.

Our approach is mainly aimed to produce two genuine LG_{01} beams with the predetermined polarization structure and efficiency about 40-45 % for each beam.

The experimental setup is shown in Fig. 2. The He-Ne laser generates a linearly polarized Hermite-Gaussian beam HG_{10}. It is diffracted on DOE (binary holographic elements) with the following phase distribution:

$$\varphi = -\varphi_0 \text{sign} \cos\left(2xy + \alpha x\right).$$

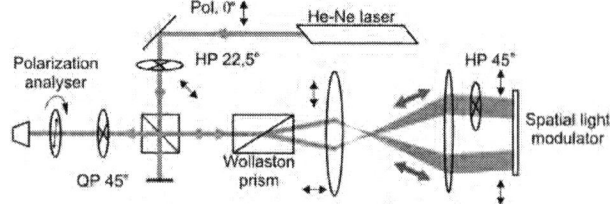

Fig. 1. Scheme of the typical experimental setup [5].

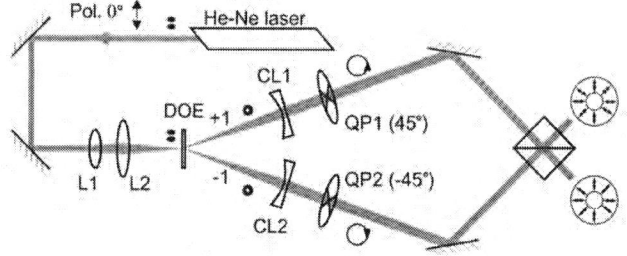

Fig. 2. Schematic layout of experimental setup for generate vector beams by means special DOE.

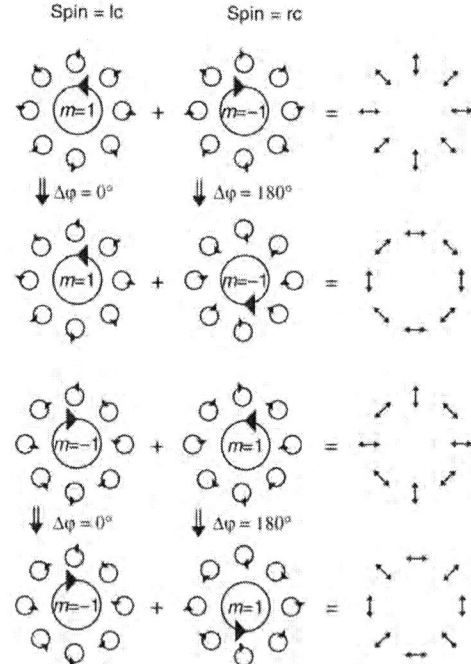

Fig. 3. Superposition of LG beams with opposite left (lc) and right (rc) -circular polarization states and opposite helical indices [5].

The choice of the constant φ_0 is explained by the condition of the minimum zero-order beam energy. This element has been fabricated at the Image Processing System Institute of RAS.

In ± 1 order we have $\pm 2xy$ phase distribution. After diffraction in this DOE we have two LG_{01} beams with opposite topological charge and residual astigmatism [6]:

978-1-4244-7043-3/10 $26.00 © 2010 IEEE   74

$$\iint HG_{10}\left(\xi,\eta\right)e^{-i\left(x\xi+y\eta\right)\pm 2i\xi\eta}\,d\xi d\eta = C\left\{\begin{array}{l} LG_{0,1}e^{-i\frac{xy}{4}}\text{ in }+1\text{ order} \\[2mm] LG_{0,-1}e^{i\frac{xy}{4}}\text{ in }-1\text{ order}\end{array}\right\},$$

where C is some constant.

The DOE diffraction efficiently in ± 1 order is about 80-90 %. After the conversion the beams are transformed into circularly polarized beam with the opposite sign of rotation by means of two quarter-wave plates. The residual astigmatism is compensated by means of two rotated cylindrical lenses. The obtained beams are summarized in a splitter cube. Hence we obtained two LG_{01} beams with the predetermined polarization structures. The summation schemes and polarization structure of the resulting light fields are shown in Fig. 3 [5]. The experimental results agree with theoretical ones.

This work was supported by the RAS program "Physical and engineering research of semiconductor lasers aimed at top parameters achievement" and by the Educative Scientific Division of the Lebedev Physical Institute.

REFERENCES

[1] A.V. Nesterov, V.G. Niziev and V.P. Yakunin, "Generation of high-power radially polarized beam", *Appl. Opt.*, vol. 29, no. 15, pp. 2234–2239, 1990.

[2] K.I. Willig , S.O. Rizzoli, V. Westphal, R. Jahn, S.W. Hell, "STED microscopy reveals that synaptotagmin remains clustered after synaptic vesicle exocytosis", *Nature*, vol. 440, pp. 935–939, 2006.

[3] Y. Kozawa and S. Sato, "Optical trapping of micrometer-sized dielectric particles by cylindrical vector beams", *Optics Express*, vol. 18, no. 10, pp. 10828–10833, 2010.

[4] P. Serafimm, P. Sprangle, B. Hafizi, "Optical guiding of a radially polarized laser beam for inverse Cherenkov acceleration in a plasma channel", *IEEE Trans. Plasma Sci.*, vol. 28, no. 4, pp. 1190-1193, 2000.

[5] C. Maurer, A. Jesacher, S. Fürhapter, S. Bernet, M. Ritsch-Marte, "*Tailoring of arbitrary optical vector beams*", New Journal of Physics, vol. 9 (78), pp. 1-20, 2007.

[6] E. Abramochkin and V. Volostnikov, "Beam transformations and nontransformed beams", *Optics Communications*, vol. 83, no. 1-2, pp. 123-135, 1991.

Shack-Hartmann wavefront sensor – advantages and disadvantages

Alexis Kudryashov, Vadim Samarkin, Alex Alexandrov, Julia Sheldakova, Valentina Zavalova

Moscow State Open University, Moscow, Russia
E-mail: kud@activeoptics.ru

(Invited Paper)

Abstract: In this paper we discuss the ability of Shack-Hartmann wavefront sensor to be used in the closed loop adaptive system. For sure we do not discuss the possibility of wavefront sensor to measure aberrations of laser beam. But at the same time we consider Shack-Hartmann wavefront sensor to be not the best device in any adaptive optical system.

To correct for the phase aberrations an adaptive optical system is usually used. The standard adaptive optical system consists of a wavefront corrector, an electronic control unit, some sensor to analyse the beam phase, and the software to determine the control voltages to be applied to the corrector (Fig. 1).

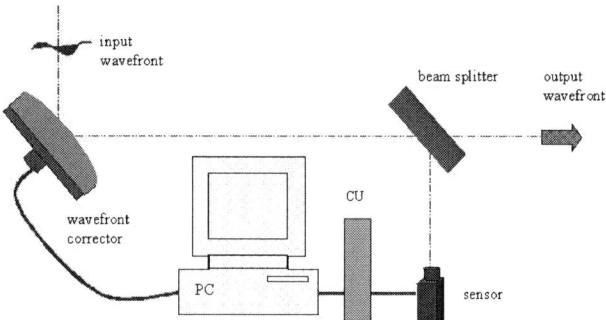

Fig. 1. Principle scheme of Adaptive Optical System

As a wavefront corrector we use bimorph deformable mirror[1, 2]; this type of correctors allows to compensate for the low-order laser beam aberrations. To control for the mirror we usually use phase conjugate algorithm and Shack-Hartmann wavefront sensorОшибка! Закладка не определена. to measure the wavefront aberrations. But we found out that sometimes the result of correction is not ideal and we can improve it manually.

The principle of Shack-Hartmann wavefront sensor is based on determination of local slopes ΔS of a distorted wavefront relative to the reference flat wavefront[3]. Laser beam is divided into a number of beamlets by a two-dimensional sub-apertures of the lenslet array. Each subaperture provides a separate focus on the detector. The position of each spot is displaced by local wavefront aberrations. The local slope of the wavefront on each subaperture of the lenslet array is proportional to the displacement of the focal spot center ΔS.

The idea of closed-loop is to minimize the displacement of real focal spots from the reference ones (Fig.); merit function could be represented as $\Phi = \min\left\{\sum_i \Delta S^2\right\}$, i – number of lenslet subapertures.

As a first step of correction response functions b should be measured. They are represented in terms of displacement of the focal spots $b_{j\,x}^{\,i} = \Delta x_i / u_{0\,j}$ and $b_{j\,y}^{\,i} = \Delta y_i / u_{0\,j}$; here $u_{0\,j}$ is the unit voltage used to measure response functions, j = number of electrode.

Then the displacements of the focal spots for the wavefront to be corrected are measured and following equation is written:

$$S_i = \left|\begin{matrix}\Delta x_i \\ \Delta y_i\end{matrix}\right| = \sum_{j=1}^{N} u_j \cdot b_j^i \; ; \text{ here } u \text{ is an array of voltages.}$$

As a last step the voltages to be applied to minimize the displacements are calculated using least square method:

$$\min\left\|\vec{S} - b \cdot \vec{u}\right\|^2, \; u = \|B\| \cdot S, \; B = (b^\mathrm{T} b)^{-1} b^\mathrm{T}.$$

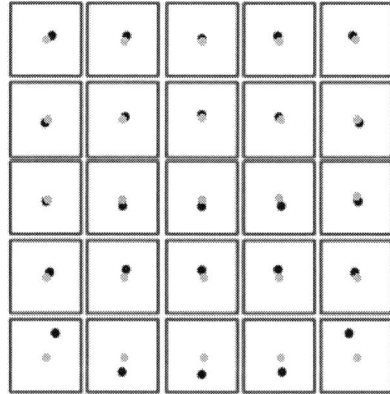

Fig. 2. View of Hartmann image with reference (grey) and real (black) spots

The problem of displacements minimization is that the minimal $\Phi = \min\left\{\sum_i \Delta S^2\right\}$ does not always correspond to the best wavefront. One of the reasons is following: different combination of displacements can lead to equal value of Φ but different wavefront amplitude. Four examples of such behavior are shown on Fig. 3; merit function Φ is the same for all four linear cases on right picture – displacement of all spots is equal, the only directions of movement differ. P-V for case #1 is 12 while for case #4 P-V=1. So from the point of view of minimization algorithm all four examples are equal to each other while the reconstructed wavefronts have different P-V.

P-V, a.u.

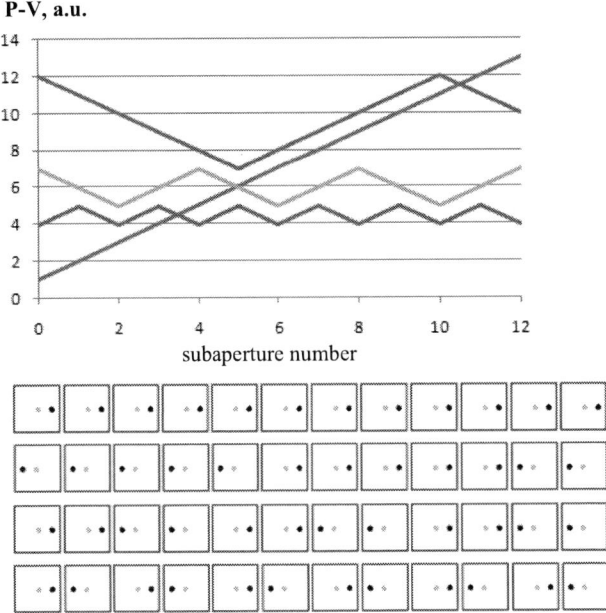

subaperture number

Fig. 3. Examples of focal spot shifts

This problem we face when you start using Shack-Hartmann wavefront sensor in a closed loop adaptive optical system. The standard algorithm is organised in such a way the it minimizes merit function Φ and as we showed before this function does not means the minimum of phase distortion. And adaptive system thus dopes not perfectly correct for the income aberrations of the wavefront.

REFERENCES

1. A.V.Kudryashov, V.I.Shmalhausen, "Semipassive bimorph flexible mirrors for atmospheric adaptive optics applications", *Opt. Eng* **35**(11), pp. 3064-3073, 1996
2. A.V.Kudryashov, V.V.Samarkin, A.Aleksandrov, "Adaptive optical elements for laser beam control", *Proc. SPIE* **4457**, pp. 170-178, 2001
3. V.Ye.Zavalova, A.V.Kudryashov, "Shack-Hartmann wavefront sensor for laser beam analyses", *Proc. SPIE* **4493**, pp. 277-284, 2002.

Nonlinear Solitary Waves in Positive-Negative Refractive Media

(Invited *P*aper)

Andrei Maimistov
Department of Solid State Physics and Nanosystems,
National Nuclear Research University Moscow Engineering Physics Institute,
Moscow, Russia, 115409,
Department of Physics and Technology of Nanostructures / REC Bionanophysics,
Moscow Institute of Physics and Technology (State University),
Dolgoprudnyi, Russia,
Email: aimaimistov@gmail.com

Abstract—**Propagation of the fundamental and third harmonic solitary waves in medium characterized by negative refraction index at the frequency of fundamental wave and by positive refractive index at the third harmonic frequency and simultons formation are considered. Gap soliton in a novel kind of nonlinear coupler with one channel or both channels fabricated from nonlinear medium having negative refraction index is found.**

I. Introduction

Recent progress in nanofabrication has lead to the design of new materials with highly unusual optical properties [1], [2], [3], [4], [5], of which negative refraction is an example. Negative refraction occurs in media in which the wave vector of the electromagnetic wave is antiparallel to the Poynting vector [6], [7], [8]. In the particular case when the real parts of the dielectric permittivity and magnetic permeability in the medium simultaneously take on negative values in some frequency range, the property of negative refraction will appear. The existence of such media was demonstrated experimentally rst in the microwave and then in the near-infrared ranges [2], [3], [4], [5].

The unusual property of negative refractive index (NRI) materials reveal themselves most prominently when a wave passes through, or is localized near, an interface between such a material and a conventional dielectric. New wave propagation phenomena can also be expected when the refractive index of the same medium can be positive in one spectral region and negative in another. We can refer to these cases as *positive-negative refractive media* (PNRM).

The present technological level does not yet allow for the fabrication of 3D materials of suf cient size and small enough losses for experiential veri cation of the effects described above [12], [13]. However, a considerable effort aimed at loss reduction and improvement of nanofabrication technology gives hope that the considered device will be manufactured.

II. Third harmonic generation

As example of the nonlinear phenomena in PNRM the third harmonic generation (THG) will be considered. THG is associated with four wave interaction of the type $\omega_1 + \omega_1 + \omega_1 \rightarrow \omega_3 = 3\omega_1$. Let us consider the THG where the frequency of the fundamental wave (pump wave) ω_1 is located in the NRI spectral region, and the third harmonic frequency ω_3 is located in the positive refractive index spectral region. In this case group velocities of pump and harmonic $\mathbf{v}_1$ and $\mathbf{v}_3$ are oppositely directed, i.e., $\mathbf{v}_1 \cdot \mathbf{v}_3 < 0$. However, the phase velocities are approximately equal. This phase match condition results in effective transformation of he pump wave into third harmonic wave. The self-modulation and cross-modulation effects are ignored. In this case THG equations after normalization read as [11],

$$
\begin{aligned}
ie_{1,\zeta} + (\sigma/2)e_{1,\tau\tau} - e_2 e_1^{*2} &= 0, \\
ie_{3,\zeta} + i\delta e_{3,\tau} - (\beta/2)e_{3,\tau\tau} - \Delta e_3 + e_1^3 &= 0,
\end{aligned}
\tag{1}
$$

Where ζ and τ are normalized time and coordinate, $\delta = Lt_p^{-1}(v_1^{-1} + v_3^{-1})$ is normalized group velocity mismatch; L is dispersion length, $\Delta = (3k_1 - k_3)L$ is normalized phase mismatch, k_1 and k_3 are wave numbers of the pump wave and harmonic wave. Parameter δ takes into account the walk-off effect for pump and harmonic pulses that is due to the difference of the group velocities' directions for the interaction waves. It should be pointed out that this parameter can not be zero.

In the large-mismatch limit of THG, the system of equations (1) can be reduced into the quintic nonlinear Schrödinger equation:

$$
ie_{,\zeta} + (\sigma/2)e_{,\tau\tau} - (1/2\Delta)|e|^4 e = 0,
\tag{2}
$$

where

$$
e_1(\tau, \zeta) = e(\tau, \zeta), \quad e_3(\tau, \zeta) = \Delta^{-1}e^3(\tau, \zeta).
$$

The quintic nonlinear Schrödinger equation has solitary pulse like and periodic cnoidal wave solutions. Solitary wave solution exists under condition of $\sigma = 1, \Delta < 0$ or $\sigma = -1, \Delta > 0$. Solution of the equation (2) read as:

$$
e(\tau, \zeta) = a(\tau + \sigma\Omega\zeta)\exp\left(-i\Omega\tau + \frac{i\sigma}{2}(p^2 - \Omega^2)\zeta\right),
$$

where

$$a^2(\eta) = \sqrt{\frac{3p^2}{2}} \frac{2|\Delta|}{\cosh(2p\eta)}$$

The solitary waves are parameterized by Ω and p.

Under mismatch condition $\Delta \approx 0$ steady state solutions of equations (1) was found [11], [14]. The approach is similar to that was exploited in [15] under consideration of quadratic solitons in nonlinear PNRM.

The proposition that there is no phase modulation leads to represent complex envelopes of the interacting waves in the following form

$$e_1 = a(\tau - \zeta/V_1)\exp(iK\zeta + i\Omega\tau),$$
$$e_3 = a(\tau - \zeta/V_3)\exp(i3K\zeta + 3i\Omega\tau),$$

where a and b are real values. The steady state waves for both frequencies must be propagating as single one. It means that the group velocities is same, $V_1 = V_3 = V$, and real amplitude of pump wave a is proportional to real amplitude of harmonic b. Let be $b = fa$. The system of the equations (1) reduces to a single equation for the real amplitude a, if the $V^{-1} = \sigma\Omega$, where parameters $\Omega = \delta/(\sigma + 3\beta)$, and is determined by formula

$$K = \sigma\left(4\beta\Omega^2 - 3\delta\Omega - \Delta\right)/(3\sigma + \beta).$$

Solving equation for a we found the eld amplitudes

$$|e_1(\tau,\zeta)| = \frac{\sqrt{p/f}}{\cosh[\sqrt{p}(\tau - \zeta/V - y_0)]}, \qquad (3)$$

$$|e_3(\tau,\zeta)| = \frac{\sqrt{pf}}{\cosh[\sqrt{p}(\tau - \zeta/V - y_0)]}, \qquad (4)$$

where y_0 is a constant of integration, $f^2 = \sigma/\beta$ and $p = 2\sigma K + \Omega^2$. For $\sigma = \beta = -1$ we can nd $\Omega = -\delta/4$, $K = \left(\delta^2 - \Delta\right)/8$ and $p = \left(4\Delta - 3\delta^2\right)/16$. The velocity of this steady state two component solitary wave V_s dependents on the particular group velocities . Under $\sigma = \beta = -1$ this velocity is read as

$$V_s = 4v_1v_3/(v_1 - 3v_3)$$

It should be note that non steady state solitary waves propagate in opposite directions. The formation of the steady state wave requires accessory investigation.

III. Nonlinear Oppositely-directed Coupler

It is well known that two closely located waveguides can be coupled due to the tunneling of light from one waveguide to the other. A coupler using tunneling, fabricated from materials with a positive refractive index, preserves the direction of light propagation, and for this reason it is named a *directed coupler*.

If one of the waveguides of the coupler is fabricated from a material with a negative refractive index, this device acts as a mirror. The radiation entering one waveguide leaves the device through the other waveguide at the same end but in the opposite direction. For this reason, this device can be called the *oppositely-directed coupler*.

In the normalized variables the system of equations describing the evolution of the elds in two tunnel connected waveguides in the normalized form [10], [16] read as

$$i\left(\frac{\partial}{\partial\zeta} + \frac{\partial}{\partial\tau}\right)e_1 - \delta e_1 + e_2 + r_1|e_1|^2 e_1 = 0,$$

$$i\left(\frac{\partial}{\partial\zeta} - \frac{\partial}{\partial\tau}\right)e_2 + \delta e_2 - e_1 - r_2|e_1|^2 e_1 = 0, \qquad (5)$$

where δ is phase mismatch parameter, r_1 and r_2 are parameters of nonlinear self-interaction due to Kerr nonlinearity of waveguides.

If the electric eld strength in the waveguides is suf ciently small, and nonlinear effects are insigni cant, we can put $r_1 = r_2 = 0$. It leads to system of equations determines the behavior of linear waves in the considered device. Using the Fourier transformation dispersion relation can be found. It reads as

$$\omega(k) = \delta \pm \sqrt{1 + k^2}.$$

Thus, the gap exists in the spectrum of linear waves whose width for $k = 0$ is $\Delta\omega = 2$. Harmonic waves with frequencies in this gap do not propagate through a system of two coupled waveguides with opposite signs of the refractive index. It has been shown [9] that due to this gap the *nonlinear* oppositely directed coupler is bistable.

In oppositely-directed couplers the group velocities of the wave packets propagating in connected waveguides are oppositely directed, while the phase velocities have the same value and direction. If the carrier frequency of a wave is localized inside the gap, then some part of the incident radiation is re ected, while another part of the radiation penetrates and decays inside the waveguide structure. An increase in the electromagnetic wave intensity, for which the nonlinear properties of the medium become noticeable, leads to the formation of coupled wave packets (*gap solitons*) that simultaneously travel through both waveguides [10]. The analytical solution of the system of equation (5) can be obtained by use the proposition that the steady state waves should be propagating as single pulse over both waveguides with group velocities V_g. The gap soliton represents by following expressions

$$e_1(\xi) = |e_1(\xi)|e^{i\varphi_1(\xi)}, \qquad e_2(\xi) = |e_2(\xi)|e^{i\varphi_2(\xi)}$$

where

$$\xi = (\zeta + \beta\tau)(\sqrt{1 - \beta^2})^{-1}$$

with free parameter β and

$$|e_1(\xi)|^2 = \frac{4}{\theta(1 + \beta)\cosh 2(\xi - \xi_0)}, \qquad (6)$$

$$|e_2(\xi)|^2 = \frac{4}{\theta(1 - \beta)\cosh 2(\xi - \xi_0)}, \qquad (7)$$

$$\varphi_1(\xi) = \varphi_1(-\infty) + \left(1 - 4\frac{\theta_1}{\theta}\right)\arctan e^{2(\xi - \xi_0)}, \quad (8)$$

$$\varphi_2(\xi) = \varphi_2(-\infty) + \left(1 - 4\frac{\theta_2}{\theta}\right)\arctan e^{2(\xi - \xi_0)}. \quad (9)$$

Parameter ξ_0 is the constant phase of gap soliton. The phases $\varphi_{1,2}(-\infty)$ can be chosen $\varphi_1(-\infty) = 0$ and $\varphi_2(-\infty) = -\pi/2$.

$$\theta_1 = r_1 \left((1+\beta)\sqrt{(1-\beta)/(1+\beta)} \right)^{-1},$$

$$\theta_2 = r_2 \left((1-\beta)\sqrt{(1+\beta)/(1-\beta)} \right)^{-1},$$

$$\theta = \theta_1 + \theta_2.$$

The negative value of the parameter β corresponds to solitary wave which propagates in the direction of the axis ζ. The solitary wave characterized by positive value of the parameter β propagates in the opposite direction. Large amplitudes of the solitary waves correspond to large positive values of the parameter β (which determines pulse velocity). When parameter β is negative the gap solitons with smaller values of the parameter β correspond to smaller amplitudes, however the absolute value of velocity determined by the parameter β is larger for less powerful solitary waves.

IV. CONCLUSION

We nd the steady state solitary wave with deferent carry waves, i.e. composite wave. It is analog quadratic soliton in $\chi^{(2)}$-PNRM [15]. The formation and stability of this waves in $\chi^{(3)}$-PNRM demand study.

Numerical simulations of the gap soliton [16] shown that interaction between gap solitons depends on their relative velocities. If relative velocity of the two colliding gap solitons is large, the pulses collide almost elastically and weak radiation generates. At small relative velocities of the gap solitons, the collision is inelastic. There is a strong energy exchange between gap solitons. This interaction leads to formation of the temporarily coupled state of the interacting solitary waves. However, the problem of the bound state formation demands an additional investigation. It is possible that a long-living bound state of two solitary waves exists in the case under consideration.

It should be noted that the term *gap soliton* is often referred to as nonlinear pulses propagating in periodic structures. The waveguide structure considered here, however, is homogeneous. Thus, the existence of a gap soliton, and the bistability of continuous waves in an oppositely directed coupler, represents new effects due to the positive-negative refraction phenomenon.

ACKNOWLEDGMENT

I would like to thank E.V. Kazantseva, N.M. Litchinitser, I.R. Gabitov, S.O. Elyutin and J.G. Caputo for enlightening discussions. I thanks the Department of Mathematics of University of Arizona and Laboratoire de Mathématiques, INSA de Rouen for the support and hospitality. This work was partially supported by NSF (grant DMS-0509589), ARO-MURI award 50342-PH-MUR and State of Arizona (Proposition 301), and by the Russian Foundation for Basic Research (grant No. 09-02-00701-a).

REFERENCES

[1] R. A. Shelby, D. R. Smith, and S. Schultz, "Experimental veri cation of a negative index of refraction," *Science*, vol. 292, no. 5514, pp. 77–79, 2001.

[2] S. Linden, C. Enkrich, M. Wegener, J. Zhou, T. Koschny, and C. M. Soukoulis, "Magnetic Response of Metamaterials at 100 Terahertz", *Science*, vol. 306, no. 5700, pp. 1351–1353, 2004.

[3] V. M. Shalaev, W. Cai, U. K. Chettiar, H. Yuan, A. K. Sarychev, V. P. Drachev, and A. Kildishev, "Negative index of refraction in optical metamaterials", *Opt. Lett.*, vol. 30, no. 24, pp. 3356–3359, 2005.

[4] S. Zhang, W. Fan, N. C. Panoiu, K. J. Malloy, R. M. Osgood, and S. J. Brueckl, "Experimental Demonstration of Near-Infrared Negative-Index Metamaterials", *Phys. Rev. Lett.*, vol. 95, p. 137404, 2005.

[5] G. Dolling, C. Enkrich, M. Wegener, and C. M. Soukoulis, "Low-loss negative-index metamaterial at telecommunication wavelengths", *Opt. Lett.*, vol. 31, no.12, pp. 1800–1803, 2006.

[6] V. G. Mandel'shtam, "Group velocity in crystalline arrays", *Zh. Eksp. Teor. Fiz.*, vol. 15, pp. 475–478, 1945.

[7] V. G. Veselago, "The electrodynamics of substances with simultaneously negative values of ϵ and μ", *Sov. Phys. Usp.*, vol. 10, no.3, pp. 509–526, 1968.

[8] V. M. Agranovich, Y. R. Shen, R. H. Baughman, and A. A. Zakhidov,"Linear and nonlinear wave propagation in negative refraction metamaterials", *Phys. Rev. B.*, vol. 69, p. 165112, 2004.

[9] N. M. Litchinitser, I. R. Gabitov, and A. I. Maimistov, "Optical Bistability in a Nonlinear Optical Coupler with a Negative Index Channel", *Phys. Rev. Lett.*, vol.99, p. 113902, 2007.

[10] A. I. Maimistov, I. R. Gabitov, and N. M. Litchinitser, "Solitary Waves in a Nonlinear Oppositely Directed Coupler". *Optics and Spectroscopy*, vol. 104, no. 2, pp. 253–259, 2008.

[11] A. I. Maimistov and I. R. Gabitov, "Nonlinear optical effects in arti cial materials", *Eur. Phys. J. Special Topics*, vol. 147, no. 1, pp. 265–286, 2007.

[12] J. Valentine, S. Zhang, T. Zentgraf, E. Ulin-Avila, D. A. Genov, G. Bartal, and X. Zhang,"Three-dimensional optical metamaterial with a negative refractive index", *Nature*, vol. 455, pp. 376–379, 2008.

[13] J. Yao, Z. Liu, Y. Liu, Y. Wang, C. Sun, G. Bartal, A. Stacy, and X. Zhang, "Optical Negative Refraction in Bulk Metamaterials of Nanowires", *Science*, vol. 321, no. 5691, pp. 930–932, 2008.

[14] S. O. Elyutin, A. I. Maimistov and I. R. Gabitov, "On third harmonic generation in medium with negative refraction of pump wave", *Zh.Eks.Ther.Fiz.*, vol. 138, no. 7, 2010. (in Russian).

[15] A. I Maimistov, I. R. Gabitov, and E. V. Kazantseva, Quadratic "Solitons in Negative Refractive Index Medium", *Optics and Spectroscopy*, vol. 102, no. 1, pp. 104-112, 2007.

[16] E. V. Kazantseva, A. I. Maimistov, and S. S. Ozhenko, "Solitary electromagnetic waves propagation in the asymmetric oppositely-directed coupler", *Phys.Rev. A*, vol. 80, p. 043833 (2009).

CAOL*2010 International Conference on Advanced Optoelectronics & Lasers, 10-14 September, 2010, Sevastopol, Ukraine

Multipole model for metamaterial homogenization

A. Chipouline[1], J. Petschulat[1], C. Menzel[2], C. Rockstuhl[2], T. Pertsch[1], and F. Lederer[2]

[1]Institute of Applied Physics, Friedrich Schiller University Jena, Max-Wien Platz 1, D-07743 Jena, Germany

[2]Institute for Solid State Theory and Optics, Friedrich Schiller University Jena, Max-Wien Platz 1, D-07743 Jena, Germany

(Invited Paper)

Abstract: Homogenization of optical metamaterials is one of the fundamental problems of electrodynamics of composite materials. Here the homogenization approach using multipole expansion is analysed and compared with the phenomenological one. Applications of the developed model for different structures and various phenomena are presented.

Metamaterials are composites consisting of artificial metaatoms/metamolecules with typical sizes less than the wavelengths for which the metamaterials are considered [1]. By means of a spatial shaping of the metaatoms and their positioning inside the host matrix, averaged dispersion properties of the metamaterials can be tuned in programmable way allowing new effects and opening up applications, inaccessible with natural materials. A key point of the metamaterials, which differs them principally from the natural one (at least in a visible wavelength region) is a magnetic response on an external electric field [2, 3].

A problem of homogenization of such materials is an important task from the point of view of fundamental physics as like as for possible applications. The problem from the one side is similar to the one of homogenization of composite materials (see, for example, [4] and references herein); from the other side, standard methods of averaging of the Maxwell equations can be applied for this problem as well [5].

The averaging procedure can be performed principally by two different methods: phenomenologically and microscopically. In a phenomenological approach, polarizability of media includes dependence on wave vector (spatial dispersion) which is finally responsible for the magnetic response [6] (this approach recently has been used in order to discuss possibility of dielectric and magnetic constants introduction for metamaterials [7]). Microscopic method involves introducing of multipoles, and the magnetic response appears from the multipoles of the second order – quadrupole and magnetic dipole moments [5].

In this work we summarize the multipole expansion approach developed in [8] in order to describe analytically linear and nonlinear optical properties of metamaterials. The model requires only a minimum number of assumptions concerning the charge dynamics of the metaatoms and is inherently analytical. We show that especially the consideration of electric quadrupole multipoles (not only the magnetic dipole one) is required by basic principles and provides the possibility to adequately analyse an effective magnetic material response. To motivate the appearance of such multipole terms we applied numerical methods to

calculate the near field radiation patterns as well as far field properties. The dispersion relation of a single metamaterial layer and 3D infinitely expanded bulk metamaterials are derived as well. In order to validate our model for particular cases, we applied the formalism for the split-ring resonator and the cut-wire structure. In principle, the formalism is not restricted by any special metaatom geometry. Only the intrinsic electron dynamics in the respective metaatom must be known.

Beginning with a single cut wire metaatom, we can identify the complex near field dynamics for the magnetic resonance – see Fig.1a. An electric near field distribution – Fig. 1a - has approximately a quadrupole-like shape shown for comparison in Fig.1b. Deviations from the purely electric quadrupole field distribution are due to the superposition of the remaining dipole-like background field supported by each metal nanoparticle. The near field distribution of the same field component within the electric resonance - see Fig.1c - is perfectly described by an electric dipole field shown in Fig.1d.

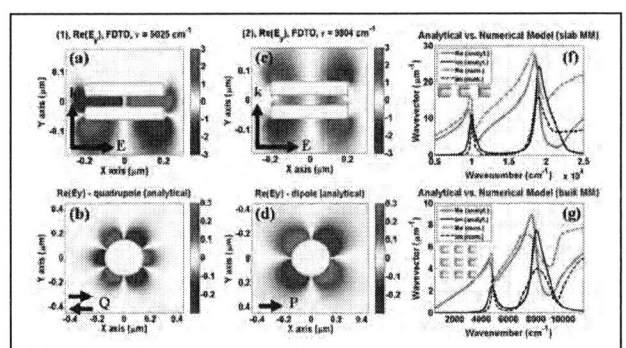

Fig. 1.The near field distribution of the cut-wire structure is shown for the magnetic **(a)** and the electric **(c)** resonances. Both show good agreement with the electric quadrupole **(b)** and the electric dipole **(d)** field distributions, respectively. The linear dispersion relation for a single layer of split ring resonators **(f)** and the corresponding bulk metamaterial **(g)** are shown.

Recently we have shown that not only the near fields exhibit higher order multipole structures, but the dispersion relation of the cut-wire structure as well as the corresponding effective parameters can be retrieved with the help of the developed formalism. In addition to the effective parameters of the single layer slab, effective parameters for the bulk

978-1-4244-7043-3/10 $26.00 © 2010 IEEE   81

alignment has been elaborated. The retrieval procedure of bulk metamaterials is an especially non-trivial task, since no material interfaces and therefore no reflection and transmission can be calculated, which is required by the conventional parameter retrieval procedure.

The next interesting feature of our formalism is the possibility to treat intrinsic multipole induced nonlinear effects [9]. As it has been shown in numerous papers, quadrupole and magnetic dipole contributions support non-linear material response. The second order nonlinear effects (e.g. second harmonic generation) for metaatoms are well described by the presented model.

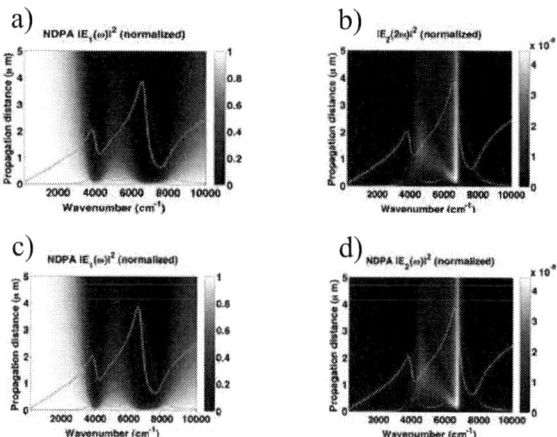

Fig. 2. Intensity (color grade) of the fundamental (a, c) and the second harmonic waves (b, d) as a function of the fundamental wavelength. "NDPA" stands for Non Depleted Pump Approximation.

Second order multipoles appear if the structures are excited from the bottom or the top due to the presence of a near field coupling between the two dipoles.

Therefore we can again use the multipole approach to explain the nonlinear interaction due to the presence of the quadruple contribution. The results of our modelling for the second harmonic generation are presented in Fig. 2 for the full model and for the nondepleted pump approximation (NDPA) as well. Moreover we discuss why the magnetic dipole influence on the nonlinear response is negligible. Finally our theoretical observations support the experimental results and the embedded discussion on the origin of the nonlinearity published recently [10].

We extend the model to more timely problems in the field of optical metamaterials. Particularly, we study the properties of planar chiral metamaterials [11] as well as metamaterial analogues of electromagnetic induced transparency (EIT) [12,13] with our model. The results of this analysis shows that the fundamental far field properties such as reflection and transmission coefficients for co- and cross-polarized light of optically active and non-active metaatoms are nicely described. By optical activity we wish to understand the ability of the metaatom to generate a signal with a cross polarized field

component. To reveal the advantage of the multipole treatment beyond the physical explanation of the observed features, we perform structural modifications to the metaatoms that enable or disable optical activity, thus changing the principle character of the metamaterial properties, while we observe that the optical response remains predictable.

We calculated then the observable far field quantities rigorously. The analytical model was also used to compute these quantities and the free parameters were optimized in a least square sense such that rigorous and analytical results match. Comparing the numerically against the analytically calculated far field intensities as well as the corresponding electric effective permittivities we observe an overall agreement between both approaches. We emphasize that the model precisely predicts also all off-diagonal components of the permittivity tensor.

We applied this description toward EIT-mimicking effects observed in similar structures and have shown that such effects can be also described by our theory.

In conclusion, we present a self consistent and analytical approach to characterize metaatoms by means of coupled oscillators. Since we observe a nice quantitative agreement with rigorous simulations we regard this analytical tool as a potential routine to support the actual research on metamaterials and to make the observed effects reducible to an intuitive and well known physical extend.

REFERENCES

[1] T. Paul, C. Menzel, C. Rockstuhl, F. Lederer, "Advanced optical metamaterials," *Advanced. Materials*, vol. 22, pp. 2354-2357, 2010.

[2] C. Soukoulis, J. Zhou, "The science of negative index materials," *J. Phys.: Condens. Matter*, vol. 20, pp. 1-7, 2008.

[3] V. Shalaev, "Optical negative index metamaterials" *Nature Photonics*, vol. 1. pp. 41-47, 2007.

[4] D. Bergman, "The dielectric constant of a composite material – a problem in classical physics", Physics Report, vol. 43, n. 9, pp. 377-407, 1978. A. P. Vinogradov, "Electrodynamics of composite materials", URSS, Moscow, 2001 [in Russian]

[5] P. Mazur and B. Nijboer, "On the statistical mechanics of matter in an electromagnetic field, I. Derivation of the Maxwell equations from electron theory", Physica XIX, pp. 971-986, 1953.

[6] V. Agranovich, Yu. Gartein, "Spatial dispersion and negative refraction of light," *Physics-Uspekhi*, vol. 49, no. 10, pp. 1029-1044, 2006.

[7] C. Menzel, T. Paul, C. Rockstuhl, T. Pertsch, S. Tretyakov, F. Lederer, „Validity of effective material parameters for optical fishnet metamaterials," Phys. Rev. B 81, pp. 035320, 2010.

[8] J. Petschulat, C. Menzel, A. Chipouline, C. Rockstuhl, A. Tünnermann, F. Lederer, and T. Pertsch, "Multiple approach to metamaterials", *Phys. Rev. A* vol. 78, 043811.

[9] J. Petschulat, A. Chipouline, A. Tünnermann, and T. Pertsch, C. Menzel, C. Rockstuhl, and F. Lederer "Multipole nonlinearity of metamaterials", *Phys. Rev. A* **80**, 063828, 2009.

[10] N. Feth, S. Linden, M.W. Klein, M. Decker, F.B.P. Niesler, Y. Zeng, W. Hoyer, J. Liu, S.W. Koch, J.V. Moloney, and M. Wegener, "Second-harmonic generation from complementary split-ring resonators," *Opt. Lett.* 33, 1975-1977 (2008).

[11] J. Valentine, S. Zhang, T. Zentgraf, G. B. E Ulin-Avila, Dentcho A. Genov, and X. Zhang, *Nature* 455, 376, 2008.

[12] J. Valentine, J. Li, T. Zentgraf, G. Bartal, and X. Zhang, *Nature Mat.* 8, 568 (2009).

[13] A. Al_u and N. Engheta, *Phys. Rev. Lett.* 102, 1 (2009).

Nonlinear coupling in plasmonic structures

(Invited Paper)

José R. Salgueiro
Departamento de Física Aplicada
Universidade de Vigo
As Lagoas s/n, 32004 Ourense, Spain
Email: jrs@uvigo.es

Yuri S. Kivshar
Nonlinear Physics Center
Research School of Physics and Engineering
Australian National University
Canberra ACT 0200, Australia
Email: ysk@internode.on.net

Abstract—**Nonlinear modes and switching characteristics of plasmonic directional couplers are studied. The structures are created by two neighbouring metal-dielectric-metal waveguides with the slots filled in a nonlinear dielectric material. The different nonlinear modes are calculated, and the FDTD numerical simulations are carried out to study the power transfer in such plasmonic couplers and evaluate the effect of optical losses on the coupler switching characteristics.**

I. INTRODUCTION

Plasmonics became a rapidly developing field of nanophotonics that may revolutionize optical integrated circuits and increase functionalities of basic operating elements in information processing [1], [2]. It is believed that incorporating metals into optical elements would allow overcoming the fundamental diffraction limits by exciting surface plasmons and squeezing light into subwavelength scales, also making the photonic circuits compatible with nowadays electronics. One of the simplest plasmonic waveguides is an interface between metal and dielectric that supports surface plasmon polaritons; however, direct excitation of plasmons at an interface is not efficient. Introducing multi-layer systems, such as metal-dielectric-metal waveguides, allows to increase substantially the propagation distance, since light propagation is tightly confined between two metal slabs, and it can even be used for efficient nanofocusing [3].

One of the key elements of integrated photonic circuits, *a directional coupler*, is based on cross-talk coupling between two neighboring waveguides. Such directional couplers for plasmonic waves have been studied theoretically and even demonstrated experimentally for different geometries [4]–[6]. Directional couplers have been also suggested for converting the modes of an optical dielectric waveguide into plasmonic modes propagating along a thin metal stripe [7].

Nonlinear plasmonic waveguides were known for many years to provide additional possibilities for the mode control [8]–[10]. It was recently shown that nonlinear slot waveguides created by a nonlinear dielectric slab sandwiched between two metals support subwavelength guided modes of different symmetries [11], including a novel type of asymmetric modes that might be important for nonlinear switching.

We now study the mode structure and switching characteristics of two coupled nonlinear plasmonic slot waveguides with nonlinear dielectric cores surrounded by metal. We find

numerically nonlinear modes of this structure, calculate the power diagram, and demonstrate a crossover between linear and nonlinear switching of gap plasmonic modes.

II. NONLINEAR MODES

We start with the Maxwell's equations for the electric $\vec{\mathcal{E}}(\mathbf{R}, t)$ and magnetic $\vec{\mathcal{H}}(\mathbf{R}, t)$ fields, where $\mathbf{R} \equiv (X, Y, Z)$ is the position vector and t the time, written in the form,

$$
\begin{aligned}
\nabla \times \vec{\mathcal{E}} &= -\mu_0 \partial_t \vec{\mathcal{H}}, \\
\nabla \times \vec{\mathcal{H}} &= \partial_t \vec{\mathcal{D}},
\end{aligned} \tag{1}
$$

where μ_0 is the vacuum magnetic permeability and $\vec{\mathcal{D}} = \epsilon_0 \epsilon \vec{\mathcal{E}}$ the electric displacement, being ϵ_0 the vacuum electric permitivity and $\epsilon(\mathbf{R})$ the dielectric function describing the medium. We consider a planar coupler formed by two dielectric cores with a nonlinear Kerr response and separated a distance d, bounded by metallic linear claddings (see Fig. 1). We also consider harmonic fields of frequency ω and scale the spatial variables as $\mathbf{r} = k_0 \mathbf{R}$, where $\mathbf{r} \equiv (x, y, z)$ and $k_0 = \omega/c$, being c the speed of light in vacuum. We assume the structure lying in the plane XZ and take z as the propagation direction. Besides, we consider TM-polarized fields (components $\mathcal{H}_y$, $\mathcal{E}_x$ and $\mathcal{E}_z$) and describe the medium by the dielectric function:

$$
\epsilon(\mathbf{r}) = \begin{cases} \epsilon_m; & \text{for claddings,} \\ \epsilon_d + \gamma(|\mathcal{E}_x|^2 + |\mathcal{E}_z|^2); & \text{for cores,} \end{cases} \tag{2}
$$

where γ accounts for the Kerr nonlinear response.

We seek stationary states of the form,

$$
\begin{aligned}
\vec{\mathcal{H}}(\mathbf{r}, t) &= \mathbf{H}(x) \exp\left[-i(\beta z - \omega t)\right]/(\mu_0 c \sqrt{\gamma}) \\
\vec{\mathcal{E}}(\mathbf{r}, t) &= \mathbf{E}(x) \exp\left[-i(\beta z - \omega t)\right]/\sqrt{\gamma}
\end{aligned} \tag{3}
$$

where β is the propagation constant. Replacing Eqs. (3) in Eqs. (1) and considering TM-fields of the form $\mathbf{H} = H_y \hat{y}$, $\mathbf{E} = E_x \hat{x} + i E_z \hat{z}$, we obtain the following z-independent system for the field components,

$$
\begin{aligned}
\partial_x H_y &= -\epsilon E_z \\
\partial_x E_z &= (1 - \beta^2/\epsilon) H_y,
\end{aligned} \tag{4}
$$

together with the relationship $\epsilon E_x = \beta H_y$.

Inside the metallic regions, the dielectric function is negative ($\epsilon = \epsilon_m < 0$) and the equations above have an analytic solution for H_y and E_z in the form of a linear combination

978-1-4244-7043-3/10 $26.00 © 2010 IEEE

CAOL*2010 International Conference on Advanced Optoelectronics & Lasers, 10-14 September, 2010, Sevastopol, Ukraine

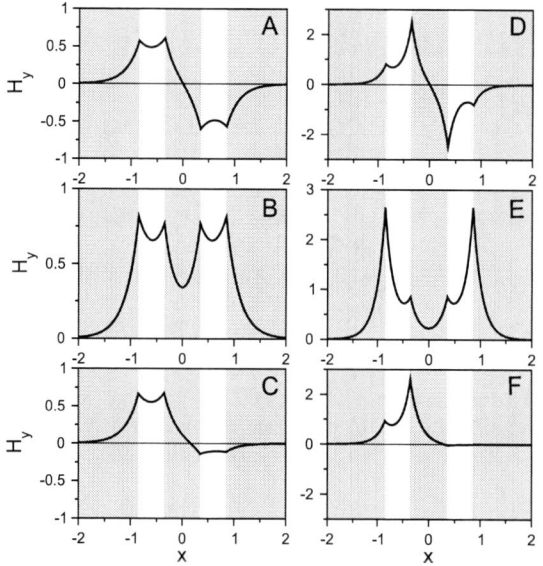

Fig. 1. Examples of TM modes of a nonlinear slot coupler for $\beta = 3.0$ (left column) and $\beta = 5.0$ (right column). The plots show the magnetic component for antisymmetric (A, D), symmetric (B, E), and asymmetric modes (C, F). The labels correspond to the points marked in Fig. 2.

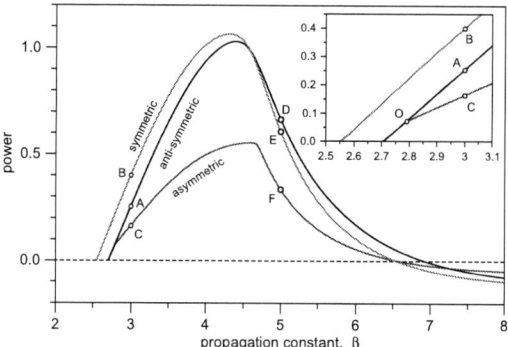

Fig. 2. Total mode power vs. propagation constant for the modes of the nonlinear plasmonic coupler. The points A, B, C, D, E, and F correspond to the modes shown in Fig. 1. Point O is the bifurcation point for the asymmetric mode.

of exponential terms containing $\exp\left(\pm\sqrt{\beta^2 - \epsilon_m}\right)$ which is properly chosen to obtain a decaying behavior at $x = \pm\infty$. Inside the cores, using Eqs. (2), (3) and the dependence $\epsilon E_x = \beta H_y$, we obtain the following cubic equation to express ϵ as a function of the field components H_y and E_z,

$$\epsilon^3 - (\epsilon_d + E_z^2)\epsilon^2 - \beta^2 H_y^2 = 0. \qquad (5)$$

It is easy to demonstrate that this equation has only one real solution which can be obtained using Cardano's method. Once this solution is replaced in Eqs. (4) the system of ODEs is numerically solved by a relaxation method imposing the continuity of fields H_y and E_z at the boundaries between claddings and cores. The solution obtained for cores is then analytically augmented to the metallic regions.

For our calculations we took $\epsilon_d = 2.25$, $\epsilon_m = -8.25$ (silver for a wavelength $\lambda \approx 480$ nm [12]), waveguide width $w = 0.5$ and separation between waveguides $d = 0.7$. The modes are characterized by their power flux, given by the z-component of the Poynting vector, $P = \int (\mathbf{E} \times \mathbf{H})\hat{z}dx = \beta \int (H_y^2/\epsilon)dx$, and the propagation constant β. Examples of modes are shown in Fig. 1 for $\beta = 3.0$ and $\beta = 5.0$ and the power versus propagation constant diagram is shown in Fig. 2.

We found three types of modes: symmetric (A, D), antisymmetric (B, E), and asymmetric (C, F). In the linear limit [13], there exist both symmetric and antisymmetric modes (A, B) guided by the slot. They are created by in-phase or out-of-phase *symmetric modes* of isolated slot waveguides. For relative small power, we observe the symmetry breaking when the new asymmetric mode (C) emerges at the bifurcation point O (see Fig. 2). This mode is composed of two antisymmetric modes (since it bifurcates from a branch of antisymmetric

solutions) but the filed amplitudes are different in two slots.

For larger powers, all those three modes evolve almost independently, but the field becomes mainly confined at the boundaries between dielectric and metallic layers in each waveguide. Such a high field concentration leads to the symmetry breaking in each of the slot waveguides, described earlier for an isolated nonlinear slot [11], so the amplitudes at the metal-dielectric surfaces become different (modes D and E). The field is shifted towards the metal cladding, so that the total power decreases, and it may even become negative due to the dominant (negative) contribution of the backward power flux concentrated in the metal (see Fig. 2), similar to a single nonlinear slot waveguide.

III. SWITCHING CHARACTERISTICS

We performed FDTD simulations to study the dynamics of the system. To account for the high dispersive metallic layers and make the algorithm numerically stable we use a classical model for cold plasmas that we introduced in Eqs. (1) writing the displacement vector as $\vec{D} = \epsilon_0 \vec{\mathcal{E}} + \partial_t \vec{\mathcal{P}}$ and modelling the polarization current $\vec{J} = \partial_t \vec{\mathcal{P}}$ by the equation [14],

$$\partial_t \vec{J} + \Gamma \vec{J} = \epsilon_0 \omega_p^2 \vec{\mathcal{E}}, \qquad (6)$$

where ω_p is the plasma frequency and Γ is the electron collision frequency that accounts for the power losses. Both parameters can be calculated considering the Drude model $\epsilon(\omega) = 1 - \omega_p^2/(\omega^2 + i\Gamma\omega)$ and using the values $\text{Re}(\epsilon) = -8.25$ and $\text{Im}(\epsilon) = 0.3$, deduced from the optical constants of silver at $\lambda \approx 480$ nm [12]. Taking $\omega = 1$, i.e. time is measured in the units of ω^{-1}, we obtain $\omega_p = 3.043$ and $\Gamma = 0.032$. On the other hand, to model the dielectric cores, we consider an instantaneous nonlinear response [15] and use Eq. (2) and (1). With this approach numerical instabilities rise after a propagation time, which is shorter the larger the power is. Anyway, for a moderate power the instabilities are delayed a time quite long enough to perfectly observe the coupling and switching characteristics. This instabilities are probably due to the simple nonlinear model we used and it is expected they will be suppressed by using a more elaborated model.

978-1-4244-7043-3/10 $26.00 © 2010 IEEE

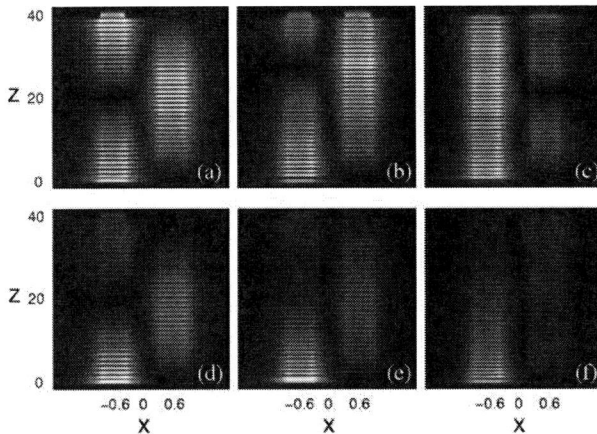

Fig. 3. Examples of mode propagation and beating in the plasmonic coupler for different input powers. In the top row we show simulations for null losses ($\Gamma = 0$): (a) linear regime, $P_0 = 0.10$, (b) intermediate regime, $P_0 = 0.20$, and (c) nonlinear regime, $P_0 = 0.25$. In the bottom row, realistic losses are assumed ($\Gamma = 0.032$): (d) linear regime, $P = 0.10$, (e) intermediate regime, $P = 0.30$, and (f) nonlinear regime, $P = 0.45$.

We initially neglected losses for simplicity. In Fig. 3(a-c) we present three images correspondent to simulations carried out for different input power to illustrate how power flux can be switched from the second core to the first one just controlling input power. The FDTD simulations were performed by exciting one of the waveguides at $z = 0$ with a *hard source* presenting the shape of the linear mode of the single-waveguide, properly scaled to the required power. After propagating for the necessary number of time-steps, further steps are calculated to time-average the Poynting vector, which is finally integrated in the transversal variable x at $z = z_b$, where z_b is the beat length, i.e. the distance at which power completely transfers to the second core under linear propagation. In Fig. 4 we show the switching curve ($\Gamma = 0$) where the normalized output power P/P_0 is plotted against the input power P_0, showing how power is completely transferred from one core to the other when power raises over a threshold. On the other hand, when realistic losses are taken into account, switching is still possible, as shown in Fig. 3(d-f) where the three corresponding

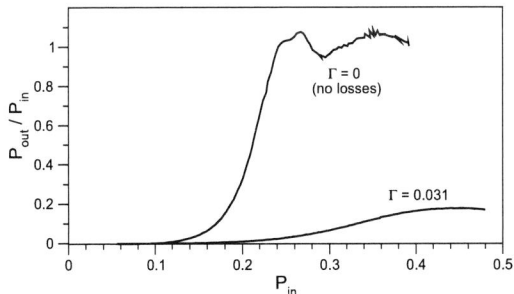

Fig. 4. Normalized output power (P/P_0) vs. input power (P_0) for the excited waveguide of the nonlinear plasmonic slot coupler at the beat length $z_b = 19.26$; shown for zero and realistic losses.

simulations for quasi-linear, intermediate and nonlinear regime are displayed. In Fig. 4 we also plot the switching curve for this case, showing the expected decrease of power. We remark that we used a particularly large distance between cores in other to achieve switching for lower power and avoid the numerical instabilities. In a practical implementation both slots in the coupler are closer to each other and consequently beat length is shorter and the effect of losses less important.

IV. CONCLUSION

We have studied nonlinear modes and switching properties of planar directional couplers created by two coupled slot metal-dielectric-metal waveguides. We have discussed nonlinear switching and reaching the conclusion that the device is still operative when realistic losses are considered.

ACKNOWLEDGMENT

We thank the Australian Research Council and the Ministerio de Ciencia e Innovación of Spain (Acción Complementaria Internacional PCI2006-A7-0561 and project MAT2008-06870 and Ramón y Cajal contract granted to JRS) for supporting this work.

REFERENCES

[1] S. A. Maier, *Plasmonics: Fundamentals and Applications*. New York: Springer, 2007.
[2] S. I. Bolzhevolnyi, V. S. Volkov, E. Devaux, J. Y. Laluet, and T. W. Ebbesen, "Channel plasmon subwavelength waveguide components including interferometers and ring resonators," *Nature*, vol. 440, pp. 508–511, 2006.
[3] R. Yang, M. A. G. Abushagur, and Z. Lu, "Efficiently squeezing near infrared light into a 21 nm-by-24 nm nanospot," *Opt. Expr.*, vol. 16, p. 20142, 2008.
[4] D. K. Gramotnev, K. C. Vernon, and D. F. P. Pile, "Directional coupler using gap plasmon waveguides," *Appl. Phys. B*, vol. 93, pp. 99–106, 2008.
[5] Z. Chen, T. Holmgaard, S. I. Bozhevolnyi, A. V. Krasavin, A. V. Zayats, L. Markey, and A. Dereux, "Wavelength-selective directional coupling with dielectric-loaded plasmonic waveguides," *Opt. Lett.*, vol. 34, pp. 810–812, 2009.
[6] T. Holmgaard, Z. Chen, S. I. Bozhevolnyi, L. Markey, and A. Dereux, "Design and characterization of dielectric-loaded plasmonic directional couplers," *J. Lightwave Technol.*, vol. 27, pp. 5521–5528, 2009.
[7] A. Degiron, S. Y. Cho, T. Tyler, N. M. Jokerst, and D. R. Smith, "Directional coupling between dielectric and long-range plasmon waveguides," *New J. Phys.*, vol. 11, p. 015002, 2009.
[8] V. M. Agranovich, V. S. Babichenko, and V. Y. Chernyak, "Nonlinear surface polaritons," *Sov. Phys. JETP Lett.*, vol. 32, pp. 512–, 1980.
[9] G. I. Stegeman, C. T. Seaton, J. Ariyasu, R. F. Wallis, and A. A. Maradudin, "Nonlinear electromagnetic waves guided by a single interface," *J. Appl. Phys.*, vol. 58, pp. 2453–2459, 1985.
[10] D. Mihalache, G. I. Stegeman, C. T. Seaton, E. M. Wright, R. Zanoni, A. D. Boardman, and T. Twardowski, "Exact dispersion relations for transverse magnetic polarized guided waves at a nonlinear interface," *Opt. Lett.*, vol. 12, pp. 187–189, 1987.
[11] A. R. Davoyan, I. V. Shadrivov, and Y. S. Kivshar, "Nonlinear plasmonic slot waveguides," *Opt. Expr.*, vol. 16, p. 21209, 2008.
[12] P. B. Johnson and R. W. Christy, "Optical constants of the noble metals," *Phys. Rev. B*, vol. 6, pp. 4370–4379, 1972.
[13] M. Conforti, M. Guasoni, and C. De Angelis, "Subwavelength diffraction management," *Opt. Lett.*, vol. 33, no. 22, pp. 2662–2664, Nov. 2008.
[14] S. A. Cummer, "An analysis of new and existing fdtd methods for isotropic cold plasma and a method for improving their accuracy," *IEEE Trans. Antennas Propagat.*, vol. 45, no. 3, pp. 392–400, Mar. 1997.
[15] R. M. Joseph and T. Allen, "Spatial soliton deflection mechanism indicated by fd-td maxwell's equations modelling," *IEEE Phot. Tech. Lett.*, vol. 6, no. 10, pp. 1251–1254, Oct. 1994.

Nonlinear Dynamics of Spiraling Laser Beams

Anton S. Desyatnikov

Nonlinear Physics Centre, Research School of Physics and Engineering,
The Australian National University, Canberra, ACT 0200, Australia

(Invited *Paper*)

Abstract: Spiraling of laser beams in nonlinear media is one of the manifestations of optical orbital angular momentum (OAM). We explore unusual dynamics of spiraling spatial solitons, including stabilization of azimuthons and quasi-periodic transformations of nonlocal solitons, as well as contribution of OAM towards suppression of catastrophic self-focusing, or collapse, in Kerr media.

During last decade, the dislocations [1] in the wave-fronts of laser beams, called optical vortices, remain subject of active research in singular optics [2, 3]. Such dislocations and, more generally, the twist of the wave-fronts, are associated with optical OAM [4], in addition to the more familiar spin angular momentum of circularly polarized waves. In nonlinear media the singular laser beams exist as self-localized vortex solitons [5], usually unstable in media with local response. The azimuthal modulational instability breaks ring-shaped vortex beams into fundamental solitons which spiral around optical axis because of OAM, this effect was observed in experiments [6-8] and it was studied theoretically [9,10].

The formation of optical spatial solitons is supported by the balance between nonlinear self-focusing and diffraction of light in two transverse dimensions [11]. The fundamental solitons have bell-shaped profiles similar to Gaussian beams and they interact similar to particles, for example creating a dynamic *spiraling* bound state, similar to the two-body system [5]. Similar approach was suggested for stabilization of vortices, by creating a ring (cluster) of fundamental solitons [12] resembling a necklace [13]. The balance between mutual attraction of solitons in the ring and effective repulsion induced by OAM requires stair-case like phase distribution, see Fig. 1.

An important link between the radially symmetric vortices and rotating soliton clusters was introduced as a general class of spatially localized self-trapped optical beams in nonlinear media, the so-called azimuthons [14]. Azimuthons can be described as vortex solitons with topological charge m and continuous N-fold azimuthal modulation, and each soliton family (m, N) includes azimuthons with negative, positive, and zero angular velocity of spiraling. A theoretical approach unifying the complete family of optical azimuthons, including vortex solitons and soliton clusters, relies on the azimuthal "discretisation" of a vortex of charge m with imposed N-fold symmetry [15]. This discretisation can be described in terms of the series expansion of a vortex phase, $\theta(\varphi) = m\varphi$, near the azimuthal position, $\varphi_n = 2\pi n/N$, of the n-th intensity peak on the ring with radius R ($n = 1...N$):

$$\theta(\varphi) = m\varphi \cong m\varphi_n + m(r/R)\sin(\varphi - \varphi_n), \quad (1)$$

here (r, φ) are the polar coordinates in the transverse plane. The *zero-order* term in Eq. (1) corresponds to a *soliton cluster* [12], i.e. a ring of N beams launched in parallel to each other with a staircase-like phase distribution. The *first-order* term describes the additional azimuthal tilt α of all lobes in the ring (initial twist). This tilt modifies the azimuthon OAM and provides generalization to the azimuthon family [14].

Recent theoretical and experimental results [15] suggest that the modulation of the power flow along the azimuthal coordinate provides an effective mechanism for the stabilization of singular spatial solitons [Fig. 2 (a-c)] in comparison to the unstable vortex solitons [Fig. 2 (d)]. It involves the reduction of the OAM of vortex beam, slowing down of azimuthon spiraling, and even inversion of the direction of rotation, see anomalous twist $\beta(\alpha)$ in Fig. 2(e).

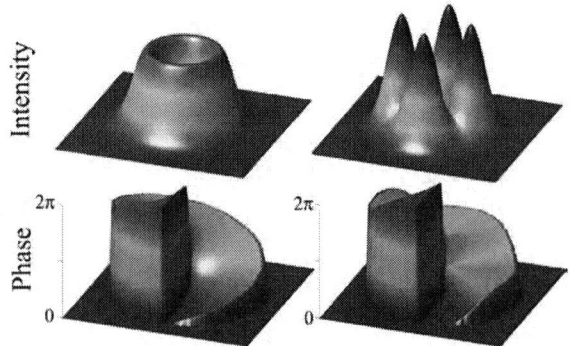

Fig. 1. Distribution of intensity (top) and phase (bottom) for a vortex soliton (left) and rotating soliton cluster (right).

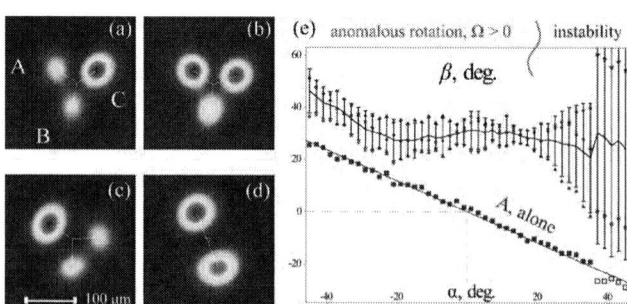

Fig. 1. Three lobe azimuthon experiments: (a-d) measured output profiles for input azimuthal phase tilt α = 1°, 15°, 29°, and 43° and (e) output angle of spatial twist $\beta(\alpha)$ compared to free propagating beam "A".

978-1-4244-7043-3/10 $26.00 © 2010 IEEE

An interesting generalization of azimuthon ideas was suggested by Lashkin for the two-dimensional "pancake-shaped" [16] as well as three-dimensional [17] Bose-Einstein condensates confined in a parabolic trap. Stable numerical solutions of the Gross-Pitaevskii equation in the rotating grid were obtained and extended later to the vector system [18].

In optics, the stabilization of vortex solitons is associated with long-range or nonlocal response of the medium [19-22]. In addition to vortex solitons, the azimuthon family has been extended to include spiraling states with multiple vortices organized into robust geometrical structures, or "vortex clusters" [23, 24]. Nonlocal media was shown to support even more general class of higher-order spatial optical solitons [25], analogous to Laguerre-Gaussian and Hermite-Gaussian linear eigenmodes, resembling various forms of soliton necklaces and soliton matrices. Surprisingly, solitons with different symmetries coexist and overlap in energy domain (power, Hamiltonian) and this overlap leads to novel effects.

As was shown in Ref. [25], the modulational instability can trigger nontrivial transformations between energetically close solitons with different symmetries through the intermediate states resembling generalized Hermite-Laguerre-Gaussian modes. The latter were studied recently [26] and the quasi-periodic transformations, resembling a self-induced mode converter, were observed in numerical simulations. Transformation dynamics of solitons with zero angular momentum, e.g. the quadrupole-type soliton in Fig. 3 (a), reveal the equidistant spectrum of spatial field oscillations typical for the breather-type solutions. In contrast, the transformations of nonlocal solitons carrying orbital angular momentum, such as 2×3 soliton matrix in Fig. 3 (b), are accompanied by their spiraling and corresponding spectra of field oscillations show mixing of three characteristic spatial frequencies. In medium with thermal nonlinearity the boundaries and the geometry of a nonlinear sample have been shown to affect the dynamics of transformations dramatically [27], changing parameters as well as symmetries of oscillations.

The OAM of self-trapped beams can also play an important role in the long outstanding problem of catastrophic self-focusing (collapse) in Kerr medium [28, 29]. Many different suggestions have been made previously on how to arrest the collapse by modifying nonlinearity while the problem remains unsolved for pure Kerr media with nonlinear correction to refractive index proportional to intensity. We recently predicted theoretically that OAM can suppress the collapse of elliptic spiraling beams [30]. Similar to the radially-symmetric beams, the elliptic beams with zero OAM collapse with the development of a sharp spike with the Towns profile [31], see Fig. 4(a), this process was experimentally observed in Ref. [32]. In contrast, if the value of the OAM σ, and the rate of spiraling, is increased above some threshold value [solid line in Fig. 4(b)], the growth of the Townes spike is strongly suppressed and the beam with super-critical power exhibits diffraction on propagation distances exceeding several times the collapse distance. In other words, the OAM strengthens the diffraction against self-focusing and suppresses the collapse. The same mechanism preserves the elliptic profile of stably rotating solitons in optical media with collapse-free nonlinearities, an example for saturable media is shown in Fig. 4(c). Similar effects are expected to occur in coherent matter waves with attractive interaction between atoms in rotating elliptic traps.

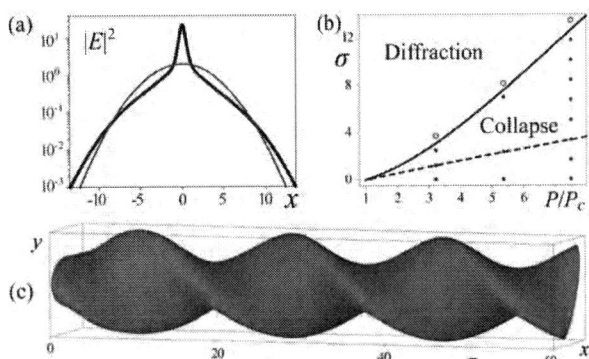

Fig. 4. (a) Collapsing beam profile in Kerr medium (thick line) and corresponding Gaussian profile (thin line). (b) Soliton domain; the solid line indicates the threshold values of (normalised) OAM σ, separating collapse from diffraction. (c) An example of a stable spiraling elliptic soliton in saturable media.

In conclusions, we present a short overview of recent theoretical and experimental results on spiraling beams carrying optical orbital angular momentum in nonlinear media. The OAM is shown to be the source of novel interesting phenomena, such as beam spiraling and symmetry transformations, as well as play important role in other effects, such as catastrophic self-focusing.

This work is supported by the Australian Research Council.

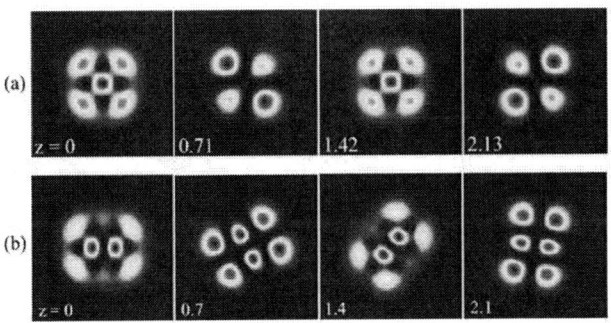

Fig. 3. Quasi-periodic transformations of nonlocal solitons; evolution dynamics along z from left to right. (a) Transformations of the quadrupole-type soliton with zero OAM. (b) Transformations accompanied by spiraling.

REFERENCES

[1] J. F. Nye and M. V. Berry, "Dislocations in wave trains", Proc. R. Soc. A **336**, 165 (1974).

[2] M. S. Soskin and M. V. Vasnetsov, "Singular optics", Progress in Optics, Vol. **42**, 219 (Ed. E. Wolf, Elsevier, 2001).

[3] M. R. Dennis, K. O'Holleran, and M. J. Padgett, "Singular Optics: Optical Vortices and Polarization Singularities", Progress in Optics, Vol. **53**, 293 (Ed. E. Wolf, Elsevier, 2009).

[4] L. Allen, M. J. Padgett, and M. Babiker, "The orbital angular momentum of light", Progress in Optics, Vol. XXXIX, 291 (Ed. E. Wolf, Elsevier, 1999).

[5] A. S. Desyatnikov, Yu. S. Kivshar, and L. Torner, "Optical Vortices and Vortex Solitons", Progress in Optics, Vol. **47**, pp. 291-391 (Ed. E. Wolf, Elsevier, 2005).

[6] V. Tikhonenko, J. Christou, and B. Luther-Davies, "Spiraling bright spatial solitons formed by the breakup of an optical vortex in a saturable self-focusing medium," J. Opt. Soc. Am. B **12**, 2046 (1995).

[7] M. S. Bigelow, P. Zerom, and R.W. Boyd, "Breakup of ring beams carrying orbital angular momentum in sodium vapor," Phys. Rev. Lett. **92**, 083902 (2004).

[8] L. T. Vuong, T. D. Grow, A. Ishaaya, A. L. Gaeta, G. W. 't Hooft, E. R. Eliel, and G. Fibich, "Collapse of optical vortices," Phys. Rev. Lett. **96**, 133901 (2006).

[9] W. J. Firth and D. V. Skryabin, "Optical solitons carrying orbital angular momentum", Phys. Rev. Lett. **79**, 2450-2453 (1997).

[10] J. P. Torres, J. M. Soto-Crespo, L. Torner, and D. V. Petrov, "Solitary-wave vortices in quadratic nonlinear media", J. Opt. Soc. Am. B **15**, 625 (1998).

[11] Yu. S. Kivshar and G. P. Agrawal, *Optical Solitons: From Fibers to Photonic Crystals* (Academic, 2003).

[12] A. S. Desyatnikov and Yu. S. Kivshar, "Rotating optical soliton clusters", Phys. Rev. Lett. **88**, 053901 (2002).

[13] M. Soljacic and M. Segev, "Integer and fractional angular momentum borne on self-trapped necklace-ring beams", Phys. Rev. Lett. **86**, 420 (2001).

[14] A. S. Desyatnikov, A. A. Sukhorukov, and Yu. S. Kivshar, "Azimuthons: Spatially modulated vortex solitons", Phys. Rev. Lett. **95**, 203904 (2005).

[15] A. Minovich, D. N. Neshev, A. S. Desyatnikov, W. Krolikowski, and Yu. S. Kivshar, "Observation of optical azimuthons", Opt. Express **17**, 23610 (2009).

[16] V. M. Lashkin, "Two-dimensional multisolitons and azimuthons in Bose-Einstein condensates", Phys. Rev. A **77**, 025602 (2008).

[17] V. M. Lashkin, "Stable three-dimensional spatially modulated vortex solitons in Bose-Einstein condensates", Phys. Rev. A **78**, 033603 (2008).

[18] V. M. Lashkin, E. A. Ostrovskaya, A. S. Desyatnikov, and Yu. S. Kivshar, "Vector azimuthons in two-component Bose-Einstein condensates", Phys. Rev. A **80**, 013615 (2009).

[19] A. I. Yakimenko, Yu. A. Zaliznyak, and Yu. S. Kivshar, "Stable vortex solitons in nonlocal self-focusing nonlinear media", Phys. Rev. E 71, 065603 (2005).

[20] D. Briedis, D. E. Petersen, D. Edmundson, W. Krolikowski, and O. Bang, "Ring vortex solitons in nonlocal nonlinear media", Opt. Express 13, 435 (2005).

[21] S. Lopez-Aguayo, A. S. Desyatnikov, Yu. S. Kivshar, S. Skupin, W. Krolikowski, and O. Bang, "Stable rotating dipole solitons in nonlocal optical media," Opt. Lett. **31**, 1100 (2006).

[22] S. Skupin, M. Saffman, and W. Krolikowski, "Nonlocal stabilization of nonlinear beams in a self-focusing atomic vapor," Phys. Rev. Lett. **98**, 263902 (2007).

[23] S. Lopez-Aguayo, A. S. Desyatnikov, and Yu. S. Kivshar, "Azimuthons in nonlocal nonlinear media," Opt. Express **14**, 7903 (2006).

[24] D. Buccoliero, A. S. Desyatnikov, W. Krolikowski, and Yu. S. Kivshar, "Spiraling multivortex solitons in nonlocal nonlinear media", Opt. Lett. **33**, 198-200 (2008).

[25] D. Buccoliero, A. S. Desyatnikov, W. Krolikowski, and Yu. S. Kivshar, "Laguerre and Hermite soliton clusters in nonlocal nonlinear media", Phys. Rev. Lett. **98**, 053901 (2007).

[26] D. Buccoliero and A. S. Desyatnikov, "Quasi-periodic transformations of nonlocal spatial solitons", Opt. Express **17**, 9608-9613 (2009).

[27] D. Buccoliero, A. S. Desyatnikov, W. Krolikowski, and Yu. S. Kivshar, "Boundary effects on the dynamics of higher-order optical spatial solitons in nonlocal thermal media", J. Opt. A **11**, 094014 (2009).

[28] L. Berge, "Wave collapse in physics: principles and applications to light and plasma waves", Phys. Rep. **303**, 259 (1998).

[29] *Self-focusing: Past and Present*, Eds. R. W. Boyd, S. G. Lukishova, and Y. R. Shen (Springer, 2009).

[30] A. S. Desyatnikov, D. Buccoliero, M. R. Dennis, and Yu. S. Kivshar, "Suppression of Collapse for Spiraling Elliptic Solitons", Phys. Rev. Lett. **104**, 053902 (2010).

[31] R. Y. Chiao, E. Garmire, and C. H. Townes, "Self-Trapping of Optical Beams", Phys. Rev. Lett. **13**, 479 (1964).

[32] K. D. Moll, A. L. Gaeta, and G. Fibich, "Self-Similar Optical Wave Collapse: Observation of the Townes Profile", Phys. Rev. Lett. **90**, 203902 (2003).

Rotating three-dimensional solitons

S. Skupin[1,2], F. Maucher[1], W. Krolikowski[3]

[1]Max Planck Institute for the Physics of Complex Systems, 01187, Dresden, Germany

[2]Friedrich Schiller University, Institute of Condensed Matter Theory and Optics, 07743 Jena, Germany

[3]Research School of Physics and Engineering, Australian National University, Canberra, ACT 0200, Australia

(Invited *Paper*)

Abstract— **We study formation of rotating three-dimensional solitons (azimuthons) in Bose Einstein condensate with attractive nonlocal nonlinear interaction. In particular, we demonstrate formation of toroidal rotating solitons and investigate their stability. The presence of repulsive contact interaction does not prevent the existence of those solutions, but allows to control their rotation.**

I. INTRODUCTION

Studies of Bose Einstein condensates (BEC) belong to one of the fastest developing research directions. In the semiclassical approach the spatial and temporal evolution of the condensates wave function is commonly described by the Gross Pitaevskii equation [1] which reflects the interplay between kinetic energy of the condensate and the nonlinearity originating from the interaction potential leading, among others, to the formation of localized structures, bright and dark solitons [2], [3]. So far the main theoretical and experimental efforts have been concentrating on condensates with contact (or hard-sphere) bosonic interaction which, in case of attraction, may lead to collapse-like dynamics. Recently, also systems exhibiting a nonlocal, long-range dipolar interaction [4] have attracted a significant attention. In particular, O'Dell *at al* [5] have recently suggested to use a series of triads of orthogonally polarized laser beams illuminating cloud of cold atoms along three orthogonal axes so that the angular dependence of the dipole-dipole nonlinear term is averaged out. The resulting nonlocal interaction potential becomes effectively isotropic of the form $1/r$. It has been already shown by Turitsyn [6] that a purely attractive "gravitational" (or Coulomb) interaction potential prevents collapse of nonlinear localized waves and gives rise to the formation of localized states - bright solitons which could be supported without necessity of using the external trapping potential. If realized experimentally such trapping geometry would enable to study effects akin to gravitational interaction. Few recent works have been dealing with this "gravitational" model of condensate looking, among others, at the stability of localized structures such as fundamental solitons and two-dimensional vortices [7]–[9].

II. MODEL

We consider a Bose-Einstein atomic condensate with the isotropic interatomic potential consisting of both, repulsive contact as well as attractive long-range nonlocal interaction contributions. Following O'Dell *et.al* [5], an attractive long-range interaction of "gravitational" form can be induced by triads of frequency detuned laser beams resulting in the following dimensionless Gross-Pitaevskii equation (GPE) for the condensate wave function $\psi(\mathbf{r}, t)$:

$$\partial_t \psi = i \quad \psi + i\Theta\psi \tag{1a}$$

$$\Theta(\mathbf{r}, t) = \int \frac{|\psi(\mathbf{r}', t)|^2}{|\mathbf{r} \quad \mathbf{r}'|} d^3r' \quad |\psi|^2 . \tag{1b}$$

The nonlinear response Θ consists of both local and nonlocal contribution. Interestingly, for the "gravitational" nonlocal interaction Θ contains no additional parameter. The ratio between local and nonlocal term is solely determined by the form of the wavefunction ψ. We will see later (Sec. IV) that for very broad solitons the local contact interaction $\sim |\psi|^2$ becomes negligible. Then the governing equation of motion is formally equivalent to the so-called Schrödinger-Newton equation, proposed by Penrose in [10].

In this work we will also consider a second, different nonlocal model, the so-called Gaussian model of nonlocality. Despite the fact that it is not motivated by a certain physical system, it serves as a popular toy model for the general class of nonlocal Schrödinger (Gross-Pitaevskii) equations in one and two dimensional problems [11]–[15]. Here, we will extend this classical model to three transverse dimensions, and moreover allow an additional local repulsive term similar to the previous case, and introduce

$$\Theta(\mathbf{r}, t) = \left(\frac{1}{2\pi}\right)^{3/2} \int |\psi(\mathbf{r}', t)|^2 e^{\frac{|\mathbf{r}-\mathbf{r}'|^2}{2}} d^3r' \quad \delta|\psi|^2 . \tag{2}$$

The additional parameter δ is necessary here to keep track of one of the two degrees of freedom of the Gaussian response, i.e. amplitude or width, which cannot be scaled out. The value of δ determines the relative strength of the local repulsive term. Note that compared to the above "gravitational" response, the interaction range of the Gaussian nonlocal response is significantly shorter due to its rapid decay for $r \to \infty$.

III. ROTATING SOLITONS

It has been shown earlier that azimuthons, i.e. multi-peak solitons with angular phase ramp, exhibit constant angular rotation and hence can be represented by straightforward generalization of the usual soliton ansatz by including an additional parameter, the angular frequency Ω [16], [17]. We write

$$\psi(r, z, \phi, t) = U(r, z, \phi \quad \Omega t)e^{iEt}, \tag{3}$$

where U is the complex amplitude and E is the normalized chemical potential, $r = \sqrt{x^2 + y^2}$ and ϕ denotes the azimuthal angle in the plane (x, y). It can be shown, that by inserting the above function into the nonlocal GPE (1) one can derive a formal relation for the rotation frequency on integrals over the stationary amplitude profiles of the azimuthon [15]. In the next Section, we will compute approximate azimuthon solutions and their rotation frequency employing a certain ansatz for the stationary amplitude profile U.

IV. VARIATIONAL APPROACH

In order to get some insight into possible localized states of the Gross-Pitaevskii equation we resort first to the so-called Lagrangian (or variational) approach [18]. In three dimensions, a reasonable ansatz for corresponding localized solutions is

$$
\begin{aligned}
\psi\left(r, z, \varphi, t\right) := & Ar \exp\left(-\frac{r^2 + z^2}{2\sigma^2}\right) e^{iEt} \\
& \times \left[\cos\left(\varphi - \Omega t\right) + ip\sin\left(\varphi - \Omega t\right)\right],
\end{aligned}
\tag{4}
$$

where parameter p varies between zero and unity. For $p = 0$ Eq. (4) describes a dipole structure consisting of two out-of-phase lobes, while for $p = 1$ it is a three-dimensional vortex, i.e. toroid-like structure with zero in the center and azimuthal (in the (x, y) plane) phase ramp of 2π.

A. The "gravitational" response

When we express the amplitude A as a function of p and σ, the localized solution (with finite amplitude) exists only if its width is greater than the critical value $\sigma_{cr}(p)$. This threshold is a consequence of competition between nonlocal and local interaction potentials. The resulting rotation frequency Ω is not sign definite, which means that we can expect both co- and counter-rotating azimuthons with respect with respect to phase. In particular, the azimuthon with the "stationary" width $\sigma_s = \sqrt{5/4\pi} \approx 0.63$ has no angular velocity. Again, this effect is due to competition between nonlocal and local contribution to Ω for small σ. The nonlocal attractive interaction leads to a positive contribution to Ω, the repulsive local interaction to a negative one. The expression for Ω *without* repulsion can be obtained by the outlined variational procedure or by asymptotic expansion $(\sigma \to \infty)$ up to $\mathcal{O}\left(1/\sigma^2\right)$. As expected, this quantity is strictly positive. Both curves Ω versus σ $(p = 0.7)$ with and without contact interaction are shown in the right panel in Fig. 1. We can see that the repulsive local interaction kicks in for $\sigma < 1.5$.

As we observe in Fig. 1, the mass behaves like $\sqrt{E}$ close to $E = 0$. The fact that the mass can become zero for $E \to 0$ is a well-known property for very long range kernels, such as the Coulomb potential in three dimensions [19]. In contrast, for shorter ranged responses (e.g., Gaussian response, see Fig. 2), the mass attains its minimum at a finite value of E.

B. The Gaussian response

For the Gaussian nonlocal response, one ends up with a similar expression for the rotation frequency. As already pointed out in Sec. II, the additional parameter δ is necessary due to an additional degree of freedom of the Gaussian response, and fixes the ratio between repulsion and attraction. Compared to the "gravitational" response, the range of this

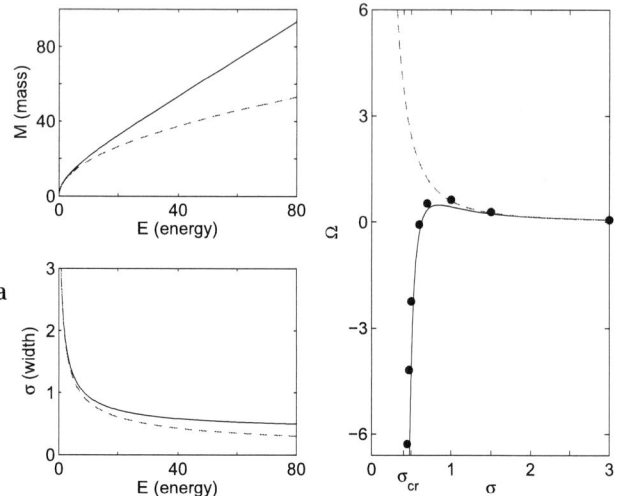

Fig. 1. (color online) The left panels show the dependence of the mass M (top) and the width σ (bottom) on the chemical potential E. Black curves show results from the variational approach including local repulsion, dashed blue curves are without contact interaction. The right panel shows the angular frequency Ω as a function of σ. Black dots denote results obtained from numerical simulations of the GPE (1). All plots are for $p = 0.7$.

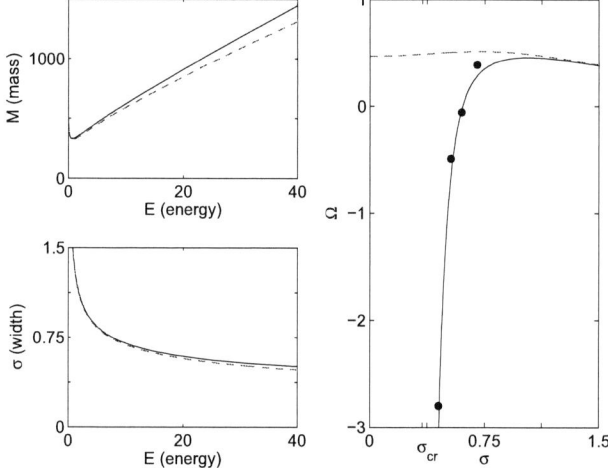

Fig. 2. (color online) Same as Fig. 1, but for the Gaussian response given in Eq. (2). Black curves are for $\delta = 0.01$, dashed blue curves without repulsion $(\delta = 0)$. All plots are for $p = 0.7$.

potential is much shorter. Hence, when considering large σ the Gaussian response acts more and more like a local attractive response and higher order solitons become unstable (see end of Sec. V).

V. NUMERICAL RESULTS

In this section, the predictions of the variational approach will be confronted with direct numerical simulations. The approximate solitons resulting from the variational approach will be used as an initial conditions to our three-dimensional code to compute their time evolution. In general, we find stable evolution, in particular the characteristic shape of the initial conditions is preserved.

CAOL*2010 International Conference on Advanced Optoelectronics & Lasers, 10-14 September, 2010, Sevastopol, Ukraine

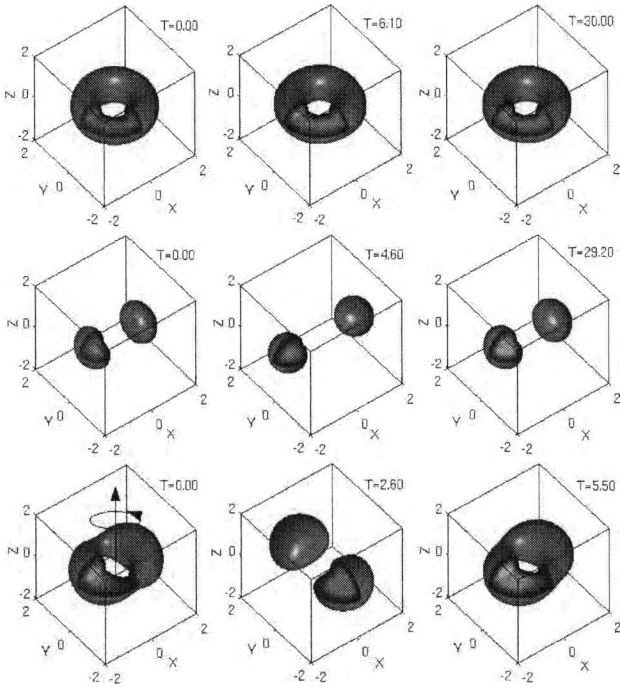

Fig. 3. (color online) Dynamics of the three-dimensional stable solitons in gravity-like BEC. Iso-surfaces of the normalized density $|\psi|^2$ are depicted for different evolution times, the interior density distribution is represented in grey-scales. The initial variational parameters used are $\sigma = 1$ and $p = 1$ (torus, iso-density surface at $|\psi|^2 = 0.76$) for the upper row, $p = 0$ (dipole, iso-density surface at $|\psi|^2 = 1.41$) for the middle one and finally $p = 0.7$ (azimuthon, iso-density surface at $|\psi|^2 = 0.86$). The sense of the rotation ($\Omega = 0.64$) is indicated by the arrows.

In Fig. 3 we illustrate the temporal evolution of three-dimensional solitons for the "gravitational" response, i.e., solutions to Eq. (1). This first two rows present the classical stationary soliton solutions torus and dipole, respectively. Due to imperfections of the initial conditions obtained from the variational approach we observe slight oscillations upon evolution, in particular for the dipole solutions (second row). Those oscillations are not present if we use numerically *exact* solutions (obtained from an iterative solver [20]) as initial conditions (not shown). In the last row of Fig. 3 we show the evolution of an azimuthon ($p = 0.7, \sigma = 1$). The rotation of the amplitude profile is clearly visible. Again we observe radial oscillations due to the imperfect initial condition, but the solution is robust.

Our variational calculations predict that for small width σ, when repulsive interaction comes into play, the sense of azimuthon rotation changes. In particular, we found a "stationary" width σ_s at which the rotation frequency Ω vanishes. Indeed, full model simulations confirm this property (see black dots in Fig. 1). Hence, we propose that tuning the strength of contact interaction in experiments allows to control the azimuthon rotation. Furthermore, we observe that very narrow azimuthons ($\sigma \to \sigma_{cr}$) have negative Ω and rotate very fast. This may be interesting for potential experiments, since the duration of BEC experiments is restricted to typically several hundreds of milliseconds.

Concerning the Gaussian nonlocal response, we find very similar evolution scenarios. However, there are some important differences: The Gaussian response has a much shorter range than the "gravitational" one. For large σ the Gaussian kernel acts like a attractive local response. As a consequence, higher order solitons become unstable in the sense that the two humps spiral out. We observe unstable evolution in numerical simulations for $\sigma \gtrsim 0.9$ at $p = 0.7$: For increasing sigma the resulting convolution term Θ [Eq. (2)], which is responsible for the self-trapping, becomes smaller in amplitude and asymmetric in the rotation plane, which eventually leads to destabilization of the azimuthon.

REFERENCES

[1] F. Dalfovo, S. Giorgini, L. P. Pitaevski, and S. Stringari, "Theory of bose-einstein condensation in trapped gases," *Rev. Mod. Phys.*, vol. 71, pp. 463–521, 1999.

[2] L. Khaykovich, F. Schreck, G. Ferrari, T. Bourdel, J. Cubizolles, L. D. Carr, Y. Castin, and C. Salomon, "Formation of a matter-wave bright soliton," *Science*, vol. 296, pp. 1290–1293, 2002.

[3] K. E. Strecker, G. B. Partridge, A. G. Truscott, and R. G. Hulet, "Formation and propagation of matter-wave soliton trains," *Nature London*, vol. 417, p. 150, 2002.

[4] K. Goral, K. Rzazewski, and T. Pfau, "Bose-einstein condensation with magnetic dipole-dipole forces," *Phys. Rev. A*, vol. 61, p. 051601(R), 2000.

[5] D. S. O'Dell, S. Giovanazzi, G. Kurizki, and V. M. Akulin, "Bose-einstein condensates with 1/r interatomic attraction: electromagnetically induced gravity," *Phys. Rev. Lett.*, vol. 84, pp. 5687–5690, 2000.

[6] S. K. Turitsyn, "Spatial dispersion of nonlinearity and stability of multidimensional solitons," *Theor, Mat. Fiz.*, vol. 64, pp. 797–801, 1985.

[7] I. Papadopoulos, P. Wagner, G. Wunner, and J. Main, "Bose-einstein condensates with attractive 1/r interaction: The case of self-trapping," *Phys. Rev. A*, vol. 76, p. 053604, 2007.

[8] H. Cartarius, T. Fabcic, J. Main, and G. Wunner, "Dynamics and stability of bose-einstein condensates with attractive 1/r interaction," *Phys. Rev. A*, vol. 78, p. 013615, 2008.

[9] A. Keles, S. Sevincli, and B. Tanatar, "Ground-state properties, vortices, and collective excitations in a two-dimensional bose-einstein condensate with gravity like interatomic attraction," *Phys. Rev. A*, vol. 77, p. 053604, 2008.

[10] R. Penrose, "Quantum computation, entanglement and state reduction," *Phil. Trans. R. Soc. (London) A*, vol. 356, pp. 1927–1939, 1998.

[11] O. Bang, W. Krolikowski, J. Wyller, and J. J. Rasmussen, "Collapse arrest and soliton stabilization in nonlocal nonlinear media," *Phys. Rev. E*, vol. 66, p. 046619, 2002.

[12] W. Krolikowski, N. Nikolov, D. Neshev, O. Bang, J. J. Rasmussen, and J. Wyller, "Modulational instability and solitons in nonlocal nonlinear media," *J. Opt. B.*, vol. 6, p. 288, 2004.

[13] D. Buccoliero, A. S. Desyatnikov, W. Krolikowski, and Y. S. Kivshar, "Laguerre and hermite soliton clusters in nonlocal nonlinear media," *Phys. Rev. Lett.*, vol. 98, p. 053901, 2007.

[14] D. Buccoliero, S. Lopez-Aguayo, S. Skupin, A. Desyatnikov, O. Bang, W. Krolikowski, and Y. S. Kivshar, "Spiraling solitons and multipole localized modes in nonlocal nonlinear media," *Physica B*, vol. 394, pp. 351–356, 2007.

[15] S. Skupin, M. Grech, and W. Krolikowski, "Rotating soliton solutions in nonlocal nonlinear media," *Opt. Express*, vol. 16, pp. 9118–9131, 2008.

[16] A. S. Desyatnikov, A. A. Sukhorukov, and Y. S. Kivshar, "Azimuthons: Spatially modulated vortex solitons," *Phys. Rev. Lett.*, vol. 95, p. 203904, 2005.

[17] D. V. Skryabin, J. M. McSloy, and W. J. Firth, "Stability of spiralling solitary waves in hamiltonian systems," *Phys. Rev. E*, vol. 66, p. 055602(R), 2002.

[18] B. A. Malomed, "Variational methods in nonlinear fiber optics and related fields," *Prog. Opt.*, vol. 43, pp. 71–191, 2002.

[19] J. Fröhlich, T. Tsai, and H. Yau, "On the point-particle (newtonian) limit of the non-linear hartree equation," *Communications in Mathematical Physics*, vol. 225, pp. 223–274, 2002.

[20] S. Skupin, O. Bang, D. Edmundson, and W. Krolikowski, "Stability of two-dimensional spatial solitons in nonlocal nonlinear media," *Phys. Rev. E*, vol. 73, p. 066603, 2006.

978-1-4244-7043-3/10 $26.00 © 2010 IEEE

CAOL*2010 International Conference on Advanced Optoelectronics & Lasers, 10-14 September, 2010, Sevastopol, Ukraine

Symmetry-induced forces on phase singularities

A. Ferrando[1] and M.A. García-March[2]

[1]Interdisciplinary Modeling Group, *InterTech*. Departament d'Òptica. Universitat de València. Spain.
[2]Department of Physics. Colorado School of Mines. Golden, Colorado. USA.

(Invited *Paper*)

Abstract: We show the existence of external forces acting on phase singularities whose origin can be attributed to the presence of short-term discrete-symmetry potentials. These special forces can break highly charged phase singularities into single-charged ones and provide them with non-zero orbital angular momentum even when the potential no longer acts.

Although a superfluid or a superconductor have a different microscopic nature, their collective properties are similar and they exhibit vortex solutions characterized by their topological charge. In both cases, mean field equations describing extended coherent collective states admit vortex solutions [1]. In mean field theory, the ground state of such systems can be described by a constant wave function corresponding to a state of uniform density. In this uniform background, vortex solutions in the form of dark spots of zero amplitude can appear. The field amplitude for such solutions vanishes at these points around which the energy flux is quantized (discretized). These points are phase singularities since phase is not defined at the point of zero amplitude. It has been proven long ago that in these physical systems phase singularities present a particle-like behavior. Phase singularities can interact among them and they can be also influenced by the presence of effective external forces (induced by the modification of the uniform background field) as if they were point-like particles. These systems are repulsively nonlinear (self-defocusing in optical language) in their collective properties (mean field). This macroscopic self-defocusing nonlinearity is the result of the microscopic repulsive interaction among the atomistic constituents of the superconductor or the superfluid. Thus, in these systems atomic interaction is an essential ingredient. Particle-like behavior and, thus, external control of phase singularities is therefore linked to nonlinearity or, equivalently, to the existence of atomic interactions.

However, the possibility of external control of phase singularities can be also achieved in optical and Bose-Einstein condensate (BEC) systems. Both systems can be described by means of the Nonlinear Schrödinger Equation (NLSE) or Gross-Pitaevsky equation in BEC language. Both systems admit vortex solutions in the self-focusing and self-defocusing regimes (which are experimentally accessible in optics and BEC.) It has been proven that the presence of discrete-symmetry potentials can be used to manipulate the phase-singularity structure of stationary nonlinear solutions of the NLSE in the self-focusing regime [3]. In Fig. 1 we show a nontrivial phase structure with multiple phase singularities. The value of the central on-axis charge and the distribution of off-axis charges are dictated by the order of the 6th-fold rotational symmetry of the potential (see lower figure in Fig. 1). According to group theory arguments, on-axis charges are restricted by the order of symmetry [4]. For 6th-fold symmetry only ±1 or ±2 central charges are allowed. Fig. 1 and Fig. 2 show configurations with central charge +1 and +2 respectively.

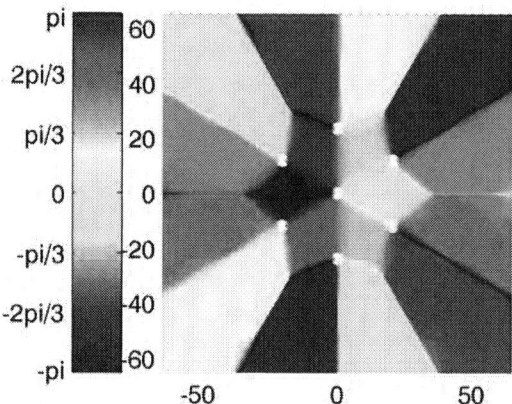

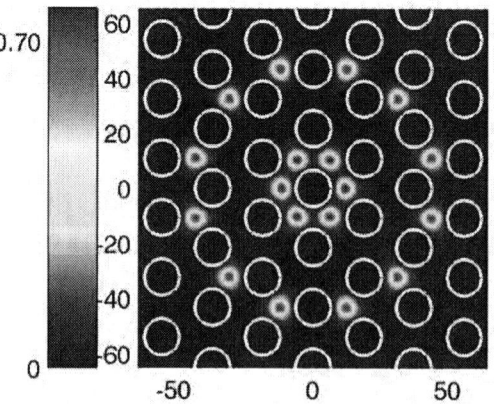

Fig. 1. Phase (upper figure) and amplitude (lower figure) of a nonlinear stationary solution exhibiting multiple singularities spatially ordered according to the 6th-fold symmetry of the potential (white circles in lower figure). Central singularity has charge +1.

978-1-4244-7043-3/10 $26.00 © 2010 IEEE

CAOL*2010 International Conference on Advanced Optoelectronics & Lasers, 10-14 September, 2010, Sevastopol, Ukraine

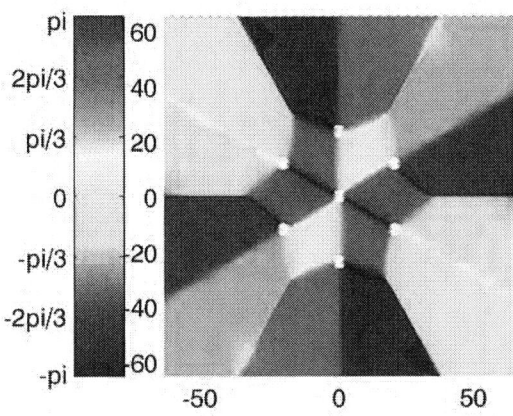

Fig. 2. Nonlinear stationary solution with multiple singularities with 6th-fold symmetry for the same system as solution in Fig. 1 but with central charge +2.

The fact that the symmetry of the potential can be used to engineer the structure of phase singularities in nonlinear stationary solutions suggests that propagation and, thus, the very dynamics of phase singularities can be influenced by the finite rotational symmetry of the external potential. In Ref. [5] it was shown that asymptotic stationary nonlinear vortex states with central charge -1 could be dynamically achieved from initial vortices of charge +3 by properly tuning a potential with 4th rotational order. The propagation issue of transmuting a high charge phase singularity into a low charge one is of great interest to understand the dynamics of phase singularities. A careful and systematic study of the process of "decay" of high charge phase singularities in self-focusing nonlinear media was performed in [6].

Fig. 3. Typical snapshot of the decay process of a vortex with topological charge 3 induced by a potential with discrete rotational symmetry of 4th order. It is clearly seen the central -1 charge and a cluster of four leaving +1 charges moving outwards.

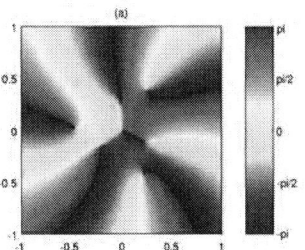 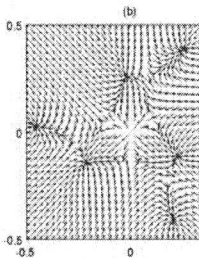

Fig. 4. Left figure: snapshot of the phase for the decay of a vortex with topological charge +5 due to the presence of a discrete-symmetry potential with $N=3$. The initial charge decays into a central singularity with charge -1 and two clusters of single-charged outwards singularities. Right figure: representation of the dual field of the gradient of the phase.

In Fig. 3. we show a typical snapshot of the decay of a +3 charged vortex after instantaneously switching on a discrete rotational potential of 4th order. Before this switching the absence of the potential permitted the existence of a vortex configuration exhibiting full rotational symmetry and a central phase singularity with topological charge +3. In this way, the appearance of the 4th order potential breaks the full rotational invariance of the system. This analysis shows that after the sudden introduction of a symmetry-breaking potential, the central +3 charge breaks into an on-axis central -1 charge and a cluster of four single-charged phase singularities following well defined trajectories in the outward direction. The initial and final central charges are related to the following simple topological selection rule [6]: $l = m + k_0 N$ ($k_0 \in Z$), where l and m are the initial and final topological charges of the on-axis (central) phase singularities, respectively, and N is the order of rotational symmetry of the breaking potential. Additionally, as shown in [4], the finite rotational order of the symmetry-breaking potential induces an upper bound for the permitted topological charges: $|m| \le N/2$ (for even N) or $|m| < N/2$ (for odd N). The topological charge selection rule along with the later cutoff condition univocally define the value of the final central charge m and of the integer k_0. So that, given the initial topological charge l and the order of discrete rotational symmetry of the potential N, m and k_0 are perfectly determined. In Ref.[5] it is also shown that the value of the natural number $|k_0|$ determines the number of ejecting clusters. In Fig. 4, it is shown how the action of a discrete-symmetry potential with rotational symmetry of order $N=3$ "breaks" a $l = 5$ vortex generating a central phase singularity with charge $m = -1$ and $|k_0| = 2$ clusters of single-charged charges moving outwards. The topological selection rule is therefore fulfilled $5 = -1 + 2 \times 3$. Nevertheless, it is also possible to generate a more sophisticated phase singularity dynamics, including the generation of vortex-antivortex pairs,

978-1-4244-7043-3/10 $26.00 © 2010 IEEE   93

by playing with the particular form of the symmetry breaking potential. We can see an example of this type of decay in Fig. 5. In this case, $l = 5$ and $N = 4$, so that the selection rule is satisfied with $m = +1$ and $k_0 = 1$. However, the selection rule does not forbid the generation of vortex-antivortex pairs: $5 = 1 + 1 \times 4 + 4 - 4$. This phenomenon is exactly what is observed in Fig. 5., in which explicit trajectories of all phase singularities are depicted.

An important feature of phase singularities in all these decay processes is that they behave like point-like particles. A well-defined trajectory and charge can be attributed to each phase singularity at any step of the decay process. The fact that trajectories in Fig. 5 are not straight lines clearly indicates that effective forces are induced on phase singularities by the introduction of the symmetry breaking potential. Effective forces must appear that break the initial high charge vortex and that bend singularity trajectories. In this sense, this particle-like behavior is analogous to that observed in superfluid or superconductors but in self-focusing media, i.e., in the presence of attractive and not repulsive interactions. However, numerical simulations suggest that the decay pattern mechanism induced by the breaking of rotational symmetry, despite occurring in a self-focusing medium, is essentially lineal.

In order to prove the previous statement rigorously we have performed an analytical calculation of the behavior of phase singularities in the absence of nonlinearity. In this calculation a $N = 4$ symmetry breaking potential is assumed to act instantaneously on a $l = 3$ charged vortex. This calculation permits to construct the dynamics of phase singularities as point-like sources by writing their equations of motion. Effective forces, induced by the instantaneous potential, can be calculated. In Fig. 6 we represent the trajectories (red lines) evaluated using these equations of motion. It can be proven that orbital angular momentum is transferred to leaving charges even when the external symmetry breaking potential no longer acts.

Note how red lines bend close to the origin thus indicating angular momentum transfer at the beginning of the process. Fig. 6 shows that trajectories are perfectly described by the particle equation of motion. Symmetry-induced forces appearing in this equation are therefore special and counterintuitive since no interaction nor external potential is active when angular momentum transfer to pahse singularities is still active.

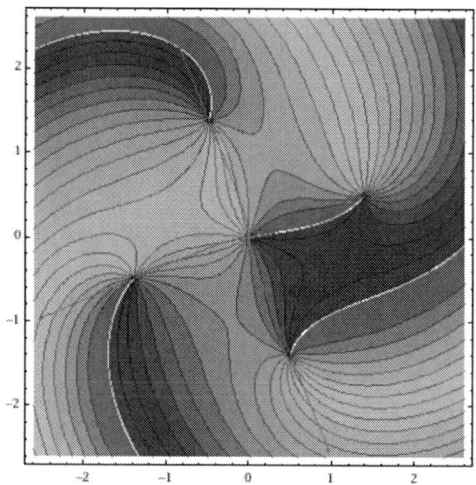

Fig. 6 Trajectories of phase singularities calculated using the particle equation of motion.

This work was partially supported by the Government of Spain Contract No. TIN2006-12890.

REFERENCES

[1] L.M. Pismen, *Vortices in Nonlinear Fields: From Liquid Crystals to Superfluids. From Non-Equlibrium Patterns to Cosmic Strings.* Oxford University Press.1999.

[2] M.A. García-March, A. Ferrando, M. Zacarés, S. Sahu, and D.E. Ceballos-Herrera, "Symmetry, winding number, and topological charge of vortex solitons in discrete-symmetry media," Phys. Rev. A 79, 053820 (2009).

[3] A. Ferrando, M. Zacarés, and M.A. García-March, "Vorticity Cutoff in Nonlinear Photonic Crystals," Phys. Rev. Lett.. 95, 043901 (2005).

[4] A. Ferrando, M. Zacarés, and M.A. García-March, "Vortex Transmutation," Phys. Rev. Lett.. 95, 123901 (2005).

[5] M. Zacarés, M.A. García-March, J. Vijande, A. Ferrando, and E. Merino, "Topological charge selection rule for phase singularities," Phys. Rev. A 80, 043812 (2009).

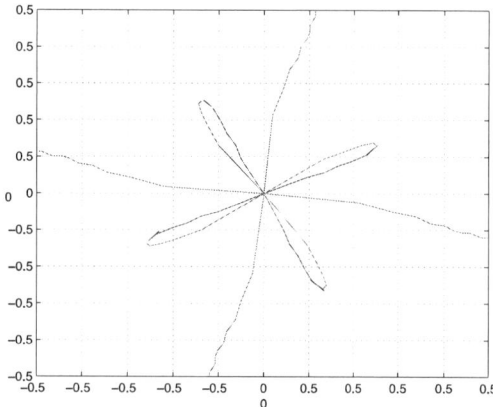

Fig. 5 Trajectories of phase singularities showing the decay of a vortex of charge +5 induced by a $N = 4$ symmetry breaking potential.

CAOL*2010 International Conference on Advanced Optoelectronics & Lasers, 10-14 September, 2010, Sevastopol, Ukraine

Optical Nanoprobing via Spin-Orbit Interaction of Light

K. Y. Bliokh[1], O. Rodríguez-Herrera[1], D. Lara[2], E. A. Ostrovskaya[3], C. Dainty[1]

[1]Applied Optics, School of Physics, National University of Ireland, Galway, Galway, Ireland
[2]The Blackett Laboratory, Imperial College London, SW7 2BW, London, United Kingdom
[3]ARC Centre of Excellence for Quantum-Atom Optics and Nonlinear Physics Centre, Research School of Physics and Engineering, The Australian National University, Canberra ACT 0200, Australia

(Invited *Paper*)

Abstract: High-NA optical microscopy is accompanied by strong spin-orbit interaction of light, which translates fine information about the nano-size specimen to the non-uniform polarization distribution of the outgoing light. We observe angular momentum conversion and spin-Hall effect of light, which are highly sensitive to the position of the nanoparticle.

In the past few years spin-orbit interaction (SOI) of light was intensively studied in connection with the spin-Hall effect in inhomogeneous media [1,2] and conversion of the angular momentum (AM) of light upon focusing and scattering [3-5]. SOI phenomena exhibit intrinsic subwavelength-scale coupling between polarization and position of light. This makes them highly attractive for nano-optical applications, but difficult for detection in the paraxial far field. In isotropic media the subwavelength-scale SOI becomes conspicuous only in highly non-paraxial fields tightly focused by high numerical aperture (NA) lenses or scattered by small particles. In these cases, its observation requires additional interaction with testing particles [3] or near-field measurements [4,5].

Here we propose and demonstrate experimentally a far-field polarimetric imaging system that employs strong SOI of light and space-variant AM conversion via a nano-particle, Fig. 1 [6,7]. An 80nm gold particle located near the focus between two high-NA lenses interacts with the non-paraxial focal field inside the system and dramatically modifies polarization distribution of the outgoing paraxial light.

The experimental and theoretical results are presented in Figs. 2 and 3 which clearly show characteristic nonuniform patterns in the output polarization [7]. First, when the scatterer is located precisely at the focal point and illuminated by circularly-polarized light, the Stokes parameters S_1 and S_2 possess four-fold symmetric patterns (Fig. 2) typical for the *spin-to-orbit AM conversion of light* upon scattering [8]. Strikingly, the sub-diffraction-limit displacement of the particle by the distance $\lambda/3$ along, e.g., the x-axis induces dramatic changes in the output polarization patterns, breaking their symmetry along the orthogonal y-axis. Figure 3 shows drastic displacement-induced separation of the two spin states (positive and negative S_3) in the linearly x-polarized light, i.e., the *giant spin-Hall effect of light*. Instead of a tiny shift of the light position caused by switching between different spin states, which is typical for the usual spin-Hall effect [1,2], we

observe *macroscopic* redistribution of spins caused by the *subwavelength* displacement of the scatterer.

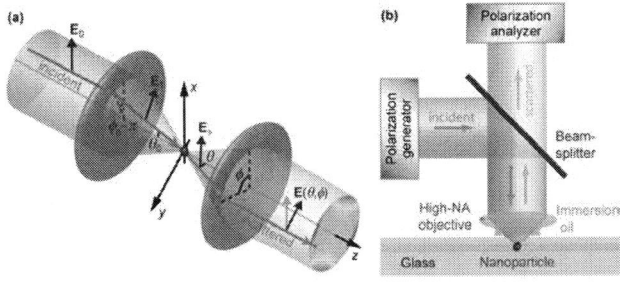

Fig. 1. Schematics of the microscopy of a nanoparticle, which involves the SOI of polarized light: (a) The "lens-scatterer-lens" system underlying the theoretical model; (b) the experimental setup in the reflection configuration, where the focusing objective is also used as a collector lens and a beam splitter separates incident and scattered light.

We describe the system by means of a simple analytical model based on the geometrical transformations of light in the system. It relates the output space-variant polarization to the input state of light via the Jones matrix $\hat{J}(\theta, \phi, \rho_s)$ which depends on the spherical angles of the exit pupil and dimensionless coordinates of the nano-particle, $\rho_s = k(x_s + iy_s)$. In the spin basis of circular polarizations it takes the form [7]

$$\hat{J} = \frac{A}{\sqrt{\cos\theta}}\begin{pmatrix} a & -be^{-2i\phi} \\ -be^{2i\phi} & a \end{pmatrix} - i\frac{B\sin\theta}{2\sqrt{\cos\theta}}\begin{pmatrix} \rho_s e^{-i\phi} & \rho_s^* e^{-i\phi} \\ \rho_s e^{i\phi} & \rho_s^* e^{i\phi} \end{pmatrix}$$

where A and B are aperture-dependent constants, $a = \cos^2(\theta/2)$, $b = \sin^2(\theta/2)$. Here the first term describes the spin-to-orbital AM conversion at $\rho_s = 0$ (Fig. 2), whereas the second term accounts for the spin-Hall effect caused by the displacement of the particle (Fig. 3).

978-1-4244-7043-3/10 $26.00 © 2010 IEEE

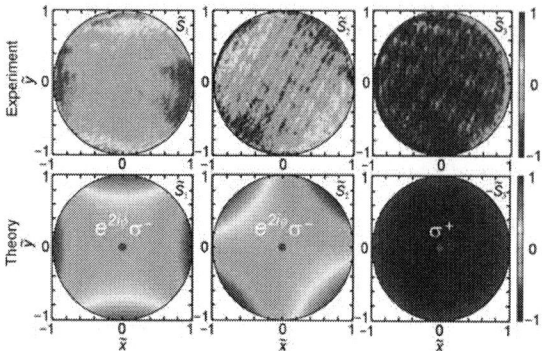

Fig. 2. Spin-to-orbital AM conversion of light when the nanoparticle is located precisely in the focal point. Experimentally measured and theoretically calculated from Eq. (1) distributions of the normalized Stokes parameters are shown for the case of circularly R-polarized incident light. We use normalized coordinates and the sign difference in S_3 between the theory and experiment arises from the helicity flip in the reflection configuration.

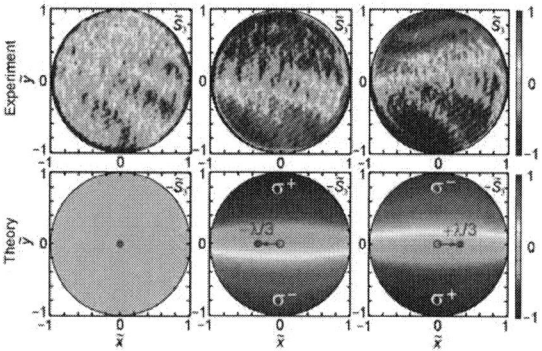

Fig. 3. Giant spin-Hall effect of light caused by subwavelength displacements of the nanoparticle. Experimentally measured and theoretically calculated, spatial distributions of the SAM density S_3 are shown for the case of linearly x-polarized incident light. Left-hand, middle, and right-hand panels correspond to the positions of the particle $x_s=0$, $-\lambda/3$, $\lambda/3$, respectively.

To summarize, we have shown that far-field imaging of nanoobjects with polarized light is accompanied by strong SOI and AM conversion. This brings forth a new type of optical probing, whereby polarization degrees of freedom of light carry and reveal subwavelength information about the specimen.

REFERENCES

[1]   P. Hosten and O. Kwiat, "Observation of the spin Hall effect of light via weak measurements," *Science*, vol. 319, pp. 787-790, 2008.

[2]   K. Y. Bliokh, A. Niv, V. Kleiner, and E. Hasman, "Geometrodynamics of spinning light," *Nature Photon.*, vol. 2, pp. 748-753, 2008.

[3]   Y. Zhao, J. S. Edgar, G. D. M. Jeffries, D. McGloin, and D. T. Chiu, "Spin-to-orbital angular momentum conversion in a strongly focused optical beam," *Phys. Rev. Lett.*, vol. 99, p. 073901, 2007.

[4]   K.Y. Bliokh, Y. Gorodetski, V. Kleiner, and E. Hasman, "Coriolis effect in optics: Unified geometric phase and spin-Hall effect," *Phys. Rev. Lett.*, vol. 101, p. 030404, 2008.

[5]   D. Haefner, S. Sukhov, and A. Dogariu, "Spin Hall effect of light in spherical geometry," *Phys. Rev. Lett.*, vol. 102, p. 123903, 2009.

[6]   O. G. Rodríguez-Herrera, D. Lara, and C. Dainty, "Far-field polarization-based sensitivity to sub-resolution displacements of a sub-resolution scatterer in tightly focused fields," *Opt. Express*, vol. 18, pp. 5609-5628, 2010.

[7]   O. G. Rodríguez-Herrera, D. Lara, K. Y. Bliokh, E. A. Ostrovskaya, and C. Dainty, "Optical nano-probing via spin-orbit interaction of light," *Phys. Rev. Lett.*, vol. 104, p. 253601, 2010.

[8]   E. Brasselet, N. Murazawa, H. Misawa, and S. Juodkazis, "Optical vortices from liquid crystal droplets," *Phys. Rev. Lett.*, vol. 103, p. 103903, 2009.

Masses of Polaritons

Igor V. Dzedolik

Taurida National V. I. Vernadsky University,
4, Vernadsky Avenue, Simferopol, 95007, Ukraine

Abstract - Linear and nonlinear models of dielectric media and electromagnetic field interaction based on the polariton approach are considered. It is shown that polaritons get the mass which quantity depends on the efficiency of electromagnetic field and medium interaction. The mass of polariton can be controlled by the external electric field.

I. INTRODUCTION

The particle mass as excitation of the field has drawn attention in connection with possibility to explain the origin of the mass of an elementary particle, and also the origin of quasi-particle mass when the electromagnetic field propagates in the waveguide or resonator [1 - 5]. The requirement of the relativistic invariance of the wave phase $\omega t - \mathbf{k} \cdot \mathbf{r}$ corresponding to the particle (quasi-particle) brings to the necessity of existence of invariant four-vector $\omega^2 / c^2 - k^2 = \kappa^2 = \text{const}$. If the requirement $\kappa^2 > 0$ is completed, the velocity of the wave packet featuring the particle (quasi-particle) as the material object must be less than the velocity of light in vacuum $d\omega / dk = ck(k^2 + \kappa^2)^{-1/2} < c$ [1]. Such dispersion law $\tilde{E}^2 / c^2 - p^2 = \hbar^2 \kappa^2 = m^2 c^4$ leads to the conclusion of a "rest mass" existence $m = \hbar \kappa / c$ of the particle in its "intrinsic" frame of reference, where the particle momentum is equal zero. The rest mass m characterizes both inertial and gravitational properties of the particle.

The quasi-particles named polaritons [6 - 7] are generated at the propagation of electromagnetic field in the ionic crystals, semiconductors, amorphous dielectric and another media. The ions of crystal lattice interact with the electromagnetic field due to their transversal high-frequency oscillations, and as a result the bound states of transversal optical phonons with the photons arise. The energy of falling electromagnetic field is redistributed between mechanical oscillations of the crystal lattice and the electromagnetic field in the medium. The polaritons represent the collective field excitations in medium, and it is possible to consider them as the quantum-field particles.

In a medium or in a waveguide it is possible to define the polariton mass by the transition to its intrinsic frame of reference in which it practically is rested, because the polariton moves with the velocity less than the velocity of light in vacuum. There is a standing wave in the resonator and in this case it is possible to measure the polariton mass in the laboratory frame of reference. The polariton mass is "rest mass", that is the true mass of a particle [8] which is considered in the special theory of relativity (STR). According to STR the mass of a particle can be expressed through its energy and momentum $m = c^{-2}(\tilde{E}^2 - c^2 p^2)^{1/2}$. The polariton mass is not equal zero, when the requirement $\tilde{E} = \hbar \omega \neq 0$ is completed at equating to zero the longitudinal component of wave vector $k_{\parallel} = 0$, i.e. when the velocity of polariton is equal zero in the chosen frame of reference. There is a dispersion equation for the plane wave $\mathbf{E} = \mathbf{E}_w \exp(-i\omega t + ikz)$ in the unbounded medium with permittivity $\varepsilon = \text{const}$ in the laboratory frame of reference $\varepsilon \omega^2 c^{-2} - k^2 = 0$, so the energy of a quasi-particle is $\tilde{E} = c\hbar k / \sqrt{\varepsilon}$. In this model the quasi-particle corresponds to a "photon in medium", its mass is equal zero $m = c^{-2}\tilde{E}(k = 0) = 0$.

II. LINEAR MODEL

We will consider the model of unbounded dielectric medium in which the mechanical oscillations propagate because the atoms are bounded by electromagnetic field and the field changes the dipole moment of the atom. Suppose the relation of the medium polarization vector and the vector of volume element displacement of medium is linear. Then in the homogeneous and isotropic dielectric medium or unbounded crystal medium of cubic system, where the external electric field $\mathbf{E}_0 = \text{const}$ is present, we can obtain the dispersion equation for polaritons by the method [6]

$$\omega_{\pm} = \left\{ \frac{1}{2}\left(\Omega_{\parallel}^2 + \frac{c^2 k^2}{\varepsilon_{\infty} \pm 4\pi\chi_2 E_0} \right) \pm \left[\frac{1}{4}\left(\Omega_{\parallel}^2 + \frac{c^2 k^2}{\varepsilon_{\infty} \pm 4\pi\chi_2 E_0} \right)^2 - \frac{c^2 \Omega_{\perp}^2 k^2}{\varepsilon_{\infty} \pm 4\pi\chi_2 E_0} \right]^{1/2} \right\}^{1/2},$$

(1)

where ε_{∞} is the high-frequency permittivity of medium, $\Omega_{\perp}, \Omega_{\parallel}$ are the oscillation frequencies of transversal and longitudinal phonons, $\Omega_{\parallel} = \Omega_{\perp}\left[\varepsilon_0 \left(\varepsilon_{\infty} \pm 4\pi\chi_2 E_0\right)^{-1}\right]^{1/2}$, χ_2 is the dielectric susceptibility featuring the linear electro-optical effect of Pockels. The polariton spectrum is presented in Fig. 1.

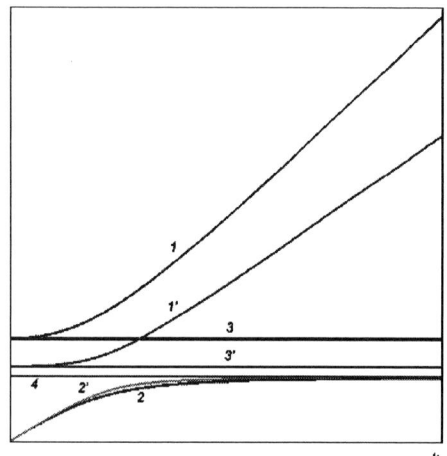

Fig. 1. The polariton spectrum in dielectric medium: curves 1, 1' are the upper branches of polaritons $\omega_+(k)$, curves 2, 2' are the lower branches of polaritons $\omega_-(k)$, straight lines 3, 3' are the longitudinal phonons $\Omega_{\parallel}$, straight line 4 is the transversal phonons $\Omega_{\perp}$.

The frequency of polaritons is equal zero $\omega_-(k = 0) = 0$ for the lower branch (minus-sign in expression (1)) at $k = 0$, that is the mass of such polariton is equal zero $m_- = 0$. But the frequency does not tend to zero for the upper branch (plus-sign in (1)) at $k = 0$, and it is equal the frequency of longitudinal phonons $\omega_+(k = 0) = \left[\varepsilon_0\left(\varepsilon_{\infty} \pm 4\pi\chi_2 E_0\right)^{-1}\right]^{1/2}\Omega_{\perp}$. The polariton mass $m_+ = c^{-2}\hbar\Omega_{\parallel} = c^{-2}\hbar\left[\varepsilon_0\left(\varepsilon_{\infty} \pm 4\pi\chi_2 E_0\right)^{-1}\right]^{1/2}\Omega_{\perp}$ of the upper branch of spectrum depends on intensity of external electric field E_0. The polariton

mass of this type is not equal zero in the laboratory frame of reference as it is caused by mechanical oscillations, instead of the electromagnetic field oscillations. The part of energy of the electromagnetic field is transported by the photons (the massless field) in the medium, and another part of the energy is bounded in the medium and transported by the polaritons (the massive field).

We can calculate the mass m_+ by measuring medium permittivity $\varepsilon_{E_0} = \varepsilon_{\infty} \pm 4\pi\chi_2 E_0$ at $E_0 = 0$ and $E_0 \neq 0$. The displacement of interference fringes depends on the phase difference $L\left(\sqrt{\varepsilon_{E_0}} - \sqrt{\varepsilon_{\infty}}\right) = N\lambda$, where L is the transparent sample length, N is the number of shift fringes in the Mach-Zehnder interferometer, λ is the wavelength in the air. Then we express the polariton mass as $m_+ = c^{-2}\hbar\sqrt{\varepsilon_0}\,\Omega_{\perp}\left(\sqrt{\varepsilon_{\infty}} + N\lambda / L\right)^{-1}$. The parameters of sample (a cube $3 \times 3 \times 3 = mm^3$ of the crystal $SrNb_2O_6 \rightarrow SrO \cdot Nb_2O_5$) are $\varepsilon_0 = 13.3$, $\varepsilon_{\infty} = 3.3$, $\Omega_{\perp} = 4.0 \cdot 10^{13}\,s^{-1}$. We've measured the number of fringe shift $N = 1$ for wavelength $\lambda = 0.633\,\mu m$ and $U_0 = 1100V$ ($E_0 = 3667V / cm$). Thus, the polariton mass is $m_+ = 0.95 \cdot 10^{-34}\,g$ in the sample.

III. NONLINEAR MODEL

Let us analyze the states of polariton field by examining the functional $U = \varepsilon_{\infty}\mathbf{E} \cdot \mathbf{E}^* + 2\pi\chi_3\left(\mathbf{E}^* \cdot \mathbf{E}\right)^2$ as the field potential in the nonlinear dielectric medium with self-focusing at $\chi_3 > 0$ or self-defocusing at $\chi_3 < 0$ properties. Suppose that the electric field has only one transversal component E, the Lagrangian of the field $L = \left(\nabla E\right)^2 / 2 - \varepsilon E^2 / 2 - \chi E^4 / 4$ is symmetrical in regard to the mirror (discrete) transformation $E \rightarrow -E$. The potential of invariable field in space $U = \varepsilon E^2 / 2 + \chi E^4 / 4$ has the global minimum $dU / dE = 0$ at $E = 0$ in the self-focusing medium (Fig. 2, curve 1).

Let us name the lowest energy state of the field as the "vacuum". In the medium the polaritons as the field excitations are absent in the "vacuum" $E = 0$. The medium with the potential $U = \varepsilon E^2 / 2 - \chi E^4 / 4$ has two degenerated maxima at the values of field $E_{max} = \pm\sqrt{\varepsilon / \chi}$ and the vacuum state at $E = 0$ (Fig. 2, curve 2). The medium with $\chi > 0$ and high-frequency permittivity $\varepsilon < 0$ can be realized in the artificial mediums and metamaterials. The field potential $U = -\varepsilon E^2 / 2 + \chi E^4 / 4$ has two degenerated minima at the values of field $E_{min} = \pm\sqrt{\varepsilon/\chi}$, and it has the local maximum at the value of field $E = 0$ (Fig. 2, curve 3). The maximum state is unstable, because it is the saddle point; therefore the system jumps in one of stable minima.

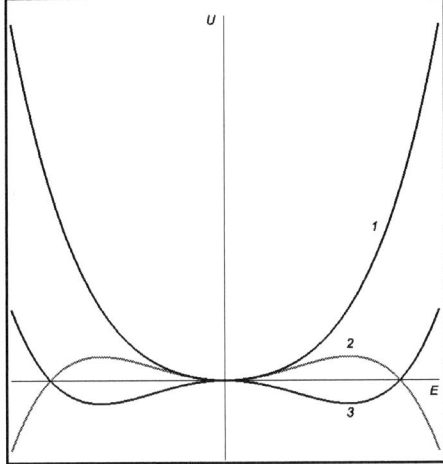

Fig. 2. The potential of the field: curve 1 in the medium with $\chi > 0, \varepsilon > 0$, curve 2 in the medium with $\chi < 0, \varepsilon > 0$, curve 3 in the medium with $\chi > 0, \varepsilon < 0$.

We can obtain the new Lagrangian $L = (\nabla E)^2 / 2 - \chi(E^2 - \varepsilon/\chi)^2 / 4$ adding the constant $\varepsilon^2 / 4\chi$, which does not change the field equations and remains symmetrical in regard to the mirror transformation $E \rightarrow -E$. But the system can be only in one of two vacuum states, either $E_+ = \sqrt{\varepsilon/\chi}$ or $E_- = -\sqrt{\varepsilon/\chi}$, i. e. the spontaneous symmetry breaking for the given system takes place. We move the system at the transformation $E \rightarrow -E$ in another vacuum. Let us present the field in the form of $E = E_1 + \sqrt{\varepsilon/\chi}$ then the Lagrangian looks like $L = (\nabla E_1)^2 / 2 - \varepsilon E_1^2 - \sqrt{\varepsilon\chi} E_1^3 - \chi E_1^4 / 4$. New field E_1 describes the excitations over vacuum $E_+ = \sqrt{\varepsilon/\chi}$ that are the quasi-particles with quadrate of mass $m^2 / 2 = \varepsilon$ (factor before quadratic term in the Lagrangian). The polariton mass in the nonlinear medium is $m = \sqrt{2\varepsilon}$.

The mass of polariton, that is the quasi-particle in dielectric medium, appears due to interaction of the electromagnetic field with medium. The appearance of polariton mass in medium is caused by its propagation with the velocity less than the velocity of light in vacuum. One can control the mass of polaritons applying the electric field into the dielectric medium. The spontaneous symmetry breaking of the system "electromagnetic field – dielectric medium" is possible at the nonlinear response of the medium with negative permittivity. The symmetry breaking of nonlinear system leads to the polariton mass arising. In the nonlinear medium the polariton mass value differs from the mass value in the linear medium.

ACKNOWLEDGMENT

The author thanks S. N. Lapayeva for the help at experiment.

REFERENCES

[1] D. I. Blokhintsev, "Fundamental particle and field," Uspekhi Phys. Nauk, vol. 42, pp. 76-92, 1950.
[2] A. S. Davydov, Theory of Molecular Excitons, Mc Graw-Hill, New York, 1962.
[3] L. A. Rivlin, "Photons in a waveguide (some imaginary experiments)," Uspekhi Phys. Nauk, vol. 167, 309-322, 1997.
[4] A. M. Gabovich, N. A. Gabovich, "How to explain the non-zero mass of electromagnetic radiation consisting of zero-mass photons," Eur. J. Phys., vol. 28, pp. 649-655, 2007.
[5] I. V. Dzedolik, "Mass of quasi-particle," Ukr. J. Phys. Opt., vol. 8, pp. 185-198, 2007.
[6] K. Huang, "On the interaction between the radiation field and ionic crystals," Proc. Roy. Soc., vol. A 208, pp. 352-365, 1951.
[7] J. J. Hopfield, "Theory of the contribution of excitons to the complex dielectric constant of crystals," Phys. Rev., vol. 112, pp. 1555-1567, 1958.
[8] L. B. Okun, "Concept of mass (The mass, energy, relativity)," Uspekhi Phys. Nauk., vol. 168, pp. 511-530, 1989.

Refraction of grey solitons at defocusing Kerr interfaces

Julio Sánchez-Curto,[1] Pedro Chamorro-Posada,[1] and Graham S. McDonald[2]

[1]Departamento de Teoría de la Señal y Comunicaciones e Ingeniería Telemática,
Universidad de Valladolid, ETSI Telecomunicación, Paseo Belén 15, Valladolid 47011, Spain
[2]Joule Physics Laboratory, School of Computing, Science and Engineering,
University of Salford, Salford M5 4WT, United Kingdom

Abstract—We present a generalized Snell's law that governs grey soliton refraction at the interface separating two defocusing Kerr media. The analysis, based on the Helmholtz theory, is valid for arbitrary angles of incidence and reveals that grey solitons undergo either external or internal refraction depending on the soliton contrast parameter.

I. INTRODUCTION

Nonlinear interfaces play a pivotal role in nonlinear science. The theory ruling the behaviour of spatial solitons at nonlinear interfaces is the cornerstone for the explanation of more complex phenomena such as the propagation in nonlinear waveguides [1], directional couplers [2] or all-optical gates [3], [4]. A vast literature on the subject lasts for three decades, since the pioneering works devoted to linear/nonlinear interfaces [5], [6], [7] till the recent studies on interfaces separating quadratic [8] or photorefractive [9] media. Those phenomena associated with nonlinear interfaces such as the formation of nonlinear surface waves [6], [10], [11], the emission of multisoliton patterns [12], [13], [14] and the existence of giant Goos Hänchen shift [15] have been analyzed in a large variety of media and interface configurations. Besides the theoretical works, experiments on nonlinear interfaces have been also carried out to verify analytical predictions [16], [17].

A review of the literature on nonlinear interfaces may conclude that two main features are found in a large number of works. First, most studies have relied on the Nonlinear Schrödinger (NLS) equation to describe soliton evolution in nonlinear media. Since the paraxial approximation is assumed in the NLS, the validity of the analysis is limited to vanishingly small angles of incidence or refraction. Such is the case, for instance, of the successful *particle-like* model of Aceves et al. [18], [19] where the complicated dynamical evolution of a beam at nonlinear interfaces is described by a simple Newtonian model. Secondly, the study of nonlinear interfaces has focused on bright solitons, so that the analysis of dark solitons has been only considered in a few number of works [20], [21].

Nonlinear interfaces have an inherent nonparaxial character which manifests, for instance, when a soliton acquires a large angle of propagation due to the phenomenon of external refraction induced by the nonlinear interface. This sort of nonparaxiality, associated to the propagation of broad beams (in relation to its wavelength) with large angles in relation to the reference axes, is properly described in the framework of the Helmholtz theory [22], [23]. Unlike the NLS, the Nonlinear Helmholtz (NLH) equation [24], [22] preserves rotations and removes previous angular limitations. The analysis of nonlinear interfaces based on the Helmholtz theory thus collects the full angular content of the problem which is summarized in a generalized Snell's law [25], [26]. Initially reported for addressing bright soliton refraction at the interface separating two focusing Kerr media [25], it has been recently extended to govern soliton refraction in defocusing Kerr media [27].

II. GENERALIZED SNELL'S LAW FOR GREY SOLITONS

Our analysis of grey soliton evolution at defocusing Kerr interfaces is based on the NLH equation

$$\kappa \frac{\partial^2 u}{\partial \zeta^2} + j\frac{\partial u}{\partial \zeta} + \frac{1}{2}\frac{\partial^2 u}{\partial \xi^2} - |u|^2 u = \left[\frac{\Delta}{4\kappa} - (1-\alpha)|u|^2\right] H(\xi)u \quad (1)$$

whose derivation for a TE optical field satisfying a Helmholtz equation can be found in [26]. $u(\xi, \zeta)$ is the complex envelope of a forward propagating beam evolving along the normalized transverse $\xi = 2^{1/2}x/w_0$ and longitudinal $\zeta = z/L_D$ coordinates. w_0 is the waist of a reference Gaussian beam with diffraction length $L_D = kw_0^2/2$ and $\kappa = 1/k^2w_0^2$ is a nonparaxiality parameter relating the beam width in relation to the number of wavelengths ($\lambda = 2\pi/k$) in the full width $2w_0$ of a reference Gaussian beam [22], [23]. $\Delta = 1 - n_{02}^2/n_{01}^2$ and $\alpha = \alpha_2/\alpha_1$ account for the linear and nonlinear refractive index mismatch at the interface, respectively. As it is displayed at both sides of the interface shown in Fig. 1, we have assumed that the refractive index of the ith medium is $n_i = n_{0i} - \alpha_i I$ where n_{0i} is the linear refractive index, α_i stands for the defocusing Kerr coefficient and I is the optical intensity.

Matching the phase of the solution solutions for focusing [22], [23], [28] and defocusing [29] Kerr media at both sides of the discontinuity one obtains a nonlinear Snell's law [25], [26], [27]

$$\gamma_\pm n_{01} \cos(\theta_{ni} + \theta_{0i}) = n_{02}\cos(\theta_{nt} + \theta_{0t}). \quad (2)$$

where θ_{ni} and θ_{nt} are the net angles of incidence and refraction of a grey soliton, respectively. They account for the total angle between the propagation direction of the soliton dip and the interface as it is illustrated by the grey arrows in Fig. 1. The interface separating two defocusing Kerr media is represented by a dotted line and has been rotated in relation to the normalized coordinates.

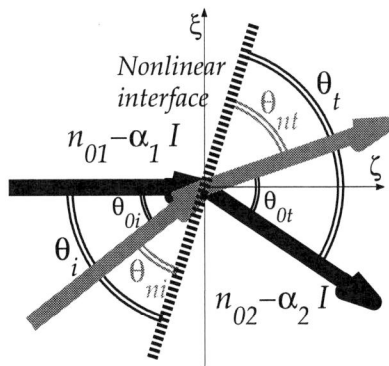

Fig. 1. Angular relationships involved in black and grey soliton refraction.

θ_{0i} and θ_{0t} represent the intrinsic angles of the incident [29] and refracted grey soliton [27] relative to the propagation direction of the background wave supporting the corresponding dark soliton. Their values are deduced from the condition supplementing Eq. (2), i.e. the preservation of the soliton grayness parameter at both sides of the interface [27]. θ_{0i} and θ_{0t} make net angles of incidence and refraction of the grey soliton differ from the ones associated to the background wave or the corresponding black soliton θ_i and θ_t, respectively. Of course, the case of bright [25], [26] and black [27] soliton refraction can be deduced from Eq. (2) provided $\theta_{0i} = \theta_{0t} = 0$.

In Eq. (2) $\gamma_\pm$ is a nonlinear correction term which has been previously calculated for bright (γ_+) [25] and black (γ_-) [27] solitons. In this case,

$$\gamma_- = \left[\frac{(1 - 4\kappa u_0^2)}{1 - 4\kappa u_0^2 \alpha (1 - \Delta)^{-1}} \right]^{1/2} \qquad (3)$$

where u_0 denote the amplitude of the background wave supporting the soliton.

The nonparaxiality parameter κ has been shown to play a pivotal role in the study of bright and black soliton refraction at nonlinear interfaces [25], [26]. The case of grey solitons is not an exception as the results shown in Fig. 2 reveal. Eq. (2) is represented for different values of the soliton contrast parameter F [29], [27] when two values of κ are used, i.e. $\kappa = 10^{-4}$ (a) and $\kappa = 10^{-3}$ (b). We assume a dark soliton $u_0 = 1$ impinging a nonlinear interface with $\alpha_2 = 4\alpha_1$ and $n_{02} = 1.0124 n_{01}$ Two different scenarios are found depending solely on κ. When $\kappa = 10^{-4}$, $\gamma_- \approx 1$ and the net angle of refraction is dictated in Eq. (2) by the relationship between the linear refractive indexes. Regardless of the value of the soliton

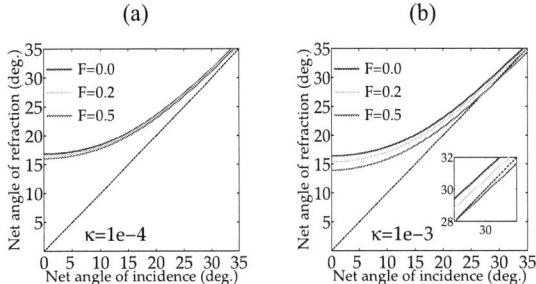

Fig. 2. Generalized Snell's law for a nonlinear interface with $\alpha_2 = 4\alpha_1$ and $n_{02} = 1.0124 n_{01}$ for $\kappa = 10^{-4}$ (a) and $\kappa = 10^{-3}$ (b).

greyness, all soliton will undergo a very similar net angle of refraction as it is shown in Fig. 2(a). This result changes when $\kappa = 10^{-3}$ as Fig. 2(b) reveals. In this case $\gamma_- \approx n_{02}/n_{01}$, so that the net angle of refraction is dependent on the intrinsic angles of the incident and refracted solitons θ_{0i} and θ_{0t}. The curve for $F = 0.5$ shows a novel feature not previously found in black or bright soliton refraction. Unlike black solitons, which experience either external or internal refraction, grey solitons may undergo both types of refraction when the net angle of incidence changes.

The inset of Fig. 2(b) also reveals that grey soliton refraction depends on F. For a fixed net angle of incidence $\theta_{ni} = 30°$, the grey soliton with $F = 0.2$ undergoes external refraction while internal refraction is achieved when $F = 0.5$. This conclusion is in excellent agreement with the results extracted from numerical simulations shown in Fig. 3. The two snapshots in Fig. 3(a) and (b) are obtained from the numerical integration of the NLH [30] and correspond to those interface and soliton parameters used in Fig. 2(b).

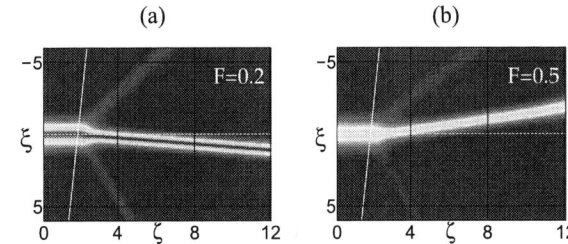

Fig. 3. Change from external (a) to internal (b) refraction when F increases. In both cases, $\alpha_2 = 4\alpha_1$, $n_{02} = 1.0124 n_{01}$ and $\kappa = 10^{-3}$.

A full description of those properties associated to the evolution of grey solitons at defocusing Kerr interfaces will be presented at the conference. Massive numerical simulations have been carried out in order to contrast the validity of the theoretical predictions.

III. CONCLUSION

In this work we have demonstrated the validity of the Helmholtz Snell's law to address not only bright or black, but also grey soliton refraction at the interface separating two Kerr media. Deduced in the framework of the Helmholtz theory, this Snell's law is valid for all possible values of any of the angles involved in contrast to the angular restrictions inherent in the paraxial theory. We have showed that a change from external to internal refraction is allowed in grey soliton refraction dependent on the angle of incidence.

ACKNOWLEDGMENT

The authors would like to thank the support provided by the Spanish Ministerio de Educación y Ciencia and Fondo Europeo de Desarrollo Regional, project TEC2007-67429-C02-01, and Junta de Castilla y León, project VA001A08.

REFERENCES

[1] D. Mihalache, M. Bertolotti, and C. Sibilia, *Nonlinear Wave Propagation in Planar Structures*, ser. Progress in Optics XXVII. Elsevier Science Publishers, 1989.

[2] A.B. Aceves, J.V. Moloney, and A.C. Newell, "Theory of light-beam propagation at nonlinear interfaces. II. Multiple-particle and multiple-interface extensions," *Physical Review A*, vol. 39, no. 4, pp. 1828–1840, 1989.

[3] J. Scheuer and M. Orenstein, "Interactions and switching of spatial soliton pairs in the vicinity of a nonlinear interface," *Optics Letters*, vol. 24, no. 23, pp. 1735–1737, December 1999.

[4] ——, "All-optical gates facilitated by soliton interactions in a multilayered Kerr medium," *Journal of the Optical Society of America B*, vol. 22, no. 6, pp. 1260–1267, June 2005.

[5] A.E. Kaplan, "Theory of hysteresis reflection and refraction of light by a boundary of a nonlinear medium," *Sov. Phys. JETP*, vol. 45, no. 5, pp. 896–905, May 1977.

[6] W.J. Tomlinson, "Surface wave at a nonlinear interface," *Optics Letters*, vol. 5, no. 7, pp. 323–325, July 1980.

[7] W.J. Tomlinson, J.P. Gordon, P.W. Smith, and A.E. Kaplan, "Reflection of a gaussian beam at a nonlinear interface," *Applied Optics*, vol. 21, no. 11, pp. 2041–2051, June 1982.

[8] I.V. Shadrivov and A.A. Zharov, "Dynamics of optical spatial solitons near the interface between two quadratically nonlinear media," *Journal of the Optical Society of America B*, vol. 19, no. 3, pp. 596–602, March 2002.

[9] A.D. Boardman, P. Bontemps, W. Ilecki, and A.A. Zharov, "Theoretical demonstration of beam scanning and switching using spatial solitons in a photorefractive crystal," *Journal of Modern Optics*, vol. 47, no. 11, pp. 1941–1957, 2000.

[10] P. Varatharajah, A.B. Aceves, J.V. Moloney, and E.M. Wright, "Stationary nonlinear surface waves and their stability in diffusive Kerr-like nonlinear media," *Journal of the Optical Society of America B*, vol. 7, no. 2, pp. 220–229, February 1990.

[11] Y.V. Kartasov, F. Ye, V.A. Vysloukh, and L. Torner, "Surface waves in defocusing thermal media," *Optics Letters*, vol. 32, no. 15, pp. 2260–2262, August 2007.

[12] N.N. Rozanov, "Nonlinear reflections and transmission of limited light beams," *Opt. Spectrosc.*, vol. 47, pp. 335–337, September 1979.

[13] E.M. Wright, G.I. Stegeman, C.T. Seaton, J.V. Moloney, and A.D. Boardman, "Multisoliton emission from a nonlinear waveguide," *Physical Review A*, vol. 34, no. 5, pp. 4442–4444, November 1986.

[14] Y.M. Aliev, A.D. Boardman, K. Xie, and A.A. Zharov, "Conserved energy approximation to wave scattering by a nonlinear interface," *Physical Review E*, vol. 49, no. 2, pp. 1624–1633, February 1994.

[15] M. Peccianti, G. Assanto, A. Dyadyusha, and M. Kaczmarek, "Nonlinear shift of spatial solitons at a graded dielectric interface," *Optics Letters*, vol. 32, no. 3, pp. 271–273, February 2007.

[16] O. Emile, T. Galstyan, A. Le Floch, and F. Bretenaker, "Measurement of the nonlinear Goos-Hänchen effect for gaussian optical beams," *Physical Review Letters*, vol. 75, no. 8, pp. 1511–1513, August 1995.

[17] E. Alvarado-Méndez, R. Rojas-Laguna, J.G. Aviña Fernández, M. Torres-Cisneros, J.A. Andrade-Lucio, J.C. Pedraza-Ortega, E.A. Kuzin, J.J. Sánchez-Mondragón, and V. Vysloukh, "Total internal reflection of spatial solitons at interface formed by a nonlinear saturable and a linear medium," *Optics Communications*, vol. 193, pp. 267–276, June 2001.

[18] A.B. Aceves, J.V. Moloney, and A.C. Newell, "Reflection, transmission, and stability characteristics of optical beams incident upon nonlinear dielectric interfaces," *Journal of the Optical Society of America B*, vol. 5, no. 2, pp. 559–564, February 1988.

[19] ——, "Reflection and transmission of self-focused channels at nonlinear dielectric interfaces," *Optics Letters*, vol. 13, no. 11, pp. 1002–1004, November 1988.

[20] S.R. Skinner and D.R. Andersen, "Stationary fundamental dark surface waves," *Journal of the Optical Society of America B*, vol. 8, no. 4, pp. 759–764, April 1991.

[21] Y. Chen, "Bright and dark surface waves at a nonlinear interface," *Physical Review A*, vol. 45, no. 7, pp. 4974–4978, April 1992.

[22] P. Chamorro-Posada, G.S. McDonald, and G. New, "Nonparaxial solitons," *Journal of Modern Optics*, vol. 45, no. 6, pp. 1111–1121, 1998.

[23] ——, "Propagation properties of non-paraxial spatial solitons," *Journal of Modern Optics*, vol. 47, no. 11, pp. 1877–1886, 2000.

[24] G. Fibich, "Small beam nonparaxiality arrests self-focusing of optical beams," *Physical Review Letters*, vol. 76, no. 23, pp. 4356–4359, June 1996.

[25] J. Sánchez-Curto, P. Chamorro-Posada, and G.S. McDonald, "Helmholtz solitons at nonlinear interfaces," *Optics Letters*, vol. 32, no. 19, pp. 1126–1128, May 2007.

[26] ——, "Nonlinear interfaces: intrinsically nonparaxial regimes and effects," *Journal of Optics A: Pure and Applied Optics*, vol. 11, p. 054015, 2009.

[27] ——, "Dark solitons at nonlinear interfaces," *Optics Letters*, vol. 35, no. 9, pp. 1347–1349, April 2010.

[28] P. Chamorro-Posada, G.S. McDonald, and G. New, "Exact soliton solutions of the nonlinear Helmholtz equation," *Journal of the Optical Society of America B*, vol. 19, no. 5, pp. 1216–1217, 2002.

[29] P. Chamorro-Posada and G.S. McDonald, "Helmholtz dark solitons," *Optics Letters*, vol. 28, no. 10, pp. 825–827, 2003.

[30] P. Chamorro-Posada, G.S. McDonald, and G. New, "Non-paraxial beam propagation methods," *Optics Communications*, vol. 192, pp. 1–12, 2001.

CAOL*2010 International Conference on Advanced Optoelectronics & Lasers, 10-14 September, 2010, Sevastopol, Ukraine

Localized states and oscillations induced by coherent interaction of waves in nonlocal media

S. Bugaychuk[1], R. Conte[2], E. Kozij[3], O. Kolesnyk[1], G. Klimusheva[1]

[1]Kharkov National University of Radio Electronics, Kharkov, Ukraine

[1]Institute of Physics of NAS Ukraine, Kiev, Ukraine

[2]LRC MESO, École normale supérieure de Cachan (CMLA);
Service de physique de l'état condense (CNRS URA 2464), Cedex, France,

[3]East Ukraine Volodymyr Dahl National University, Lugansk, Ukraine

Abstract: We show the dynamical four-wave mixing in a nonlocal medium is described by the complex Ginzburg-Landau equation. Two regimes are considered for this FWM: (i) the self-oscillations and (ii) stationary localized states in the form of sech-function for the intensity pattern in the medium volume. The applications of such FWM in all-optical switching and optical phase conjugation are examined.

I. INTRODUCTION

The property of spatial localization of the grating amplitude profile formed during dynamical four-wave mixing (FWM) in photorefractive media with nonlocal response (Fig.1) has been noted in several papers [1-4]. The damped sine-Gordon equation was obtained to describe the dynamics of the FWM in a nonlocal medium [4-6]. A regime of self-oscillations was found and investigated theoretically for such scheme [6, 7]. The first experimental observation of the spatial localization of the grating amplitude was provided in [7].

In the present paper we show that the damped sine-Gordon equation can be transformed to the cubic complex Ginzburg-Landau equation, which describes the properties of the dynamical FWM. We investigate the conditions required to get the self-oscillation regime. Also we investigate the properties of spatial localized states both for the intensity pattern and for the grating amplitude distribution that are described by a sech-function along the medium volume (the axis z). We show that the stationary distribution is controlled by the input intensity ratio. This fact gives methods to optimized different applications of the FWM. We consider here all-optical switching scheme and optical phase conjugation.

II. COMPLEX GINZBURG-LANDAU EQUATION DISCRIBED DYNAMICAL FWM IN NONLOCAL MEDIA

Transmission dynamical FWM is described by the set of four coupled-wave equations [7, 8]

$$\partial_z A_1 = -iEA_2, \partial_z \overline{A}_2 = iE\overline{A}_1, \partial_z \overline{A}_3 = -iE\overline{A}_4, \partial_z A_4 = iEA_3 \ (1)$$

and one evolution equation for the grating amplitude

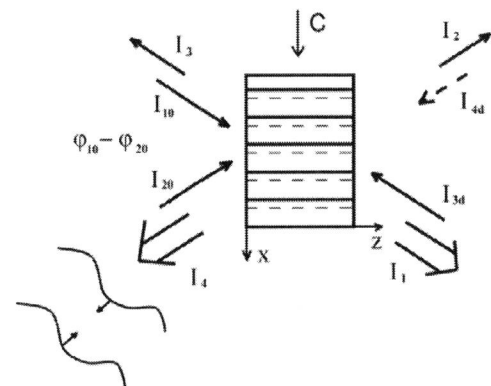

Fig. 1. The conventional scheme of the degenerate FWM in the transmission geometry. Straight lines are the maximums of the interference pattern, dash lines are the maximums of the grating amplitude. C is the polar axis. $I_{10}, I_{20}, I_{3d}, I_{4d}$ are the input waves, I_1, I_2, I_3, I_4 are the output waves. The usual assumption is $I_{4d}=0$.

$$\partial_t E = \gamma I_m / I_0 - E / \tau \qquad (2)$$

where the intensity pattern is $I_m = A_1 \overline{A}_2 + \overline{A}_3 A_4$. Here $A_j(t,z), j = 1,2,3,4$ are the slow variable amplitudes of the interacted waves normalized to the total intensity $I_0 = |A_1|^2 + |A_2|^2 + |A_3|^2 + |A_4|^2 = const$; $E(t,z)$ is the amplitude of the dynamical grating; τ is the time relaxation constant of the grating; γ is the photorefractive gain normalized to the time constant τ. In a nonlocal medium γ is a complex value.

We consider here a particular case of a pure nonlocal response when $\gamma = i\gamma_N$ is the pure imaginary. Then the system (1)-(2) is reduced to the damped sine-Gordon equation [4-8]:

$$\partial_t \partial_z u + \partial_z u / \tau - K \sin(2u + \alpha) = 0 \qquad (3)$$

978-1-4244-7043-3/10 $26.00 © 2010 IEEE   103

where $|E| = \partial_z u$ and the both coefficients K and α are depended on t. This equation has a solution in a form of bright soliton [8]:

$$I_m = |E| / (\gamma_N \tau) = -i e^{-i(t-t_0)/\tau} \cdot \tau K \sec h(2\tau K(z - z_0)) \quad (4)$$

The damped sine-Gordon equation (3) is transformed to the cubic complex Ginzburg-Landau equation [8]:

$$
\begin{aligned}
& i \frac{\partial \psi}{\partial \eta} + \frac{2}{q^3} \frac{\partial^2 \psi}{\partial \zeta^2} + \frac{4}{q} |\psi|^2 \psi \cdot e^{-2T_0/\tau - 2\Im(q)Z_0} \\
& = -i \frac{2\Im(q)}{|q|^2} \frac{k_2}{k_0} \psi
\end{aligned}
\quad (5)
$$

where the gain/loss coefficient k_2 is time depended and the wave-vector q is a complex constant. This way various solutions of the complex Ginzburg-Landau equation [9] may find their practical realizations during dynamical FWM in nonlocal media.

III. SELF-OSCILLATIONS IN MEDIA WITH STRONG RESPONSE

Stable oscillations (see Fig.2) are appeared in some area of input parameters due to influence of white noise to the phases of interacted waves:

$$\Phi(0) = \varphi_{10} - \varphi_{20} = \Phi_0 + \Lambda f(t), \Lambda \sim \pi/3 \div \pi/100 \quad (6)$$

where Λ is a scalar, $f(t)$ is a white noise. The amplitude and the period of the oscillations depend on the input conditions: the value of the coupling constant and the intensity ratio. The reason of the oscillation behaviour is the emergence of a local component of the grating that causes the changes of wave phases during their propagation; the light contrast changes with time, and the grating is erased and rerecorded repeatedly.

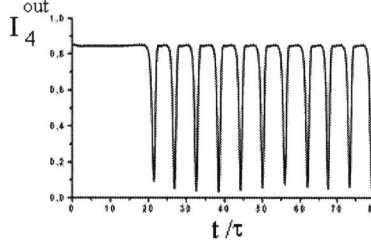

Fig. 2. The self-oscillations of the output intensity. $I_{10}/I_{20} = 3$, $I_{3d} = 0.87$, $I_{4d} = 0$, $\gamma_N d = 15$.

IV. CONTROL OF OPTICAL BEAMS

In the steady state the solution for the intensity pattern and the grating amplitude distribution has a form of the sech-function [4-5, 7]:

$$I_m = E / \gamma_N = C / \cosh(2\gamma_N C z - p),$$
$$tg(u) = \exp(2\gamma_N C z - p) \quad (7)$$

where C and p are the constants defined by the input conditions of a FWM scheme. The main parameter is u_d, which is the integral under the grating amplitude profile over the medium boundaries. The value u_d determines the diffraction efficiency:

$$u_d = \int_0^d E(z)dz, \quad (8)$$

for the two-wave mixing:

$$\eta = \sin^2[\gamma_N u_d + arctg(A_{10}/A_{20})]$$

The value u_d depends on the location of the stationary sech-function of $E(z)$ relative to the crystal boundaries $z = 0 \div d$. To characterize FWM schemes with different input conditions we introduce new parameters: the position of the grating amplitude maximum Z_0 and the half-width of the grating amplitude distribution Δ:

$$Z_0 = p/2\gamma_N C, \quad \Delta = \ln(3 + \sqrt{2})/\gamma_N C \quad (9)$$

The calculations of the value u_d in dependence of these parameters for the FWM schemes with two different input conditions are combined in Fig.3. However only a part of the presented curves will correspond to any concrete scheme. This way one can choose optimal input conditions for different applications.

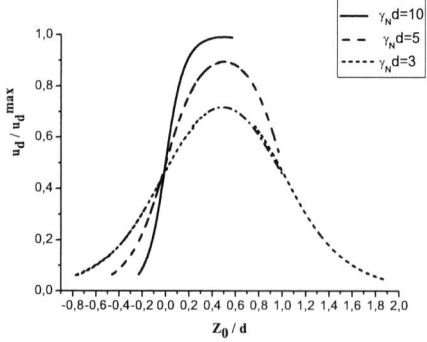

Fig. 3. The normalized value u_d calculated for the FWM with two driven beams when their input intensity ratio is changed on the 3 orders of magnitude.

Optical switching. As it is following from the Fig.3 the optical switching is realized if to change u_d from its maximum value to its minimum value with the help of a guided beam. This situation occurs on the base of the transmission two-wave mixing scheme, where input beams I_1 and I_2 record the grating. But the third beam I_{3d} is the guided one, which switch on/off the value of the u_d and by this way the diffraction efficiency (see Fig.4).

Optical phase conjugation (OPC). OPC is realized when $u_d = u_{d\max}$ over all range of changes of the input intensities. The optical scheme is on the base of the double phase conjugation mirror (DPCM) when $I_{10} = I_{4d} = 0$. The DPCM scheme is unstable. But introducing a weak beam $I_{10} \neq 0$ makes the stabilization of this scheme (see Fig.5 a,b).

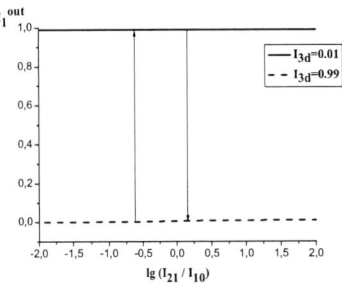

Fig. 4. All-optical switching on the base of two-wave mixing scheme. $I_{10}+I_{20}+I_{3d}=1$, $I_{4d}=0$, $\gamma_N d=10$.

The CGLE governs the spatiotemporal dynamics of the spatially localized interference pattern formed by the FWM. These properties of degenerate FWM in nonlocal media may find numerous applications, e.g. for optical phase conjugation, all-optical switching, manipulation of laser pulses, optical logic, transmission of solitary waves through fibers etc.

REFERENCES

[1] D.I. Stasel'ko, V.G. Sidorovich, "Efficiency of transformation of light beams by dynamic volume phase gratings", *J. of Technical Physics*, No.3, p. 580, 1974 (in Russian).

[2] J.H. Hong, R. Saxema, "Diffraction efficiency of volume holograms written by coupled beams", *Opt. Lett.*, Vol. 16, p. 180, 1991.

[3] A.A. Zozulya, M. Saffman, D.Z. Anderson, " Propagation of light beams in photorefractive media: fanning, self-bending, and formation of self-pumped four-wave-mixing phase conjugation geometries", *Phys. Rev. Lett.*, Vol. 73, No. 6, p. 818, 1994.

[4] M. Jeganathan, M.C. Bashaw, L. Hesselink, "Evolution and propagation of grating envelopes during erasure in bulk photorefractive media", *J. Opt. Soc. Am. B*, Vol. 12, p. 1370, 1995.

[5] A. Zozulya, V.T. Tikhonchuk, "Investigation of stability of four-wave mixing in photorefractive media", *Phys.Lett. A*, Vol. 135, p. 447, 1989.

[6] A. Bledowski, W. Krolikowski, A. Kujawski, "Temporal instabilities in single-grating photorefractive four-wave mixing", *J. Opt. Soc. Am. B*, Vol. 6, p. 1544, 1989.

[7] S. Bugaychuk, L. Kovacs, G. Mandula, K. Polgar, R. Rupp, "Nonuniform dynamic gratings in photorefractive media with nonlocal response", *Phys. Rev. E*, vol. 67, p. 046603, 2003.

[8] S. Bugaychuk, R. Conte, "Ginzburg-Landau equation for dynamical four-wave mixing in gain nonlinear media with relaxation", *Phys. Rev. E*, Vol. 80, p. 066603, 2009.

[9] I.S. Aranson, L. Kramer, "The world of the complex Ginzburg-Landau equation", *Rev. Mod. Phys.*, Vol. 74, p. 99, 2002.

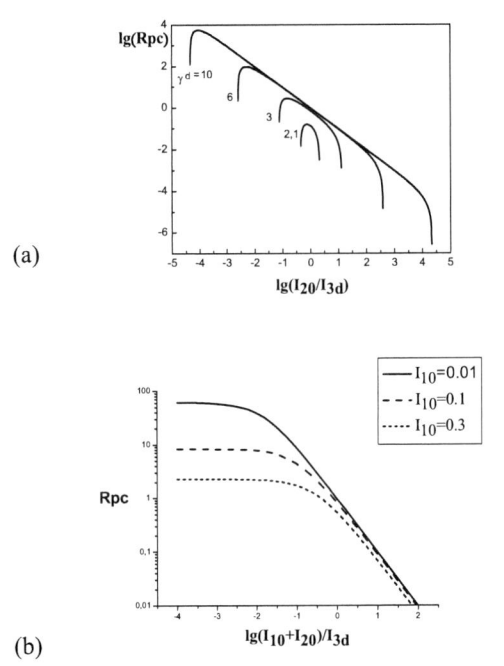

(a)

(b)

Fig. 5. The coefficient of the phase conjugation Rpc=I4(0)/I$_{20}$ in DPCM (a) and in the FWM scheme based on the DPCM scheme (b) - $\gamma_N d=6$, $I_{10}+I_{20}+I_{4d}=1$.

V. CONCLUSION

We show rigorously that a nonlinear system describing the degenerate wave mixing in a medium which possesses both a nonlocal response and relaxation is reduced to one nonlinear complex Ginzburg-Landau equation (CGLE). We develop the technique to obtain the cubic CGLE by using the reductive perturbation method for the nonlinear dynamical system described wave coupling of four waves.

Soliton based DPSK encoded sequences: Stability and dynamical properties

Jaroslaw E. Prilepsky
and Stanislav A. Derevyanko
Nonlinearity and Complexity Research Group
Aston University
B4 7ET Birmingham, UK
Emails: y.prylepskiy1@aston.ac.uk
s.derevyanko@aston.ac.uk

Sergei K. Turitsyn
Photonics Research Group
Aston University
B4 7ET Birmingham, UK
Email: s.k.turitsyn@aston.ac.uk

Abstract— We study the dynamical properties of the RZ-DPSK encoded sequences of bits, focusing on the instabilities in the train leading to the bit stream corruption. The problem is studied within the framework of the complex Toda chain model for optical solitons. We show how the bit composition of the pattern affects the initial stage of the train dynamics and explain the general mechanisms of the appearance of unstable collective soliton modes. Then we discuss the nonlinear regime using the asymptotic properties of the pulse stream at large propagation distances and analyze the dynamical behavior of the train elucidating different scenarios for the pattern instabilities.

I. INTRODUCTION

The differential phase-shift-keying (DPSK) encoding format has being actively discussed for the last couple of years [1], [2]. In particular in the return to zero (RZ) DPSK information is encoded by the phase difference between the two adjacent well separated optical pulses. Logical "ones" are encoded onto a π phase change, whereas a logical "zeros" correspond to zero phase shift. In order achieve the best quality of transmission any signal distortions must be minimized. There exists a multitude of sources leading to the signal garbling and in this paper we will be concerned with the intrachannel pulse interactions. However the study of this interaction is normally limited to the short patterns (typically triplets) and the pattern effects, i.e. collective excitations affecting the large sets of pulses are often neglected. It is already known that the density of marks in the message *does have an effect of the bit error rate* in the pulse overlapped system [3].

In this study we will be interested in how the initial bit composition of a (long) sequence of RZ-DPSK encoded pulses affects its dynamics i.e. in the rate at which pulses change its position and phase. Eventually pulses will leave their slots which leads to appearance of errors in the bit stream. We shall study these effects within the framework of the Nonlinear Schrodinger Equation (NSE). The NSE model is known to appear in the description of telecommunication systems operating at a relatively low bit rate (< 10 Gb/s) where it appears as an averaged model for the short-scale dispersion management [4].

When pulses are well separated it is often possible to reduce NSE to a system of ODEs describing propagation of

individual pulses weakly interacting via the rapidly decreasing tails. To achieve such a reduction one can use either different modifications of the collective coordinate approach [5] or variational approach [6]. Because the obtained reduced system is only valid for small pulse overlap it will therefore be suitable only for modelling the initial stages of the pulse collision. There is an important class of RZ systems where such a reduction has proven to be asymptotically exact – a train of weakly interacting classical soliton pulses in the NSE [7], [8], [9]. In this system it is possible to use the machinery of adiabatic soliton perturbation theory to reduce the dynamics of the individual parameters of each pulse to the integrable complex Toda chain (CTC).

When considering the initial stages of pulse evolution it is sufficient to linearize the CTC around the initial configuration which is determined by the bit pattern. Analyzing the eigenvalues of the emerging linear self-adjoint eigenproblem one can immediately find the unstable modes. The eigenmode with the highest increment of instability provides an estimate for the rate of the pulse distortion development and hence the maximum error-free propagation distance. We elucidate the structure of the emerging eigenvalue pattern for the linearized system versus the initial bit pattern.

Next we use the fact that the long time asymptotes of the solutions of nonlinear CTC are known [9] and the asymptotical behavior can be studied by examining the eigenvalues of another symmetric tridiagonal (generally non-Hermitian) matrix of the same dimensionality. We utilize this to provide additional long-distance estimates of the pattern stability. It is important to mention that in the normalized units the elements of both matrices depend only on the given bit pattern (and not the parameters of the system) so their eigenvalues are the universal quantities for each given pattern.

II. PROBLEM STATEMENT

NLSE with anomalous dispersion (in soliton units) reads as:

$$\frac{\partial Q}{\partial z} - \frac{i}{2}\frac{\partial^2 Q}{\partial t^2} - i|Q|^2 Q = 0, \qquad (1)$$

where $Q(t;z)$ is the optical pulse envelope in a single-mode silica fiber, z stands for the propagation distance and t is the time in the frame co-moving with the group velocity of the envelope. Time is normalized to the FWHM-time, $t/T_F \rightarrow t$, the power $W = |Q|^2$ – by the characteristic power $W_* = (\gamma L_D)^{-1} \equiv |\beta_2|/(\gamma T_F^2)$, where γ is the nonlinear coefficient in W^{-1}/km, $L_D = T_F^2/|\beta_2|$ is the dispersion length measured in km and β_2 is the dispersion parameter measured in ps^2/km. The propagation distance is also normalized to the dispersion length, $z/L_D \rightarrow z$. We deal with the input having the form DPSK-coded train of equidistant solitons having equal amplitudes (with zero overall train velocity):

$$Q(t;0) = \sum_{n=0}^{N} \frac{2\nu\,e^{i\delta_n}}{\cosh\left[2\nu(t - \xi_n)\right]}, \qquad (2)$$

and δ_n, ξ_n, and 2ν are the phase, position, and amplitude of a n-th individual soliton from the train, correspondingly; $N+1$ being the total number of solitons in the train. We suppose that $\exp(-\nu r) \ll 1$, r being the distance between the solitons.

Then we can make use of the simplified description for the dynamics of individual n-th soliton position, ξ_n, and phase, δ_n. Let us introduce a complex quantity $q_n(z)$ as:

$$q_n = -2\nu\xi_n(z) + 2n\ln 2\nu + i\left[n\pi - \delta_n(z)\right]. \qquad (3)$$

Then the dynamics of each constituent soliton of the train can be described by means of the integrable CTC [7], [8]:

$$\frac{d^2 q_n}{dz^2} = e^{q_{n+1} - q_n} - e^{q_n - q_{n-1}}, \qquad n = 1 \ldots N, \qquad (4)$$

where we have applied the renormalization: $4\nu z \rightarrow z$. The dynamics of the the complex variables q_n provides complete information about the positions, and phases of individual solitons and Eq.(4) describes the interaction induced jitters.

III. RESULTS OF THE CTC ANALYSIS

A. The initial stage of the soliton train dynamics

Many properties of evolving bit stream can be analysed by using linearization approach valid for the onset of instabilities. We assume that initially all solitons are equidistant with the common distance r and then linearize Eq. (4) around the initial configuration q^0:

$$q_n(z) = q_n^0 + w_n(z), \qquad |w_n| \ll |q_n^0|. \qquad (5)$$

The initial distribution is:

$$q_{n+1}^0 - q_n^0 = -2\nu r + 2\ln 2\nu + i\left[\pi - \Delta_1\delta_n\right],$$

$\Delta_1\delta_n \equiv \delta_{n+1} - \delta_n$, and $n = 0, \ldots, N - 1$. For an arbitrary DPSK-coded configuration we have: i) $\Delta_1\delta_n = 0$ for a "zero" bit with number $n+1$; (ii) $\Delta\delta_n = \pi$ for a logical "one" at the $n+1$-th position. Further on we define the universal z-scale: $a = \nu\exp(-r\nu)$, and then again rescale the distance – the overall scaling is $8\nu a z \rightarrow z$.

After the linearization of Eq.(4) with respect to $w_n(z)$ we arrive at the set of inhomogeneous equations:

$$\frac{d^2\mathbf{w}}{dz^2} = \hat{\Lambda}\mathbf{w} + \mathbf{f}, \qquad (6)$$

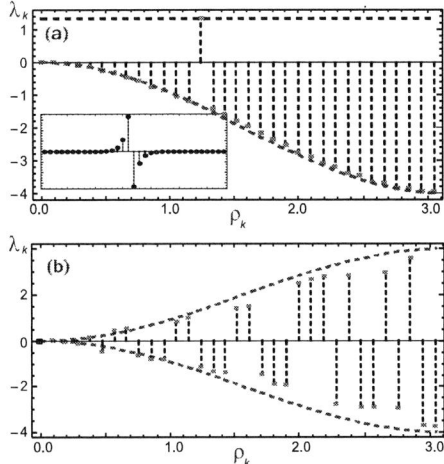

Fig. 1. Eigenvalue distribution (a) for a single "zero" posed in the center of the 32-bit pattern of "ones" (the inset shows the profile of the unstable mode) and (b) for a randomly encoded sequence of 32 bits. Dashed lines show the loci of λ_k for the pure homogeneous pattern of "ones" (a lower line) and "zeros" (an upper line).

the bold symbols stand for vector notations and hat indicates a matrix. Matrix $\hat{\Lambda}$ is real, symmetric and tridiagonal. Its elements depend on the direct bit values entering the pattern transmitted. If we introduce σ_n being equal to -1 for a zero bit and 1 for a unit one, then $\Lambda_{i-1\,i} = \sigma_i$, $\Lambda_{ii} = -\sigma_i - \sigma_{i+1}$, $\Lambda_{i+1\,i} = \sigma_{i+1}$. The driving vector is: $f_i = \sigma_i - \sigma_{i-1}$. Now one can find the set of normalized orthogonal eigenvectors $\mathbf{v}_k$ and corresponding eigenvalues λ_k of $\hat{\Lambda}$. Let us define the functions:

$$u_k(z) = \begin{cases} \gamma_k^{-1}\sin\gamma_k z & \text{for } \lambda_k < 0, \\ \gamma_k^{-1}\sinh\gamma_k z & \text{for } \lambda_k > 0, \\ z & \text{for } \lambda_k = 0, \end{cases} \qquad (7)$$

where $\gamma_k = \sqrt{|\lambda_k|}$. Then the solution of Eqs.(6) can be written by means of the $\hat{\Lambda}$ eigenvector expansion as:

$$\mathbf{w} = \sum_k U_k(z)\mathbf{v}_k, \qquad (8)$$

where $U_k(z) = f^k \int_0^z d\tau u_k(z - \tau)$, and $f^k = \mathbf{f} \cdot \mathbf{v}_k$. So, the type of bit pattern behavior depends on the signs of λ_k: when $\lambda_k > 0$, the mode is unstable and oppositely, for the negative sign the mode is stable.

The results of our analysis are the following. (i) For pure "ones" the bit pattern is completely stable with $\lambda_k = -2(1 - \cos\rho_k)$, where $\rho_k = \pi k/(N + 1)$. (ii) For pure "zeros" the pattern is absolutely unstable with $\lambda_k = 2(1 - \cos\rho_k)$, the most unstable are the modes with $\gamma_N \approx 2$. (iii) By using the derivations similar to the calculation of localized modes frequencies in the crystal lattice, see Ref.[11], it is possible to find the appearance of the unstable mode in the homogenous pattern of "ones" by introducing a single "zero" inside this originally stable pattern. This mode has a single node and is localized on the two sites bearing a "zero" bit, the corresponding eigenvalue for a single bit is:

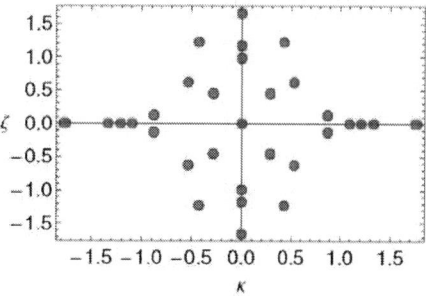

Fig. 2. Eigenvalue distribution for a Lax matrix $\hat{L}$ (divided by a) with the random bit pattern used for generating the plot in Fig.1(b).

$\lambda^* = 4/3$, see Fig.1(a). When "zeros" are rare and well separated inside the pattern of "ones", all unstable modes have the eigenvalues being very close to λ^*, and their eigenvectors are mostly localized on the pairs of "zero"-bearing sites. (iv) In the general case eigenvalue pattern consists of negative and positive λ_k lying within the interval $[-4, 4]$, see Fig.1(b).

B. Nonlinear regime and long time asypmtotics

As the CTC (4) possesses the property of complete integrability, this allows one to establish the asymptotical behavior of the given pattern at large distances. Eq.(4) can be recast in terms of the compatibility condition for the pair of matrix equation involving so-called Lax matrices [7], [8], [9], [10]. Let us define the matrix $\hat{L}$ as (up to a factor a): $L_{i-1\,i} = \sqrt{\sigma_i}$, $L_{i+1\,i} = \sqrt{\sigma_{i+1}}$, all other elements are 0. Then the complex eigenvalues, $l_k = \kappa_k + i\zeta_k$, of $\hat{L}$ give us the information about the asymptotical behavior of the CTC solution [9].

For an arbitrary bit pattern we have mixed asymptotic regime, where some κ-th coincide, and the part of solution forms bound complexes while the rest sites are asymptotically free. Fig.2 illustrates a typical situation, and by virtue of this picture one can classify the main regimes occurring in a generic bit pattern. (i) *Breathers*, corresponding to the pairs of purely imaginary eigenvalues, $l_k = \pm i\,\zeta_k$. (ii) *Moving breather pairs*, corresponding to the four complex eigenvalues, $\pm l_k$ and $\pm \bar{l}_k$, with nonzero imaginary and real parts. Each quaternion corresponds to the two bound pairs of pulses (breathers) moving in the opposite directions with the equal velocities. (iii) A set of *asymptotically free pairs* of solitons moving in the opposite directions with the equal velocities $\pm\kappa_k$. The solutions with the biggest κ_k can be used as a criterion of the timing jitter significance, i.e. for estimating the distance where the fastest solitons definitely move out of the slots.

From the symmetries of the general eigenvalue pictures there follows a *duality principle* for the asymptotic train of DPSK encoded solitons. Suppose that we have an arbitrary sequence of "zeros" and "ones", which yields a corresponding eigenvalue portrait of the matrix $\hat{L}$. Then, if one flips "zeros" and "ones", the matrix $\hat{L}$ is transformed into $i\hat{L}^\dagger$, and the eigenvalues distribution of the new pattern is obtained by flipping the distribution with respect to the bisector $\kappa = \zeta$.

IV. CONCLUSION

We have then studied how the particular symmetries imposed by the DPSK encoding affects the dynamics of a soliton pattern. In particular we have elucidated the two regimes: (i) the initial regime of exponential inflation/deflation of intrapulse distances and phase differences and (ii) asymptotical regime occurring for large distances. In both cases the study of the dynamical regime reduces to classifying the eigenvalues of a certain tridiagonal matrix which is real symmetric in the linear case and complex symmetric in the asymptotical regime. The practical applications of the method consists in determining the distances where a particular corraption of a given pattern takes place. For instance for the random 32-bit pattern used for calculations given at Figs.1 (a) and 2 we can evaluate the linear instability distance Z_{lin}, i.e. the distance before which we do not have a significant distortion of the bit pattern. Taking the intersoliton distance to be $r = 5$ and amplitudes $\nu = 1$, we arrive at the estimation: $Z_{lin} \approx 10$. From the other hand the considerations based on the nonlinear analysis and the eigenvalues of the Lax matrix $\hat{L}$, bring about the distance Z_{nln}, where the time jitter starts to garble significantly the encoded information. Using the same values of the system parameters we find: $Z_{nln} \approx 50$. So, the method proposed allows one to simplify the calculation and tremendously diminish the time needed for the estimation of the stability and distortion influence of a given DPSK-encoded pattern of solitons.

REFERENCES

[1] A.H. Gnauk and P. J. Winzer, "Optical Phase-Shift-Keyed Transmission", *J. Lightwave Technol.*, vol. 23, pp. 115-130, 2005.

[2] P.J. Winzer and R.J. Essiambre, "Advanced Modulation Formats for High-Capacity Optical Transport Networks", *J. Lightwave Technol.*, vol. 24, pp. 4711-4728, 2006.

[3] E.G. Shapiro, M.P. Fedoruk, S.K. Turitsyn, and A. Shafarenko, "Reduction of Nonlinear Intrachannel Effects by Channel Asymmetry in Transmission Lines With Strong Bit Overlapping", *IEEE Photonic. Tech. Lett.*, vol. 15, pp. 1473-1475, 2003.

[4] S.K. Turitsyn, E.G. Turitsyna, S.B. Medvedev, and M.P. Fedoruk, "Averaged model and integrable limits in nonlinear double-periodic Hamiltonian systems", *Phys. Rev. E*, vol. 61, pp. 3127-3132, 2000.

[5] A. Sanchez and A. R. Bishop, "Collective coordinates and length-scalecompetition in spatially inhomogeneous soliton-bearing equations", *SIAM Rev.*, vol. 40, 579-615, 1998.

[6] B. A. Malomed, "Variational methods in nonlinear fiber optics and related fields", *Prog. in Opt.*, vol. 43, 71-193, 2002.

[7] V.S. Gerdjikov, D.J. Kaup, I.M. Uzunov, and E.G. Evstatiev, "Asymptotic Behavior of N-Soliton Trains of the Nonlinear Schrodinger Equation", *Phys. Rev. Lett.*, vol. 77, pp. 3943-3946, 1996.

[8] V.S. Gerdjikov, I.M. Uzunov, E.G. Evstatiev, and G. L. Diankov, "Nonlinear Schrodinger equation and N-soliton interactions: Generalized Karpman-Solovev approach and the complex Toda chain", *Phys. Rev. E*, vol. 55, pp. 6039-6060, 1997.

[9] V.S. Gerdjikov, E.G. Evstatiev, and R.I. Ivanov, "The complex Toda chains and the simple Lie algebras - solutions and large time asymptotics", *J. Phys. A: Math. Gen.* **31**, 8221-8232, 1998.

[10] V.S. Gerdjikov, E.V. Doktorov, and J. Yang, "Adiabatic interaction of N ultrashort solitons: Universality of the complex Toda chain model", *Phys. Rev. E*, vol. 64, 56617 (2001).

[11] A.M. Kosevich, *The Crystal Lattice: Phonons, Solitons, Dislocations*, Wiley-VCH, Berlin, (1999).

CAOL*2010 International Conference on Advanced Optoelectronics & Lasers, 10-14 September, 2010, Sevastopol, Ukraine

Magneto-optical Kerr effect in a dielectric film with cubic optical nonlinearity on a magneto-electric slab

Yu. S. Dadoenkova[1,2], I. L. Lyubchanskii[1,2], Y.P. Lee[3], and Th. Rasing[4]

[1]Donetsk Physical & Technical Institute of the NAS of Ukraine, Donetsk, Ukraine
[2]Donetsk National University, Donetsk, Ukraine
[3]q-Psi and Department of Physics, Hanyang University, Seoul, Republic of Korea
[4]Institute for Molecules and Materials, Radboud University Nijmegen, Nijmegen, the Netherlands

Abstract: We investigate magneto-optic Kerr effect at light reflection from a nonlinear optical dielectric film on a bigyrotropic magneto-electric slab. We analyzed the optical nonlinearity and the magneto-electric coupling contributions to the Kerr rotation angles for both *s*- and *p*-polarized light at the incidence angles close to pseudo-Brewster angles.

Linear and nonlinear optical properties of magneto-electric (ME) materials are a subject of theoretical and experimental investigations [1, 2] because of their wide application in modern magneto- and opto-electronics. A lot of papers are devoted to the study of light reflection from magneto-electrics and from patterned magnetic structures with ME interaction as well as from the interfaces with ME properties [2]. Nonlinear optical phenomena like optical second harmonic generation and nonlinear optical spectroscopy of ME materials are also developed [2].

The influence of the cubic nonlinear optical susceptibility (NOS) on the reflectivity of the ME materials is studied in [3, 4]. It was shown that the cubic nonlinear optical interaction combined with linear ME coupling leads to changes in the refractive index of nonlinear media. These changes can be observed, for instance, by investigating light reflection under pseudo-Brewster angles (PBAs) [3, 4]. It can be expected that the presence of both the ME and the nonlinear optical couplings in the system results in considerable changes in the magneto-optical (MO) effects. In this communication, we investigate the MO Kerr effect at reflection of the electromagnetic waves (EMWs) from the system which is characterized by the ME and third-order NOS tensors.

Let us consider the reflection of a plane EMW from the bilayer structure composed of the nonlinear dielectric film (NDF) and the ME substrate (MES) of thicknesses d_{NL} and d_{ME}, respectively, as depicted in Fig. 1. We assume the substrate is magnetized perpendicularly to the interface and consider the polar MO configuration (MOC).

In the NDF the relations between the electric field vector **E** and the dielectric displacement vector **D** as well as between the magnetic field vector **H** and the magnetic induction **B** can be written as [5]

$$D_i = \left(\varepsilon_{ij}^{(1;0)} + \varepsilon_{ij}^{(1;NL)} \right) E_j, \qquad B_i = H_i, \qquad (1)$$

where $\varepsilon_{ij}^{(1;0)} = \varepsilon^{(1;0)} \delta_{ij}$ is linear permittivity tensor and $\varepsilon_{ij}^{(1;0)} = \chi_{ijkl}^{(3)} E_k E_l$ is the nonlinear contribution to the permittivity tensor, induced by the third-order NOS tensor $\chi_{ijkl}^{(3)}$, and δ_{ij} is delta-symbol. Superscript (1) denotes the NDF.

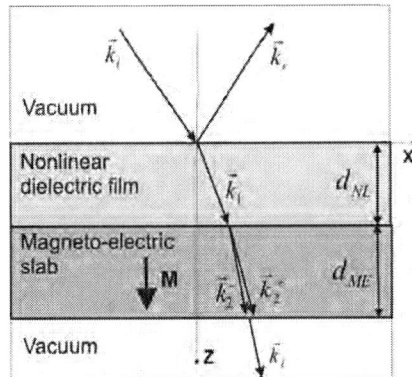

Fig. 1. Schematic of light reflection from the nonlinear film on the ME substrate in the case of polar MO configuration. Here $\vec{k}_i$, $\vec{k}_r$ and $\vec{k}_t$ denote wave vectors of incident, reflected and transmitted waves in vacuum, respectively, $\vec{k}_1$ is wave vector of transmitted wave in the NDF, $\vec{k}_2^{\pm}$ denote wave vectors in the MES.

Propagation of the EMW through the NDF is described by the wave equation [5]

$$\nabla div\, \mathbf{E} - \Delta\mathbf{E} - k_0^2 \hat{\varepsilon}^{(1;0)} \mathbf{E} = 4\pi \mathbf{P}^{NL}, \qquad (2)$$

where the nonlinear polarization at the frequency ω can be expressed as [5]

$$\mathbf{P}^{NL} = 3k_0^2 \left(2\chi_{1122}^{(3)} + \chi_{1221}^{(3)} \right) \mathbf{E} |\mathbf{E}|^2. \qquad (3)$$

Here $\chi_{1122}^{(3)}$ and $\chi_{1221}^{(3)}$ are the nonzero components of $\chi_{ijkl}^{(3)}$ [5]. The solution of Eq. (2) can be presented as

$$E_i(z) = E_i^{(0)} + 4\pi k_0^2 \int dz' G_{im}^{(1)}(z-z') P_m^{NL}(z'), \qquad (4)$$

978-1-4244-7043-3/10 $26.00 © 2010 IEEE   109

where $E_i^{(0)}$ is the solution of the homogeneous wave equation and $G_{im}^{(1)}(z-z')$ are the Green functions of the NDF.

In the MES, material relations can be written as [1]

$$D_i = \varepsilon_{ij}^{(2)} E_j + \alpha_{ij} H_j, \ B_i = \mu_{ij}^{(2)} H_j + \alpha_{ji} E_j, \quad (5)$$

where $\varepsilon_{ij}^{(2)}$ and $\mu_{ij}^{(2)}$ are the permittivity and the permeability tensors, respectively, and $\hat{\alpha}$ is the ME tensor, which is diagonal for cubic crystal: $\alpha_{ij} = \alpha \delta_{ij}$ [6]. The superscript (2) denotes the MES.

For the polar MOC, $\varepsilon_{ij}^{(2)}$ and $\mu_{ij}^{(2)}$ tensors of the MES can be presented in the following form:

$$\hat{\varepsilon}^{(2)} = \begin{pmatrix} \varepsilon_\perp & i\varepsilon' & 0 \\ -i\varepsilon' & \varepsilon_\perp & 0 \\ 0 & 0 & \varepsilon_\| \end{pmatrix}, \ \hat{\mu}^{(2)} = \begin{pmatrix} \mu_\perp & i\mu' & 0 \\ -i\mu' & \mu_\perp & 0 \\ 0 & 0 & \mu_\| \end{pmatrix}. \quad (6)$$

where ε' and μ' are linear on magnetization off-diagonal components of $\varepsilon_{ij}^{(2)}$ and $\mu_{ij}^{(2)}$ tensors, respectively. The difference in the diagonal components $\varepsilon_\|$ and $\varepsilon_\perp$, as well as the difference in $\mu_\|$ and $\mu_\perp$ results from a quadratic MO interaction and is sufficient for the effects of magnetic birefringence. For simplicity, we neglect these effects and concentrate on the influence of linear MO and ME couplings on the light reflection, i. e. we assume the diagonal components of the tensors (6) to be equal: $\varepsilon_\| = \varepsilon_\perp$, $\mu_\| = \mu_\perp$.

The reflected light magnitudes $\mathbf{E}^{(r)}$ can be written via the incident light magnitudes $\mathbf{E}^{(i)}$ using the reflection matrix $\hat{R}$:

$$\begin{pmatrix} E_p^{(r)} \\ E_s^{(r)} \end{pmatrix} = \hat{R} \begin{pmatrix} E_p^{(i)} \\ E_s^{(i)} \end{pmatrix}, \ \hat{R} = \begin{pmatrix} R_{pp} & R_{ps} \\ R_{sp} & R_{ss} \end{pmatrix}, \quad (7)$$

where subscripts p and s correspond to the p- and s-polarized EMWs, respectively.

The Kerr rotation (KR) angle ϕ^K can be calculated from the ratio of the complex magnitudes $E_s^{(r)}$ and $E_p^{(r)}$ components of the reflected field, $\kappa = \dfrac{E_s^{(r)}}{E_p^{(r)}}$, as

$$\phi^K = \frac{1}{2} arctg \frac{2\operatorname{Re}(\kappa)}{1 - |\kappa|^2}. \quad (8)$$

We make numerical estimations for an yttrium-iron garnet film as a substrate, which demonstrates giant ME coefficients [7], covered by an optical glass characterized by the third-order NOS tensor. For our calculations, we used the following material parameters for light wavelength λ=1.15 μm, and magnitude of the incident EMW E~10^8 V/m: $\varepsilon^{(1;0)} = 3.61$, $\chi^{(3)} \sim 10^{-13}$ (esu), $\varepsilon^{(2)} = 4.5796$, $\mu^{(2)} = 1$, $\varepsilon' = -2.47 \cdot 10^{-4}$ $\mu' = 8.76 \cdot 10^{-5}$, and $\alpha \sim 10^{-2}$. Thicknesses of the NDF and MES are equal and taken to be 0.5λ. Results of the numerical calculations are illustrated in Figs. 2 – 4.

Figs. 2(a) and 3(a) show diagonal components R_{ss} and R_{pp} of the reflection matrix, respectively, as functions of the incidence angle θ. Solid lines correspond to the nonlinear interaction absence in the NDF, and dotted lines depict reflection coefficients with taking into account the third-order NOS tensor components. As it was mentioned in [3, 4], diagonal reflection coefficients demonstrate several minima not reaching zero values. These minima correspond to reflection at PBAs.

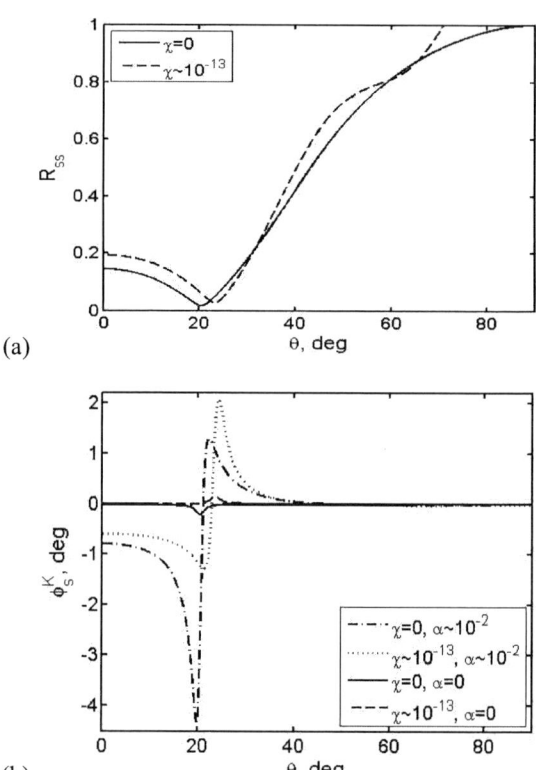

(a)

(b)

Fig. 2. Diagonal reflection coefficient R_{ss} (a) and KR angle ϕ_s^K (b) as functions of the incidence angle θ for s-polarized incident light.

Figs. 2(b) and 3(b) present KR angles ϕ_s^K and ϕ_p^K as functions of the incidence angle θ for s- and p-polarized incident light, respectively. Dash-dotted and dotted (dashed and solid) lines correspond to KR with and without taking into account the nonlinear interaction in the NDF, respectively, and with (without) taking into account the ME coupling in the MES. One can see from Figs. 2(a) and 3(a) that the nonlinear interaction in the NDF leads to the PBAs shift towards larger incidence angles for s-polarized EMW and towards smaller incidence angles for p-polarized EMW. This behavior results in the corresponding minima and maxima shifts of the KR angles at the angles of incidence close to the PBAs (see Figs. 2(b) and 3(b)). For example, ϕ_s^K decreases at the incidence angles before the PBA and ϕ_s^K increases after the PBA. From Figs.

978-1-4244-7043-3/10 $26.00 © 2010 IEEE

2(b) and 3(b) one can see that the ME coupling enhances the KR at about 20% whether nonlinear interaction presents in the NDF or not.

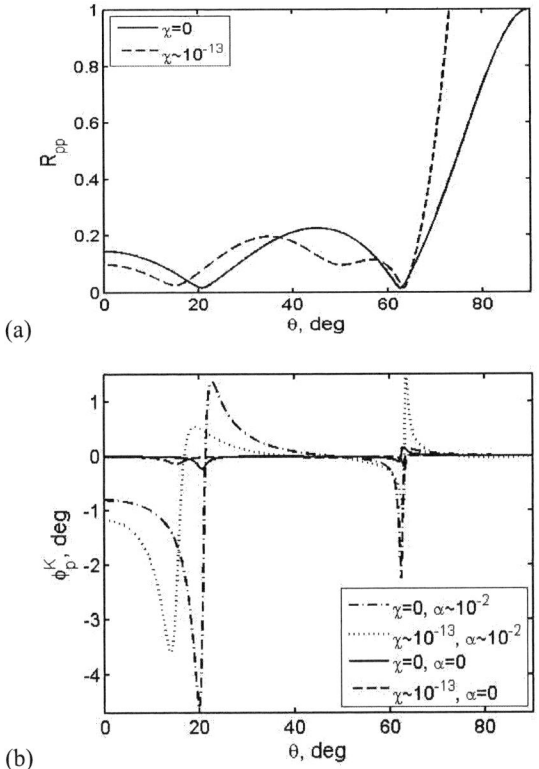

(a)

(b)

Fig. 3. Diagonal reflection coefficient R_{pp} (a) and KR angle ϕ_p^K as functions of the incidence angle θ for p-polarized incident light

Diagonal reflection coefficients for all general MOCs differ very small [3, 4] for the system under consideration, whereas the off-diagonal coefficients, which are about three orders smaller the diagonal ones, demonstrate strong dependence under the ME interaction constant. It should be noted that for the transverse MOC, when the magnetization vector **M** is perpendicular to the incidence plane, the KR takes place due to the ME coupling which results in nonzero off-diagonal components of the matrix $\hat{R}$. The KR angles for all general MOCs are the same order of value. Fig. 4 shows the difference between the KR angles $\Delta\phi^K$ for the longitudinal MOC, when the magnetization vector **M** lies in the incidence plane parallel to the interface, and polar MOC. Dotted (dashed) and dash-dotted (solid) lines show $\Delta\phi^K$ for $p(s)$-polarized EMWs with and without taking into account the third-order NOS tensor components in the NDF. One can see that difference in KR angles is about several percents and depends upon the cubic optical nonlinearity. It should be noted that reverse of the magnetization vector **M** do not leads to the KR direction

change because of small MO Kerr effect on comparison with the ME-induced one. When magnetization reversed, the KR angle decrease is about 10 %.

In conclusions, we have shown that for all general MOCs, the nonlinear optical interaction leads to sufficient changes of the KR for both s- and p-polarized EMWs at the incidence angles close to the PBAs. The ME coupling results in the KR enhancement. Thus, combined ME and nonlinear interaction in such bilayer structure allows to increase optical rotation.

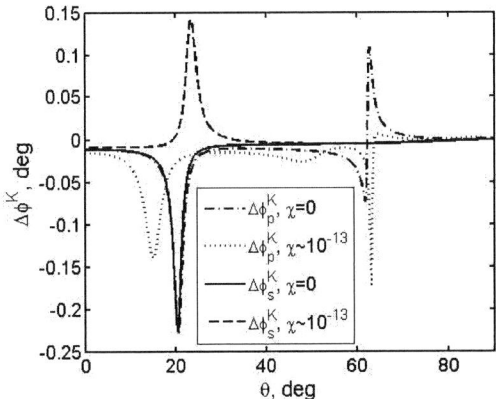

Fig. 2. Difference in KR angle ϕ^K in the cases of the polar and longitudinal MO configurations, as function of the incidence angle θ for s- and p-polarized incident light.

This work was supported via project FP7-People-2009-IRSES, No.247556 (Yu.S.D., I.L.L.), by National Research Foundation (Korea) through q-Psi (Y.P.L.) and by NWO – the Netherlands Organization for Scientific Research (Th.R.).

REFERENCES

[1] T.H. O'Dell, *The Electrodynamics of Magnetoelectric Media*, North-Holland, Amsterdam, 1970.

[2] M. Fiebig, "Revival of the magnetoelectric effect," J. Phys. D: Appl. Phys., vol. 38, R123, 2005.

[3] Yu. S. Dadoenkova, I. L. Lyubchanskii, Y.P. Lee, and Th. Rasing, "Investigation of light reflection at Brewster angles from a nonlinear optical film on a magneto-electric substrate", Eur. Phys. J. B 71, 401-406, 2009.

[4] Yu. S. Dadoenkova, I. L. Lyubchanskii, Y.P. Lee, and Th. Rasing, "Reflection of light from a nonlinear optical film on a bigyrotropic magneto-electric substrate under angles near the Brewster angles," Fizika Nizkih Temperatur (Kharkov), vol. 36, No. 6, pp. 673-679, 2010.

[5] R.W. Boyd, *Nonlinear Optics*, Academic Press, San Diego, 1992.

[6] R. R. Birss, *Symmetry and Magnetism*, North-Holland, Amsterdam, 1966.

[7] R. V. Pisarev, "Crystal optics of magnetoelectrics," Ferroelectrics, vol. 162, pp. 191-209, 1994.

CAOL*2010 International Conference on Advanced Optoelectronics & Lasers, 10-14 September, 2010, Sevastopol, Ukraine

Direct quantitative estimation of the spatial energy migration between ions Yb^{3+} in $Gd_3Ga_5O_{12}$ matrix

E.V. Ivakin, I.G. Kisialiou

Institute of physics, National academy of sciences

Nezavisimosti Ave., 68, Minsk, 220072, Belarus

Abstract-**The phenomenon of spatial diffusion of stored energy between excited and not excited ions has been experimentally studied under conditions of high concentration of dopant and high pump gradient. Direct measurements of the diffusion coefficient as well as lifetime of the excited state in laser crystal Yb^{3+}(21 at.%):$Gd_3Ga_5O_{12}$ have been conducted by the method of tunable transient gratings.**

I. INTRODUCTION

When setting the goal of our studies, we proceeded from the assumption that some well-known and deeply investigated effects of excitation energy transfer among rear-earth dopant ions in laser crystals can be accompanied with the effect of spatial diffusion which becomes especially apparent under high density of dopant in matrix and high spatial gradient of population inversion of the energetic levels. In particular, the diffusion effect can noticeably impact the output characteristics of solid state laser in laser systems with compact active media (e.g. laser microchip, fiber laser, distributed-feedback laser and so on).

To our knowledge, the first attempt of direct measurement of diffusion coefficient D_E of the energy excitation was made with ruby crystal at chromium ions concentration from 0.05 to 5% by the method of transient gratings [1]. Despite the grating spacing was varied in rather wide range of the value – from 1 to 100 μm – the influence of the excitation energy migration upon the character of diffraction signal decay was not observed. Instead, upper limit of the diffusion coefficient was estimated.

Indirect measurements of D_E in laser crystal Nd:YVO$_4$ were carried out in [2] via effect of washing out of spatial modulation of population inversion in laser crystal that was induced by the interference of two counter propagating beams in resonator. Auger upconversion and energy excitation diffusion that impact the characteristics of emission from single mode laser were considered. The waveform of emission under rapid longitudinal movement of the active media was analyzed to determine the D_E value.

In this work we have studied the temporal behavior of the diffracted signal when spacing-variable transient gratings were recorded in highly doped laser crystal Yb^{3+}:$Gd_3Ga_5O_{12}$ by light flash. The technique enabled us to perform the direct measurement of both the excitation energy diffusion coefficient as well as the lifetime of the upper lasing level with no impact of re-absorption of the emitted light.

II. THE TECHNIQUE OF MEASUREMENT

The method of high frequency transient gratings gives the opportunity of direct measurement of D_E and seems to be the most adequate to the problem to be solved. It provides high and well controllable gradients of pumping density creation in the crystal.

According to the method, two beams from a pumping laser intersect at a sample at the angle desired and give rise to spatial modulation of higher energetic level population density due to interference.

The polarisability change of ions provides the phase transient grating recording in a sample. The diffraction efficiency of this grating is probed with a second (CW or pulsed) laser.

The amplitude of active ions density at high energetic level in ytterbium doped laser crystal decreases with time (after pumping pulse is switched off) due to the following processes [2]:

1) Depopulation of the excited state of Yb^{3+} ions due to downward transitions.

2) Auger upconversion, i.e. energy exchange between two excited ions. In result, ion-donor passes to the ground state whereas ion-acceptor undergoes transition to the higher energetic level. Normally, the rapid nonradiative transitions give rise to ion-acceptor returning back to the ground state. Hence, the upconversion phenomenon leads to inversion depletion and the temperature of active crystal increases. The probability of the process grows nonlinearly as the pumping energy density rises [3].

3) The energy exchange between excited and not excited ions. Due to this process the velocity of transient grating decay depends quadratically on the frequency of spatial modulation of ions population density.

According to [1], the waveform of diffraction signal relaxation is accepted to be of single exponential character with time constant t_S:

$$\frac{1}{t_S} = \frac{2}{t_R} + 8\pi^2 D_E \frac{1}{L^2}, \qquad (1)$$

where t_R is lifetime of the excited state of Yb^{3+} ion, L is the transient grating spacing, D_E is diffusion coefficient of excitation energy.

978-1-4244-7043-3/10 $26.00 © 2010 IEEE

III. SAMPLE DESCRIPTION

We studied the crystal $Gd_3Ga_5O_{12}$, doped by active ions Yb^{3+} with the atomic concentration as high as 21%. The sample was a 1 mm thick slab with both facets being optical grade polished.

The reference value of upper lasing level lifetime is $8 \cdot 10^{-4}$ s [4]. The absorbing lines 932, 944 and 971 nm were determined via maxima of the diffracted signal which well compiled with that obtained via spectral transmittance measurement. The maximum of luminescence was observed at the wavelength 1025 nm.

We accepted the value of thermal conductivity of ytterbium doped $Gd_3Ga_5O_{12}$ crystal to be about 7.5 $Wm^{-1}K^{-1}$ at high doping concentration according to [5].

IV. EXPERIMENTAL

Since the diffusion coefficient was expected to be rather small, high spatial gradient of refraction index has been produced via small values of L=11.7, 1.76, 1.1, 0.87 and 0.76 μm in order to find out the excitation energy migration phenomenon. For the adequate organizing of research and correct interpretation of experimental results the following specific aspects were taken into account:

1) Negligibly small relaxation time of a thermal grating in observed signal of diffraction which is governed by thermal diffusivity of matrix. Really, with thermal conductivity 7.5 $Wm^{-1}K^{-1}$ the value of thermal diffusivity was estimated to be about 0.04 cm^2/s. It gives the lifetime of thermal grating of 1 to 10 μm spacing lying within interval from 3 to 300 ns.

2) Bulk nature of a transient grating and necessity to meet Bragg condition.

3) Pump energy dependence of t_S, which is typical for upconversion phenomenon.

The experimental setup is schematically shown in Fig. 1. CW radiation (633 nm) from He-Ne laser (a) served as probe beam. The excitation pulse of 8 ns duration at 971 nm was derived from the output of Ti-Sapphire laser (b) pumped with radiation from Q-switched Nd:YAG laser. The outlet beam was split with a diffraction beam splitter (c) to yield the two excitation pulses that crossed at the sample. Focusing lenses (d) were used to provide sufficient excitation energy density in the sample. Folding optical wedges and 90° prisms (e) ensured the proper angle between excitation beams as well as probe angle of incidence of CW beam.

Since the $Gd_3Ga_5O_{12}$ is transparent to the probe wavelength, the experiments were performed in transmission geometry, as shown in Fig. 1. It was necessary to meet Bragg diffraction condition and to find out the proper angle of probe beam incidence, especially when high frequency grating was recorded in the slab.

The first order diffraction beam was detected with a photomultiplier tube. Signal of diffraction was then digitized and stored with digital oscilloscope Tektronix (300 MHz bandwidth).

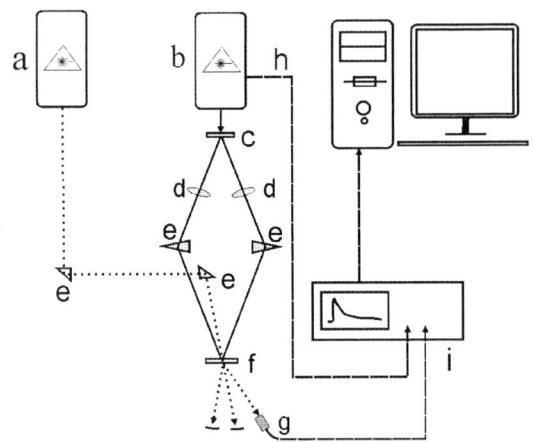

Fig.1. Experimental setup for real-time transient grating experiments with laser crystal slabs: CW probe laser (a), excitation laser (b), grating-based beam splitter (c), focusing lenses (d), folding prisms (e), sample (f), photomultiplier tube (g), triggering (h), digital oscilloscope (i).

Fading waveforms of diffraction were detected at various grating spacing. Pumping energy density in the plane of sample was held constant at approximately 0.1 J/cm^2. Each waveform exhibited the decay close to single exponential form as it is shown in Fig. 2.

In addition, the detection of sample luminescence in both visible and near infrared ranges was carried out in order to compare luminescence duration with the lifetime of transient grating. To do this the input aperture of optical fiber coupled with photomultiplier tube was fixed in vicinity of the sample. The emitted radiation from the sample was selected by proper spectral filters.

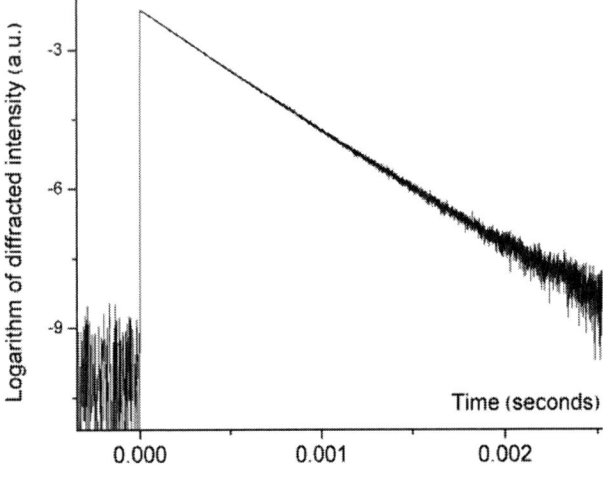

Fig.2. Typical time-resolved signal of diffraction recorded at L = 0, 87 μm

V. RESULTS AND DISCUSSION

Given in Fig. 3, the resulting plot demonstrates a total of various diffraction fading waveforms recorded at different grating spacing (L). As active medium is normally operating at high temperature, the measurements were carried out at room and at elevated sample temperature. Graphs are plotted as linear functions of t_s^{-1} versus L^{-2} in accordance with Eq. (1).

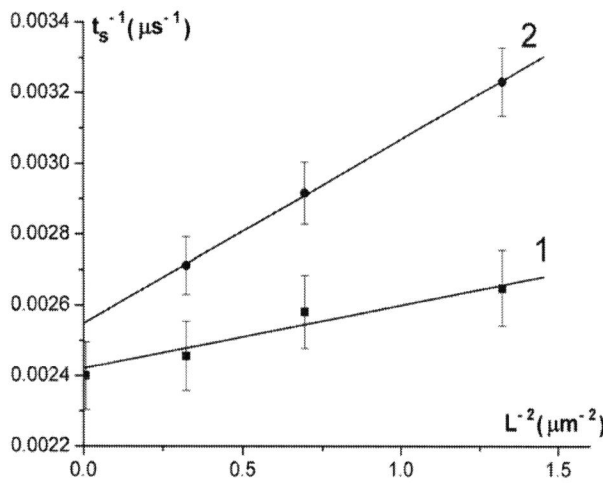

Fig.3. Reverse time of diffracted signal relaxation vs. L^{-2} at sample temperature about 30^0 C (1) and at 200^0 C (2).

The slope of lines in Fig. 3 gives a coefficient of excitation diffusion. The intersection with reverse-time axis ($L \rightarrow \infty$) corresponds to grating fading due to depletion of excited state of ytterbium ions. The results of measurements are given in Table 1.

It results from measurements that diffusion coefficient D_E increases with sample heating. One of the reasons is presumably connected with the increase of probability and rate of energy-exchange events between ions.

The obtained lasing level lifetime in $Yb^{3+}:Gd_3Ga_5O_{12}$ crystal corresponds to the reference value t_R [4] at room temperature with uncertainty less than 5%. Thermal quenching and reduction of lasing level lifetime are observed according to the Table 1.

TABLE 1
Diffusion coefficient and excited level lifetime of Yb^{3+} ion

Sample temperature, ^{0}C	D_E, cm^2/s	t_R, s
≈ 30	$2.4 \cdot 10^{-8}$	$8.25 \cdot 10^{-4}$
≈ 200	$6.6 \cdot 10^{-8}$	$7.85 \cdot 10^{-4}$

The luminescence from excited sample both in visible and infrared ranges was observed in the experiment.

A weak green luminescence with the peak at approximately 540 nm is presumably caused by upconversion with participation of ytterbium ions as well as little amount of holmium ions impurity [6]. The time-resolved waveform of this green luminescence exhibits some rise in the beginning, which confirms the assumption of its origin.

Infrared emission with the peak at 1025 nm corresponds to downward transitions of excited ytterbium ions into the ground state. However a measured duration of this luminescence (up to 2 ms) noticeably exceeded the value given in [4]. The same phenomenon of radiation trapping was also observed and described in [5].

It is known that measurement of excited-state lifetime in laser crystals where absorption and emission bands partially overlap would be a problem because of re-absorption phenomenon. Special means based on significant reduction of emitting volume have been developed to decrease the influence of this factor [7, 8].

However, t_R in (1) is a true excited state lifetime because luminescence and its re-absorption do not contribute into detected signal of diffraction. Determination of lasing level lifetime in crystal via diffraction fading could be considered as alternative coherent technique with respect to conventional methods based on detection of luminescence.

ACKNOWLEDGMENT

This work is partially supported by the Program of National Fundamental and Applied Scientific Research (Grant "Fotonika-3.03") and by National Scientific and Technological Program (Grant No 1.14).

REFERENCES

[1] H.J. Eichler, J. Eichler, J. Knof, and Ch. Noack, "Lifetimes of laser-induced population density gratings in ruby," *Phys. Stat. Sol.* (a), vol. 52, p.p. 481–486, 1979.

[2] L. Meilhac, G. Pauliat and G. Roosen, "Determination of energy diffusion and of Auger upconversion constants in a Nd:YVO₄ standing-wave laser," *Opt. Com.*, vol. 203, p.p. 341-347, 2002.

[3] Marie-France Joubert, "Photon avalanche upconversion in rare earth laser materials," Optical Materials, vol.11, p.p.181-203, 1999.

[4] M.I. Belovolov, E.M. Dianov, M.I. Timoschechkin, L.V. Barashov, A.M. Belovolov, M.A. Ivanov, N.P. Morosov, A.M. Prokhorov, K.M.Timoschechkin, "Room temperature cw Yb:GGG laser operation at 1.038-µm", OSA Conference on Lasers and Electro-Optics Europe 1996 (CLEO Europe 96), technical digest, p. 43.

[5] S. Chenais, F. Druon, F. Balembois, P. Georges, A. Brenier and G. Boulon, "Diode-pumped Yb:GGG laser: comparison with Yb:YAG" *Optical Materials*, vol. 22, p.p. 99-106, 2003.

[6] A.M. Belovolov, M.I. Belovolov, E.M. Dianov, V.V. Dudin, M.I. Timoschechkin, "CW lasing in Yb^{3+}:GGG crystal pumped at 0.925 µm", *Quantum Electronics*, vol. 36 (7), p.p. 587-590, 2006.

[7] H. Kühn, S.T. Fredrich-Thornton, C. Kränkel, R. Peters and K. Petermann, "Model for the calculation of radiation trapping and description of pinhole method," *Optics Letters.* – Vol. 32,. – p.p.1908–1910, 2007.

[8] V.E. Kisel, A.E. Troshin, V.G. Shcherbitsky, N.V. Kuleshov, A.A. Pavlyuk, F. Brunner, R. Paschotta, F. Morier-Genoud and U. Keller, "Luminescent and lasing characteristics of heavily doped Yb3+:KY(WO₂)₄ crystals", *Quantum Electronics*, vol. 36, p.p. 319-323, 2006.

Liquid crystal defects : nonlinear optics and optical vortices

E. Brasselet

Centre de Physique Moléculaire Optique et Hertzienne, CNRS, Université Bordeaux 1

351 cours de la libération, 33405 Talence Cedex, France

e.brasselet@cpmoh.u-bordeaux1.fr

(Invited *Paper*)

Abstract: Liquid crystals can self-assemble into various phases characterized by well-defined orientational ordering of their crystalline axis. Exception may happen at locations in space where the orientation is ill-defined, which correspond to liquid crystal defects. In addition, liquid crystals are well-known to be sensitive to external fields and light is no exception. In particular, Gaussian light beams can generate singular liquid crystal reorientation defects owing to the giant optical orientational nonlinearities. Such a singular optical reordering of liquid crystals can be used to generate light beams carrying on-axis phase singularity that are usually called optical vortices.

Liquid crystals can self-assemble into various phases characterized by well-defined orientational ordering of their crystalline axis and are well-known to be sensitive to external fields. Light is no exception. Indeed the high birefringence and low elastic constants of liquid crystals confer to them genuinely high optical nonlinearities that are essentially driven by their orientational degree of freedom [1]. A famous example is the optical Fréedericksz transition in the early 80's that corresponds to the light induced optical reordering of thermotropic liquid crystals above a laser intensity threshold [2], which has been the subject of many research activities since then [3].

Experimentally, the most studied case corresponds to a nematic liquid crystal film under the course of fundamental Gaussian light beams. In such a film the molecular orientation is locally defined by the unit vector **n** (called director) and the case of perpendicular director alignment with respect to the plane of the film (called homeotropic alignment, see Fig.1(a), where **n**₀ refers to the director at rest) is a common situation. From the theoretical point of view, the plane wave approximation has been a main assumption to describe quantitatively the nonlinear optical orientational processes for a long time and a mature theoretical toolbox has become available only recently [4]. However, until now, optical reordering of nematics has been experimentally discussed, and theoretically described, in the framework of smooth spatial reorientation profile (i.e., spatial distribution of the molecular axes free from orientational singularities) generated by smooth optical fields (i.e., light beams free from optical singularities). The generation of singular orientational patterns in liquid crystals using Gaussian beams is nevertheless possible, as shown very recently [5]. Indeed, the optical torque of a focused fundamental Gaussian light impinging at normal incidence on a homeotropic nematic film possesses radial and azimuthal components that are responsible for the generation of azimuthal and radial elastic distortions of the liquid crystal, respectively [see Fig.1(b)]. When the incident polarization is circular, the total distorted director spatial pattern is the superposition of two cylindrically symmetric modes : the radial one [see Fig.1(c)] and the azimuthal one [see Fig.1(d)], respectively.

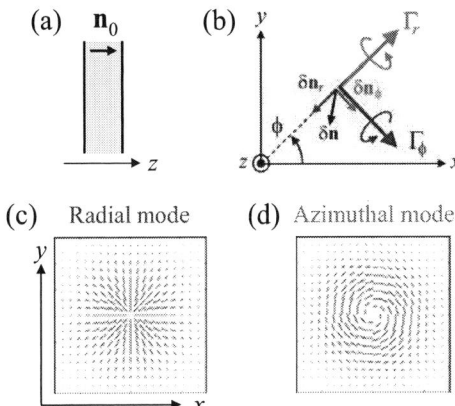

Fig. 1. (a) Homeotropic nematic liquid crystal film at rest, where the molecular orientation is given by the director **n**₀ that defines the normal to the film. (b) Radial (Γ_r) and azimuthal (Γ_ϕ) components of the optical torque that generate azimuthal ($\delta\mathbf{n}_\phi$) and radial ($\delta\mathbf{n}_r$) components of the reoriented director field. (c) and (d) : typical radial and azimuthal component of the distorted director spatial pattern $\delta\mathbf{n}$ in the (x,y) plane of a film illuminated by a circularly polarized fundamental focused Gaussian beam.

This illustrates the possibility to generate liquid crystal defect patterns from an initially homogeneously aligned sample. Interestingly, we note that the total distorted director field is chiral, as shown theoretically in Fig.2(a) and Fig.2(b) for right- and left-handed incident circular polarizations, respectively. Such a spin-dependent chirality of the singularly reoriented nematic film is retrieved in experiments by looking at conoscopic cross-polarized images of a probe beam, as shown in Fig.2(c) and Fig.2(d), respectively. Also, we notice that these polarization sensitive elastic distortions in liquid crystals may be viewed as the two-dimensional (2D) analog,

978-1-4244-7043-3/10 $26.00 © 2010 IEEE

i.e. nematic homeotropic film geometry, of previously observed chiral conoscopic patterns in a three-dimensional (3D) case, i.e. nematic radial droplet geometry [6]. However, there is a main difference between these 2D and 3D situations : in the present case the defect is artificially created by light through a nonlinear process whereas in the case of a radial droplet the defect is a natural one, which is located in the center of the droplet.

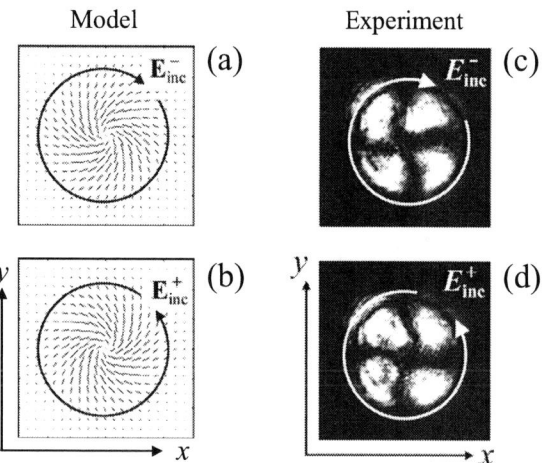

Fig. 2. (a) and (b) : calculated total distorted director field in the plane of a nematic homeotropic film singularly reoriented by a right- and left-handed circularly polarized Gaussian beam (see Fig.1). (c) and (d) : twisted conoscopic Maltese cross under crossed polarizers observed experimentally, which correspond to panel (a) and (b), respectively.

The use of singularly patterned liquid crystal systems has been recently revealed to be useful to generate light beams having polarization-dependent helical wavefronts [7], usually called optical vortices. Such beams possess on-axis phase singularity of the form $\exp(i\ell\phi)$, where ℓ is called the topological charge and ϕ is the azimuthal angle, and find many applications in imaging, quantum information or micromanipulation, see for example [8]. If several realizations of continuously 2D patterned liquid crystal systems have been already demonstrated [7,9,10] they all rely on artificially created patterns without reconfiguration capabilities and, general, restricted operating range (e.g. acceptance angle, wavelength). Recently, the extension to 3D patterned systems has been obtained using radial nematic droplets, which lead to unique optical vortex generators operating at any incidence angle and wavelength [11].

In the present case, the 2D patterned system relies on the optical orientational nonlinearity of liquid crystals, which offers the possibility to control, both in space and time, the generation of an optical vortex. The idea is to fabricate, on-demand, a liquid crystal defect that will operate as an optical vortex generator. This is possible owing to the charge one topological defect of the spatial distribution of the optical axis

[see Fig.2(a,b)]. Indeed, an efficient spin-to-orbital light angular momentum conversion can take place in that case [7]. This is illustrated in Fig.3 where the intensity [Fig.3(a)] and phase [Fig.3(b)] profile of the contra-circular component of the output field (with respect to an incident circularly polarized probe plane wave) is calculated for a typical singular pattern as shown in Fig.2(a,b). It appears that the phase profile is clearly identified as a charge two optical vortex ($\ell = 2$), see Fig.3(b). As expected, the on-axis intensity is zero, as shown in Fig.3(a). The calculation is performed using Jones formalism for the propagation of a plane wave on the distorted director field obtained from [5].

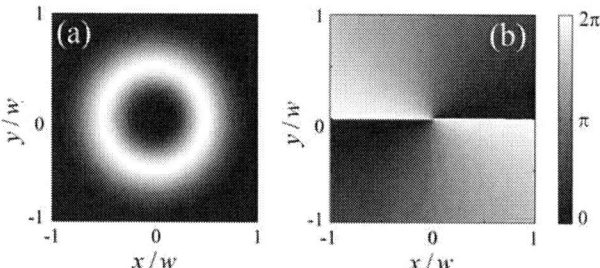

Fig. 3. Calculated intensity (a) and phase (b) profile of the output contra-circular component of an incident circularly polarized probe plane wave on a laser induced continuously 2D patterned nematic film (see Fig.2) where w is the characteristic spatial scale in the transverse plane.

Experimentally, these predictions are retrieved following the set-up detailed in [5], but the additional insertion of a collimated probe beam whose incident polarization state is set as circular. Then, the output probe beam is analyzed on the circular polarization basis using a quarter wave plate and a polarizer. The typical obtained intensity profile is shown in Fig.4(a), which exhibits a doughnut shape. Although this agrees with predictions shown in Fig.3(a), the null on-axis intensity does not ensure the presence of a phase singularity. The latter is unambiguously retrieved from the interference with a reference Gaussian beam, as shown in Fig. 4(b). In this figure, the double spiral fringes structure demonstrates the generation of a charge two optical vortex.

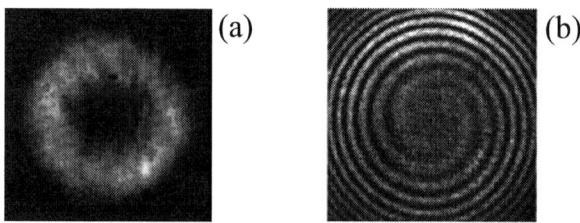

Fig. 4. Experimental counterpart of Fig.3 : (a) intensity profile. (b) double spiral fringes from the interference between the obtained optical vortex and a Gaussian beam reference that identifies a helical phase profile with topological charge two.

The concept of singular optical reordering of liquid crystals allows generating orientational defects within initially homogeneous anisotropic media. Such defects can be switched on and off at will by activation of the pump beam responsible for the optical reorientation. The charge one topological defect of the resulting spatial distribution of the optical axis can be used for the on-demand generation of optical vortices, both in time and space, through an optical spin-orbit coupling process activated by optical nonlinearities.

REFERENCES

[1] I. C. Khoo, "Nonlinear optics of liquid crystalline materials", Phys. Rep. **471**, 221–267 (2009).

[2] N. V. Tabiryan, A. V. Sukhov and B. Ya. Zel'dovich, "The orientational optical nonlinearity of liquid crystals", Mol. Cryst. Liq. Cryst. **136**, 1–139 (1986).

[3] G. Demeter and D. O. Krimer, "Light-induced dynamics in nematic liquid crystals – a fascinating world of complex nonlinear phenomena" Phys. Rep., **448**, 133–162 (2007).

[4] E. Brasselet, B. Piccirillo and E. Santamato, "Three-dimensional model for light-induced chaotic rotations in liquid crystals under spin and orbital angular momentum transfer processes", Phys. Rev. E **78**, 031703 (2008).

[5] E. Brasselet, "Singular optical manipulation of birefringent elastic media using nonsingular beams" Opt. Lett. **34**, 3229–3231 (2009).

[6] E. Brasselet, N. Murazawa, S. Juodkazis and H. Misawa, "Statics and dynamics of radial nematic droplets manipulated by laser tweezers", Phys. Rev. E **77**, 041704 (2008).

[7] L. Marrucci, C. Manzo and D. Paparo, "Optical spin-to-orbital angular momentum conversion in inhomogeneous anisotropic media" Phys. Rev. Lett. **96**, 163905 (2006).

[8] D. L. Andrews, "Structured Light and Its Applications: An Introduction to Phase-Structured Beams and Nanoscale Optical Forces", Academic Press-Elsevier, Burlington, 2008.

[9] H. Ren, Y.-H. Lin and S.-T. Wu, "Linear to axial or radial polarization conversion using a liquid crystal gel" Appl. Phys. Lett. **89**, 051114 (2006).

[10] S. C. McEldowney, D. M. Shemo, R. A. Chipman and P. K. Smith, "Creating vortex retarders using photoaligned liquid crystal polymers", Opt. Lett. **33**, 134–136 (2008).

[11] E. Brasselet, N. Murazawa, H. Misawa and S. Juodkazis, "Optical vortices from liquid crystal droplets", Phys. Rev. Lett. **103**, 103903 (2009).

Nonlocal interactions of solitons in bias-free nematic liquid crystals

Ya. V. Izdebskaya
Nonlinear Physics Center, Research School of Physics and Engineering,
The Australian National University, Canberra ACT 0200, Australia

(Invited *Paper*)

Abstract: With an emphasis on experimental observations, we review the state of the art in the area of optical spatial solitons in unbiased nematic liquid crystals, outlining their basic and applied features and discussing recent results, including soliton interaction with curved surfaces and observation of dynamical instability of two counterpropagating solitons.

Optical spatial solitons, i.e. self-localized light beams in nonlinear media, are excellent building blocks for all-optical signal processing [1]. Recent, the highly nonlocal and nonresonant molecular response of undoped nematic liguid crystals (NLC) has enabled the demonstration of stable spatial solitons (or *nematicons*) and their interactions at milliwats power levels [2].

In the majority of previous experimental studies with nematicons the liquid-crystal cells were biased by an external electric field; the bias allows for robust control over the orientation of NLC molecules at rest and, therefore, it provides access to fine tuning of nonlinearity as well as nonlocality [3]. However, nematicons also exist in unbiased NLC cells with the molecules pre-aligned in the bulk of the planar cell, as seen in Fig. 1(a,b), by the anchoring conditions at the boundaries. The nonlinearity, induced by small tilt of molecules by laser light, is sufficiently strong to observe self-focusing effects at milliwatt input power, provided the input beam is extraordinary polarized, i.e. with the electric field and molecular director coplanar with the plane (*x, z*), parallel to the cell interfaces. Because of the optical anisotropy of the liquid crystalline molecules and the resulting birefringence of nematic, the light beams propagating in NLC walk-off the direction of their wave vector. Earlier studies of spatial solitons in NLC considered the propagation dynamics of individual nematicons [2] as well as nonlocal interactions of two co-propagating [4] and counter-propagating (CP) nematicons [5,6].

In this paper we review our recent experimental results of the interaction spatial solitons in a planar bias-free NLC cell [7-9]. We use custom made NLC cells, as shown in Fig. 1(a,b), without external bias, exploring different optical regimes depending on the boundary conditions as well as input power, polarization, and focusing of laser light. We investigate in detail the power-dependent nonlocal interaction of two identical mutually incoherent CP nematicons and demonstrate the existence of a novel class of *vector nematicons* consisting of two self-trapped beams propagating in opposite directions. Depending on parameters of the setup, we observe both, the stable stationary states and the dynamic instabilities of two CP

beams, in the form of splitting and spatial entanglement. The interaction of spatial optical solitons with curved dielectric surfaces is studied in NLC layer with injected air-bubbles of different size; it allows for large-angle bending of nematicons as well as total internal reflections preserving non-diffracting character of self-trapped beams.

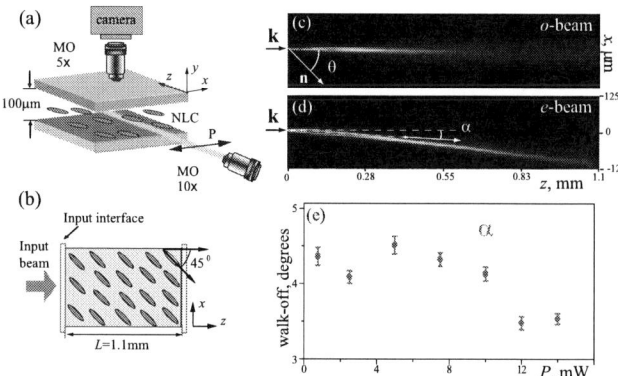

Fig. 1. (Color online) (a) Perspective view and (b) top view of the NLC planar cell. The ellipses indicate the orientation of the molecules in the plane (*x, z*). MO - microscope objectives; P - polarizer. Experimental results: (c) Linear diffraction of an *o*-beam ($P = 2,5$ mW) and (d) self-trapping and walk-off of the *e*-solitary beam ($P = 14$ mW). (e) Walk-off angle vs power, as defined in (d) [7,8].

DYNAMICS OF COUNTER-PROPAGATING NEMATICONS

One of the simplest processes in nonlinear optics leading to a variety of complex nonlinear phenomena is the mutual interaction of two CP optical beams. Numerous concepts in nonlinear optics, such as phase conjugation, Bragg reflection by volume gratings, wave-mixing in photorefractive media, are based on this geometry. Nevertheless, the simple CP geometry can give rise to an extremely complicated and sometimes counterintuitive dynamical behaviour, including both mutual beam self-trapping and the formation of stationary states, as well as complex spatiotemporal instabilities [6].

Earlier fundamental concepts motivated the studies of spatial optical solitons created by two beams propagating in of two CP beams can lead to the formation of a novel type of *vector soliton* [6], for both coherent and incoherent interactions.

978-1-4244-7043-3/10 $26.00 © 2010 IEEE

CAOL*2010 International Conference on Advanced Optoelectronics & Lasers, 10-14 September, 2010, Sevastopol, Ukraine

In experiments we use a planar cell of length 1.1 mm consisting of two polycarbonate slides spaced by 100µm and filled with NLC 6CHBT. The inner surfaces of the slides are unidirectionally rubbed in order to align the NLC molecular director $\mathbf{n}$ in the plane (x, z) at 45° with respect to the z-axis. Such boundary conditions in (x, z) are analogous to the bulk pre-alignment in (y, z) by an external bias voltage. Two additional glass slides are attached perpendicularly to z from two opposite sides of the cell in order to define air-NLC input/output interfaces, avoiding lensing and depolarization due to NLC meniscus. A beam from a cw laser ($\lambda = 532$ nm) with wave-vector $\mathbf{k}$ parallel to the z-axis is coupled into the cell using a 10× objective, resulting in an input waist $w = 3$ µm. A halfwave plate controls the input polarization state and the nematicon evolution is monitored with a CCD camera collecting the light scattered out of the top slide.

First, we study the propagation of a single beam for various excitations. Figures 1(c,d) show the experimental results for ordinary (o) and extraordinary (e) beams. We observe that the o-beam diffracts, whereas the e-beam experiences self-focusing at $P > 1$ mW and forms self-localized state, a nematicon. However, due to high birefringence in our geometry, the nematicon propagates with walk-off with respect to z-direction. The decay of the walk-off angle with the input power is plotted in Fig. 1(e).

In order to investigate the propagation dynamics and interaction of CP solitons we launched beams from opposite sides of the same cell by focusing them with 10X micro-objectives. When the two inputs propagate towards each other with a finite impact parameter r [the offset, see dashed lines in Fig. 2(a,A)], we observe that two e-beams form nematicons and bend in opposite transverse directions due to walk-off. Since the walk-off depends on the input power, the interaction of CP nematicons depends on both, the offset r and the input power P; below we explore these dependencies in more details. Selected experimental results for the interaction of CP solitons are presented in Fig. 2(a). For a given offset r and small powers the beams do not interact, as in Fig. 2(a,A). With further decrease offset r (or increase of input power P) the two CP beams still remain spatially separated but the attraction between them leads to additional bending, clearly visible in Fig. 2(a,B) for $r = 130$ µm. We reveal the absence or presence of attraction by blocking one of the beams and observing the other one which relaxes (in time) to its independent (unperturbed) trajectory. The bound state or vector of two CP nematicons is formed for specific parameters r and P, as visible in Fig. 2(a,C). On the other hand, the sequence of Figs. 2(a,C) and 2(a,D) for $r = 105$ µm shows how the bound state can be destroyed by increasing the power, demonstrating in Fig. 2(a,D) the development of CP instability which splits otherwise stable vector nematicon.

The results of numerous experiments are summarized in Fig. 2(b) by distinguishing the regions of qualitatively different interaction scenarios described above. Importantly, in all our experiments the beams are launched parallel to the optical axis z with equal powers P and, since the interaction is *phase-independent*, the key parameters governing the dynamics are

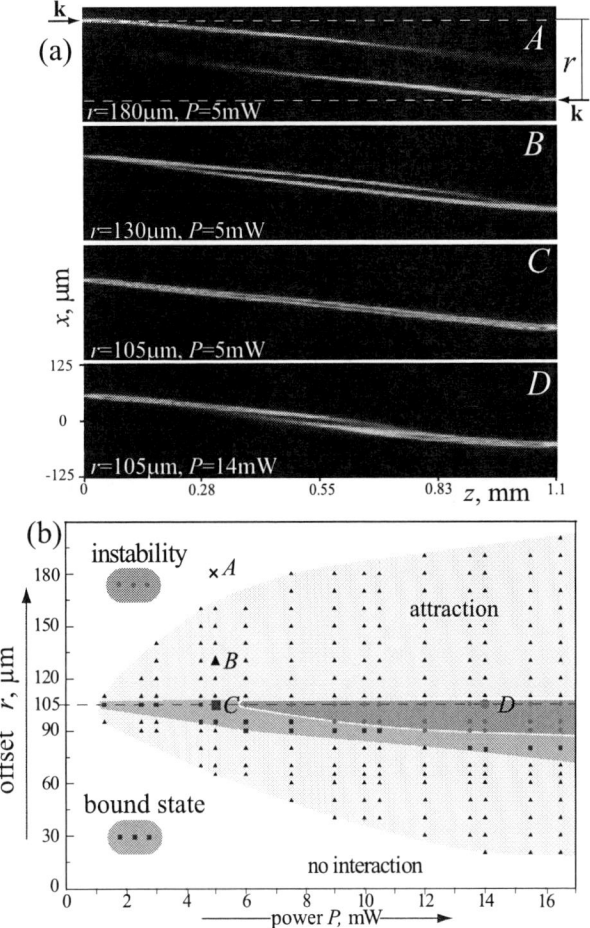

Fig. 2. (Color online) (a) Interaction of CP solitons for various input separations r: (A) $r = 180$ µm, no interaction; (B) $r = 130$ µm, soliton attraction; (C) $r = 105$ µm, bound state with a single common waveguide; (D) $r = 105$ µm, instability. The input power in each beam is $P = 5$ mW in (A-C) and $P = 14$ mW in (D). (b) Scenarios of CP soliton interaction shown on the plane (P, r). The points A-D correspond to the regimes shown in (a) [8].

the offset r and power P. For the power below a threshold value, no interaction is observed (blank area) and each nematicon's trajectory does not bend when the CP soliton is launched, see Fig. 2(a,A). The threshold is offset-dependant and it can be seen as the border of the gray domain of attraction marked with black triangles in Fig. 2(b). With increase of power we enter this domain of mutual attraction and observe the bending of two trajectories towards each other while the propagation remains stationary in time [see Fig. 2(b, B)]. The vector nematicons [Fig. 2(b,C)] are formed in the green area with blue square markers, and the instability [Fig. 2(b,D)] is observed in a narrow region marked with red circles in Fig. 2(b).

978-1-4244-7043-3/10 $26.00 © 2010 IEEE   119

SOLITON INTERACTION WITH CURVED SURFACES

The efficient and tunable routing of nematicons can be achieved by reflection or refraction of nematicons at the interface between two nonlinear media, each controlled by an independent external bias or by anchoring at the boundaries. Similarly, the localized nonlinear defects, induced by an additional "control" light beam, can be repulsive or attractive, depending on the induced change of the refractive index (negative or positive). However, if the boundary conditions are linear, e.g. fixed by the anchoring of NLC molecules, the interface became repulsive.

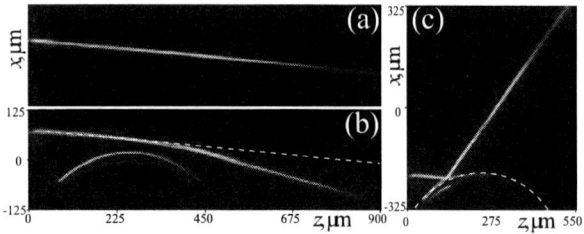

Fig. 3. (Color online) Experimental results of the nematicon interaction with curved dielectric surfaces. For better comparison we show the image of unperturbed nematicon in (a). The soliton-surface interaction in (b) bends soliton trajectory towards the curved NLC-air interface, the later visible as a blue curved contour. The dashed line in (b) corresponds to the trajectory of unperturbed soliton in (a). (c) Total internal reflection of a nematicon from a curved dielectric surface (dashed line shows the air bubble with radius $R = 270$ μm) [9].

To induce curved linear/nonlinear interface, we used the syringe and injected into the NLC layer micro-scale air bubbles. From the side view of the cell, such extended defects have cylindrical shape with the generatrices orthogonal to planar cell boundaries. Similar to experiments above, the nematicons are generated by monochromatic beam focused on the NLC layer by the 10x microobjective into a laser spot of several microns in the sample. The formed soliton travels in the NLC along the direction of Poynting vector with a significant spatial walk-off angle due to the large birefringence of 6CHBT liquid crystal, see Fig. 3(a) and dashed line in Fig. 3(b). In our experiments we kept the input power of the nematicon $P = 3,5$ mW constant while varying the relative distance between soliton and the defect.

Finally, when the target soliton trajectory crosses the bubble surface, as in Fig. 3(c), the soliton experiences the total internal reflection, observed earlier in different settings. Unlike the cases studied earlier, however, in our geometry the nematicon steering angle may exceed 90^0. Surprisingly, despite very different propagation directions after reflections, the solitons propagate similar to rays of light, with hardly visible differences of their widths and profiles.

CONCLUSIONS

We experimentally investigated generation and power-dependence of interacting counter- propagating spatial solitons in unbiased nematic liquid crystals. We demonstrated that two counter-propagating nematicons can form vector solitons with two components propagating in opposite directions, as well as experience dynamic instabilities. We performed series of experiments varying two key parameters, the transverse separation of CP nematicons r and their excitation power P, and identified the parameter domains where different dynamics can be observed. Also, we studied experimentally the interaction of spatial optical solitons with curved dielectric surfaces and demonstrated that, by varying the impact parameter, the soliton beam can be efficiently steered by using its attractive interaction with the surface. We have observed experimentally the total internal refraction of the soliton beam from an interface between air and liquid crystal with the soliton deviation angle exceeding 90^0.

Acknowledgements. The author is grateful to V. Shvedov, A. S. Desyatnikov, Yu. S. Kivshar, W. Krolikowski, G.Assanto, and M. Beli`c for essential contributions. This work was supported in part by the Australian Research Council through the Discovery Project.

REFERENCES

[1] Yu. S. Kivshar and G. P. Agrawal, *Optical Solitons: From Fibers to Photonic Crystals* (Academic Press, 2003).

[2] M. Peccianti, G. Assanto, A. De. Luca, C. Umeton, and I. Khoo, "Electrically assisted self-confinement and waveguiding in planar nematic liquid crystal cells," Appl. Phys. Lett. 77, 7-9 (2000).

[3] M. Peccianti, C. Conti, and G. Assanto, "Interplay between nonlocality and nonlinearity in nematic liquid crystals," Opt. Lett. 30, 415-417 (2005).

[4] M. Peccianti, K. A. Brzdiakiewicz, and G. Assanto, "Nonlocal spatial soliton interactions in nematic liquid crystals," Opt. Lett. 27, 1460-1462 (2002).

[5] J. Henninot, J. Blach, and M. Warenghem, "Experimental study of the nonlocality of spatial optical solitons excited in nematic liquid crystal," J. Opt. A 9, 20-25 (2007).

[6] See, e.g., the recent review: M. S. Petrovi`c, M. R. Beli`c, C. Denz, and Yu. S. Kivshar, "Counterpropagating optical beams and solitons," Las. Photon. Rev., in press (2010); preprint: arXiv:0910.4700.

[7] Ya. V. Izdebskaya, V. Shvedov, A. S. Desyatnikov, W. Krolikowski, G. Assanto, and Yu. S. Kivshar, "Incoherent interaction of nematicons in bias-free liquid-crystal cells," J. of the Eur. Opt. Soc. - Rapid Publ. 5, 10008-4 (2010).

[8] Ya. V. Izdebskaya, V. G. Shvedov, A. S. Desyatnikov, W. Z. Krolikowski, M. Belic, G.Assanto, and Yu. S. Kivshar, "Counterpropagating nematicons in bias-free liquid crystals," Opt. Express 18, 3258 (2010).

[9] Ya. V. Izdebskaya, V. G. Shvedov, A. S. Desyatnikov, W. Z. Krolikowski, and Yu. S. Kivshar, "Soliton bending and routing induced by interaction with curved surfaces in nematic liquid crystals," Opt. Lett., 35, 10, 1692-1694 (2010).

Reversible optical nonreciprocity in periodic structures with liquid crystals

(Invited Paper)

Andrey E. Miroshnichenko*, Etienne Brasselet[†], Yuri S. Kivshar*,

*Nonlinear Physics Centre and Centre for Ultra-high bandwidth Devices for Optical Systems (CUDOS),
Australian National University, Canberra ACT 0200, Australia
[†]Centre de Physique Optique Moléculaire et Hertzienne, Université Bordeaux 1, CNRS,
351 Cours de la Libération, 33405 Talence Cedex, France

Abstract— We suggest a novel approach to achieve reversible nonreciprocal optical response in a periodic photonic structure with a pair of defects, one of them being a nonlinear liquid crystal. We show that nonreciprocal effects can be reversed by changing the wavelength due to change of localization properties of defect modes.

I. INTRODUCTION

Optical reciprocity relates input and output waves, and is usually associated with time-reversal symmetry and phase-conjugation [1]. It can be even traced back to the third Newtonian law of equality of action and reaction. The standard approach to derive an optical reciprocity theorem is to start directly from the Maxwell's equations, which for the first time was done by Lorentz [2]. This results was further generalized for scattering by an obstacle by de Hoop [3], which is valid under rather general assamptions, which relates the reciprocity to existence of some symmetry for reverse energy flow in an optical system. In addition to that, there were several attempts to derive principles of reversibility of light beam based on energy balance relations, but their failure was demonstrated by Knittl [4]. The well-known example of nonreciprocal response if the magneto-optical (Faraday or Kerr) effect of a circular polarized light in gyrotropic materials [5]. Applied magnetic field breaks the time-reversal symmetry, which leads to nonresiprocal behaviour.

Another kind of reciprocity holds for light scattering by stratified media [6]. Namely, the reciprocity of transmission and reflection for the same input from opposite directions. Originally this problem was studied by Stokes [7] for light scattering at a plane interface, for which he derived reciprocity relations for scattering amplitudes. It can be shown that under quite general conditions transmission through a stratified medium is reciprocal. However, the reflection is not. The phases of the reflected waves in both directions are different. The difference can be further enhanced by using asymmetrically located absorbing layers [8]. It led to generalized reciprocity relations for reflected amplitudes derived by Agarwal and Gupta [9], which underpin the role of spatial symmetry in optically active media. There are other approaches which demonstrate a nonreciprocal response without magneto-optical effect. The most common one is the use of nonlinearity in systems with broken spatial symmetry [1]. It was successfully

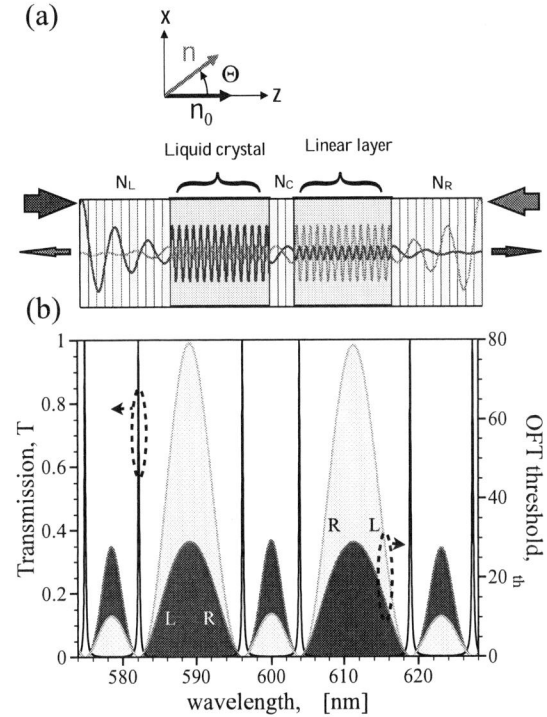

Fig. 1. (Color online) Schematic view of an one-dimensional photonic structure with asymmetrically placed NLC and dielectric defect layers: (a) director **n** describes the local molecules orientation inside liquid crystal. (b) Transmission T and normalized threshold intensity ρ_{th} for light sent in opposite directions.

applied to achieve all-optical diode and directional coupler operation.

We study light propagation in a one-dimensional periodic structure with defect layers [see Fig. 1(a)]. We are interested in different optical responses for the light impinging from the left or right. In general, the scattering setup implies that the light field distribution inside the structure is an asymmetric one. The introduction of defects will enhance this asymmetry even further. Due to specific properties of the transfer matrix the transmission in both directions will be the same regardless the field distribution inside [1]. But if we chose an optically sensitive material as the defect layer the situation may drastically change. To demonstrate this idea we introduce nematic liquid crystal (NLC) defects and utilize optical Fréederickz transition phenomenon [10], [11], [12]. It will allow to change the effective refractive index of LC beyond a certain power threshold of the light, which is inversionally proportional to the electric field intensity inside the liquid crystal cell. As a result, the threshold will be different for the light impinging from different directions due to asymmetric field distribution [13]. Due to threshold modulation the nonreciprocal response can be reversed by proper choosing the wavelength. It makes the whole structure colour sensitive with the possibility to invert the nonreciprocal response.

II. DESCRIPTION OF THE MODEL

We consider one-dimensional periodic structure consisting of alternating layers of SiO_2 and TiO_2. For dielectric layers of thicknesses $d_{SiO_2} = 103$ nm, and $d_{TiO_2} = 64$ nm there is a bandgap in the visible range between 520 and 710 nm. Without loss of generality, we consider a NLC defect layer with typical refractive indices $n_\perp = 1.5$ and $n_{||} = 1.7$, and normalized elastic constants $K_1/K_3 = 2/3$ and $K_2/K_3 = 1/2$.

The dynamics of the NLC is usually described by the director $\mathbf{n}$, which defines the local molecular orientation, and corresponding equations are obtained in a standard manner using the free energy density [10]. For homeotropic alignment and incident plane-wave linearly polarized along the x-axis, the director can be described by a single angle Θ [see Fig.1 (b)] that only depends on space coordinate z and time t following

$$\mathbf{n}(z,t) = \{\sin\Theta(z,t), 0, \cos\Theta(z,t)\} . \quad (1)$$

Due to homeotropic anchoring boundary conditions $\Theta(0,t) = \Theta(L,t) = 0$ of the NLC layer of thickness L the angle Θ can be uniquely expand onto Fourier series [14]. Near threshold dynamics can be effectively described by using projection technique with just few modes retained.

For convenience, we introduce normalized intensity $\rho = I/I_{lin}$, where $I_{lin} = \pi^2 cK_3\epsilon_{||}/(\epsilon_a\epsilon_\perp L^2)$ is the OFT threshold for linearly polarized light [10]. The time t and spatial coordinate z are normalized following $\tau = t/\tau_{NLC}$, with $\tau_{NLC} = \gamma_1 L^2/\pi^2 K_3$, and $\xi = z/L$. In addition, we define the total phase delay between the extraordinary and ordinary waves with refractive indices $n_e(\xi,\tau) = \left[\epsilon_\perp\epsilon_{||}/(\epsilon_{||}\cos^2\Theta(\xi,\tau) + \epsilon_\perp\sin^2\Theta(\xi,\tau))\right]^{1/2}$ and $n_\perp$, respectively,

$$\Delta(\tau) = k_0 L \int_0^1 (n_e(\xi,\tau) - n_\perp)d\xi . \quad (2)$$

This integral quantity is a measure of the overall director reorientation, which is accessible to experiments.

To study the scattering properties of the one-dimentional layered structure with anisotropic defect layers, we utilize the transfer matrix approach by using the Berreman's 4×4 matrix formalism, where Maxwell's equations can be written as follows

$$\frac{d\Psi}{dz} = ik_0\mathbf{D}(z)\Psi , \quad (3)$$

with $\Psi = (E_x, H_y, E_y, -H_x)^T$ and the matrix $\mathbf{D}(z)$ only depends on dielectric properties [10].

For calculation purposes, the whole structure is divided into thin layers with constant permittivity tensor that define a finite discrete set of matrices $\mathbf{D}_n$ [15], [16]. The resulting matrix is a product of matrix exponents of all layers, $\mathbf{P} = \prod \exp(ik_0\mathbf{D}_n)$, which links the scattering transmitted $(\Psi^{(t)})$ amplitude to the incident $(\Psi^{(i)})$ and reflected $(\Psi^{(r)})$ ones,

$$\Psi^{(t)} = \mathbf{P}\left(\Psi^{(i)} + \Psi^{(r)}\right) . \quad (4)$$

The input power of the light at the threshold is inversionally proportional to the averaged intensity of the electric field inside liquid crystal cell $\rho_{th} \sim 1/\langle|E_x|^2\rangle_{LC}$ [16]. On the other hand, the threshold power is strongly dependent on the input wavelength and number of the periodic layers. All those dependencies can be understood in terms of the photonic band gap, where the light distribution inside the structures is significantly affected by the Bragg's reflection mechanism. As a result, at the defect modes the threshold is extremely low due to electric field amplification, and becomes relatively large in between two nearest defect modes, because of exponential field intensity decay within the band gap region. If we now place the NLC defect asymmetrically inside the periodic structure the threshold value immediately become different for light impinging from left or right due to asymmetric light confinement. It turns out, that from one side of the structure it requires less input power to achieve the critical field intensity inside the liquid crystal cell than from another. Above the threshold the effective refractive index of NLC will change with the reorientation of the molecules resulting in variation of the optical properties of the whole structure. In other words, the system can be switched from reciprocal to nonreciprocal by using the input light intensity only, which makes this kind of structures quite unique.

The situation becomes even more interesting when we introduce the second defect [see Fig. 1(b)]. We consider a dielectric defect layer, which is identical to the liquid crystal layer at rest, i.e. whose refractive index is the same as ordinary one of NLC, $n_{def} = n_\perp$, and of the same width. By putting these two defects, NLC and dielectric ones, symmetrically inside the periodic structures, with $N_L = N_R$, will result in restoration of the mirror symmetry of the whole structure at rest. Due to this symmetry the distribution of the field intensity exactly at the defect modes will be symmetric as well. For all other conditions the field distribution will be asymmetric due to the scattering boundary conditions. Thus, one can imaging the following situation. Let the field be localized mostly in one of the defects for the light impinging from one side. Then, the field distribution will be reversed for the light sent from the opposite direction, because of the reflection symmetry. In other words, the averaged field

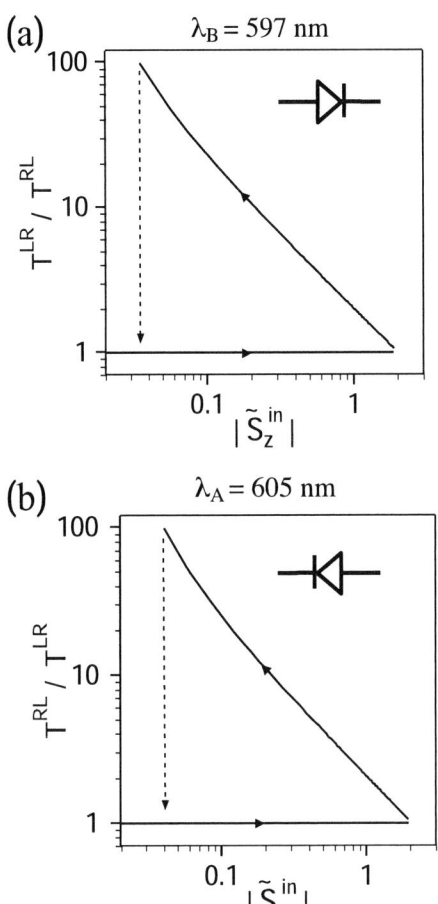

Fig. 2. (Color online) Ratio of the transmission field in opposite directions (a) $T^{\mathrm{RL}}/T^{\mathrm{LR}}$ for wavelength $\lambda = 605$nm, and (b) $T^{\mathrm{LR}}/T^{\mathrm{RL}}$ for wavelength $\lambda = 597$nm. Both plots are nearly identical which illustrates the reversability of the ONR unidirectionality.

III. Conclusion

We have demonstrated that periodic structures with asymmetrically placed a nematic liquid crystal defect may exhibit optical nonreciprocity. Asymmetric field distribution inside the photonic band gap results in different optical Fréedericksz thresholds for light impinging for the left or right. The variation of the liquid crystal optical properties above the transition alternates the optical properties of the whole structure. Therefore, the difference of threshold values from left and right implies that there is an interval of input intensities where nonreciprocal response can be observed. It can be clearly seen in difference of dynamics of the system for the same input intensity in both directions, where the system switches or not to a highly excited state. Due to alternation of the sign the thresholds difference of the given structure with two defects the nonreciprocal response alternates as well. And is some situation it can be completely reversed. This tunability of the nonreciprocal response makes the structure very attractive for future all-optical switching applications.

References

[1] R. J. Potton, "Reciprocity in optics," *Reports on Progress in Physics*, vol. 67, no. 5, pp. 717–754, 2004.

[2] H. A. Lorentz, *Versl. Kon. Akad. Wetensch. Amsterdam*, vol. 4, p. 176, 1895.

[3] A. de Hoop, "A reciprocity theorem for the electromagnetic field scattered by an obstacle," *Applied Scientific Research, Section B*, vol. 8, no. 1, pp. 135–140, 12 1960/12/01/.

[4] Z. Knittl, "The principle of reversibility and thin film optics," *Opt. Acta*, vol. 17, pp. 33–45, 1962.

[5] A. Zvezdin and V. Kotov, *Modern Magnetooptics and Magnetooptical Materials*, ser. Condensed Matter Physics Series. Taylor & Francis, 1997.

[6] M. Nieto-Vesperinas and E. Wolf, "Generalized stokes reciprocity relations for scattering from dielectric objects of arbitrary shape," *J. Opt. Soc. Am. A*, vol. 3, no. 12, pp. 2038–2046, 1986.

[7] G. G. Stokes, *Cambridge Dublin Math. J.*, vol. 4, p. 1, 1849.

[8] A. Armitage, M. S. Skolnick, A. V. Kavokin, D. M. Whittaker, V. N. Astratov, G. A. Gehring, and J. S. Roberts, "Polariton-induced optical asymmetry in semiconductor microcavities," *Phys. Rev. B*, vol. 58, no. 23, pp. 15 367–15 370, Dec 1998.

[9] G. S. Agarwal and S. D. Gupta, "Reciprocity relations for reflected amplitudes," *Opt. Lett.*, vol. 27, no. 14, pp. 1205–1207, 2002.

[10] H. L. Ong, "Optically induced freedericksz transition and bistability in a nematic liquid crystal," *Phys. Rev. A*, vol. 28, no. 4, pp. 2393–2407, Oct 1983.

[11] N. V. Tabiryan, A. V. Sukhov, and B. Y. Zel'dovich, "Orientational optical nonlinearity of liquid crystals," *Molecular Crystals and Liquid Crystals*, vol. 136, pp. 1 – 139, 1986.

[12] I. C. Khoo, "Nonlinear optics of liquid crystalline materials," *Phys. Rep.*, vol. 471, no. 5-6, pp. 221 – 267, 2009.

[13] A. E. Miroshnichenko, E. Brasselet, and Y. S. Kivshar, "Reversible optical nonreciprocity in periodic structures with liquid crystals," *Appl. Phys. Lett.*, vol. 96, no. 6, p. 063302, 2010.

[14] E. Brasselet, T. V. Galstian, L. J. Dubé, D. O. Krimer, and L. Kramer, "Bifurcation analysis of optically induced dynamics in nematic liquid crystals: circular polarization at normal incidence," *J. Opt. Soc. Am. B*, vol. 22, no. 8, pp. 1671–1680, 2005.

[15] A. E. Miroshnichenko, I. Pinkevych, and Y. S. Kivshar, "Tunable all-optical switching in periodic structures with liquid-crystal defects," *Opt. Express*, vol. 14, no. 7, pp. 2839–2844, 2006.

[16] A. E. Miroshnichenko, E. Brasselet, and Y. S. Kivshar, "Light-induced orientational effects in periodic photonic structures with pure and dye-doped nematic liquid crystal defects," *Phys. Rev. A*, vol. 78, no. 5, p. 053823, 2008.

intensity inside the NLC defect layer will be different for light impinging from different directions. As a consequence of that, the light-induced Fréedericksz transition threshold will be different as well, leading to nonreciprocal response. Our numerical results support this idea. For simulations we took equal number of layers $N_L = N_R = 5$ from left and right, and $N_D = 1$ periodic layer between two defects. Note here, that the difference of the thresholds from both side alternates its sign. It is exactly zero at the defect modes, indicating that both thresholds become identical, because of the restoration of the reflection symmetry, and, consequently, of the reciprocity. Such alternation of sign originates from the interplay between even and odd defect modes supported by such structures. The difference of the thresholds suggests that there are such values of the input power when the system starts to change its optical properties for light sent from one direction, and remains at rest for light sent from another direction [see Fig. 2].

Fast Near-Infrared CdHgTe:V:Mn Photorefractive Material for Optical and Biomedical Applications

Yuriy P. Gnatenko[1], Petro M. Bukivskij,[1] Yuriy P. Piryatinski,[1] Ivan O. Faryna,[1] Mykhaylo S. Fur'yer[1],
Oleg A. Shigiltchoff,[1] Roman V. Gamernyk[2], N. Kukhtarev[3], T. Kukhtareva[3]

[1]Institute of Physics of NASU, Kyiv, Ukraine
[2]Lviv National University, Lviv, Ukraine
[3]Alabama A&M University, Huntsville, Alabama, USA

Abstract: We describe spectroscopic investigations of $Cd_{1-x}Hg_xTe:V:Mn$ crystals. It was shown that the electronic processes in this material are fast and occur in nanosecond range. These materials are very sensitive up to 1800 nm. The nature of the electronic transitions which determine their NIR photosensitivity was established.

At present there is a strong need for fast near-infrared (NIR) photorefractive materials and devices with the potential applications in the areas of optical communications, optical storage, image recognition, optical computing and optoelectronics, including the elaboration of novel adaptive sensors for laser systems of remote ultrasonic inspection of the quality of different industrial goods and highly sensitive laser vibrometers for the detection of vibration over a wide frequancy spectrum [1, 2, 3, 4]. Last decade the photorefractive materials are also successfully used for the development of a high-resolution, non-destructive contact-free holographic method for image of biological tissue [5]. The photorefractive holography allows to investigate human tissue physiology from millimeters to centimeter below the tissue surface by producing real microscopic analysis of the fine structure of the holographically reconstructed-time depth resolved 3D-dimenshional imaging from different depths within a biological object. Thus this technique allows to obtain a new kind of medical imaging based on light instead of tissue-damaging X-rays. Light in scattering media such as most human tissue is both scattered and adsorbed within object. This directly limits the maximum thickness of biological tissue through which it is possible to image. In order to minimize absorption it is necessary to use a transmission window for biomedical diagnostics : the NIR wavelength from 800 nm to 1700 nm [6]. Use this wavelength region allows to reduce scattered coefficient of tissue. By using NIR light it is possible to resolve features of tissue more than an order of magnitude than ultrasound and to provide imaging resolution as fine as 10 μm.

At first, a photorefractive rhodium-doped barium titanate crystal (bulk ferroelectric crystal) was used for photorefractive holography [6]. This crystal have a high sensitive. However its response time, i.e. the time taken for grating being written in the photorefractive medium to reach its maximum steady-state value, was limited to several second. For many biomedical applications it is desirable to image moving or evolving samples. Thus it is necessary to use the photorefractive sensors with a fast response time. Last years the photorefractive multiple wells (MQW) are used for the depth-resolved holographic biomedical imaging which have a response time smaller than 1 ms. In this case the depth resolved images can be viewed in real-time by using fiber coupled laser diodes with center wavelength ranging from 760 nm to 850 nm. These photorefractive sensors are composed of GaAs and AlGaAs layers on a GaAs substrate and are about 1 μm thick. They require typical electric field of 6-10 kV/cm and have a optical bandwidth of 830-850 nm. This is problem since, as it was above mentioned, it is necessary to use longer wavelength (>850 nm) to reduce scattering and absorption in biological tissue. In the case of the MQWs as photorefractive imaging system there is another problem. It is well known[4] that the response time of the photorefractive materials is reverse proportional to the total average intensity in the signal and reference beams. It allows to reduce a response time by increasing the intensity of reference beam using high-power laser with low spatial coherence. Since the scattered signal beam is very low intensity (~1μW) it is possible to achieve reduce of response time by increasing the intensity of reference beam. However the power incident on the MQW sensor need to be limited to about 1 mW to avoid damage due to excessive photocurrents. It limits the application of the photorefractive holography in microscopy and endoscopy as well as for the imaging of dynamics living biological subjects and other evolving objects. It should be also noted that the optical quality of the photorefractive materials because unwanted scattering of the read-out beam by heterogeneities in the medium becomes the dominant source that determines the achievable dynamic range of photorefractive sensors and their performance. The use of photorefractive materials of very optical quality allows to increase the sensitivity (i.e. the minimum power level of coherent scattered signal light required to record an image) and thus the imaging depth through scattered media such as biological tissue. It has been shown [7, 8, 9, 10, 11, 12] that an efficient method of obtaining the information on the energy and crystal structure of deep impurity centers and structural defects for CdTe-based semiconductor crystals involves complex low-temperature investigation of their optical and photoelectric properties. In order to characterize the optical quality of the $Cd_{1-x}Hg_xTe:V:Mn$ crystals we used the measurements of their optical properties.

978-1-4244-7043-3/10 $26.00 © 2010 IEEE

CAOL*2010 International Conference on Advanced Optoelectronics & Lasers, 10-14 September, 2010, Sevastopol, Ukraine

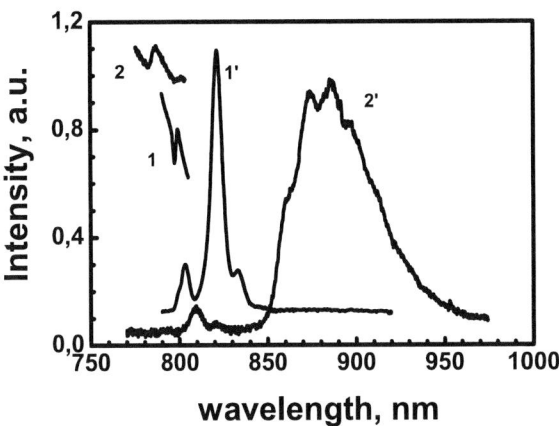

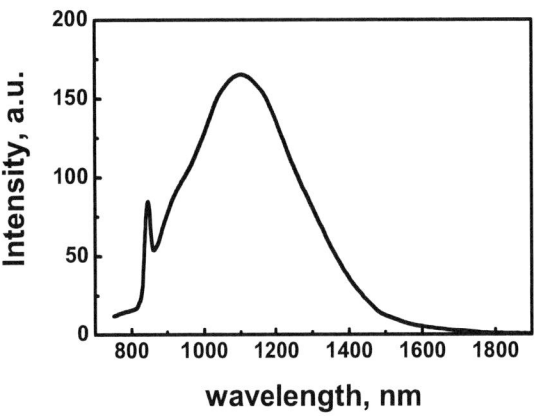

Fig.1. The photoluminescence and exciton reflection spectraa for $Cd_{1-X}Hg_XTe$:V:Mn (X=0.0183, curves 1 and 1') and $Cd_{1-X}Hg_XTe$:V (X=0.0075, curves 2 and 2') crystals at T=4.5 K

Fig. 2 Photoconductivity spectrum for $Cd_{1-X}Hg_XTe$:V:Mn (X=0.0183) crystals at T=4.5 K.

In Fig. 1 the excitonic photoluminescence and reflection spectra for the different crystal samples are shown at 4.5 K. An analysis of the observed photoluminescence spectra reveal an an exciton emission A^0X-line at a short-wavelength region. The free excitons occur in reflection spectra. It should be noted that the exciton reflection band of $Cd_{1-X}Hg_XTe$:V crystals is shifted in the long-wavelength region as compared to the CdTe:V crystal [4, 8, 9, 11, 12]. The fact that low-temperature exciton reflection and photoluminescence spectra are observed indicates that the $Cd_{1-X}Hg_XTe$:V:Mn crystals are of fairly good optical and crystal quality. In the case of $Cd_{1-X}Hg_XTe$:V (X=0.0075) the semi-width of the exciton reflection band is considerably larger than for the Mn-codoped $Cd_{1-X}Hg_XTe$:V crystal.

The decreasing of the semi-width of exciton band which take place as a result of the Mn-codoping is caused by the decreasing of the intrinsic defects. In this case the improving of the optical quality of crystals and the increasing of the resistance of such materials take place. Both factors are very important for the obtaining of very effective semiconductor photorefractive materials.

The characteristic peculiarity of the $Cd_{1-X}Hg_XTe$:V crystal is the appearance of the intensive PL bands in the spectral range 840-940 nm which are due to the donor-acceptor recombination with the acceptor complexes of $(V^{2-}+D^+)^-$ type. There is a direct correlation between the photoluminescence bands in this spectral range and the dislocation density [3, 4, 12]. The bands increase in intensity on the dislocation density. Therefore the Mn-codoping also leads to the strong decreasing of the dislocation in the $Cd_{1-X}Hg_XTe$:V:Mn crystals.

Information on the energy position of the defects relative to the crystal energy bands as well as on the photosensitivity spectral range may be obtained by measuring the photoconductivity (PC) spectrum, which for $Cd_{1-X}Hg_XTe$:V:Mn crystal is shown in Fig. 2 at 300 K.

The appearance of the intensive band in the short-wavelength region near 860 nm is caused by the phototransitions from the discrete level of a single charged cadmium vacancy (V^-_{Cd}) [14] to the conduction band. The PC spectrum also shows a broad band in the long-wavelength spectral range which is caused by the phototransitions from the main $^4T_1(F)$ state to the different spin-orbit compounds of the excited $^4T_1(P)$ state of the V^{2+} ions. The later state is a resonance with the conduction band. The long-wavelength edge of the PC impurity band is due to the direct photoionization of V^{2+} ions and the transition to the other excited $^3A_2(F)$-state which also is in the conduction band. For the $Cd_{1-X}Hg_XTe$:V:Mn crystal (X=0.0083) the long-wavelength edge of a broad band extends up to 1800 nm.

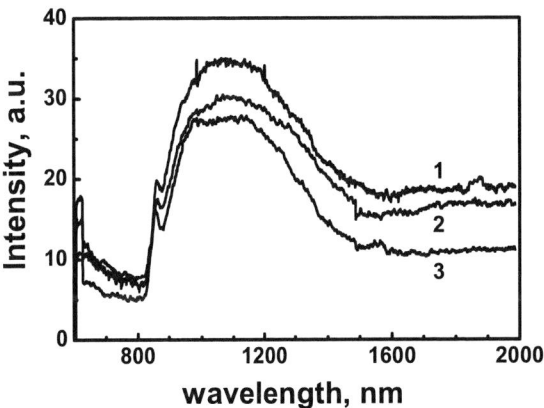

Fig. 3 Spectral dependence of the intensity of pulse photocurrent for $Cd_{1-X}Hg_XTe$:V (X=0.0075) crystals at T=300 K. Curves 1-3 correspond to a delay time of 5.0, 15.0 and 40 ns, respectively.

978-1-4244-7043-3/10 $26.00 © 2010 IEEE

The very important characteristic of the semiconductor photorefractive materials is their speed which is determinated by the electronic processes in the crystals and depends on their defect structure. Recently[16], we elaborated a new method, i.e. a time-resolved photoelectric spectroscopy (TRS).Use this method let us establish the nature and role of the impurity centers and intrinsic defects participating in the trapping and detrapping processes of electrons in the CdTe:V crystals. In this work we use this effective technique for the studies of the $Cd_{1-x}Hg_xTe$:V:Mn crystals. The TRS technique was presented in detail in [16]. For the excitation of the crystal a pulse nitrogen laser was used. The pulse nitrogen laser freaquency correspond to 100 Hz. The pulse duration was 9 ns and the pulse power of laser emission was 5 kW.

Fig. 3 represents the dependence of the PPC intensity for the delay time t_d=4.5 ns on the energy of stationary monochromatic illumination for $Cd_{1-x}Hg_xTe$:V:Mn crystal. It was shown that for CdTe:V crystal the wide band near 1100 nm is caused by the photoionization transitions of V^{2+} as a result of the autoionization of electrons from $^4T_1(P)$ excited state to the conduction band [13].

This is main mechanism of the electron realization from the deep impurity centers. In the case of $Cd_{1-x}Hg_xTe$:V:Mn crystals the long-wavelength wing of spectral dependence of PPC is protracted up to 1800 nm for It results obtained shown that the electronic processes in $Cd_{1-x}Hg_xTe$:V:Mn alloys with the participation of the impurity centers and intrinsic defects are high-speed and correspond to the nanosecond time region.

In summary, the results described here demonstrate that the $Cd_{1-x}Hg_xTe$:V:Mn crystals are very promising NIR photorefractive materials for optical communications and for photorefractive holography to investigate dynamics living biological subjects and other evolving objects.

Acknowledgements

This work has been supported by Grant No UKE2-2855-KV-07 of U.S. Civilian esearch and Development (CRDF).

REFERENCES

[1] T.O. Santos, J.C. Launay, J. Frejilich, "Photo-electromotive-force from volume speckle pattern vibration with large amplitude", *J. Appl. Phys.*, vol. 103, no. 11, note 113104.1-113104.6, 2008.

[2] I. Lahiri, L.J. Pirak-Nolte, D.D. Nolte, M.R. Melloch, R.A. Kruger, G.D. Bacher and M.V. Klein, "Laser-based ultrasonic detection using photorefractive quantum wells", *Appl. Phys. Lett.*, vol. 73, no. 8, pp. 1041-1043, 1998.

[3] Yu.P. Gnatenko, M.S. Brodyn, I.O. Faryna, P.M. Bukivskij, O.A. Shigilchoff and M.S. Fur'yer, R.V. Gamernyk, N. Kukhtarev, and T. Kukhtareva, "Energy structure and micromechanism of photo-electromotive force effect in V-doped CdTe crystals", *Phys. Stat. Sol.* (a), vol. 204, no. 7, pp. 2431-2440, 2007.

[4] Yu.P. Gnatenko, P.M. Bukivskij, Yu.P. Piryatinski, I.O. Faryna, O.A. Shigiltchoff, and R.V. Gamernyk, "Spectroscopic study of V-doped $Hg_{0.018}Cd_{0.981}Mn_{0.001}Te$ bulk crystals as near-infrared materials for optical applications", *Appl. Phys. Lett.*, vol. 95, no. 11, p. 112109, 2009.

[5] C. Dunsby, Y. Gu and P.M.W. French, "Single-shot phase-stepped wide-field coherencegated imaging", *Optics Express*, vol. 11, no. 2, pp. 105-115, 2003.

[6] N.P. Barry, R. Jones, S.C. Hyde, J.C. Dainty, and P.M.W. French, "High background holographic imaging using photorefractive barium titanate", *Electronics Letters*, vol. 33, no. 20, p. 1732-1733, 1997.

[7] Yu.P. Gnatenko, A.O. Borshch, N. Kukhtarev, T. Kukhtareva, I.O. Faryna, V.I. Volkov, P.M. Bukivskij, R.V. Gamernyk, V.I. Rudenko, S.Yu. Paranchych, L.D. Paranchych, "Optical, photoelectric, and photorefractive properties of Ti-doped CdTe crystals", *J. Appl. Physics,* vol. 94, no. 8, pp. 4896-4903, 2003.

[8] Yu.P. Gnatenko, I.O. Faryna, P.M. Bukivskij, O.A. Shigiltchoff, R.V. Gamernyk, S.Yu. Paranchych and L.D. Paranchych, "Optical and photoelectric properties of vanadium-doped $Cd_{1-x}Hg_xTe$:V crystals" *J. Phys.: Condens. Matter*, vol. 14, no. 29, pp. 7027-7033, 2002.

[9] Yu.P. Gnatenko, I.O. Faryna, P.M. Bukivskij, O.A. Shigiltchoff, R.V. Gamernyk, S.Yu. Paran-chych, "Nature and energy structure of impurity and intrinsic defects n V-doped $Cd_{1-x}Hg_xTe$:V crystals", vol. 20, no. 5, pp. 378, *Semicond. Sci. Technol.*, vol. *20*, no. 5, pp. 378-388, 2005.

[10] Yu.P. Gnatenko, Yu.P. Piryatinski, P.M. Bukivskij, D.D. Kolendryckyj, O.A. Shigilchoff and R.V. Gamernyk, "Time-resolved photoelectric spectroscopy of photorefractive CdTe:V crystals", *J. Phys.: Condens. Matter*, vol. 18, no. 42, pp. 9603-9613, 2006.

[11] R.V. Gamernyk, Yu.P. Gnatenko, P.M. Bukivskij, P.A. Skubenko, and V.Yu. Slivka, "Optical and photoelectric spectroscopy of photorefractive $Sn_2P_2S_6$", *J. Phys .: Condens. Matter*, vol. 18, no. 23, pp. 5323-5333, 2006.

[12] Yu.P. Gnatenko, R.V. Gamernyk, I.A. Farina and P.I. Babii, "Deep impurity states and intrinsic defects in photorefractive $Cd_{1-x}Fe_xTe$ crystals", *Physics of the Solid State*, vol. 40, no. 4, pp. 564-568, 1998.

[13] Yu.P. Gnatenko, I.A. Farina, R.V. Gamernyk, V.S. Blashkiv, and A.S. Krochuk, "Impurity states of vanadium in cadmium and zinc telluride", *Semiconductors,* vol. 30, no. 11, pp. 1027-1030, 1996.

CAOL*2010 International Conference on Advanced Optoelectronics & Lasers, 10-14 September, 2010, Sevastopol, Ukraine

Nonlinear LBO and BBO crystals for extreme light sources

A.E. Kokh[1], T.B. Bekker[1], K.A. Kokh[1], N.G. Kononova[1], V.A. Vlezko[1], Ph. Villeval[2], S. Durst[2], D. Lupinski[2]

[1]Institute of Geology and Mineralogy SB RAS, 3, Koptyug ave., 630090, Novosibirsk, Russia
a.e.kokh@gmail.com
[2]CRISTAL LASER, Nancy, France

Abstract: Using the approach of heat filed symmetry change and rotation LBO crystals of more than 1.5 kg of weight were grown from MoO_3-containing melt-solution. High quality optical elements up to 65 mm in diameter were produced. At the same time with this approach we obtained high quality BBO crystals about 400 g using $NaBaBO_3$ flux. Elements with apertures up to 30x30 mm^2 were produced. The progress in the dimensions of grown crystals provides the opportunity to use them in laser systems of multi petawatt and exawatt power level.

Due to good operating performance LiB_3O_5 crystals (LBO) are widely used for frequency conversion in the visible and near UV regions. Compared to KDP, the LBO have more preferable operating characteristics in particular higher nonlinearity, angular tolerance and damage threshold as well as smaller absorption and larger thermal conductivity. But its applying in broad-aperture high power laser systems was limited due to difficulties to grow crystals with big sizes. According to the Curie principle, rhombic LBO crystals were grown in the heat field of the same symmetry by Kyropoulos method.

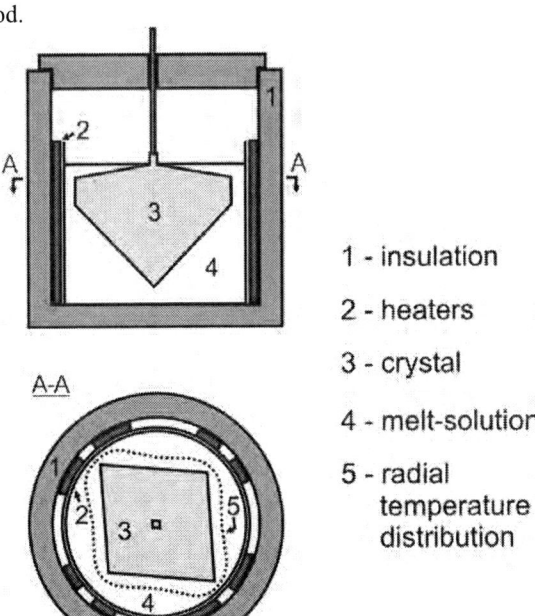

1 - insulation
2 - heaters
3 - crystal
4 - melt-solution
5 - radial temperature distribution

Fig. 1. Sketch of the growth zone.

Fig. 1 presents the sketch of the growth zone. Eight separate heaters around the crucible are responsible for a specific temperature distribution in the horizontal section. Electrical connection of the heaters is realized through power commutator which in turn provides switched on/off state of any heater for any time period. In this work a sequential switching of four from eight heaters was used: 1234->2345->3456... If the delay between switchings is high, an effective homogenization of the melt-solution may be achieved (Fig. 2a). Before the start of crystal growth the delay of switching is decreased to 3 sec, so the convection becomes similar to that if all heaters would be switched on at the same time (Fig. 2b). Crystal growth was performed from Li_2O-B_2O_3-MoO_3 melt-solution with molar ratio between oxides as 1:1.29:1.71, correspondingly.

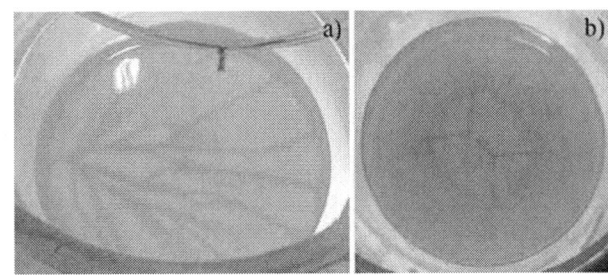

Fig. 2. Convective patterns on the surface of melt-solution; delay between switching of the heaters: (a) – 20 min, (b) – 3 sec.

But since imperfections in heaters and insulation result in nonsymmetrical radial temperature distribution, this provides in fact a significant progress in the growth technology. By means of correction of switching delay for some group of heaters, it is possible to shift the cold point on the melt surface in the geometric center of the crucible which is highly important for large diameter crucibles. Grown crystals (Fig. 3) possesses high quality with optical absorption for most of the samples less than 5 ppm/cm at 1064 nm. Currently the maximum diameter of NLO element cut from the crystals is 65 mm [1].

978-1-4244-7043-3/10 $26.00 © 2010 IEEE

CAOL*2010 International Conference on Advanced Optoelectronics & Lasers, 10-14 September, 2010, Sevastopol, Ukraine

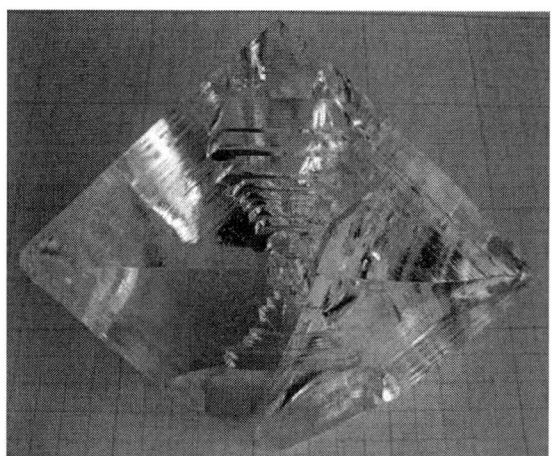

Fig. 3. As grown LBO crystal 1558g in weight.

In the optical experiments of Commissariat a l'Energie Atomique (Bordeaux, France), 217 J pulse of second harmonic wave (λ=527 nm) was achieved with ~92% efficiency [2].

Therefore, progress in LBO crystal growth and results of experiments on the transformation of laser radiation of high energy have risen the status and prospects of the crystal to a new level. Currently there are many debates of replacing the family of KDP crystals in high-power laser systems for more efficient LBO crystals, which will provide substantially higher levels of output power.

BBO crystals occupy a special niche in nonlinear optics. Despite the active search for new crystals for the near-UV range, BBO is still a widely-used commercial crystal. Its potential for powerful wide-aperture laser systems operating in the femtosecond regime remains very high. Due to the phase transition at 925°C, β-BaB$_2$O$_4$ crystals are commonly grown from high temperature solution. In high-temperature solution methods solvent must provide extended primary crystallization region of β-BaB$_2$O$_4$ and reduce the viscosity and glass-forming ability of the melt.

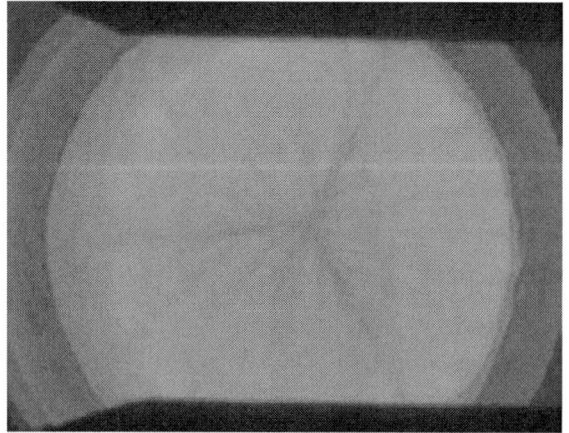

Fig. 4. Surface of the melt-solution under L$_3$ heat field.

Also, it must be of low volatility, possess stable properties and provide very weak interaction with the crystal to exclude the possibility of poisoning. Main solvents used for β-BaB$_2$O$_4$ crystal growth are compositions in BaO-B$_2$O$_3$-Na$_2$O system; NaF; mixed oxide-fluoride compositions.

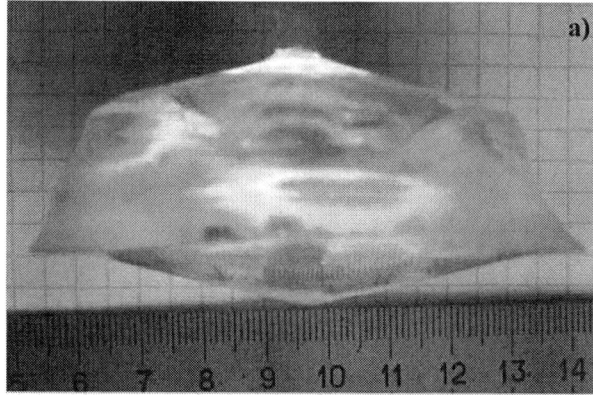

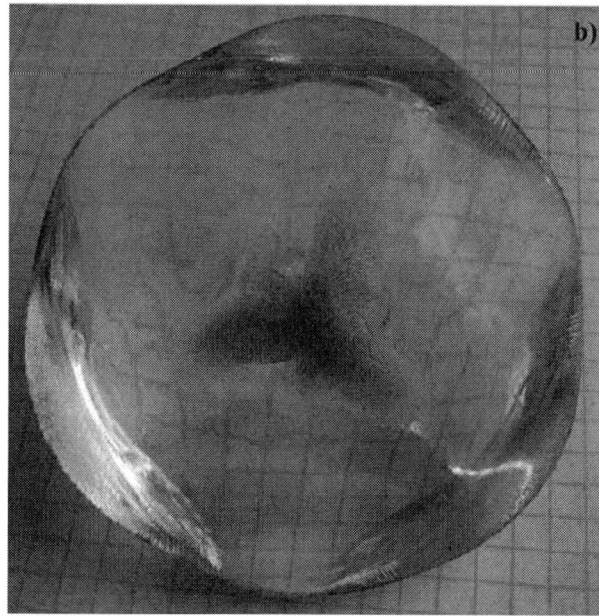

Fig. 5. As grown BBO crystal 400 g in weight; side (a) and bottom (b) views.

Because of high melt viscosity, growth of inclusion free crystal is quite complicated. So reproducible production of crystals for large aperture elements is still the aim to achieve.

In our prior publications we have shown the efficiency of β-BaB$_2$O$_4$ (BBO) crystal growth in the heat field of three-fold axis symmetry by top-seed solution method (TSSG) [3, 4]. In such heat field liquid body splits into three convective cells, in which the thermal-gravitational flows moved up-and-down and formed three intensive flows at the surface merging in the centre (Fig. 4). This convective regime provides a permanent renewal of high temperature solution in the region near crystallization interface. In order to involve the whole body of

978-1-4244-7043-3/10 $26.00 © 2010 IEEE   128

the high temperature solution in the convective motion it is very important to achieve vertical temperature distribution with the 'hot points' at the lower part of the growth crucible. At the same time temperature of the latter at the surface of high temperature solution must be always higher than crystallization temperature. In order to accomplish such temperature distribution two-zone heating furnace with three heating sectors in each zone has been developed. Load commutator included in thermoregulation system executes power distribution on each growth sector during the growth run.

The $BaO-B_2O_3-Na_2O$ system remains basic for growing BBO crystals. Using $NaBaBO_3$ compound as a solvent [5], we obtained high-quality crystals about 400 g of weight with the convex form of the crystallization front which is an evidence of the stability of the crystallization process (Fig. 5). Diameter of the crucible was 100 mm. Currently the maximum aperture of NLO element cut from the crystals is 30x30 mm^2 (Fig. 6).

Fig. 6. NLO element from BBO crystal, dimensions: 30x30xx6 mm^3

This work is supported by RFBR grant #09-02-12261-ofi_m in Russia and in France by OSEO Innovation, contract # A0610001L, by MIPI and by "Région de Lorraine", contract #A0610002L.

REFERENCES

[1] A. Kokh, N. Kononova, G. Mennerat et al, " Growth of high quality large size LBO crystals for high energy second harmonic generation", *J Crystal Growth*, vol. 312, pp. 1774-1778, 2010.

[2] Dr. Gabrial Mennerat – private communications.

[3] A.E. Kokh, N.G. Kononova, P.W. Mokruchnikov, "An azimuthal pattern of heat field in β-BaB₂O₄ crystal growth" *J Crystal Growth*, vol. 216, pp. 359-362, 2000.

[4] A.E. Kokh, V.N. Popov, T.B. Bekker et al, "Melt-solution BBO crystal growth under change of the heat field symmetry and its rotation" *J Crystal Growth*, vol. 275, pp. e669-e674, 2005.

[5] Fedorov, P.P., Kokh, A.E., Kononova, N.G., Bekker, T.B., "Investigation of phase equilibria and growth of BBO (β-BaB₂O₄) crystals in BaO-B₂O₃-Na₂O ternary system" *J Crystal Growth*, vol. 310, pp. 1943-1949, 2008.

CAOL*2010 International Conference on Advanced Optoelectronics & Lasers, 10-14 September, 2010, Sevastopol, Ukraine

Kaleidoscope laser properties
and new optical fractal contexts

J. M. Christian[1], G. S. McDonald[1], A. S. Heyes[1], J. G. Huang[2]

[1]Materials & Physics Research Centre, University of Salford, Greater Manchester M4 4WT, UK
[2]School of Engineering and Mathematical Sciences, City University London, London EC1V 0HB, UK

Abstract: We present the first detailed account of kaleidoscope laser modes where the equivalent Fresnel number N_{eq} and magnification M may assume arbitrary values. Properties of these *linear* fractal eigenmodes are explored through extensive numerical computations. Considerations are extended to demonstration and analyses of new contexts for spontaneous *nonlinear* optical fractals.

I. KALEIDOSCOPE FRACTAL LASERS

Unstable cavity lasers exhibit a plethora of bizarre and potentially exploitable phenomena that have fascinated researchers over the last four decades. In particular, the innate capacity of such simple systems for generating complex light patterns continues to attract wide and sustained interest. Within earlier collaborations [1,2], we discovered that the linear eigenmodes of one-dimensional (1D) and two-dimensional (2D) unstable resonators are fractals – patterns that exhibit proportional levels of detail spanning many length scales. The fractality of these self-reproducing mode profiles was shown to originate from the interplay between round-trip cavity magnification, M, and periodic diffraction at the defining aperture of the system (e.g., a small feedback mirror) [3].

Kaleidoscope lasers are intuitive generalizations of classic unstable resonators to fully-2D transverse geometries, where the defining aperture has the shape of a regular polygon [3]. The non-orthogonal edges of this element have a profound impact on the structure of the cavity eigenmodes, which exhibit a striking level of beauty and complexity (see Fig. 1). Most obviously, N-sided regular-polygon boundary conditions impose N-fold rotation symmetry on the intensity pattern. Kaleidoscope lasers offer enormous prospects for inspiring novel photonic device designs and architectures. For example, fractal light may prove to be more efficient for probing, scanning, writing and ablation applications in various technological, medical, research and industrial contexts.

Here, we present the first detailed analysis of kaleidoscope fractal lasers through accommodation of arbitrary equivalent Fresnel number N_{eq} and magnification M. All previous analyses have been restricted to regimes where either $N_{eq} = O(1)$ (in which case conventional ABCD matrix modelling can be deployed [4]) or $N_{eq} \gg O(1)$ (where asymptotic approximations can be used [5]). Our approach is based on a fully-2D generalization of Southwell's Virtual Source method

[6], and exploits exact (Fresnel) mathematical descriptions of the constituent diffraction patterns [7].

One key advantage of our technique is that a single calculation allows one to access entire families of modes; another is that any particular mode may be computed to any desired accuracy. We also quantify the convergence properties of kaleidoscope laser modes (eigenvalue spectra and mode patterns themselves in the limit $N \to \infty$, where the feedback mirror becomes circular). The analysis employs a technique

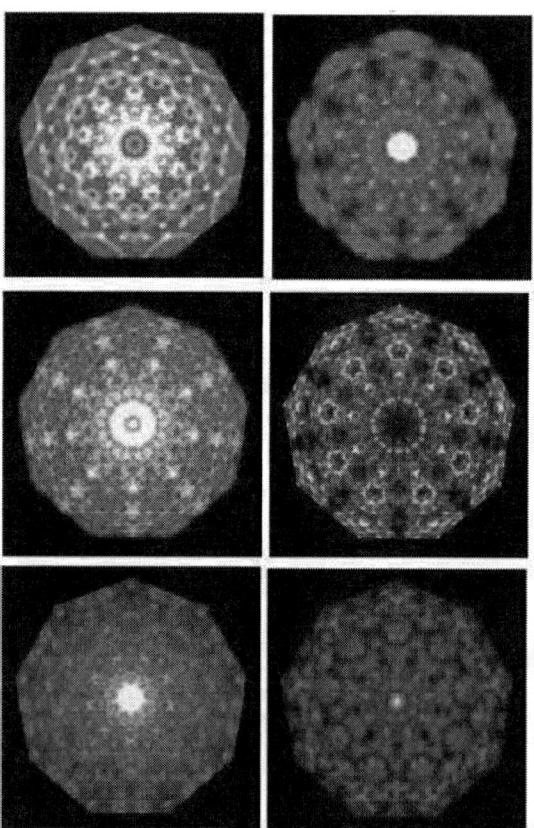

Fig. 1. Examples of lowest-order (column 1) and first-order (column 2) kaleidoscope fractal modes. Parameters are $N = 9$ (nonagonal feedback mirror) and magnification $M = 1.5$. Top row: $N_{eq} = 10$; middle row: $N_{eq} = 20$; bottom row: $N_{eq} = 30$.

978-1-4244-7043-3/10 $26.00 © 2010 IEEE 130

CAOL*2010 International Conference on Advanced Optoelectronics & Lasers, 10-14 September, 2010, Sevastopol, Ukraine

that unfolds each unstable cavity into an equivalent sequence of $N_s = \log(250N_{eq})/\log(M)$ virtual apertures. Any eigenmode can then be constructed from a weighted superposition of the edge-waves diffracted by each aperture, plus a plane-wave contribution. In scaled units, the mode pattern $V(\mathbf{X})$ is given by

$$V(\mathbf{X}) = e_0 \left[\frac{E_{N_s+1}(\mathbf{X_c})}{\alpha^{N_s}(\alpha-1)} + \sum_{m=1}^{N_s} \alpha^{-m} E_m(\mathbf{X}) \right], \quad (1)$$

where $\mathbf{X}$ denotes an appropriate set of transverse coordinates, $\mathbf{X_c}$ is any point on the boundary of the feedback mirror, and $E_m(\mathbf{X})$ is the edge-wave pattern arising from the mth virtual aperture [7].

Since our interest lies with calculating the field over the feedback mirror, $e_0 = +1$ if $\mathbf{X}$ lies inside this domain, and $e_0 = 0$ otherwise. The weighting factor α, which plays the role of the mode eigenvalue, is obtained by finding the roots of an associated polynomial equation. A single virtual-source calculation provides information for determining the lowest-loss mode and all the higher-order modes.

When $N \to \infty$, the feedback mirror becomes circular and the cavity essentially has only a single transverse dimension. This limit has been investigated by Berry under the assumption $N_{eq} \gg O(1)$, and only for the lowest-loss mode [8]. For cavities with *arbitrary* N_{eq} and M, this type of fully-2D convergence phenomenon does not lend itself to asymptotic analysis; indeed, it can only be truly addressed via numerical computation. We will present, to the best of our knowledge, the first in-depth treatment of the circular limit of families of kaleidoscope laser modes. It was found that this is a far more subtle problem than might first be imagined.

II. NEW OPTICAL FRACTAL CONTEXTS

In later work [9], we examined whether *nonlinear* systems might give rise to spontaneous spatial fractals. It is well-known that nonlinear systems can exhibit a Turing instability in which homogeneous states develop large-amplitude emergent patterns from the smallest of fluctuations. However, such patterns are typically "simple", i.e. characterized by a *single* dominant length scale that corresponds to a single minimum in a modulational instability threshold plot. A range of nonlinear systems actually present a hierarchy of local Turing minima in their stability curves. It was proposed [9] that systems with such a multi-Turing property were capable of spontaneous spatial fractal formation. This hypothesis was tested, and confirmed true, for a single feedback mirror context involving a transient and diffusive nonlinear medium.

Here, we extend investigations to examine whether fractal pattern formation may arise from the one of the simplest possible configurations – *the instantaneous interaction of two counter-propagating waves in a non-diffusive Kerr medium*. This system [10,11] is known to present a multi-Turing modulational instability characteristic (see Fig. 2).

It will be shown, for the first time, that this fundamental configuration can also give rise to spontaneous fractal patterns. While we focus here on optical contexts, the implications of our findings extend to wave interactions in other (e.g. fluid and plasma) systems that are governed by the same pair of universal coupled nonlinear equations.

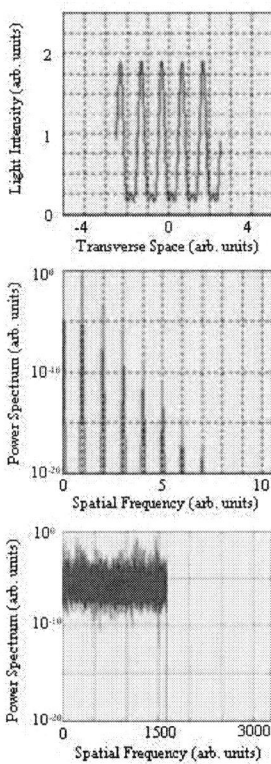

Fig. 3. Pattern formation in counter-propagating waves within a focusing Kerr medium. Top: spontaneous emergence of a simple transverse pattern. Middle: power spectrum of a simple pattern (involving a single dominant frequency and its harmonics). Bottom: power spectrum of a spontaneous spatial fractal (with a sharp cut-off at the carrier frequency).

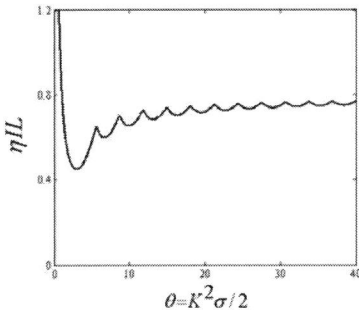

Fig. 2. A modulational instability threshold curve for two counterpropagating waves in a Kerr medium. The vertical axis is proportional to wave intensity, I, while the horizontal axis is involves the transverse wavenumber of the modulation, K. Other parameters are σ (a transverse scale parameter), L (the longitudinal interaction length), and $\eta = \pm 1$ (for a focusing or defocusing Kerr effect, respectively).

978-1-4244-7043-3/10 $26.00 © 2010 IEEE

We will summarise investigations into the dependence of pattern formation character on various system parameters and confirm fractality of the resulting patterns through deployment of various appropriate fractal analysis measures.

Fig. 3 shows typical results for the counter-propagating waves system. By either: considering wave intensities close to the value dictated by the global instability minimum, or allowing for some spatial filtering mechanism in the system, simple patterns emerge. For higher wave intensities, and without strong spatial filtering, the system spontaneously generates fractal patterns with structure across decades of scale. The data shown in Fig. 3 corresponds to a nonlinear phase shift per medium transit of $\eta IL = 0.8$. A simple pattern emerges when strong filtering is employed and fractal formation occurs when no spatial filtering is present. This result is consistent with the detail of the threshold curve plotted in Fig. 2.

More generally, it is found that depending on the precise parameters spontaneous spatial fractals can be either static (i.e.

stationary) or dynamic in time. Also, for this particular system, each fractal generated has proved to be of a statistical and area-filling character (see Fig. 4).

Finally, we will briefly summarize our most recent results concerning *spatial pattern formation in incoherent systems*. The leading works in this field [12,13,14] proposed that spatial pattern formation required either a perfectly coherent system or one involving a sufficiently non-instantaneous nonlinearity. However, subsequent work examining *temporal* pattern formation in fibre optic systems [15] reported that patterns may form irrespective of medium response time.

Two questions have been addressed concerning *spatial* pattern formation in incoherent systems involving instantaneous nonlinearity. Firstly, we have examined whether coherent spatial patterns emerge from initial states of high incoherence. Secondly, spatial pattern formation has been studied in systems that are maintained in states of partial incoherence. We will report on our finding that, within both of these contexts, spatial pattern formation is relatively robust.

REFERENCES

[1] G.P. Karman, and J.P. Woerdman, " Fractal structure of eigenmodes of unstable-cavity lasers," *Opt. Lett.*, vol. 23, pp. 1909-1911, 1998.

[2] G.P. Karman, G.S. McDonald, G.H.C. New, and J.P. Woerdman, "Fractal modes in unstable resonators," *Nature*, vol. 402, pp. 138, 1999.

[3] G.H.C. New, M.A. Yates, G.S. McDonald, and J.P. Woerdman, "Diffractive origin of fractal resonator modes," *Opt. Commun.*, vol. 193, pp. 261-266, 2001.

[4] G.S. McDonald, G.P. Karman, G.H.C. New, and J.P. Woerdman, "Kaleidoscope laser," *JOSA B*, vol. 17, pp. 524-529, 2000.

[5] M. Berry, C. Storm, and W. van Saarloos, "Theory of unstable laser modes: edge waves and fractality," *Opt. Commun.*, vol. 197, pp. 393-402, 2001.

[6] W.H. Southwell, "Virtual-source theory of unstable resonator modes," *Opt. Lett.*, vol. 6, pp. 487-489, 1981.

[7] J.G. Huang, J.M. Christian, and G.S. McDonald, "Fresnel diffraction and fractal patterns from polygonal apertures," *JOSA A*, vol. 23, pp. 2768-2774, 2006.

[8] M. Berry, "Fractal modes of unstable lasers with polygonal and circular mirrors," *Opt. Commun.*, vol. 200, pp. 321-330, 2001.

[9] J.G. Huang, and G.S. McDonald, "Spontaneous optical fractal pattern formation," *Phys. Rev. Lett.*, vol. 94, art. no. 174101, 2005.

[10] W.J. Firth, and C. Pare, "Transverse modulational instabilities for counterpropagating beams in Kerr media," *Opt. Lett.*, vol. 13, pp. 1096-1098, 1988.

[11] W.J. Firth, A. Fitzgerald, and C. Pare, "Transverse instabilities due to counterpropagation in Kerr media," *J. Opt. Soc. Am. B*, vol. 7, pp. 1087-1097, 1990.

[12] D. Kip, M. Soljacic, M. Segev, E. Eugenieva, and N. Christodoulides, "Modulational instability and pattern formation in spatially incoherent light beams," *Science*, vol. 290, pp. 495-498, 2000.

[13] T. Schwartz, J.W. Fleischer, O. Cohen, H. Buljan, M. Segev, and T. Carmon, "Pattern formation in a ring cavity with temporally incoherent feedback," *J. Opt. Soc. Am. B*, vol. 21, pp. 2197-2205, 2004.

[14] T. Schwartz, T. Carmon, H. Huljan, and M. Segev, "Spontaneous patter formation with incoherent white light," *Phys. Rev. Lett.*, vol. 93, art. no. 223901, 2004.

[15] A. Sauter, S. Pitois, G. Millot, and A. Picozzi, "Incoherent modulational instability in instantaneous nonlinear Kerr media," *Opt. Lett.*, vol. 30, pp. 2143-2145, 2005.

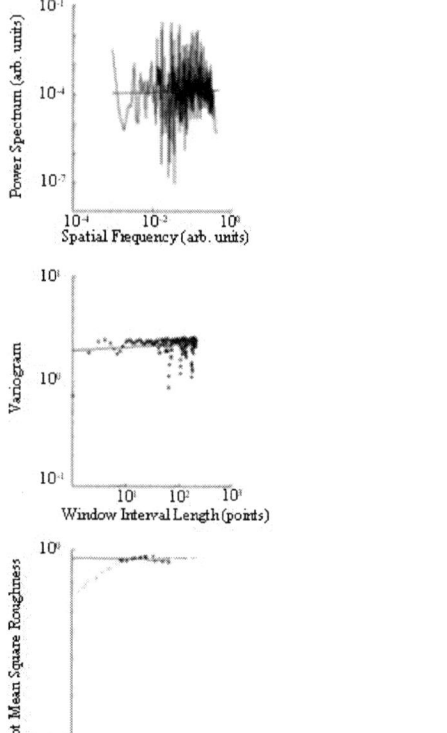

Fig. 4. Fractal analyses of spontaneous multi-scale pattern formation when $\eta IL = 0.8$. Top: power spectrum fractal dimension determination, middle: the variogram method, bottom: roughness-length fractal dimension. Each measure confirms an area-filling curve of dimension close to 2 (for frequencies below the carrier wave cut-off frequency, and at length scales above the carrier wavelength).

CAOL*2010 International Conference on Advanced Optoelectronics & Lasers, 10-14 September, 2010, Sevastopol, Ukraine

Modification of the Power Radiated by an Electrical Dipole in the Presence of a Thin Dielectric Disk

M. V. Balaban[1], R. Sauleau[2], A. I. Nosich[1]
[1]Institute of Radio-physics and Electronics NASU, Kharkov, Ukraine
[2]IETR, Universite de Rennes 1, Rennes, France

Abstract: We consider the problem of electromagnetic waves emission by a horizontal electrical dipole in the presence of a thin high-contrast dielectric disk. To find the solution of the problem we use analytical-numerical scheme which is based on the dual integral equations and the method of analytical regularization. We present the normalized radiation patterns and total radiated power for different values of geometrical and material parameters.

Electromagnetic wave scattering by thin dielectric disks is interesting for many reasons: such a disk is met as a part of printed antennas; it is used as a simplified model of a tree leave [1]; thin few-micron radius disks are used as resonators of semiconductor lasers with ultralow thresholds [2]. Moreover, this scattering problem is of great interest for the optical community because it can be used to study the spontaneous emission in disk-shape nanosize dielectric particles. As known, the change in the power radiated by a dipole (or any other source) in the presence of a dielectric particle is equivalent to the modification of the spontaneous emission rate of a quantum-mechanical atom in the same particle [3].

Many analytical high-frequency approximations and direct computational methods have been used for the scattering analysis. However, high-frequency methods fail to reproduce fine details of the wave field and have uncertain domains of applicability. Direct computational methods meet other problems like large-size matrices to be inverted, low convergence of solution, and hence huge computational time.

Therefore in this study we use a rigorous numerical method, which is fast and accurate for any disk size and needs moderate computer resources; this method has been presented in [4].

Consider the problem of diffraction of the electromagnetic field emitted by a horizontal electrical dipole from a dielectric disk of radius a and thickness τ (Fig. 1). Introduce dimensionless cylindrical coordinates ($\rho = r / a, \varphi, \zeta = z / a$) with origin in the center of the disk. Assume that the dipole is located at the height d above the disk center ($\zeta_0 = d / a$). Denote total field as a sum of the scattered and the incident fields: $E = E_{in} + E_{sc}$, $H = H_{in} + H_{sc}$, where the incident field corresponds to the dipole in the free space. The scattered field has to satisfy the homogeneous Maxwell equations out of the sources and the disk, and the following generalized boundary conditions at the disk median section [5]:

$$(E_{tg}^+ + E_{tg}^-) = 2Z_0 R \cdot \vec{n} \times (H_{tg}^+ - H_{tg}^-),$$
$$Z_0(H_{tg}^+ + H_{tg}^-) = -2Q \cdot \vec{n} \times (E_{tg}^+ - E_{tg}^-).$$

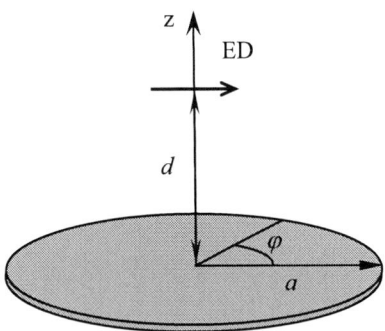

Figure 1. Problem geometry

Here, Z_0 is the free-space impedance, and R and Q are the electric and magnetic resistivities which are given by $R = iZ / 2 \cot(\sqrt{\varepsilon_r \mu_r} k\tau / 2)$, $Q = R / Z^2$ in case of $k\tau \ll 1$ and $|\varepsilon_r \mu_r| \gg 1$, Z is the relative impedance of the disk material, $k = \omega / c$ is the wavenumber, ε_r and μ_r are the relative permittivity and permeability, respectively. On the rest part of the disk plane the components of the total field are continuous. Also, the components of the scattered field must satisfy the 3-D radiation condition and the condition of local integrability of power [6].

To find the solution of the scattering problem, we use the method of analytical regularization [9] applied to the dual integral equations (IEs) [7,8] in the Fourier-Hankel spectral domain. It enables us to obtain the Fredholm second kind IEs for the unknowns that are the images of the jumps and the average values of the normal to the disk field components. This guarantees the existence of solution, which can be found numerically using any reasonable disctretization. We truncate the interval of integration form $(0, \infty)$ to $(0, N)$, where $N > ka + 1$ and apply the Nystrom method with the Gauss-type higher-order quadratures to solve it numerically.

Some of preliminary results of computations are presented in Figs. 2 to 4. Fig. 2(a) shows the dependences of the total power radiated by the elementary electrical dipole in the presence of thin dielectric disk for three different types of high-contrast dielectric materials with $\varepsilon_r = 10 + i$; $100 + i$; $1000 + i$ and $\mu_r = 1$. The power is normalized by the same quantity for the

978-1-4244-7043-3/10 $26.00 © 2010 IEEE

CAOL*2010 International Conference on Advanced Optoelectronics & Lasers, 10-14 September, 2010, Sevastopol, Ukraine

elementary electrical dipole in free space. One can see resonance nature of the normalized radiated power. That correspond to two different resonance types which are became more clear for the materials with low dielectric losses ($\tan \delta = 0.01$ and 0.001). First type of resonances (marked by 1, 2, 3, 4 and 5 in Fig 2(a)) are radial resonances in the dielectric disk. Radiation patterns of the incident and total fields at these points are presented in Fig. 3(a)-3(e).

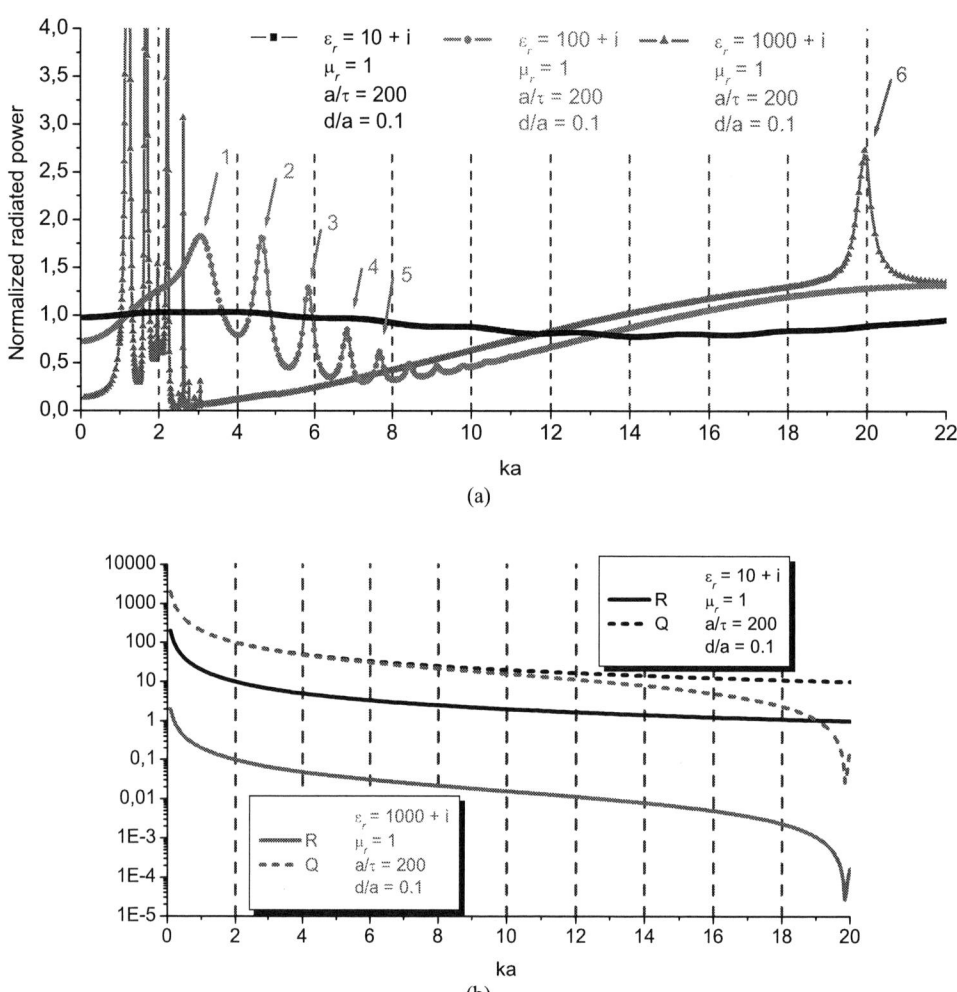

(a)

(b)

Figure 2. Normalized radiated power (a) and resistivities dependences as a functions of the normalized radius of the disk ka

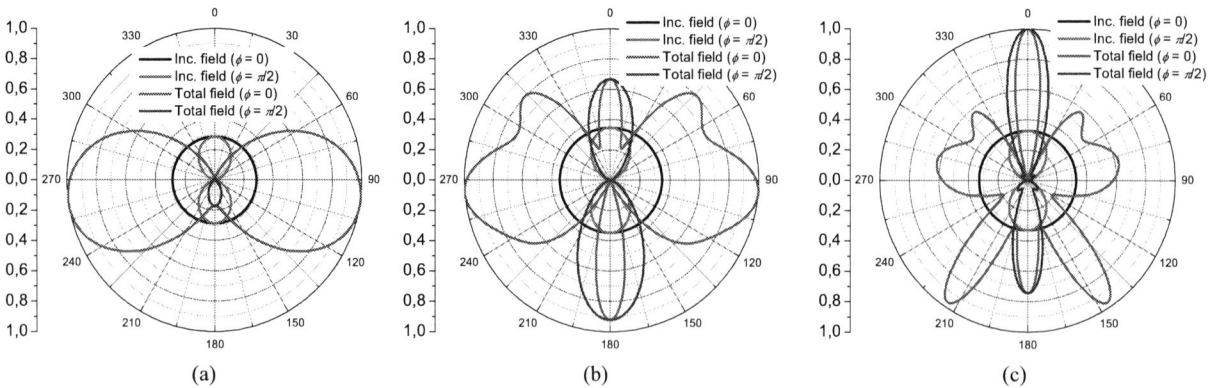

(a)   (b)   (c)

978-1-4244-7043-3/10 $26.00 © 2010 IEEE   134

CAOL*2010 International Conference on Advanced Optoelectronics & Lasers, 10-14 September, 2010, Sevastopol, Ukraine

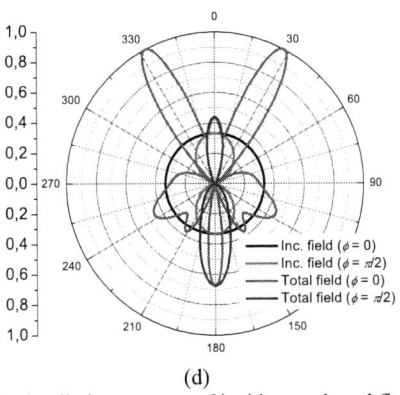 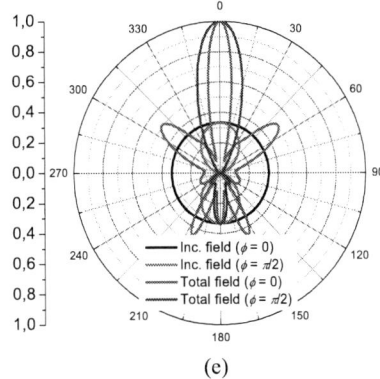

(d)   (e)

Figure 3. Normalized radiation patterns of incident and total field at points $ka = 3.043$ (a); $ka = 4.6386$ (b); $ka = 5.836$ (c); $ka = 6.883$ (d); $ka = 7.681$ (e) for dielectric disk material $\varepsilon_r = 100 + i$; $\mu_r = 1$ and geometry parameters $a./\tau = 200$; $d/a = 0.1$

A resonance of the other type appears only if the permittivity is large enough. It is marked by 6 in Fig. 2(a). This resonance corresponds to the disk thickness equal to half-wavelength in the dielectric material ($k\tau\sqrt{\varepsilon_r\mu_r} = \pi$) and is connected to a zero of the electric and magnetic resistivities (see Fig 2(b)). Therefore this is a dielectric layer transversal resonance. Radiation patterns of the incident and the total fields at this point are presented in Fig. 4.

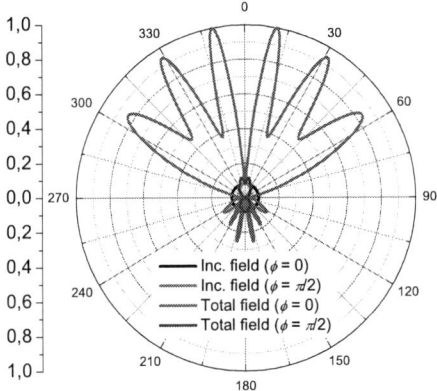

Figure 4. Normalized radiation patterns of incident and total field at point $ka = 19.95$ for dielectric disk material $\varepsilon_r = 1000 + i$; $\mu_r = 1$ and geometry parameters $a/\tau = 200$; $d/a = 0.1$

ACKNOWLEDGEMENT

This work has been partially supported by the European Science Foundation via the "Newfocus" network project.

REFERENCES

[1.] M. A. Karam, A. K. Fung, "Electromagnetic scattering from a layer of finite length, randomly oriented, dielectric circular cylinders over a rough interface with application to vegetation," *Int. J. Remote Sensing*, vol. 9, pp. 1109-1134, 1988.

[2.] E. I. Smotrova, A. I. Nosich, T. M. Benson, and P. Sewell, "Cold-cavity thresholds of microdisks with uniform and non-uniform gain: Quasi-3D modeling with accurate 2D analysis," *IEEE J. Sel. Topics Quantum Electron.*, vol. 11, no. 5, pp. 1135–1142, Sep./Oct. 2005.

[3.] L. Rogobete, C. Henkel, "Spontaneous emission in a subwavelenght encironment by boundary integral equations," *Phys. Rev. A.* vol. no 70, 2004.

[4.] M.V. Balaban, R. Sauleau, T.M. Benson, A.I. Nosich, "Dual integral equations technique in electromagnetic wave scattering by a thin disk," *Progress in Electromagnetic Research B*, 2009, vol. 16, pp. 107-126.

[5.] E. Bleszynski, M. Bleszynski, T Jaroszewicz, "Surface-integral equations for electrmagnetic scattering from impenetrable and penetrable sheets", *IEEE Antennas Propag. Mag.*, vol. 35, pp. 14-24, 1993.

[6.] I. Braver, P. Fridberg, K. Garb, I. Yakover, "The behavior of the electromagnetic field near the edge of a resistive half-plane", *IEEE Trans Antennas Propag.*, vol. 36, pp. 1760-1768, 1988.

[7.] S.M. Ali, W.C. Chew, J.A. Kong, "Vector Hankel transform analysis of annular-ring microstrip antenna", *IEEE Antennas Propagat.*, vol. 30, no 4, pp. 637-644, July 1982.

[8.] N.Y. Bliznyuk, A.I. Nosich, A.N. Khizhnyak, "Accurate computation of a circular-disk printed antenna axysimmetrically excited by an electric dipole", *Microwave and Optical Technology Lett.*, 2000, vol. 25, no 3, pp. 211-216.

[9.] A.I. Nosich, "Method of analytical regularization based on the static part inversion in wave scattering by imperfect thin screens", *J. Telecommunications and Information Technology*, Warsaw: NIT Press, 2001, no 3, pp. 72-79.

978-1-4244-7043-3/10 $26.00 © 2010 IEEE   135

Adaptive optical system with water-cooled bimorph deformable mirror

V.V.Samarkin[1], M.E.Dryagin[2], A.V.Kudryashov[1], P.N.Romanov[2]

[1]Moscow State Open University, Shatura, Russian Federation

[2]Active Optics NightN Ltd., Moscow, Russian Federation

Abstract: The water cooled deformable mirror of the bimorph type is intended for beam correction in the CW solid state laser with output power of 10 kW level. An operation of the adaptive system using Shack-Hartmann wavefront sensor and deformable mirror is based on the phase conjugating algorithm. The first results will be presented in this report.

Any adaptive system consists of three basic elements that determine the properties of such a system – active deformable mirror, wavefront sensor and control unit that implements the feedback between corrector and sensor. Adaptive systems are now widely used for correction and formation of the laser beam.

A choice of each element is made in terms of the real parameters of the laser beam such as peak and average power, beam aperture, wavefront distortions etc. There are special peculiarities of the use of the adaptive systems in high power lasers. The wavefront of the output beam here is highly aberrated because of thermal deformation, inhomogeneities of the optical pumping, self-focusing, non-linear effects and so on. Deformable mirror should have enough damage threshold for high power beam and the surface should be thermally stabilized.

A water-cooled deformable mirror of the bimorph type was developed and investigated. The corrector consists of silicon substrate firmly glued to a plate actuator disk made from piezoelectric ceramic (PZT) [1]. The outer surface of the PZT disc is divided in 48 controlling actuators, which have the shape of a part of a sector. It's a design of the passive cooling, when the cooling comes not from inside of the passive substrate like copper mirror [2], but from periphery surface [3]. Here the surface is thermally contacted with the condenser which temperature is always constant (20^0C). The material of the substrate is Si because it has a good thermal conductivity and it more thermally fits to PZT material. Photos of such deformable mirrors are given in the Fig.1.

Some estimation was made for the deformable mirror when 1 kW CW laser beam illuminates the mirror surface. Aperture of the beam is 80 mm, while the aperture of the mirror is 100 mm. Difference of temperature for the central and periphery area is less than 3^0C. This difference produces a small curvature about 500 m, which can be compensated by low enough voltage (10 V) on the 1st actuator of the mirror.

Dielectric coating ensures 99.9% of the reflectivity for the wavelength $\lambda=1.06$ µ. This coating was not damaged by CW power density 50 kW/cm^2, it was a maximal available density while testing the coating.

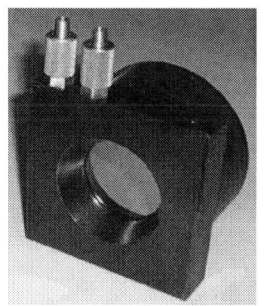

Fig.1. Photos of the water cooled deformable mirrors for beam control and correction in: a) 1 kW output power CW Nd:YAG laser; b) 4 kW output power CW CO$_2$ laser.

The water-cooled deformable mirror is intended for intracavity correction of the CW solid state Nd:YaG laser with output beam power 10 kW. An operation of the adaptive system Shack-Hartmann wavefront sensor and deformable mirror is based on the phase conjugating algorithm. The first results will be presented in this report.

REFERENCES

[1] A.Kudryashov, V.Shmalhausen, "Semipassive bimorph flexible mirrors for atmospheric adaptive optics applications", *Opt. Eng.* Vol.35, pp.3064-3073, 1996

[2] A. V. Kudryashov and V. V. Samarkin, "Control of high power CO2 laser beam by adaptive optical elements," Opt. Commun., vol. 118, pp. 317–322, 1995.

[3] J.V.Sheldakova, V.V.Samarkin, A.V.Kudryashov, "Correction of the radiation of 1kW CW diode pumped Nd:Yag laser", *Proc. SPIE*, vol. 6101, pp.61010B, 2006.

A theorem for the propagation of electromagnetic waves through a multilayered structure consisting of left handed material and dielectric

Muin F. Ubeid and Mohammed M. Shabat

Department of Physics, Faculty of Science, Islamic University of Gaza
P.O. 108, Gaza, Gaza Strip, Palestinian Authority
Fax: + 790 82860800; Tel: +970 82860700; mail: shabat@mail.iugaza.edu

Abstract: We prove the law of conservation of energy for propagation of electromagnetic waves through a multilayered structure consisting of LHM and dielectric slabs inserted in vacuum. Numerical method is presented to calculate the reflected and transmitted powers of the structure. The obtained results satisfy the law of conservation of energy.

Materials of simultaneously negative permittivity ε and permeability μ are called left-handed materials (LHMs) or metamaterials [1, 2]. Regular materials are called right handed materials (RHMs).

In this paper we consider a structure consisting of LHM and dielectric slabs inserted between two semi-infinite dielectric media. A plane polarized wave is obliquely incident on it. We use Maxwell's equations and match the boundary conditions for the electric and magnetic fields of the incident waves at each layer interface. Then we solve the obtained equations for the unknown parameters to calculate the reflection and transmission coefficients. We take into account the frequency dependence of permittivity and permeability of the LHM. We propose a numerical proof for the law of conservation of energy given by [3].

We consider four regions each with permittivity ε_ℓ and permeability μ_ℓ, where ℓ represents the region order (see fig. 1). The electric field in each region is [4, 5]:

$$\vec{E}_\ell = \left(A_\ell e^{ik_{\ell z}z} + B_\ell e^{-ik_{\ell z}z}\right) e^{i(k_{\ell x}x - \omega t)} \, \hat{y} \qquad (1)$$

To find the corresponding magnetic field $\vec{H}_\ell$, we start with

Maxwell's equation $\quad \bar{\nabla} \times \vec{E}_\ell = -\partial \vec{B}\big/\partial t$, substituting

$\vec{B} = \mu_\ell \vec{H}_\ell$ and solving for $\vec{H}_\ell$ yield:

$$\vec{H}_\ell = \frac{1}{\mu_\ell \omega}\left[\begin{array}{l} (A_\ell k_{\ell x} e^{ik_{\ell z}z} + B_\ell k_{\ell x} e^{-ik_{\ell z}z})\hat{z} + \\ (-A_\ell k_{\ell z} e^{ik_{\ell z}z} + B_\ell k_{\ell z} e^{-ik_{\ell z}z})\hat{x} \end{array} \right] e^{i(k_{\ell x}x - \omega t)} \qquad (2)$$

Where A_ℓ and B_ℓ are the amplitude of forward and backward traveling waves, $k_\ell = n_\ell \omega\big/c$ is the wave

vector inside the material and n_ℓ is the refractive index of it.

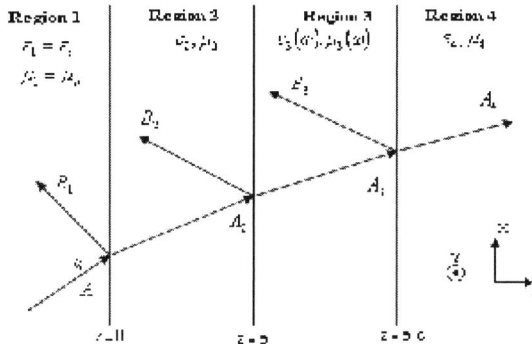

Fig. 1. Wave propagation through a structure consisting of a pair of dielectric and metamaterial embedded between two dielectric semi-infinite media.

Matching the boundary conditions for $\vec{E}$ and $\vec{H}$ fields at each layer interface, that is at z=0, $E_1 = E_2$ and $H_1 = H_2$ and so on. This yields six equations with six unknown parameters [3,4]. Letting $A_1 = 1$ and solving the obtained equations for the unknown parameters enables us to calculate the reflection and transmission coefficients B_1 and A_4 [4]. The reflected power R and the transmitted power T are given by [4]:

$$R = B_1 B_1^{\,*}\,,\, T = A_4 A_4^{\,*} \qquad (3)$$

Where $B_1^{\,*}$ and $A_4^{\,*}$ are the complex conjugate of B_1 and A_4 respectively.

The law of conservation of energy is given by [3]:

$$R + \left(k_{4z}\big/k_{1z}\right)T = 1 \qquad (4)$$

Consider Regions 1 and 4 are free spaces ($n_1 = n_4 = 1$), Region 2 is any dielectric ($n_2 = 1.883$). Region 3 is a non-dispersive metamaterial with ε_3 and μ_3 used as appeared in [6-8]:

$$\varepsilon_3(\omega) = 1 - \frac{F_e \omega_{ep}^2}{\omega^2 - \omega_{eo}^2} \quad , \quad \mu_3(\omega) = 1 - \frac{F_m \omega_{mp}^2}{\omega^2 - \omega_{mo}^2} \quad (5)$$

where ω_{ep} and ω_{mp} are the electric and magnetic plasma frequencies, ω_{eo} and ω_{mo} are the electric and magnetic resonance frequencies. F_e and F_m are the scaling filling parameters.

We have used the following parameters in (5) appearing in [4]: $\omega_{mp} = 2\pi 10.95\,GHz$, $\omega_{mo} = 2\pi 10.1\,GHz$, $F_m = .26$, $\omega_{ep} = 2\pi 13.3\,GHz$, $\omega_{eo} = 2\pi 10.3\,GHz$, $F_e = .37$. In this case, the frequency range in which $\varepsilon_3(\omega)$ and $\mu_3(\omega)$ are negative extends from 10.3 up to 11.4 GHz. Let $\omega = 2\pi 11.0\,GHz$, then: $\mu_3 = -.641 + .03i$, $\varepsilon_3 = -3.39 + .05i$, $k_{4z}\big/k_{1z} = 1$, $B_1 = -.55315627 - .3208659517i$, $A_4 = -.2465788325 - .7281909568i$, R=.408936809, T = .591063190. In this case the left hand side of (4) is equal to .99999999 which verifies the law of conservation of energy. Figs. 2 and 3 show the transmitted and reflected powers respectively as a function of frequency.

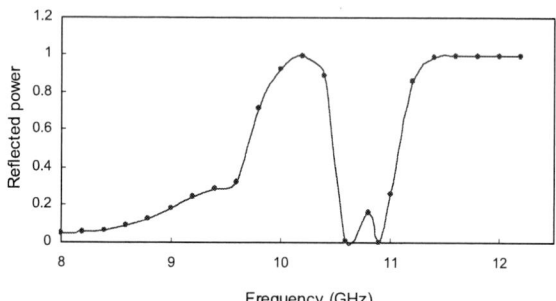

Fig. 3. : Reflected power against frequency

A theorem has been formulated for the propagation of electromagnetic waves through multilayered structures consisting of a pair of LHM and dielectric slabs inserted between tow semi-infinite media. The followed method is based on Maxwell's equations and matching the boundary conditions for the electric and magnetic fields at each interface layer. The frequency dependence of permittivity and permeability of the LHM has been taken into account. The reflected and transmissted powers have been calculated and the law of conservation of energy has been verified numerically.

REFERENCES

[1] V.G. Veselago, The electrodynamics of substances with simultaneously negative values of ε and μ, Soviet Phys Uspekhi vol. 10, pp. 509-514, 1986 .

[2] R.A. Shelby, D.R. Smith and S. Schultz, Experimental verification of a negative index of refraction, science vol. 292, pp. 77-79, 2001.

[3] H. Cory and C. Zach, Wave propagation in metamaterial multi-layered structures, Microwave and Optical Tech. vol. 40, pp. 460-465, 2004.

[4] R.A. Shelby, Thesis (PhD.), University of Calefornia, San Diego, Microwave Experiments with Left-Handed Materials, Bell and Howell Information and Learning Company 2001.

[5] J.A. Kong, Theory of Electromagnetic waves, EMW Publishing, 2005.

[6] H. Oraizi and A. Abdolali, Mathematical formulation for zero reflection from multilayered metamaterial structures, IET Microw. Antennas Progpag., vol. 3, pp. 987-996, 2009.

[7] J.B. Pendry, A.J. Holden, W.J. Sewart, I. Youngs,Extremely low frequency plasmons in metallic mesostructure, Phys. Rev. Lett. vol. 76, pp. 4773-4776, 1996.

[8] B. Pendry A.J. Holden, D.J. Robbins, W.J. Stewart, Low frequency plasmons in thin wire structures, Phys. Condens. Matter vol. 10, pp. 4785-4809, 1998.

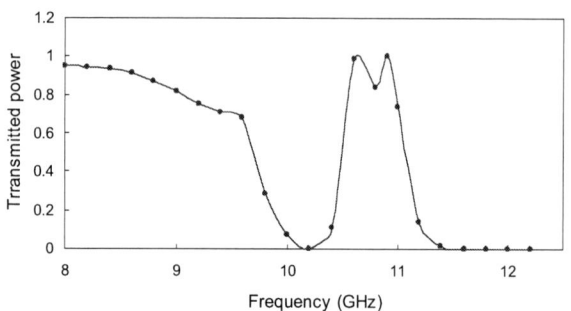

Fig. 2 Transmitted power against frequency.

Theoretical and experimental investigations compact slab guide resonators

V. A. Saetchnikov[1], E. A. Tcheriavskaia[1]

[1]Belarusian State University, Minsk, Belarus

Abstract: Investigation of compact slab guide resonators with variable geometry, guide material and surface quality by automated scanning interferometer with high precision positioning system has been performed. Obtained optimization criteria are used to improve parameters of compact (including dimensions up to less then some cm) slab guide and multi pass laser resonators. On the other hand very high sensitivity to resonator surface quality and internal media properties allows to use such a structure as instrumentation for development of optical and fiber optical (chemical, biomedical etc.) sensors.

INTRODUCTION

Waveguide lasers have been studied for long time and have proved useful over a wide range of applications. Recent generation of such type gas lasers based on slab resonator geometry had extremely high efficiency combined with a compact design, allowed to get CW output power till some kilowatts and are widely used now in industry. Wide area of electrode surface forming a RF-discharge was found can influence rather strongly the output parameters of such a device. Moreover sometimes even small variation of resonator geometry or electrode surface can change drastically both optical parameters of resonator and output of the complete laser. To get a detailed information on the complex influence of electrodes on the optical parameters of slab laser resonator a very precise measurements of such a parameter as optical losses on the level of 10^{-5} under variation of a resonator geometry on the level of some microns are necessary. Development of experimental equipment which allows to perform such measurements and advanced computer models were the aim of our investigation.

EXPERIMENTAL EQUIPMENT AND MATERIALS

Experimental measurement of optical losses in waveguide and slab laser resonators have been performed by automated scanning interferometer, adapted for automated measurements under noisy and low intensity detected signal. Sealed off CO_2 laser (Spectral CTM-300) with controlled output power (Power Meter OPHIR AN/2) and a state of polarization was used as source of excitation. Piezo – elements (PI R-841.40) were controlled by Multifunction Waveform Synthesizer (ROHDE&SCHWARZ AFS) and Piezo –driver/control unit (PI P864.00). Signal from fast detector (PEM-L-3) with self made amplifier was digitized by 14-bit ADC, filtered by a low pass filter and after additional smoothing and averaging was used to estimate optical losses by a standard procedure. Developed system of positioning allowed to vary geometry of the resonator and electrodes with the precision till some micrometers (minimal step 0.5 µ). The influence of a material, type of a covering and quality of a surface of electrodes on optical parameters of slab guide resonators have been investigated. Alumina, copper, aluminium and their mixture were used as a material for electrodes. Alumina was used also as a dielectric layer for electrode surface. Electrode surface quality was varied by special instrument which allowed to get both periodical and random profile of the surface roughness. Profile deepness of the electrode surface was varied from 2 till 20 micrometers.

RESULTS AND DISCUSSION

Multimode model, modified for variable curvature radius of mirrors has been used to calculate optical losses as an eigenvalue problem of a system. Irradiation output from a waveguide travelling twice in the gap between a waveguide and both mirrors also twice inside a waveguide and then appears in the same point.

To calculate optical losses on resonator complex matrix of the complete roundtrip M is calculated taking into account losses on all elements of resonator. Then eigenvalue problem is solved. To take into account variation of Gaussian beam in the gap between a waveguide and mirrors and mirror curvature complex parameter q is used. Entering the waveguide beam has new waist and curvature radius of a wave front. Transmission of the beam can be divided into two parts: before and after reflection. Then elements of the complex matrix of a mode EH_{mn} coupling into the mode $EH_{m'n'}$ through free space modes TEM_{pq} in the waveguide – mirror gap are calculated. For the second gap it is necessary to calculate a similar matrix.

Propagation through the waveguide is taking into account with the help of two additional matrices. Besides all matrix multiplication should be made in the proper turn. That is if we start one end of the waveguide the first coupling matrix should be multiplied on the waveguide propagation matrix, then the second coupling matrix and the waveguide propagation matrix once more.

Variation of an electrode material is simulated by real an imagine parts of the complex refractive index. When absolute values of this index increases average level of optical losses becomes higher and they are less sensitive to the resonator geometry as will be shown below.

978-1-4244-7043-3/10 $26.00 © 2010 IEEE

On the basis of a model described above influence of a rectangular geometry resonator on optical losses have been performed. Periodical behavior of this value due to variation of mode resonance conditions is more pronounced for smaller mirror curvature. It is necessary to point out that optical losses in such a system depend strongly on resonator geometry. And the most critical parameter is an electrode gap.

Compare resonance frequencies of the slab guide resonator its parameter when optical losses are minimized can be evaluated. Minimum mirror – guide gap under minimal losses depends mainly on frequency coincidence between axial modes and transverse modes of the same axial mode as for the Case I resonator. Moreover a small variation of this gap when the electrode gap is close to optimum correlate with resonant frequency varied due to resonator length variation. For curved mirrors the influence of this resonance on losses is more pronounced due to additional variation of a wave front curvature on the guide aperture.

Experimental results are well correlate with calculated data represented above, however only qualitatively. The reason of this discrepancy will be discussed below. Now we would like to pay attention that a field structure inside a waveguide resonator is different from both pure waveguide and free space mode structure. Thus mutual coupling of waveguide and free space modes is the most impotent factor when a mirror – guide gap is large. Calculated and experimentally measured data for a resonator with internal diaphragm make obvious, that the results can not be explained by only suppression of the higher order modes because under different mirror – electrode gaps optical losses of a fundamental mode in the resonator with a diaphragm can be both more and less then without it.

Coming back to discrepancy of experimental and theoretical results we would like to discuss the main to our mind factor. Previous investigations have shown that an absolute value and behavior of optical losses depend drastically on the quality of electrode surface. New measurements also correlate with this conclusion. Thus nearly ideal surface (roughness on the level of some nm) are not optimal to get minimal losses. The best results can be obtained under periodical roughness with the deepness of the order of a wavelength. When the profile is not periodical average level of losses increases and they depend less on the resonator geometry.

Comparison of experimental and calculated data for electrodes with the roughness deepness of around 10 µ allowed to conclude that relative variation of losses are considerably less then calculated for ideal electrode surface. But taking into account a statistical variation of the electrode gap much better correlation of calculated and experimental data can be obtained.

CONCLUSION

The generalized model of a slab guide resonator allowing to take into account several guide and free space modes have been developed. Optical losses and mode structure of such a resonator have been evaluated as a solution of the eigenvalue problem for the complex round trip matrix of the resonator.

The change of a material of electrodes within the framework of the developed model is described by a variation of the real and imaginary part of the refractive index. Numerical model was applicable for actual geometry of scanning interferometer (curved mirrors, symmetric and asymmetric slab resonator, varied gap between mirrors and guide, internal elements).

Automated measurement system based on scanning interferometer of the middle IR range have been developed. Positioning system allowed to vary geometry of the resonator and electrodes with the precision till some micrometers. The influence of a material, type of a covering and quality of a surface of electrodes on optical parameters of slab guide resonators have been investigated. Alumina, copper, aluminium and their mixture were used as a material for electrodes. Alumina was used also as a dielectric layer for electrode surface. Electrode surface quality was varied by special instrument which allowed to get both periodical and random profile of the surface roughness. Profile deepness of the electrode surface was varied from 2 till 20 micrometers.

Significant (two, three times and more) variation of optical losses in the resonator under small (down to several micrometers) change of its geometry has been found out experimentally. Replacement of a material or use of dielectric covering of the electrode surfaces has been followed not only the change of an average level of optical losses due to complex index of refraction, but also rather complex variation of their dependence on geometry of the resonator. It was shown experimentally that absolute value and character of dependence of optical losses on resonator geometry can change drastically with variation of the roughness of electrode surface. For example nearly smooth surfaces (with a the deepness of the roughness of some nm) are not optimum for minimal optical losses. The best results can be obtained for a structure of a surface of electrodes representing periodic or random structure (depending on electrode material or its dielectric covering) with depth of around a radiation wavelength.

Obtained optimization criteria are used to improve parameters of compact (including dimensions up to less then some cm) slab guide and multi pass laser resonators. On the other hand very high sensitivity to resonator surface quality and internal media properties allows to use such a structure as instrumentation for development of optical (interferometer) and fiber optical (chemical, biomedical etc.) sensors. This research is under development now.

ACKNOLEGEMENTS

Investigations were supported by International Bureau of WTZ DLR, project BLR 00/003.

The contribution of evanescent waves into the power transfer

A.B. Katrich

Kharkov National University, Kharkov, Ukraine

Abstract: Theorems of Fourier transform cannot be applied to prove that evanescent waves do not contribute into the power passed through the initial plane because the Fourier spectrum of the boundary value is by the most part singular. Their contribution is considered in details in the case of linear source. It is shown that evanescent waves transfer the power in both directions and do not contribute into the total passed power.

The statement [1] that "on the average no energy is carried away from the initial plane by evanescent waves" is based on the proof which uses Prseval's or Fubini's theorem [2,3]. It has been shown recently [4] that the Fourier spectrum of the boundary value (cross-section of the field by any pane) is singular if only in the far-field the angular distribution function in this plane is not equal to zero. However singular functions do not meet validity conditions of mentioned theorems [5] and known proof [2,3] holds only when the boundary value has continuous spectrum. Thus, the statement in [1] must be proved by some other way.

The proof given in [4] concerns only propagating waves and the case when the far-field angular distribution $D(\theta,\varphi)$ has no zeroes. In general case the boundary value $U(r,\pi/2,\varphi)$ can be expanded into Fourier series in the plane $z=0$ ($\theta=\pi/2$) in polar coordinates since it is an analytical and periodical function

$$U\left(r,\frac{\pi}{2},\varphi\right) = U_{r<r_0}(r,\varphi) + U_{r>r_0}(r,\varphi),$$

(1)

$$U_{r<r_0}(r,\varphi) = \sum_n f_n(k_0 r) e^{in\varphi}$$

$$U_{r>r_0}(r,\varphi) = G(k_0 r)\sum_n c_n e^{in\varphi},$$

(1a)

where $r=(x^2+y^2)^{1/2}$, c_n are Fourier coefficients of the expansion of the angular distribution $D(\pi/2,\varphi)$, and the radius r_0 is so large that Eq. (1a) is valid with acceptable accuracy in the region $r>r_0$. Now the spectrum of $U(r,\pi/2,\varphi)$ is

$$A(\rho,\psi,z) = e^{ik_0 mz}\sum_n(-i)^n\left[g_n(\rho)+c_n d_n(\rho)\right]e^{in\psi}$$

(2)

where $\rho=(p^2+q^2)^{1/2}$, $\tan\psi=q/p$ and

$$g_n(\rho) = \int_0^{r_0} f_n(k_0 r) J_n(k_0 r\rho) r\, dr$$

(3)

$$d_n(\rho) = \int_{r_0}^{\infty} e^{ik_0 r} J_n(k_0 r\rho)\, dr$$

(4)

if only integrals exist and sums in Eq. (2) converge. Functions $f_n(k_0 r)$ are defined inside the circle $r<r_0$ and so are finite and continuous. Thus, $g_n(\rho)$ are continuous and absolutely integrable [5] and the first sum in Eq.(2) converges always. The angular distribution is a periodic and analytical function, which Fourier expansion converges always. So, the second sum in Eq. (2) will converge if only coefficients d_n do not rise when n increases.

The region of interest is the vicinity of possible singularity, that is the circle $\rho=1$ where $k_0 r\rho \gg 1$. Using the asymptotic expression for $J_n(k_0 r\rho)$ and formula 3.761.2,7 [6], after simplification

$$d_n(\rho) = a\Gamma\left(\tfrac{1}{2},-iu\right) + b\Gamma\left(\tfrac{1}{2},-iv\right),$$

$$u = k_0 r_0(1+\rho), \quad v = k_0 r_0(1-\rho),$$

$$a = \frac{(-i)^n}{k_0\sqrt{2\pi\rho(1+\rho)}}, \quad b = \frac{(i)^{n+1}}{k_0\sqrt{2\pi\rho(1-\rho)}}$$

(5)

In the vicinity of the circle $\rho=1$ the argument u of the incomplete gamma-function $\Gamma(1/2,-iu)$ is $u>2k_0 r_0 \gg 1$ and the argument $|v|\ll u$. Incomplete gamma-functions are of the order $u^{-1/2}$ and $|v|^{-1/2}$ for large arguments (8.357 [6]) and $\Gamma(1/2,-iv)$ is approximately equal to $\pi^{1/2}$ when $|v|\ll 1$ (8.354, [6]). Since $u \gg |v|$, the first term in expression for d_n in Eq. (5) is negligible comparing with the second one and the product $(-i)^n d_n(\rho)$ in Eq.(2) does not depend on n

$$(-i)^n d_n(\rho) \approx \frac{i\Gamma(\tfrac{1}{2},ik_0 r_0(\rho-1))}{k_0 m\sqrt{\pi}},$$

(6)

where $2\rho(1-\rho)\approx 1-\rho^2=m^2$ is used. The incomplete gamma-function of a small argument is about $\pi^{1/2}$ and Eq. (6) represents the spectrum of the free space Green function in the vicinity of singular points. In Eq.(2) it can be taken outside the second sum that becomes equal to $D(\pi/2,\psi)$ and Eq.(2) can be rewritten in the form

$$A(\rho,\psi,z) = e^{ik_0 mz}\left[\sum_n(-i)^n h_n(\rho)e^{in\psi} + \frac{i}{k_0 m}D(\tfrac{\pi}{2},\psi)\right],$$

(7)

where into $h_n(\rho)$ are included $g_n(\rho)$ and the continuous difference between the second term in Eq. (1) and the second sum in Eq.(2). Thus, singular points at the circle $\rho=1$ can be only of the order m^{-1}, they are irregular and exist always if

978-1-4244-7043-3/10 $26.00 © 2010 IEEE

only $D(\pi/2,\psi)$ is not equal to zero for all angles in the plane. The boundary value can be now written in the form

$$U\left(R,\frac{\pi}{2},\varphi\right) = U_c(R,\varphi) + G(k_0R)D(\frac{\pi}{2},\varphi), \qquad (8)$$

where $U_c(R,\varphi)$ is a continuous and absolutely integrable function.

In the two-dimensional case

$$A(p,z) = \int_{-x_0}^{x_0} U(x,z)e^{-ik_0xp}\,dx + \left[D(\tfrac{\pi}{2})+D(-\tfrac{\pi}{2})\right]\int_{x_0}^{\infty} H_0^{(1)}(k_0R)\cos(k_0xp)\,dx$$

where $D(\pm\pi/2)$ are magnitudes of the angular distribution $D(\varphi)$ in the initial plane. The part of the spectrum determined by the first integral is the continuous absolutely integrable function. Since $k_0x_0 \gg 1$ the asymptotic expression for Hankel function can be used and when $z=0$

$$\int_{x_0}^{\infty} H_0^{(1)}(k_0x)\cos(k_0xp)\,dx \approx \frac{\Gamma(\tfrac{1}{2},ik_0x_0(1-p))}{k_0m\sqrt{\pi}} \qquad (9)$$

In the vicinity $|p|=1$ the incomplete Gamma-function is about $\pi^{1/2}$ and the integral is equal to the spectrum of Hankel function. Thus, the result is the same: the spectrum is discontinuous if only points $D(\pm\pi/2)$ are not equal to zero.

The total power passed through any plane $z=$const is the normal component of the Poynting vector

$$S_z = C\,\mathrm{Re}\left[E_xH_y^* - E_yH_x^*\right] = S_z^p + S_z^e \qquad (10)$$

integrated over this plane. Here C is a constant and superscripts p and e denote power density components determined by propagating and evanescent waves. So,

$$S_z^p = C\,\mathrm{Re}\left[E_x^pH_y^{p*} - E_y^pH_x^{p*}\right] \qquad (11)$$

includes only the contribution from propagating modes and

$$S_z^e = C\,\mathrm{Re}[E_x^eH_y^{p*} - E_y^eH_x^{p*} + E_x^pH_y^{e*}$$
$$\qquad - E_y^pH_x^{e*} + E_x^eH_y^{e*} - E_y^eH_x^{e*}] \qquad (12)$$

is a sum of contribution from only evanescent modes and mixed terms. When only the contribution into the total power is considered then from Eq. (12) all products of two absolutely integrable functions can be excluded even if this term has the real part. But terms where at least one function has singular spectrum cannot be removed.

Unfortunately the general proof that the integral from S_z^e over the plane is always equal to zero cannot be done by conventional way. To answer the question on the role of evanescent waves in the power transfer consider the simple example with the line source placed along the axis y with field components $E_y=-iH_0^{(1)}(k_0R)/4$, $H_x=-zH_1^{(1)}(k_0R)/(4R)$, $H_z=xH_1^{(1)}(k_0R)/(4R)$, $R=(x^2+z^2)^{1/2}$. The normal to the plane (x,y) the Pointing vector component is

$$S_z = -C\,\mathrm{Re}\left[E_yH_x^*\right]$$
$$= \frac{zC}{16R}\left[J_1(k_0R)N_0(k_0R) - J_0(k_0R)N_1(k_0R)\right] \qquad (13)$$
$$= \frac{zC}{8\pi k_0R^2},$$

because the combination of Bessel and Neumann functions is the Wronskian [5]. Integration with respect to x yields the constant total power $P_0=C(8k_0)^{-1}$ which doesn't depend on z, i.e. is the conservation value on propagation. Spectra of Bessel functions and functions $N_0(k_0R)$ and $N_1(k_0R)/(k_0R)$ in Eq.(13) have singularity at the frequency k_0. Partition of field components into propagating and evanescent parts yields

$$E_y(x,z) = E_y^p(x,z) + E_y^e(x,z) = -\frac{i}{4}H_0^{(1)}(k_0R)$$

$$H_x(x,z) = H_x^p(x,z) + H_x^e(x,z) = -\frac{z}{4R}H_1^{(1)}(k_0R)$$

$$S_z^p = -C\,\mathrm{Re}\left[E_y^pH_x^{p*}\right] \qquad (14)$$

$$S_z^e = -C\,\mathrm{Re}\left[E_y^eH_x^{p*} + E_y^pH_x^{e*} + E_y^eH_x^{e*}\right]$$

where

$$E_y^p(k_0R) = -\frac{i}{4}J_0(k_0R) + \tfrac{1}{4}I_1; \quad E_y^e(k_0R) = -\tfrac{1}{4}I_2;$$

$$H_x^p(k_0R) = -\frac{z}{4R}J_1(k_0R) + \frac{i}{4}I_3; \; H_x^e(k_0R) = \frac{i}{4}I_4;$$

and integrals are

$$I_1 = \frac{2}{\pi}\int_0^1 \frac{\sin(k_0m^pz)}{m^p}\cos(k_0xp)\,dp$$

$$I_2 = \frac{2}{\pi}\int_1^{\infty} \frac{\exp(-k_0m^ez)}{m^e}\cos(k_0xp)\,dp$$

$$I_3 = \frac{2}{\pi}\int_0^1 \cos(k_0m^pz)\cos(k_0xp)\,dp \qquad (15)$$

$$I_4 = \frac{2}{\pi}\int_1^{\infty} \exp(-k_0m^ez)\cos(k_0xp)\,dp$$

where $m^p=(1-p^2)^{1/2}$; $m^e=(p^2-1)^{1/2}$. Relative contributions of propagating and evanescent parts into the total power are $\varepsilon^{p,e}=P^{p,e}/P_0$

$$\varepsilon^p = k_0\int_0^{\infty}\left[J_0(k_0R)I_3 + \frac{z}{R}J_1(k_0R)I_1\right]dx$$

$$\varepsilon^e = k_0\int_0^{\infty}\left[\frac{z}{R}J_1(k_0R)I_2 - J_0(k_0R)I_4\right]dx \qquad (16)$$

and the sum $\varepsilon^p+\varepsilon^e$ must be equal to 1. Note that integrands in Eq.(15) and Eq.(16) are not absolutely integrable and the order of integration in Eq. (16) cannot be changed.

Behavior of ε^p can be considered analytically when z is small. Integrals I_1 and I_3 can be calculated using approximation $\sin(k_0m^pz) \approx k_0m^pz$ and $\cos(k_0m^pz) \approx 1$: $I_1 \approx 2z\sin(k_0x)/\pi x$;

$I_3 \approx 2\sin(k_0 x)/\pi k_0 x$. Integration yields $\lim_{z\to 0}(\varepsilon^p)=1$ and evanescent waves in the case of linear source really do not

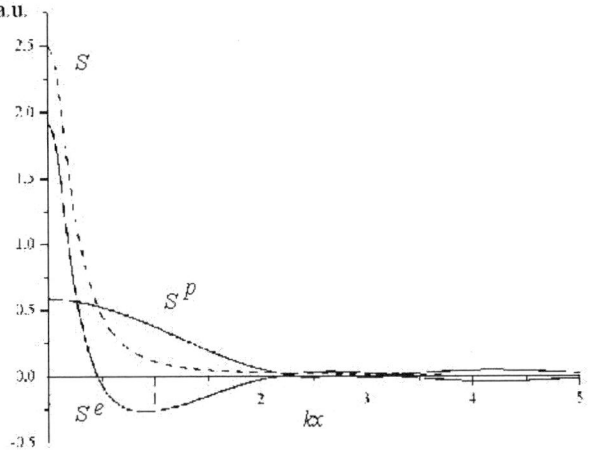

Fig. 1

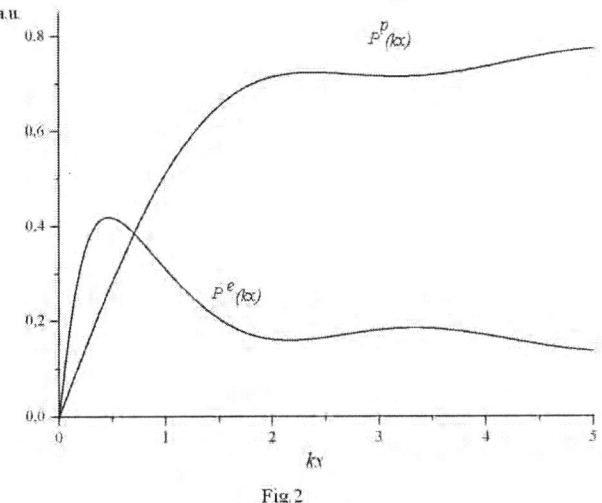

Fig.2

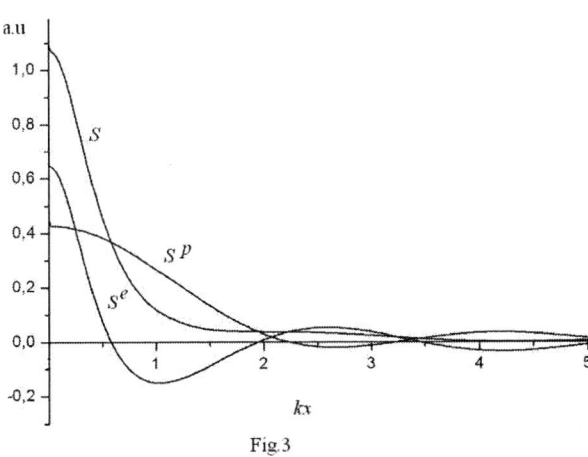

Fig.3

contribute into the total power passed through the initial plane. When z is not small integrals have no analytical expressions and relative contributions have been calculated numerically. Normalized distributions $S_z^p(k_0 x)$ and $S_z^e(k_0 x)$ and their sum S_z are shown in Fig.1, $k_0 z=0.25$ and Fig. 3, $k_0 z=0.5$. Both functions oscillate and the power is transferred in both directions similarly by propagation and evanescent waves. However integrated value of the contribution rises for propagated waves and decreases for evanescent ones as it is shown in Fig.2, $k_0 z=0.25$. When the upper limit of the numerical integration range over $k_0 x$ is expanded, the value ε^p tends to 1.0 and ε^e to zero.

The angular distribution $D(\theta,\varphi)$ is the spectral amplitude function $D(\alpha,\beta)$ for real values of polar angles α, β [7] where it has been proved that the spectral amplitude function is an entire function of two complex variables if only the field is generated by the continuous scattering object. In physical space this condition is satisfied: physical scattering objects and primary sources are continuous and finite. So, neither $D(\alpha,\beta)$ nor $D(\theta,\varphi)$ can be equal to zero within a finite solid angle [5]. In a plane the angular distribution can be identically equal to zero but such a case is rather exception then a rule and any rotation of a plane, however small, will restore the spectrum singularity.

Thus, the role of evanescent waves is redistribution of passed power and the level of power density carried by them depends on the size of the object that generate the field and position of the plane. The averaged value of the power transferred through the plane by evanescent waves is rather always equal to zero but the proof of this must be done by some other way then it was done before.

REFERENCES

[1] M. Born, E. Wolf, *Principles of Optics*. 7th edn. (Cambrige University Press, New York, 1999).

[2] L.E.R. Petersson, G.S. Smith, "Role of evanescent waves in power calculations for counterpropagating beams," J. Opt. Soc. Am. A **20**, 2378-2384 (2003).

[3] S. R. Seshadri, „Dynamics of the linearly polarized fundamental Gaussian light wave," J. Opt. Soc. Am. A **24**, 482-492 (2007).

[4] Katrich A.B., "On the propagation of evanescent waves". Proc. of the 4th International Conference on Advanced Optoelectronics and Lasers (CAOL'2008). Yalta, 2008, p. 304-306

[5] G.A. Korn, T.M. Korn, *Mathematical Handbook*. 2nd ed. (McGrow-Hill, New York, 1968).

[6] I.S. Gradshtein, I.M. Ryzhik, *Tables of Integrals, Series and Products*. Academic Press, NY,1980.

[7] E. Wolf, M.Nieto-Vesperinas, "Analyticity of the angular spectrum amplitude of scattered field and some of its consequences," *J. Opt. Soc. Am.* A **2**, 886-890, (1985).

CAOL*2010 International Conference on Advanced Optoelectronics & Lasers, 10-14 September, 2010, Sevastopol, Ukraine

Simulation of Lasing Modes in a Kite-Shaped Microcavity Laser

E. I. Smotrova, *Member, IEEE* and A. I. Nosich, *Fellow, IEEE*

Institute of Radiophysics and Electronics of National Academy of Sciences of Ukraine, Kharkov, Ukraine

Abstract: We consider the lasing modes in a thin kite-shaped active microcavity as solutions to the 2-D linear eigenproblem for the Maxwell equations with exact boundary and radiation conditions. This problem is reduced to the set of Muller's integral equations with smooth and integrable kernels discretized using the adequate quadrature formulas. The eigenvalues are found numerically as the roots of the corresponding determinantal equation. The results of the study of several modes are presented.

I. INTRODUCTION

A serious drawback of circular microdisk lasers is the low directionality of light emission because any whispering-gallery mode with azimuth index m has $2m$ identical beams in the disk plane. Directionality can be conveniently quantified with the aid of *directivity* as a ratio of power emitted into the main beam direction to the total power averaged over all directions. Then any mode of a circular cavity with $m > 0$ has directivity equal to 2. It is evident that improvement of the directionality needs a distortion of the cavity shape from the circle. As such a deformed shape, we will study the kite cavity whose contour can be characterised using a smooth function.

II. EIGENVALUE PROBLEM FOR ACTIVE MICROCAVITIES

Denote the interior domain of a two-dimensional (2-D) model of an active dielectric (non-magnetic) microcavity as D_i, its closed contour as Γ, and the outer domain as D_e. Consider a function $U(x,y)$, which is either the E_z or the H_z field component. When simulating a microlaser, we are interested in real-valued pairs of numbers (k,γ) generating non-zero functions U solving, off Γ, the Helmholtz equation $(\Delta + k^2 v^2)U = 0$ with a piecewise-constant effective refractive index v equal to $v_i = \alpha_i - i\gamma$ ($\gamma > 0$) in D_i, and $v_e = \alpha_e$ in D_e. Here, the following two-side boundary conditions are required on Γ: $U^e = U^i$ and $\eta_e \partial U^e / \partial n = \eta_i \partial U^i / \partial n$, where the superscripts "i,e" refer to the corresponding domains, $\eta_{i,e} = 1$ (E-polarisation) or $\eta_{i,e} = 1/v_{i,e}^2$ (H-polarisation), and $\vec{n}$ is the outward normal vector to Γ. Furthermore, the time-averaged electromagnetic energy must be locally integrable to prevent source-like field singularities, and the Sommerfeld radiation condition must be satisfied at infinity. This is the Lasing Eigenvalue Problem (LEP) that we have introduced in

[1] and systematically used in [2]-[5] to study the modes in various active cavities.

In this paper, we consider a kite-shaped microcavity (Fig.1).

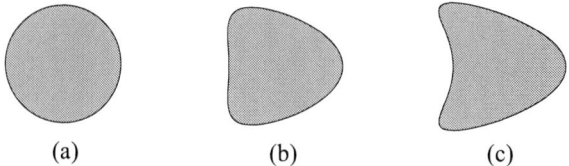

(a)   (b)   (c)

Fig. 1. Geometries of microcavity for different values of parameter δ: (a) $\delta = 0$, (b) $\delta = 0.3$, (c) $\delta = 0.5$.

For the contour Γ representation, we use the following smooth (i.e. infinitely continuously differentiable) function, where $t \in [0, 2\pi]$:

$$\mathbf{r}(t) = \{x(t), y(t)\},$$
$$x(t) = \cos t + \delta \cos 2t - \delta, \quad y(t) = \sin t \quad (1)$$

Here, δ is the contour shape parameter, so that if $\delta = 0$ then (4) turns to a circle. Note that the values of $\delta < 0.29$ provide convex contours and the true kite shape appears only if δ takes larger values.

III. MULLER'S INTEGRAL EQUATIONS

Introduce the Green's functions $G_j(R) = (i/4)H_0^{(1)}(kv_j R)$ of the homogeneous media, where $j = i, e$, $R = |\vec{r} - \vec{r}'|$ is the distance between the points $\vec{r}$ and $\vec{r}'$, and $H_0^{(1)}(\cdot)$ is the Hankel function. After applying the second Green's formula to the functions $G_j(\vec{r}, \vec{r}')$ and U_j, using boundary conditions, and taking into account the properties of single-layer and double layer potentials, we obtain two integral equations as

$$\varphi(\vec{r}) + \int_\Gamma \varphi(\vec{r}')A(\vec{r}, \vec{r}')dl' - \int_\Gamma \psi(\vec{r}')B(\vec{r}, \vec{r}')dl' = 0,$$

$$\frac{\eta_i + \eta_e}{2\eta_e}\psi(\vec{r}) + \int_\Gamma \varphi(\vec{r}')C(\vec{r}, \vec{r}')dl' - \int_\Gamma \psi(\vec{r}')D(\vec{r}, \vec{r}')dl' = 0, \quad (2)$$

978-1-4244-7043-3/10 $26.00 © 2010 IEEE   144

where dl' is the element of the arc on Γ, $\varphi(\vec{r}) = U_i(\vec{r})$ and $\psi(\vec{r}) = \partial U_i(\vec{r}) / \partial n$, $\vec{r} \in \Gamma$.

Here, the kernel functions are

$$A(\vec{r}, \vec{r}') = \partial G_i(\vec{r}, \vec{r}') / \partial n' - \partial G_e(\vec{r}, \vec{r}') / \partial n',$$

$$B(\vec{r}, \vec{r}') = G_i(\vec{r}, \vec{r}') - \eta_i / \eta_e G_e(\vec{r}, \vec{r}') \quad (3)$$

$$C(\vec{r}, \vec{r}') = \partial^2 G_i(\vec{r}, \vec{r}') / \partial n \partial n' - \partial^2 G_e(\vec{r}, \vec{r}') / \partial n \partial n',$$

$$D(\vec{r}, \vec{r}') = \partial G_i(\vec{r}, \vec{r}') / \partial n - (\eta_i / \eta_e) \partial G_e(\vec{r}, \vec{r}') / \partial n \quad (4)$$

Note that the kernel functions $A(t, \tau)$ and $D(t, \tau)$ are continuous, and the kernel functions $B(t, \tau)$ and $C(t, \tau)$ have logarithmic singularities.

IV. DISCRETIZATION OF INTEGRAL EQUATIONS

One of the most efficient discretization techniques is the method of quadratures, also known as the Nystrom method [6]-[8]. This latter method is based on the replacement of the integrals with approximate sums using the appropriate quadrature formulas. As some of the kernel functions have logarithmic singularities, it is convenient to represent all of the kernels in (3) and (4) in such a way that these singularities are extracted [7],[8]. Then the integrals are approximated by two different quadrature rules for the regular and singular parts with the same equidistant set of points $t_p = \pi p / N$, $p = 0, 1, \ldots, 2N - 1$. Namely, we use a trigonometric quadrature rule for the parts with logarithmic singularities and a trapezoidal rule for the regular parts [5]. By evaluating the integrals from (1) for each $t = t_s$ with the aid of the quadrature rules, we obtain a determinantal equation for the eigenvalues. A secant-type iterative method [2] is further used to find the eigenvalues numerically from this equation.

V. NUMERICAL RESULTS

For simplicity, we consider the kite contour as a continuous deformation of the circle, and therefore denote the modes of the kite cavity using the notations of their limiting forms in the circle with a prefix "quasi". In Fig. 2, we present the dependences of the lasing frequency and threshold gain of the doublet of low-order modes quasi-H_{21} on the kite contour deformation parameter, δ. The initial values of thresholds of these modes are quite high. If δ gets larger, the frequencies of both modes grow up. Unlike this, only one of the modes (odd with respect to the x-axis) has the threshold that grows up monotonically while another (even mode) displays a maximum of threshold around $\delta = 0.3$.

The shape of the far-field radiation pattern changes very dramatically with the growth of δ. In Fig. 3, the patters of the modes quasi-H_{21} are shown for the maximum value of δ of the studied range, i.e. $\delta = 0.5$. As visible, the directivity can be both larger and smaller than the circular-cavity value 2.

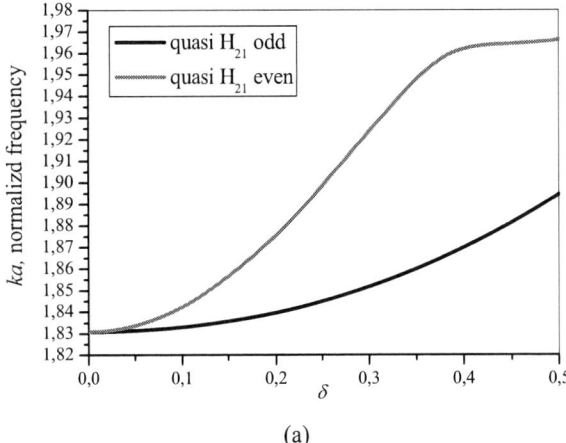

(a)

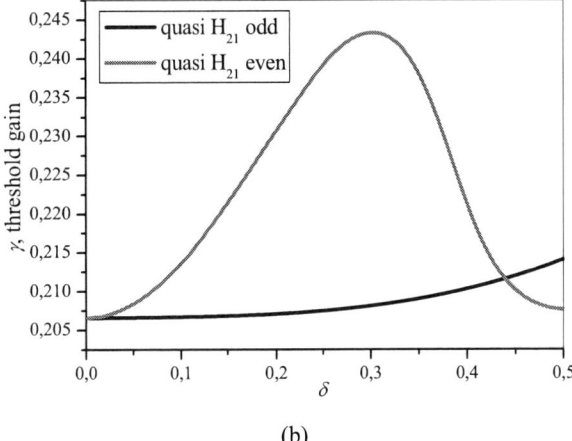

(b)

Fig. 2. Normalized lasing frequency (a) and threshold gain (b) as a function of the kite deformation parameter δ, for the doublet of quasi-$H_{2,1}$ modes, $\alpha = 2.63$, $N = 50$

In Fig. 4, presented are the lasing frequencies and thresholds for the doublet of modes quasi-H_{51} that display the features of the whispering-gallery modes at $\delta = 0$. Their initial thresholds are some 10 times lower than those of the quasi-H_{21} modes.

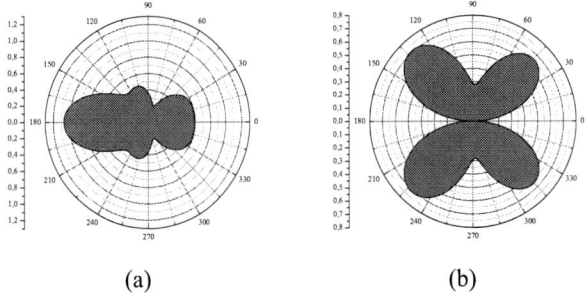

(a)   (b)

Fig. 3. Normalized far-field emission patterns for $\delta = 0.5$, $\alpha = 2.63$, and $N = 50$: (a) quasi-$H_{2,1}$ even, directivity = 3.43; (b) quasi-$H_{2,1}$ odd, directivity = 1.98.

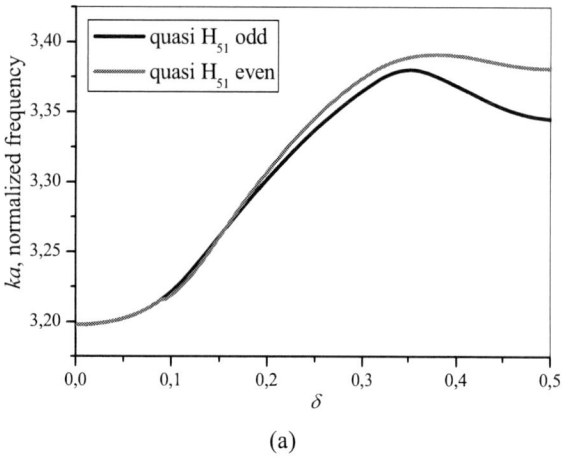

(a)

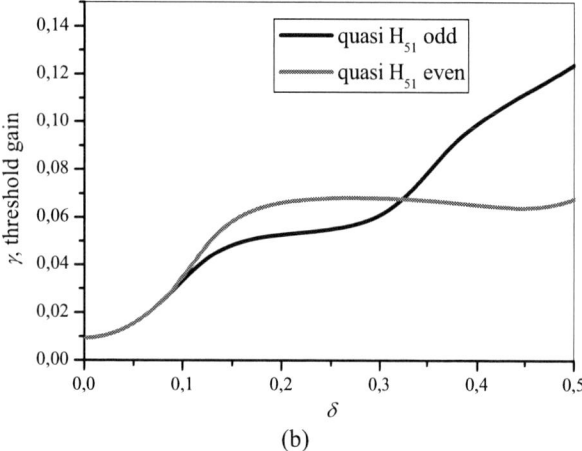

(b)

Fig. 4. The same as in Fig. 2 however for the modes quasi-H_{51}.

The even-type mode of this doublet also show a maximum threshold value around $\delta = 0.2$. In Fig. 5, the far-field patters of the modes quasi-H_{51} are shown for $\delta = 0.5$. As visible, at such deformation their directivities are larger than for the low-order modes quasi-H_{21}. Note that the number of emission lobes is no more the same as in the circular cavity.

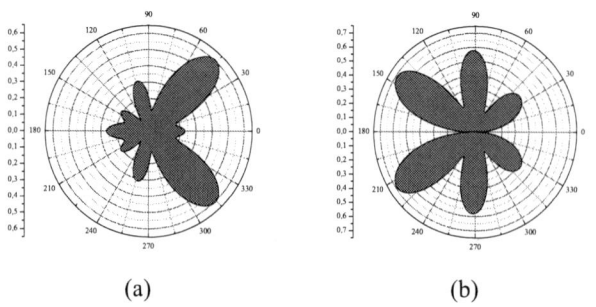

(a)   (b)

Fig. 5. The same as in Fig. 2 however for the modes (a) quasi-H_{51} even, directivity = 3.71; (b) quasi-H_{51} odd, directivity = 2.74

VI. CONCLUSIONS

We have presented preliminary results of the LEP-based numerical analysis of the lasing modes in the 2-D model of a kite-shaped thin microcavity laser. This shape is attractive as it enables one to study the variations of the near and far-field modal patterns, and also the lasing frequencies and threshold gains for a variety of shapes changing smoothly from a circle to a "boomerang" cavity. As one can see, even small deformations of this sort can lead to considerable changes of the emission patterns and thus provide greater directivities.

ACKNOWLEDGMENTS

This work was supported in part by the National Academy of Sciences of Ukraine via the Young Scientist Scholarship to the first author, the European Science Foundation via research network "Newfocus", and the Ministry of Education and Science, Ukraine via project M/146-2010.

REFERENCES

[1] E.I. Smotrova, A.I. Nosich, "Mathematical study of the 2D lasing problem for the whispering-gallery modes in a circular dielectric microcavity," *Optical and Quantum Electronics*, 2004, vol. 36, no 1-3, pp. 213-221.

[2] E.I. Smotrova, *et al.*, "Cold-cavity thresholds of microdisks with uniform and non-uniform gain: quasi-3D modelling with accurate 2D analysis," *IEEE J. Selected Topics Quant. Electron.*, vol. 11, pp. 1135-1142, 2005.

[3] E.I. Smotrova, et al., "Ultralow lasing thresholds of the pi-type supermodes in cyclic photonic molecules composed of sub-micron disks with monopole and dipole modes," *IEEE Photonics Technology Letters*, vol. 18, no 19, pp. 1993-1995, 2006.

[4] E.I. Smotrova, *et al.*, "Lasing frequencies and thresholds of the dipole-type supermodes in an active microdisk concentrically coupled with a passive microring," *J. Opt. Society America A*, vol. 25, no 11, pp. 2884-2892, 2008.

[5] E.I. Smotrova, *et al.*, "Optical fields of the lowest modes in a uniformly active thin sub-wavelength spiral microcavity," *Optics Letters*, vol. 34, no 24, pp. 3773-3775, 2009.

[6] L. Wang, J.A. Cox, A. Friedman, "Modal analysis of homogenious optical waveguides by the boundary integral formulation and the Nyström method," *J. Opt. Soc. Am. A*, vol. 15, no. 1, pp.92-100, 1998.

[7] D. Colton, R. Kress, *Inverse Acoustic and Electromagnetic Scattering Theory*, Springer, Berlin, 1998.

[8] J. Tsalamengas, "Exponentially converging Nystrom methods applied to the integral/integro-differential equations of oblique scattering hybrid wave propagation in presence of composite cylinders of arbitrary cross section," *IEEE Trans. Antennas Propag.*, vol. 55, no 11, pp. 3239-3250, 2007.

Space confinement of random lasing

Vasil P.Yashchuk, Olga A. Prygodiuk

Kyiv National Taras Shevchenko University, Kyiv, Ukraine

Abstract: Lasing properties of dyed random media under different excited region size and the medium thickness were investigated. Random lasing threshold, spectrum wavelength and intensity are appreciably dependent both on the excited region size and the medium thickness. It proves both lengthwise and crosswise spatial confinement of random lasing.

Recently the stimulated emission research under multiple scattering has been under the detailed interest of several scientific groups [1-3]. It mainly conditioned by specific features of the radiation process behavior in these media. This results in peculiar phenomena observation as photon localization and associated noncoherent backscattering, random lasing and stimulated Raman scattering. Due to lasing occurrence possibility such media are named random lasers. These lasers do not demand any cavity use so they are of great interest due to the construction simplicity being useful for their producing and operation.

Owing to the specific way of positive feedback formation under multiple scattering the random lasing obtain some essential features. Particularly the lasing parameters should strongly depend on any changes of the medium scattering properties. Thus the active volume size where the lasing occurs should influence the lasing formation. This work is devoted to random lasing phenomenon investigation under spatial confinement in both longitudinal and crosswise directions according the pump incidence.

Random active media (random lasers) were fabricated from solid $R6G$ polyvinyl acetate solutions with silica scattering particles (diameter d~1μm). Weight concentration of the particles was 30% (C_{part}~10^{12}cm^{-3}). The solid solutions were prepared by mixing of alcohol solutions of polyvinyl acetate and $R6G$ dye with silica and follow alcohol evaporation from the mixture. All the samples were cut from the obtained blanks.

The random lasers were pumped with the second harmonic of quasi single-mode Q-switched Nd^{+3}: YAG (λ=532nm, τ~15ns). The spectra were registered in backward to pump incidence direction. We investigated the spectra and radiation energy of the random laser depending on its thickness and pump beam diameter at the random laser surface.

Fig.1 presents the dependency of radiation spectrum width versus pump intensity I_p. Different curves correspond to different pump beam diameters d_p that defines the crosswise size of excited area in random laser. The beam diameter was changed by distance variation between the investigated sample and focusing lens.

As it was defined earlier [4] the random lasing does not characterizes with abrupt lasing threshold in terms of conventional lasing. Random lasing appears as enhanced

luminescence saturation. But we proposed to define the random lasing threshold as the inflection point on the I_p – dependency of the spectrum bandwidth $\Delta\lambda$. The threshold defined in the described way is presented in the Fig. 1 Inset. As can be seen from the pictures the random lasing threshold I_{th} growths significantly under the d_p decreasing lower the defined value $d_\perp$ ~ 150-200μm.

Confinement of the longitudinal size of the random laser was provided by the sample thickness h variation. It was shown [5] the I_{th} is also depends on the confinement. Data on Fig.2 demonstrates the nonlinear dependency of the peak intensity of random lasing spetrum I_{RL} on the thickness h: the dependency is nonmonotonic and characterizes with maximum when sample thickness is about $d_\parallel$ ~ 150-200μm.

The obtained results can be interpreted by defining the size of radiation localization within scattering medium [6]. Equivalence of the $d_\parallel$ and $d_\perp$ values demonstrate the appropriateness to introduce the characteristic size of photons trajectories in the strongly scattering medium. It proves that effective radiation trajectories of the random lasing origin are confined in random medium.

The confinement size is equal in both geometrical direction $d_\parallel = d_\perp = d_c$ (Fig. 2b) and it influences spectra parameters: spectrum intensity, spectrum width [5], and spectrum peak position (Fig.3). It appears in essential difference of the dependency on working size (as pump beam diameter or

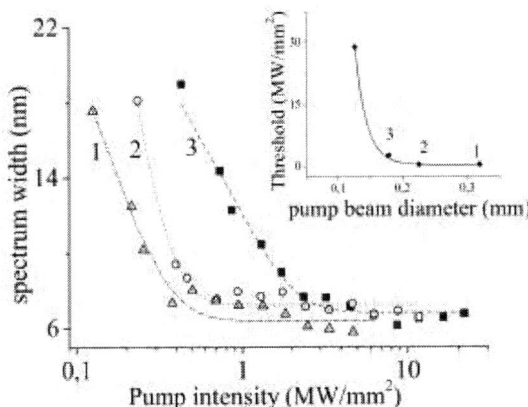

Fig. 1. The radiation spectrum width dependency on pump intensity at pump beam diameter 1 — 0.3mm, 2 — 0.2mm, 3 — 0.15mm. Inset: the corresponding random lasing threshold value versus the pump beam diameter.

978-1-4244-7043-3/10 $26.00 © 2010 IEEE

CAOL*2010 International Conference on Advanced Optoelectronics & Lasers, 10-14 September, 2010, Sevastopol, Ukraine

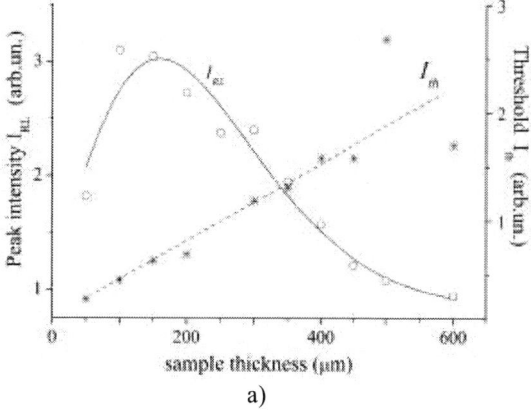

a)

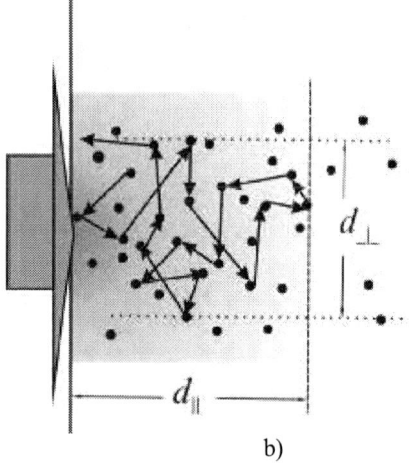

b)

Fig. 2. a) Lasing intensity in the spectrum maxima and random lasing threshold dependence on the sample thickness. b) The random lasing confinement sizes are equal in lenthwise and crosswise direction.

sample thickness variation within the radiation confinement size and out of it.

It is noticeable that sample thinning influences the I_{th} in a different way than the crosswise pump beam narrowing. Fig.2a demonstrates the linear decreasing of the I_{th} throughout the h change range in spite of the lasing intensity decreasing within the confinement region ($h < d_c$). The answer is boundary conditions whose influence growth under the sample thinning. As it was shown in detail [5], positive feedback in random laser is formed in two ways: the first feedback mechanism is multiple scattering within active volume. The second is reflection on the sample surfaces. The mean reflection coefficient is equal to the angle-averaged energy reflection coefficient of unpolarized light:

$$\overline{R} = \frac{1}{2\pi} \int_0^{2\pi} d\phi \int_0^{\pi/2} \frac{1}{2}\left(r_\perp^2(\theta) + r_\parallel^2(\theta)\right)\sin\theta d\theta,$$

where θ — the incidence angle on the surface, $r_\perp(\theta)$, $r_\parallel(\theta)$ are Fresnel energy reflection coefficients for light being

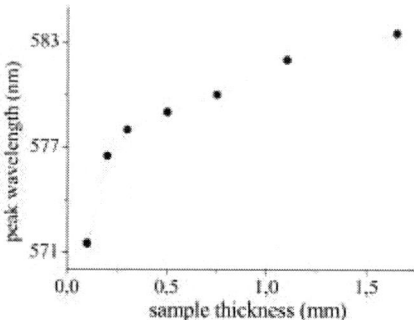

Fig. 3. Lasing spectrum peak wavelength dependence on the sample thickness.

linearly polarized in the incidence and perpendicular plane correspondingly. For the typical refraction coefficient value of polymer $n = 1.5$ it achieves $\overline{R} = 57\%$.

When the sample is thick enough only the front surface contributes into the feedback formation. The surface influence minimization or cancellation by sample immersion causes I_{th} appreciable increasing in this case. In Fig.4 it is exhibited in insignificant spectrum widening. When sample thickness is lower than the confinement size $h < d_c$ the sample immersion mostly suppressed random lasing as surfaces reflection is determinative mechanism of feedback formation.

Thus under sample thinning within the confinement region one feedback formation mechanism is progressive replaced with the other more effective one. It results in monotonical threshold decreasing all over $h < d_c$ region. The intensity I_{RL} decreasing for thin samples (Fig.2a) testifies pump absorption efficiency lowering: due to small sample thickness significant part of pump transmits through the sample.

The crosswise confinement is "unambiguous": when excited region d_p is wider than the confinement size d_c radiation is amplified all over the trajectory and the lasing origins. Otherwise a part of effective radiation trajectory is out of the

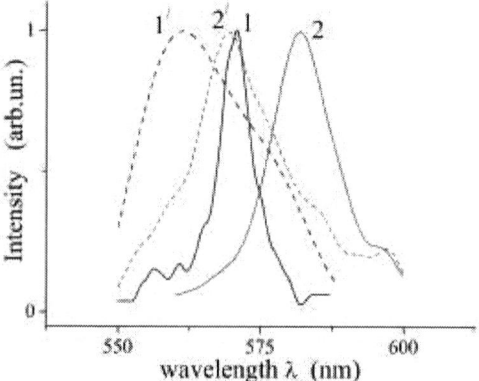

Fig. 4. Lasing spectra reveal shortwave shift when the sample is immersed: the sample thickness is 0.1mm (1 and 1′) and 0.4mm (2, 2′).Dotted lines — immersed samples, and solid lines the samples are in the air.

978-1-4244-7043-3/10 $26.00 © 2010 IEEE   148

excited region and losses suppress random lasing.

We showed [7] that main mechanism influencing the lasing spectrum position in the investigated samples is luminescence reabsorption and reemission. The mechanism is sensitive to any factor influencing radiation path length within the sample. The most influential sample parameters having an effect on path length are sample thickness and scattering efficiency. Thus the h-dependence of spectrum wavelength is conditioned by the reabsorption efficiency change. Referring to Fig. 3 the dependency of lasing wavelength on sample thickness allow us realize the wavelength tuning within 13nm by simple change of sample thinning [5]. The tuning efficiency, determined as $\Delta\lambda/\Delta h$ is significantly higher within confinement region than out of it.

On the basis of obtained results we proposed the way to optimize random lasing conditions and lasing wavelength tuning [5]. We showed the most effective tuning of random lasing wavelength for random laser thickness is within the size of confinement. It allows effectively improve feedback efficiency of random laser by random laser parameter optimization conforming the working size with the confinement size. The obtained results demonstrate the appropriateness to introduce the characteristic size of photons trajectories in the strongly scattering medium, allow determining the optimal sizes of the random laser and defining its optimal pump requirements.

REFERENCES

[1] N. M. Lawandy, R. M. Balachandran, a. S. L. Gomes & E. Sauvain "laser action in strongly scattering media", nature, 368, p.436-438, 1994.

[2] H. Cao, "Lasing in random media", waves of random media, 13, R1-R39, 2003.

[3] Sushil Mujumdar, Marilena Ricci, Renato Torre, and Diederik S.Wiersma "Amplified Extended Modes in Random Lasers", Phys.Rev.Lett., vol. 93, no. 5, pp.053903-1 – 053903-4, 2004.

[4] E. Tikhonov, Vasil P.Yashchuk, O.Prygodjuk, V.Bezrodny, "The Dye Activated Polymers with Added Fine Scattering Particles as New Lasing Media", Solid State Phenomena,vol. 94, pp.295-298, 2003.

[5] E.O.Tikhonov, Vasil P.Yashchuk, O.A.Prygodjuk, V.I.Bezrodny, Multiple Scattering Active Media as Small-size Frequency Tunable Sources of Stimulated Emission, Solid State Phenomena, vol. 99-100, p.77-82, 2004.

[6] V. P.Yashchuk, M.V.Zhuravsky, O.A.Prygodiuk, "Numeric simulation and experimental research of the spatial size effect of random laser lasing", Izvestija RAN, vol.70, no.9, pp.1318-1322, 2006.

[7] E.O.Tikhonov, Vasil P.Yashchuk, O.A.Prygodjuk, V.I.Bezrodny, Influence of fine-dispersed dielectric particles on dye luminescence in a polymer matrix, Solid State Phenomena, vol. 99-100, pp.153-156, 2004

CAOL*2010 International Conference on Advanced Optoelectronics & Lasers, 10-14 September, 2010, Sevastopol, Ukraine

Fast Adaptive Optical System for Laser Beam Correction

A.L.Rukosuev, A.G.Alexandrov, A.V.Kudryashov, V.V.Samarkin

"Active Optics NightN" Ltd., 115407, Sudostroitelnaya 18 bld. 5, Moscow, Russia, tel. (499) 618-81-71

Abstract: Considered adaptive optical system is intended for correction of laser wavefront aberrations, having characteristic frequency 50-70 Hz. The system consists of the following basic elements: stacked-actuators adaptive mirror, fast Shack-Hartmann wavefront sensor and a control system which includes computer under OS Windows and the fast control unit for adaptive mirror. The results of modeling of the correction process are presented and the interrelation between frequency of samples and accuracy of correction is revealed. Recommendations about future improvement of the system work are given.

The closed loop adaptive optical system can be used for correction of various wave fronts, for example, wave front of the powerful femtosecond and petawatt lasers, and as for correction of wave front of the radiation which have passed through turbulent atmosphere. The wave front in femtosecond laser facilities changes extremely slowly, the change period can reach several hours. Petawatt systems generate impulses with period from tens minutes till several hours. Thus, on use of adaptive optical system in such lasers it is not imposed essential restrictions on speed. In this case, from the point of view of time, use of adaptive optical system for correction of the radiation which passed through turbulent atmosphere is critical. The system variant is considered, allowing to correct for aberrations with characteristic frequency of 60-70 Hz.

The closed loop adaptive optical system [1] is constructed under the classical scheme (fig. 1) and consists of a stacked-actuator adaptive mirror, the control driver for adaptive mirror, Shack-Hartmann wave front sensor based on CCD or CMOS camera and the operating computer working under OS Windows.

Owing to the nature, the adaptive system constructed under the given scheme is discrete that predetermines some delay between measurement and applying the correcting voltage to a mirror. Correction process can be divided in time into following characteristic intervals: 1) integration of the image by matrix CCD or CMOS camera; 2) transfer of the image to the computer; 3) processing of the image and calculation of correcting voltage; 4) transfer of the correcting voltage array to the control driver 5) applying of voltage to a mirror; 6) a mirror deflection according to the exposed voltage. Thus to carry out in parallel the given processes is impossible, since before reception of the new image the mirror should change wavefront. Thus, the system will work with clock frequency $1/(t1+t2+t3+t4+t5+t6)$.

The mathematical model describing correction of wavefront of the sinusoidal form has been developed for calculation of key parameters of correction. On fig. 2 the original sinusoid, result of correction is presented at clock frequency of system 10 times of a sinusoid surpassing frequency, and result of correction at clock frequency of system 50 times of a sinusoid surpassing frequency.

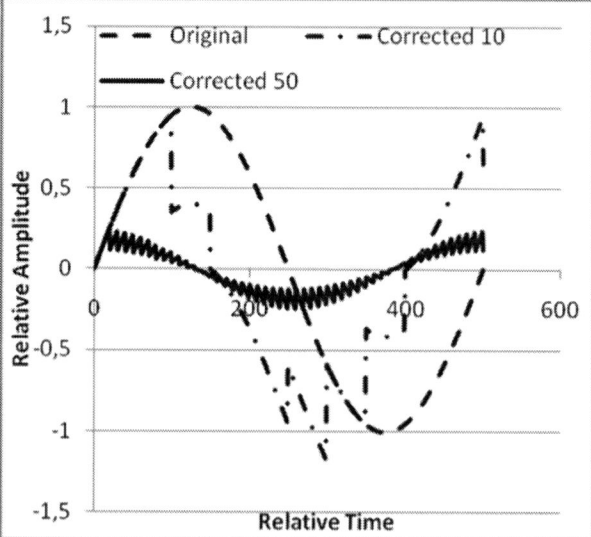

Fig. 2. Correction with 10 and 50 relative sample rate.

Fig. 3 shows dependence of accuracy of correction on clock frequency of system. It is visible, that at 50 multiple

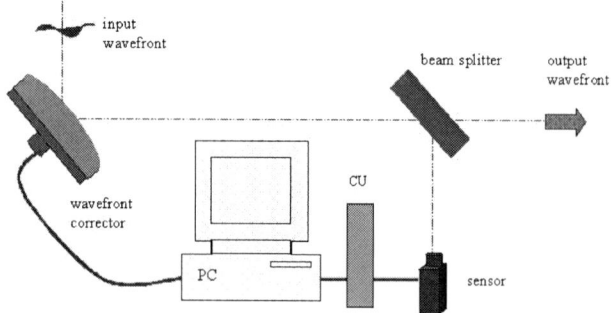

Fig. 1. Closed-loop adaptive optical system.

978-1-4244-7043-3/10 $26.00 © 2010 IEEE

excess of clock frequency of system potential accuracy of correction does not exceed 20 %.

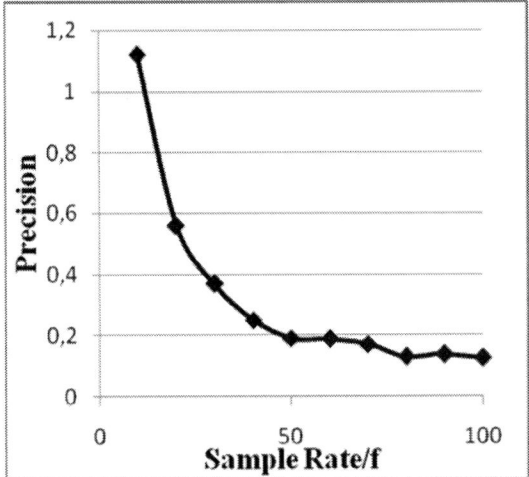

Fig. 3. Correction accuracy depending on relative sample rate.

REFERENCES

[1] A.L.Rukosuev, A.Alexandrov, V.Ye.Zavalova, V.V.Samarkin, A.V.Kudryashov, "Adaptive optical system based on bimorph mirror and Shack-Hartmann wavefront sensor", *Proc. SPIE 4493, pp. 261-268, 2002.*

CAOL*2010 International Conference on Advanced Optoelectronics & Lasers, 10-14 September, 2010, Sevastopol, Ukraine

Optimizing energy performance of «zig-zag» slab lasers in form of flat truncated prisms

I.S. MANAK, V.V. ZUKOWSKI, M.S. LEANENIA

Belarusian State University, Minsk, Belarus

Abstract: Considered the optimal power characteristics of «zig-zag» lasers in the form of a flat truncated prism with a zig-zag course of the ray in the three-mirror stable resonator, depending on the design parameters. It is noted that this design is characterized by the additional loss of radiation on the mirrors of the resonator, contributing to the generated radiation specific power decreasing.

One possible solution of the problem of laser's miniaturization is to develop designs where the effective interaction of active particles and radiation in the cavity can be achieved. These include the lasers with an active element in the form of a flat truncated prism (slab-laser) with a zig-zag course of the ray («zig-zag» lasers) in the three-mirror stable resonator (Fig. 1a) [1]. Slab-lasers are tech. Flat truncated prism (Fig. 1b) is cut out and properly handled from the active substance in the solid phase, for example, crystal aluminum-yttrium garnet or glass doped with neodymium ions. Two plane-parallel surfaces of the prism, separated by a distance t, can be used for pumping the active substance. Particularly promising in this case are the laser diodes or LED matrix on the basis of ternary compounds of $Ga_{1-x}Al_xAs$, which spectrum can be effectively managed by changing the component composition of the semiconductor. The absorption spectrum of neodymium has a broad band at a wavelength of ~ 0.8 µm, and therefore the perfect alignment of the absorption spectrum of Nd^{3+} and the spectrum of healing $Ga_{1-x}Al_xAs$-matrix can be achieved [2]. The other two flat surfaces of truncated prism 1 and 2 are inclined to each other at an angle α. The surfaces are carefully polished. Then they are covered with high-reflective mirrors. Output semitransparent mirror 3 is inclined at an angle φ to one of these surfaces. The optical beam propagating in the active element is perpendicular to the mirror 3, is reflected by mirrors 2 and 1 in turn until the angle of incidence on one of the mirror does not become zero. Then the beam is reflected in the opposite direction and goes through the mirror 3. Thus the zig-zag course of the beam in the optical resonator is realized. The number of possible reflections of optical beams in such system provides by simple formula $N = 1 + \varphi / \alpha$ [3]. It is easy to get a requirement imposed on the angles α and φ of plane truncated prisms: $\varphi \geq \alpha$ and $\varphi / \alpha = n$, where n - integer.

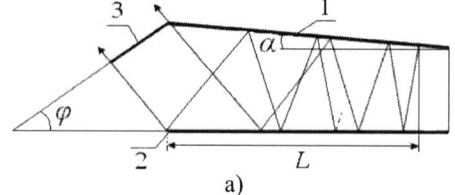

a)

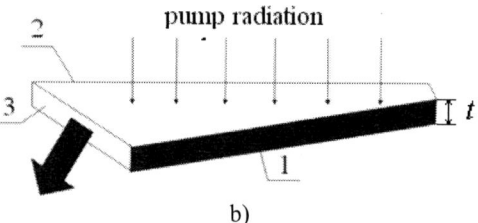

b)

Fig.1. Optical scheme of three-mirror laser resonator and the ray path in it (a) and planar laser geometry of the active element (b)

Although the height of the prism varies with the distance L, however, owing to the smallness of the angle α, in the first approximation we can assume that the height of the prism is equal to d, where d - the maximum distance between the planes 1 and 2. The expression that relates the length of the ray l which passes an optical beam in the active medium in a single pass from point A to point B, with angles α and φ, has the form [1]:

$$l = d\left\{\frac{1}{\cos\phi} + \cos(\alpha+\phi)\sum_{k=1}^{\phi/\alpha}\{\cos[\phi-(k-1)\alpha]\cdot\cos(\phi-k\alpha)\}^{-1}\right\}\cdot \quad (1)$$

Value for the length L of mirror 1, at which a repetition of the trajectory of the beam in the opposite direction is realized, is the following:

$$L = d[ctg\alpha - \cos(\phi+\alpha)/\sin\alpha]. \quad (2)$$

The volume V of the active element filled with beam radiation generated in the «zig-zag» laser is calculated by the formula:

$$V = td^2\left[\frac{\cos^2\alpha - \cos^2(\phi+\alpha)}{\sin 2\alpha} + \frac{\cos^2\phi - \cos^2 2\phi}{\sin 2\phi}\right]. \quad (3)$$

When the propagation of the laser beam is almost orthogonal to the axis of the crystal, the effective length l of one beam pass in such resonator is much higher than its length L. At angles $\varphi < 5^0$ and $\alpha < 0.5^0$ ratio of the length of the beam passes in the active medium l to the total length L is more than $l/L > 15$. In addition the ratio l/L does not depend on the magnitude of the output aperture d.

978-1-4244-7043-3/10 $26.00 © 2010 IEEE

For a laser with a Fabry-Perot resonator radiation specific power per unit volume is calculated by the formula [4]:

$$S^{sp} = v\frac{1}{2L}\ln\frac{1}{r_1r_2}(k_0 - \rho - \frac{1}{2L}\ln\frac{1}{r_1r_2})(\rho + \frac{1}{2L}\ln\frac{1}{r_1r_2})^{-1}\beta^{-1}. \quad (5)$$

Here v - velocity of light in the active medium, k_0 – gain coefficient, β - nonlinearity parameter, $k_r = \frac{1}{2L}\ln\frac{1}{r_1r_2}$ - utility losses caused by the escape of radiation from the resonator with length L through the mirror with reflection coefficients r_1 and r_2, ρ - coefficient taking into account the absorption losses and scattering in the matrix.

For «zig-zag» laser (Fig.1), formula (5) is modernized with regard to the fact that in the active medium revealed additional losses while reflection of radiation on the mirrors 1 and 2 with the same average reflection coefficients r. They can be taken into account by introducing instead of the coefficient ρ, which takes into account the absorption losses and scattering in the matrix, in formula (5) parameter [3]:

$$\rho^* = \frac{2N-1}{2l}\ln\frac{1}{r} + \rho. \quad (6)$$

Taking into account (6), the average in time radiation specific power density of «zig-zag» laser generated per unit volume of the active substance, is calculated by the formula:

$$S^{sp} = v\frac{1}{2l}\ln\frac{1}{r_3}(k_0 - \rho^* - \frac{1}{2l}\ln\frac{1}{r_3})(\rho^* + \frac{1}{2l}\ln\frac{1}{r_3})^{-1}\beta^{-1}. \quad (7)$$

Here $k_r = \frac{1}{2l}\ln\frac{1}{r_3}$ - utility losses caused by the escape of radiation from the resonator through the output mirror with reflection coefficient r_3, l - the length of the optical beam in the active medium from point A to point B (Fig. 1).

Differentiating the specific power S^{sp} of «zig-zag» laser on $\frac{1}{2l}\ln\frac{1}{r_3}$ and equating the derivative to zero we obtain:

$$\left(\frac{1}{2l}\ln\frac{1}{r_3}\right)_{opt} = \sqrt{k_0\rho^*} - \rho^*. \quad (8)$$

This expression allows us to determine the optimal value of efficiency losses for a given values of k_0 and ρ, as well as the optimal value of r_3 for given values of k_0, ρ and l, providing maximum output power of «zig-zag» laser:

$$r_3^{opt} = \exp[-2l(\sqrt{k_0\rho^*} - \rho^*)]. \quad (9)$$

Optimized by the reflective coefficient of the output mirror expression for calculating the power density S^{sp} of «zig-zag» laser radiation is determined by the formula:

$$S_{opt}^{sp} = vk_r^{opt}(k_0 - \rho^* - k_r^{opt})(\rho^* + k_r^{opt})^{-1}\beta^{-1}. \quad (10)$$

Figure 2 shows the dependence of power, generated per unit volume of a neodymium glass laser, and the coefficient of energy losses of ρ^* versus the angles α and φ for the active medium, characterized by the parameters $\beta = 0.1\cdot10^7\text{cm}^3/\text{J}$, $\rho = 0.01\text{cm}^{-1}$, $k_0 = 0.15\text{cm}^{-1}$, and the design of the resonator with the values of $d = 1\text{cm}$ and $r = 0.99$.

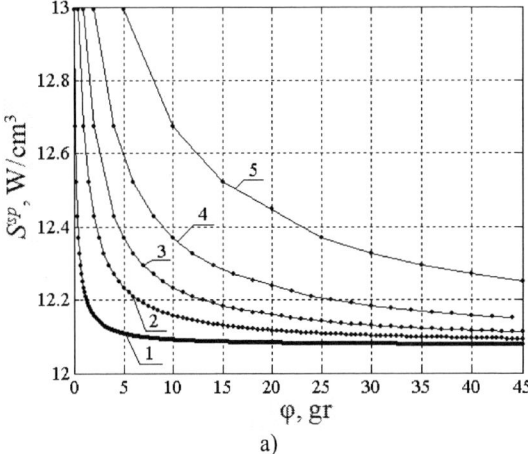

a)

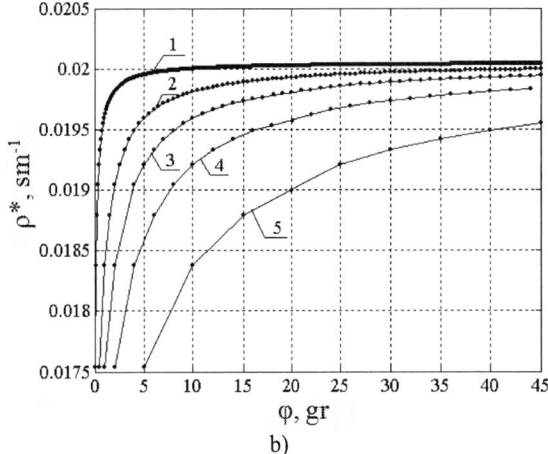

b)

Fig. 2. The dependence of the power of S^{sp} (a) and energy losses ρ^* (b) on the angle φ at angles $\alpha = 0.1^0$ (1), $\alpha = 0.5^0$ (2), $\alpha = 1^0$ (3), $\alpha = 2^0$ (4), $\alpha = 5^0$ (5)

As shown in Figure 2, the fall of power density S^{sp} by 7% relative to the maximum value ($S^{sp} = 13\text{W/cm}^3$ for $N = 2$) as well as increasing of energy loss ρ^* is due to harmful radiation losses while reflection on the mirrors 1 and 2 [5].

Note that to estimate the approximate value of the beam path length l in the active element may be used simple in comparison with (1) expression:

$$l \approx N \cdot d. \qquad (11)$$

The effective length of the radiation beam passage in the active medium can be calculated by formula (11) with a small relative error $\delta < 5\%$ for angles $\alpha < 5^0$ and $\varphi < 25^0$. Then the loss factor ρ^* and the specific power S^{sp} can be determined by the expressions where the dependence on the angles α and φ is represented by a single parameter N:

$$\rho^* = \frac{2N-1}{2Nd}\ln\frac{1}{r}+\rho, \qquad (12)$$

$$S^{sp} = vk_r^{opt}(k_0 - \rho^* - k_r^{opt})(\rho^* + k_r^{opt})^{-1}\beta^{-1}. \qquad (13)$$

Figure 3 shows the dependence of specific output power S^{sp} and energy losses ρ^* in the active element, calculated by formulas (12) and (13), versus the reflection coefficient r of mirrors 1 and 2 and the number of reflections N at d=1cm for the active medium on the glass doped with Nd^{3+}.

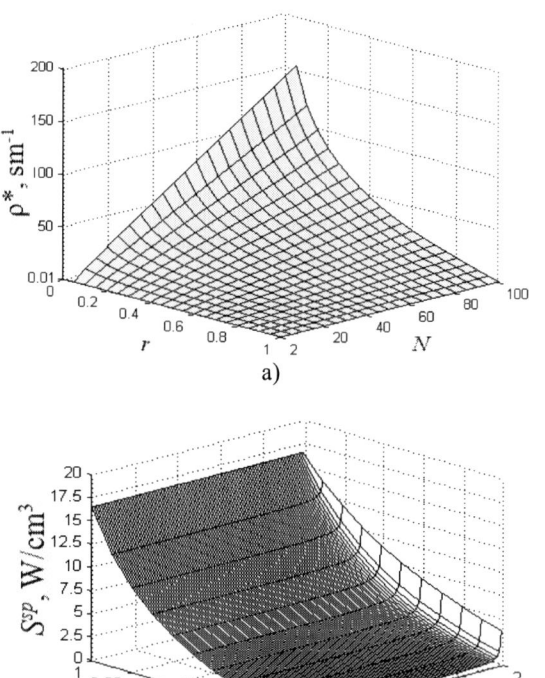

a)

b)

Fig. 3. The dependence of the power S^{sp} and energy losses ρ^* on the reflection coefficient r of the mirrors 1 and 2 and the number N of reflections

The fall of power S^{sp} and rise in loss are greater with decreasing the reflection coefficient r, as shown in Figure 3. Beginning with some values of N for a given value of r the rate of specific power S^{sp} decreasing approaches zero. The optimum value of the reflection coefficient r of mirrors 1 and 2 is r=1, at which the maximum value of specific power can be achieved. However, even when using high-reflective coatings, the maximum-possible reflection coefficient is $r \approx 0.998$. That places high demands on the quality of manufactured constructions.

Despite the drop in power density due to losses while reflection on mirrors 1 and 2, due to design features of a laser resonator «zig-zag» lasers are promising sources of miniature lasers, because their inherent high quality of laser beam by reducing the impact of technological and thermal inhomogeneity and increased length of one passage of radiation in the active element [6].

REFERENCES

[1] L.N. Orlov, Y.I. Nekrashevich, G.I. Zheltov, V.V. Zhukovsky. Patent 6244 Republic Belarus. "Solid-state laser", Publ. 30.06.2004, Official bulletin, № 2 (41), p. 236, 2004

[2] I.S. Manak, V.V. Zhukovsky, M.S. Leanenia, "Compact solid-state flat «zig-zag» lasers," Instrumentation: status and prospects. IX international scientific-technical conference, NTU "KPI", Kiev, PP. 55-56, 2010.

[3] V.V. Zhukovsky, M.S. Leanenia, I.S. Manak. "Compact solid-state lasers with a zig-zag course of radiation in the active element," Electronics-info, №7, pp. 55-58, 2008

[4] Methods of calculating optical quantum generators. Volume 1., By ed. B.I. Stepanov, Mn., 385 pp., 1967

[5] I.S. Manak, V.V. Zhukovsky, M.S. Leanenia, "Dependence of the energy characteristics of «zig-zag» lasers from the design of three-mirror resonator," Instrumentation: status and prospects. IX international scientific-technical conference, NTU "KPI", Kiev, PP. 54-55, 2010.

[6] I.S. Manak, V.V. Zhukovsky, M.S. Leanenia, "Physical basis of reducing the overall dimensions of laser sources," Electronics-Info, № 7, pp. 76-77, 2009

Microchip laser with active output mirror

V.V. Kiyko, V.I. Kislov, E.N. Ofitserov

Prokhorov General Physics Institute, Russian Academy of Science, Moscow, Russia

Abstract: Presented is an operator model of a microchip laser with an active output mirror based on Fabry-Perot interferometer. Considered are apodizing properties of a mirror-interferometer composed of two spherical, semi-transparent reflectors with variable gap between those. Proposed is an operator model describing the mirror under diffraction approximation. Results of numerical simulation of the micro-chip laser with variable thermal lens and active output mirror are presented. It is shown that use of an active interferometer as an output cavity mirror enables to control mode composition and power of output radiation at divergence close to diffraction limit.

The feasibility of control over temporal and spatial characteristics of micro-chip laser radiation is limited. Especially problematic is radiation control over such passive Q-switched lasers operating in a high frequency mode. Radiation temporal characteristics (repetition rate and pulse duration) control is realized by varying of pumping power. Simultaneously, owing to change of the thermal lens, also inevitably occurs variation of spatial characteristics of output radiation. One of the ways for microchip laser output parameters control is based on application of an output mirror with controlled reflection coefficient [1, 2].

In this work presented is an operator model of a micro-chip laser with output mirror on the basis of Fabry-Perot interferometer with non-flat mirrors and variable gap.

This approach is especially preferred for application in small-size and microchip lasers with small diameter of the output beam.

In the modeling of such type of a mirror it is necessary to take into account beam diffraction at each pass through the interferometer meanwhile simplified geometrical models may results in essential errors. Applicability boundary of the geometrical model is determined by an expression $N_{eff} = N_{int}/F \gg 1$ (where N_{eff} is effective Fresnel number, $N_{int} = d^2/8\lambda l$ is Fresnel number of the interferometer, d is beam diameter, λ is radiation wavelength, F is number of interfering beams, l is size of the gap). Comparative analysis of geometrical and diffraction models is conducted by means of developed operator approach to the interferometer description. Within the frame of this approach reflection coefficient across the field is determined by a relation

$$R_{int} = \left(R_1 - \left(R_2 |T_1|^2 H_{int} \right) / \left(E - R_1^* R_2 H_{int} \right) \right) \exp\left(i\Psi \right)$$

(where $R_{1,2}$ are of the first and the second reflectors reflection coefficients taken into account along direction of falling onto the interferometer field, H_{int} is propagation operator per round

trip, Ψ is phase addition, E is unit operator, T_1 is transparency of reflector 1).

Some results of calculations of averaged over aperture power reflection coefficient R_{int}^2 of the mirror-interferometer are presented in Fig. 1, 2. Here incident on the mirror radiation was presented as corresponding to several lowest transverse modes of a stable cavity.

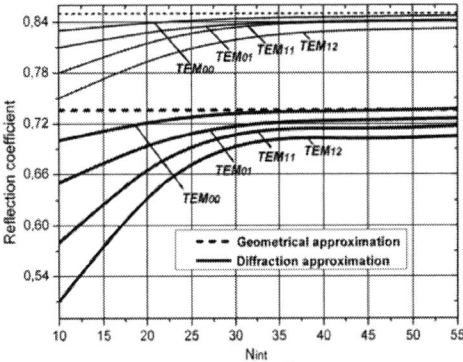

Fig.1 Reflection coefficient R_{int}^2 of the interferometer vs Fresnel number (thin lines: $\Delta l = 0{,}05\lambda$; thick lines: $\Delta l = 0{,}03\lambda$; curvature radius is -0,07m; $d = 100\,\mu m$; $R_1 = R_2 = 0{,}88$; $\lambda = 1{,}064\,\mu m$).

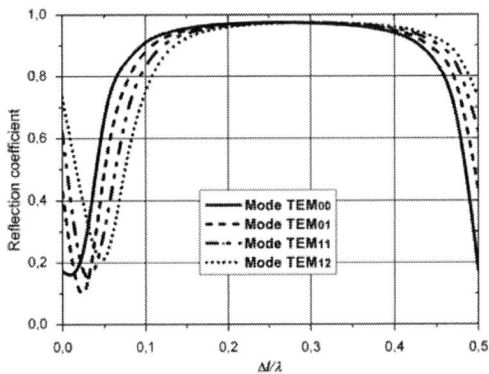

Fig.2 Reflection coefficient R_{int}^2 of the interferometer vs gap mismatch Δl (mirror 1 is flat; mirror 2 curvature radius is -0,06m; $l_0 = 100\,\mu m$; $\Delta l = \lambda/40$; $d = 200\,\mu m$; $R_1 = R_2 = 0{,}85\,\mu m$; $\lambda = 1{,}064\,\mu m$)

978-1-4244-7043-3/10 $26.00 © 2010 IEEE

Within the frame of the operator model of the laser eigenvalues γ and eigenfunctions U are being found from equation:

$$R_{int} H_{rez} U = \gamma U$$

where: U describes field distribution at the output mirror of the cavity; R_{int} is the interferometers reflection coefficient; H_{rez} is propagation operator per round trip for the resonator. In the calculations applied was matrix description of the operators.

The model enables to optimize characteristics of system resonator and mirror-interferometer taking into account thermal lens (TL) and requirements on output radiation characteristics. One of the cavity basic characteristics is squared modulus of eigenvalue $|\gamma|^2$ (EV) [1, 3]. Some results of calculations of the laser system characteristics are presented in Fig. 3, 4. Here the initial cavity is composed of flat mirrors and the interferometer is of output cavity mirror and spherical semi-transparent mirror. In further, the cavity is called stable or unstable corresponding to conventional terms used for description of optical resonators. In Fig. 3 depicted are of $|\gamma|^2$ vs number of mode for plane-parallel cavity without interferometer and with interferometer of the microchip laser. Involved are cases without TL and with TL. System parameters: Output mirror curvature radius is $\pm 0,06m$; $\lambda = 1,064 \mu m$ is radiation wavelength; $l = l_0 + \Delta l$ is interferometer length; $l_0 = 100\lambda$; Δl is interferometer mismatch varying within limits from 0 to $\lambda/2$; laser output aperture $d = 200 \mu m$; $R_1 = R_2 = 0,8$; TL focal length was $6mm$; cavity length - $5,5mm$; active element refraction index is 1.82.

From analysis of calculations results and Fig.3 follows that with TL (curve 2) the resonators becomes multimode. With this, approximately, divergence of output radiation of cavity with TL is higher than one of cavity without TL by a factor of 2 - 2.5. Use of interferometer as an output mirror (curves 3, 4) enables to enter the system into almost single mode operation. Characteristics of system including cavity and mirror-interferometer depends essentially on controlled mismatch of the interferometer Δl (Fig.4). Parameter ω is proportional to the angle of energy divergence Θ of output radiation in the far zone at the level 0.5 for dominating mode, $\Theta = 2\lambda\omega/\pi d$; $|\gamma|^2$ is square of modulus EV for the dominating mode and for nearest one to it. In the region of mismatch $(0,35 - 0,45)\lambda$ dominates mode TEM$_0$, in the region of mismatch $(0,18 - 0,22)\lambda$ dominates mode TEM$_1$. Discontinuities of ω in Fig. 4 are caused by change of number of dominating mode at the interferometer mismatch. Magnitude $|\gamma|^2$ (and, correspondingly, laser output radiation power) depends essentially on the interferometer length. Thus $|\gamma|^2$ for TEM$_0$ mode varies within range 0.58 - 0.90. Radiation divergence in this case varies not higher than for 25%.

Conducted numerical investigations of system of a microchip laser with thermal lens demonstrate that use of an active cavity output mirror based on Fabry-Perot interferometer enables to control mode composition and power of laser output radiation.

This work was supported in part by Russian Foundation for Basic Research under grant (№ 09-02-00343).

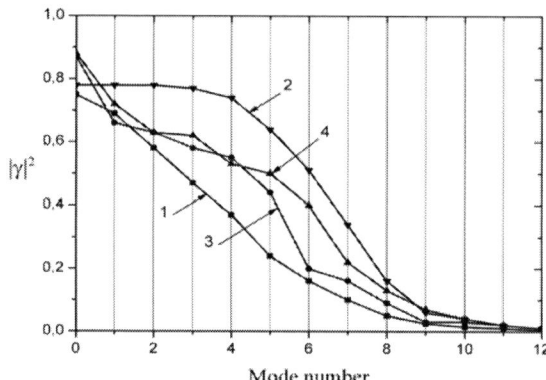

Fig. 3. Dependence's of $|\gamma|^2$ vs mode number (1 - plane-parallel cavity without TL; 2 - cavity with TL; 3 - cavity with stable interferometer and TL; 4 - cavity with unstable interferometer and TL).

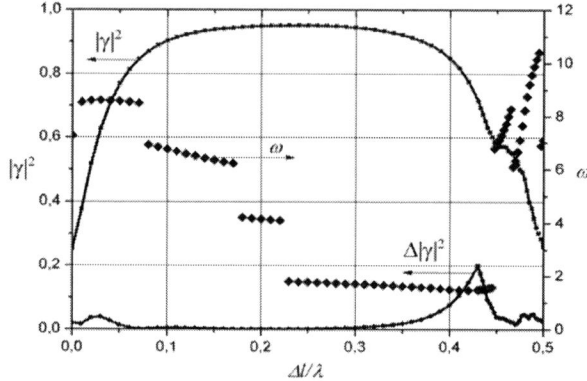

Fig. 4. Dependence's $|\gamma|^2$, $\Delta|\gamma|^2$ and ω vs mismatch $\Delta l/\lambda$.

References

[1] O. Svelto Principles of Lasers, 4-th ed., 604p., 1998.

[2] S. De Silvestri, P. Laporta, V. Magni, and O. Svelto *Optics Letters*, vol. 12, pp. 84-86, 1987.

[3] A.G Fox. and T. Li. B*ell Syst. Tech. J.,* vol. 40, p.453, 1961.

Hill-climbing algorithm for adaptive optical system with Shack-Hartmann sensor

J.Sheldakova[1], A.Rukosuev[1], P.Romanov[1], A.Kudryashov[2], V.Samarkin[2]

[1]Active Optics NightN Ltd., Moscow, Russia
[2]Moscow State Open University, Moscow, Russia
sheldakova@nightn.ru
+7 (499) 6188171

Abstract: We describe multi-dither adaptive optical system based on Shack-Hartmann wavefront sensor. RMS of the wavefront is minimized to get the best correction of the aberrated laser beam. Hill-climbing algorithm is used to determine voltages to be applied to the electrodes of a bimorph mirror which is used as a wavefront corrector.

Presented closed-loop adaptive optical system is used to correct for the wavefront aberrations and consists of bimorph deformable mirror [1],[2] with electronics and Shack-Hartmann wavefront sensor [3]. To control for the mirror we usually use phase conjugate algorithm. But we found out that sometimes the result of correction is not the best.

The principle of Shack-Hartmann wavefront sensor is shown on Fig.1. Measurements are based on determination of local slopes ΔS of a distorted wavefront relative to the reference flat wavefront. Laser beam is divided into a number of beamlets by a two-dimensional subapertures of the lenslet array. Each subaperture provides a separate focus on the detector. The local slope of the wavefront on each subaperture of the lenslet array is proportional to the displacement of the focal spot center ΔS.

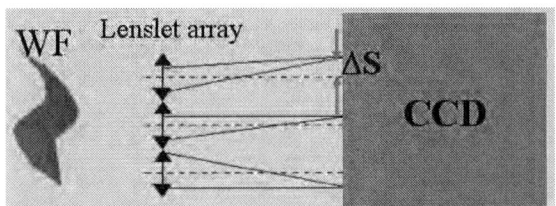

Fig. 1. Group delay in the chirped mirror described in the text.

Phase-conjugated algorithm allows us to minimize the displacement of real focal spots from the reference ones; so closed-loop system minimizes $\sum_i \Delta S^2$, i – number of lenslet subapertures.

As a first step of correction response functions b should be measured. They are represented in terms of displacement of the focal spots $b_{j\,x}{}^i = \Delta x_i / u_{0\,j}$ and $b_{j\,y}{}^i = \Delta y_i / u_{0\,j}$; here $u_{0\,j}$ is the unit voltage used to measure response functions, j = number of electrode.

Then the displacements of the focal spots for the wavefront to be corrected are measured and following equation is written:

$$S_i = \begin{vmatrix} \Delta x_i \\ \Delta y_i \end{vmatrix} = \sum_{j=1}^{N} u_j \cdot b_j^i \; ; \text{ here } \boldsymbol{u} \text{ is an array of voltages.}$$

As a last step the voltages to be applied to minimize the displacements are calculated using least square method:

$$\min \left\| \vec{S} - b \cdot \vec{u} \right\|^2 , \; u = \| B \| \bullet S, \; B = (b^T b)^{-1} b^T .$$

The problem of displacements minimization is that the minimal $\sum_i \Delta S^2$ does not always correspond to the best wavefront. Fig.2 shows the example of "bad" correction – the resulting wavefront is far from ideal. Moreover we can improve this result manually without any closed-loop systems (Fig.3).

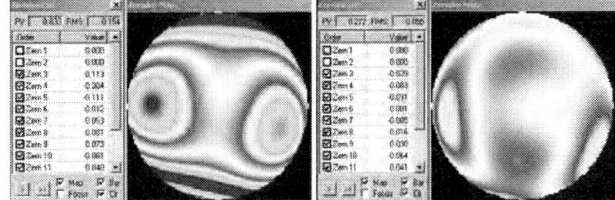

Fig. 2. Wavefront before (left) an after (right) correction

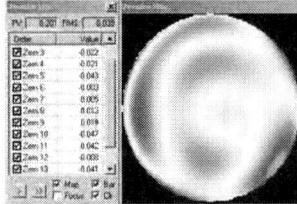

Fig. 3. Wavefront after manual correction

To correct for the residual aberrations after phase-conjugation we could use hill-climbing algorithm. As a criterion of good correction of the wavefront aberrations a wavefront amplitude $P\text{-}V$ or RMS could be used.

To minimise RMS of the wavefront with hill-climbing algorithm first we apply the probe signal to one electrode of the mirror. If after that *RMS* becomes better we continue to change voltage on the same direction on this electrode. And do it until the function becomes worse. Then we apply the probe signal to the next electrode and do the same with this electrode and with all other electrodes. This is a quite simple procedure but it takes rather long time.

Fig.4 shows the result of correction with hill-climbing algorithm. First we used phase-conjugation and improved RMS of the wavefront from RMS=0.467μ to RMS=0.088μ. Then after 10 minutes of hill-climbing use we got RMS=0.047μ; after 20 minutes the result was improved to RMS=0.039μ.

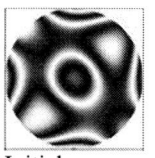 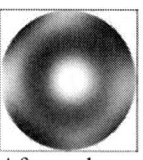 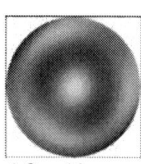 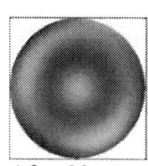

Initial	After phase-	After 10	After 20
P-V=2.451μ	conjugation	minutes of	minutes of
RMS=0.467μ	P-V=0.505μ	hill-climbing	hill-climbing
	RMS=0.088μ	P-V=0.269μ	P-V=0.207μ
		RMS=0.047μ	RMS=0.039μ

Fig. 4. Wavefront changes while correction

Let us estimate how fast the system based on hill-climbing algorithm could be.

1. As we use standard camera in our wavefront sensor – the sensor is the main limiting element in our case (30 Hz – 30 frames per second).

2. One electrode needs about 5 steps to get better *RMS*.

3. One loop of hill-climbing algorithm includes 5 steps * Number of electrodes:

for 17-elements mirror one loop is 85 steps – 3 seconds

for 32-elements mirror one loop is 160 steps – 5.5 seconds

for 60-elements mirror one loop is 300 steps – 10 seconds

4. Hill-climbing algorithm requires at least 100 loops to get reasonable result:

for 17-element mirror –300 seconds – 5 minutes

5. The more electrodes we have – the more loops of hill-climbing we need:

for 60-element mirror – 200 loops - 2000 seconds – about 35 minutes

6. In case of large initial aberrations (more then P-V=1μ) we need more loops and much more time.

So the system based on hill-climbing method allows to correct for static or slow-changing aberrations. Phase-conjugation could be used to speed up the procedure and find an initial approach for hill-climbing method.

Future work:
- To speed up hill-climbing method.
- To try more approaches.
- To find better criterion for laser beam correction.

REFERENCES

[1] A.V.Kudryashov, V.I.Shmalhausen, "Semipassive bimorph flexible mirrors for atmospheric adaptive optics applications", *Opt. Eng* **35**(11), pp. 3064-3073, 1996

[2] A.V.Kudryashov, V.V.Samarkin, A.Aleksandrov, "Adaptive optical elements for laser beam control", *Proc. SPIE* **4457**, pp. 170-178, 2001

[3] V.Ye.Zavalova, A.V.Kudryashov, "Shack-Hartmann wavefront sensor for laser beam analyses", *Proc. SPIE* **4493**, pp. 277-284, 2002.

CAOL*2010 International Conference on Advanced Optoelectronics & Lasers, 10-14 September, 2010, Sevastopol, Ukraine

Design of the interference structures taking into account absorption by thin-film materials

Ya.Bondarchuk, H.Petrovska

Department of Photonics, Lviv Polytechnic National University, 12 Bandera Str., 79013 Lviv, Ukraine
Phone: 032-2582581, e-mail: galina_petrovska@mai.ru

Abstract. **In this paper, the design method of polychrome laser interference mirrors is proposed. This method reduces the total absorption in the mirrors at all wavelengths, while ensuring optimum transmission at each wavelength.**

I. INTRODUCTION

It is known that the implementation of the integrated regime of generation in lasers polychrome is achieved using broadband interference mirror with maximum reflection at wavelength λ_0, that corresponds to centre spectrum range of the polychrome radiation. However, maximization of the integral power in such lasers is possible only at provide optimal oscillating mode for each transition, in particular, for those wavelengths, whose contribution to total power is most significant.

At the same time, know that for each wavelength is optimal transmittance of the exit mirror τ_i, whose value depends on the amplification and total losses in the resonator for this wavelength [1]. The optimal transmittance for all wavelengths provide practically impossible using broadband interference mirror with maximum reflectance in the centre spectrum range of the polychrome radiation. In addition, it is necessary take into account range properties of the interference coatings in terms of spectral absorption in them for multiline lasers. The spectral absorption, which has the minimum value at resonant wavelength λ_0 of the interference structure and deviation from it for 20÷25%, can increase ten times (see Fig. 1).

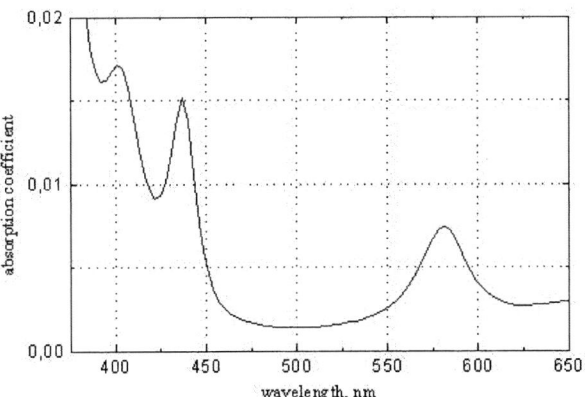

Fig. 1. Spectral dependence absorptance $A(\lambda)$ of the 15-layers interference coating $ZrO_2–SiO_2$ (λ_0=500 nm)

The transmittance (reflectance) for wavelength of the oscillation lines are used as control parameters of the synthesized coatings at design the cavity interference mirrors of the lasers polychrome. Ignoring specific spectral dependence

absorptance of the real interference mirrors in area of high reflectance (Fig. 1) results into significant loss of integrated power for the low-power lasers and formation intracavity thermal effects for powerful lasers [2].

Hence, the search new approach in the design of cavity mirrors for lasers with the wide spectrum of radiation corresponding to the range properties of the interference structures is topical problem. The various technical solutions are offer for low power and power lasers, since the effects of absorption in the multilayer mirrors affect differently on exit radiation characteristics.

II. DESIGN CAVITY MIRRORS OF THE LOW-POWER LASERS

Absorption in the mirrors of the low-power lasers not results marked thermal effects, but significantly reduces output power and makes impossible the generation of the weak transitions at the critical values. The maximum power in such lasers, where the excess gain over intracavity losses are small, traditionally achieved by strict compliance with the maximum reflection of the "nontransmitting mirror" and optimum transmission of the exit mirror. The optimal transmittance of the exit mirror at each wavelength depends from the gain and total losses in the resonator for a given wavelength. The absorption of the mirror coatings is significant portion of these losses, thus is necessary consider the spectral dependence absorptance $A(\lambda)$ of the interference structures. The maximum integral power of the low-power lasers will achieved when parameters exit mirror are satisfied of the following minimum goal function:

$$M = \sum_{i=1}^{N} \left(\tau_{ionm} - \tau_i \right)^2 \qquad (1)$$

Since, the optimum transmittance of the exit mirror is defined as [3]

$$\tau_{ionm} = \sqrt{2G_{0i}} - A_i - \sigma, \qquad (2)$$

than

$$M = \sum_{i=1}^{N} \left(\sqrt{2G_{0i}} - A_i - \tau_i - \sigma \right)^2. \qquad (3)$$

The minimum goal function for nontransmitting mirror is:

$$M = \sum_{i=1}^{N} \left(A_i + \sigma \right)^2, \qquad (4)$$

where G_{0i} is unsaturated gain mirror for i wavelength,
A_i and τ_i are absorptance and transmittance of the mirror for i wavelength, correspondingly,

978-1-4244-7043-3/10 $26.00 © 2010 IEEE

159

σ is weakly dispersive compound of the intracavity losses (scattering of the laser irradiation),

N is number wavelengths.

The compact low-power air-cooled Ar^+-lasers are characteristic representatives of the lasers polychrome. There is length of the discharge gap not exceed 150 mm (diameter $d \le 1$ mm), discharge current to 10 A. The commercial models of such lasers realize four lines generation λ_1=514.5 nm, λ_2=488 nm, λ_3=476.5 nm, λ_4=496.5 nm correspondingly. Constants that define G_{0i} for laser lines of the ionized argon, were investigated and determined with reasonable accuracy [3] at which calculate the optimal values of the transmittance of the exit mirrors each transition. However, in our work these parameters were determined experimentally using calibrated losses method for compact lasers ЛГН-410 and ЛГН-52. The results of the measurements indicated in Fig. 2. The curves shows that optimal transmittance are 514.5 nm – 1÷2%; 488.0 nm – 2÷4%; 476.5 nm – 0.5÷2.5%; 496.5 nm – 0.5÷1.5% for the investigated wavelengths.

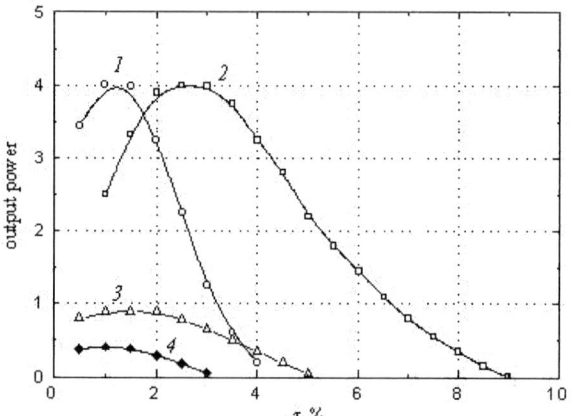

Fig. 2. Dependences of the output power from the transmittance of the exit mirror for wavelengths
1 – 514.5 nm; 2 – 488 nm; 3 – 476.5 nm; 4 – 496.5 nm

The optical thicknesses of layers used as design parameters at development of the numerical algorithm mirrors design on the assumption of the conditions their technological effectiveness. The filming materials selected from the list used in mass production (ZnS, MgF_2, Na_3Al_6, ZrO_2, HfO_2, and SiO2). The design interference coating conducted with subject to dispersion of the refractive indices and absorption of thin-film materials [4], as substrate used a sintered quartz KB.

Note that the optimal structure of coating for materials ZnS–MgF_2 and ZnS–Na_3Al_6, are obtained as result synthesis correspond to higher values of the parameters M (3) and (4) than for ZrO_2–SiO_2 and HfO_2–SiO_2. The first pair of materials no longer considered. Table 1 shows two pairs of coatings that best correspond to the minimum functional (3) and (4).

Fig. 3 represents spectrum characteristics of the transmittance and absorptance of the synthesized interference mirror $S_0(0.7H,1.3L)^70.7H$.

Table 1

Synthesized structures mirrors of the low-power argon lasers

Type of the mirror	Thin-film structure	Material H	Material L	λ_0, nm
"exit"	$S_0(0.7H,1.3L)^70.7H$	ZrO_2	SiO_2	510
"nontransmitting"	$S_0(0.7H,1.3L)^{14}0.7H$	ZrO_2	SiO_2	510
"exit"	$S_0(0.7H,1.3L)^70.7H$	HfO_2	SiO_2	510
"nontransmitting"	$S_0(0.7H,1.3L)^{14}0.7HB$	HfO_2	SiO_2	510

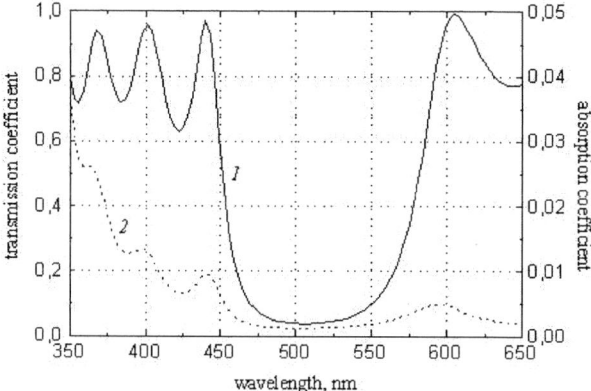

Fig. 3. Spectrum characteristics of the transmittance and absorptance of the synthesized unequal thickness interference coating $S_0(0.7H,1.3L)^70.7H$ λ_o=510 nm

Results comparison of the mirrors characteristics (see Table 2) show that synthesized structure allows achieved practically twice-lower absorption at provide optimal transmission for the reference wavelengths.

Table 2

Comparison characteristics of the synthesized structure $S_0(0.7H,1.3L)^70.7H$ and standard structure $S_0(H,L)^7L$ for materials ZrO_2 and SiO_2

Wavelength, nm	Transmission,%			Absorption, %	
	Optimal value	Standart structure	Synthesis structure	Standard structure	Synthesis structure
476,5	0,5÷2,5	2,4	2,6	0,27	0,13
488,0	2÷4	1,8	2,4	0,25	0,14
496,5	0,5÷1,5	1,6	2,1	0,24	0,12
514,5	1÷2	1,97	2,0	0,25	0,12

The value of the parameter M (4) "nontransmitting mirrors" the standard coating is M_{st}=1.37, and synthesized unequal thickness coating is M_{synth}=0.26 that indicate on its significant optimization.

Synthesized mirrors made by electron beam evaporation in vacuum [5]. Averaged results of the mirrors in open cavities represented in the Table 3.

Table 3.

Results use of the interference mirrors optimized in the industrial low-power argon lasers

Type of the active element	L, mm	d, mm	I, A	Passport power, mW	Achieved power, mW
ЛГН-410	150±1	$2^{-0,1}$	10	50	75
ЛГН-522	90±1	$1^{-0,1}$	8	25	38

Thus, the achieved results show the effectiveness of the chosen way to improve the basic characteristics of the output irradiation of the low-energy argon laser polychrome.

III. DESIGN CAVITY MIRRORS OF THE POWER LASERS

The absorption in the mirrors is crucial in terms of photothermal intracavity processes in the power laser, where typically the significant gain is over the excess losses [2]. Until recently, the only effective way to minimize the thermal effects in ionic power lasers was use as material of the mirror substrate optical quartz glass. Low value coefficient of thermal expansion optical quartz glass grades КУ-1, КУ-2, КВ can reduce surface thermodeformations of working surfaces of the mirrors. However, a production thermoagreed durable vacuum-tight elements node of the constructions of the power gas lasers polychrome using quartz optics is rather problematic. The effective solution of this problem is the use as material of the mirror substrates colorless optical glass and minimization of the absorbed power in the mirror coatings. The structure of the thin-film coating the following minimum functional must be satisfied:

$$M = \sum_{i=1}^{N} B_i \left(\frac{G_{0i}}{\tau_i + A_i} - 1 \right) \cdot A_i , \qquad (5)$$

where B_i is parameter of the saturation i wavelength.

As the filming materials for the mirrors of the power lasers used refractory oxides ZrO_2, HfO_2, SiO_2. The interference structure of the cavity mirrors with spectral characteristics, providing minimum functional (5) were synthesized for serial lasers ЛГН-402, ЛГН-406.

The optimum transmittance of the mirrors τ_i for all transitions laser Ar^+ were determined experimentally by calibrated loss method and their values taken equal 514.5 nm – 2÷5%; 488.0 nm – 5÷8%; 476.5 nm – 2÷6%; 496.5 nm – 1÷4%. Results syntheses of the coatings given in Table 4.

Production of the mirrors with the thin-film structures given in Table 4, conducted using vacuum technology of the electron-beam deposition of the thin films. The active elements of the ЛГН-402, ЛГН-406 and ЛГН-502 lasers with obtained mirrors, are produced. Investigations their output

characteristics in compare with active elements of the standard optics (see Table 5), are conducted.

Table 4.

Synthesized structures mirrors of the power argon lasers

Type of the mirror	Thin-film structure	Material H	Material L	λ_0, nm
"exit"	$S_0(0.7H,1.3L)^6 0.7H$	ZrO_2	SiO_2	510
"nontransmitting"	$S_0(0.7H,1.3L)^{14} 0.7H$	ZrO_2	SiO_2	510
"exit"	$S_0(0.8H,1.2L)^6 0.8H$	HfO_2	SiO_2	510
"nontransmitting"	$S_0(0.8H,1.2L)^{14} 0.8H$	HfO_2	SiO_2	510

Table 5

Results use of the interference mirrors optimized in the industrial power argon lasers

Type of the active element	L, mm	d, mm	I, A	Passport power, mW	Achieved power, mW
ЛГН-406	500	$2,5^{-}_{0.15}{}^{+0.25}$	35	5	7
ЛГН-402	500	$2,5^{-}_{0.15}{}^{+0.25}$	35	4	5,8
ЛГН-502	500	$2^{-}_{0.15}{}^{+0.25}$	35	2	2,6

Laboratory test of the lasers with active elements showed full compliance with the spatial and temporal characteristics of output radiation specifications. As the good time it should be noted that the development mirrors began, the series works for creation the active elements with no glued vacuum-tight couplings. Its allow thermodegassing active elements on the stage of pumping providing achievement of the parameters higher reliability gas-discharge lasers polychrome.

[1]. Ya.M. Bondarchuk, Ya.O. Dovgyy // Theoretical and technology aspects synthesis of the laser mirrors. Physics and Chemistri of Solid State. Vol.3, №2,-2002.- P.199-214. (In Ukraine)

[2]. Ya.M. Bondarchuk, H.A. Petrovska // Design optical resonators of the power continuous lasers. Bulletin Lviv Polytechnic National University, Electronics, №455.- 2002.-P.60-64. (In Ukraine)

[3]. Laser handbook edited by A.M. Prokhorov, Vol 1. (In Russian)

[4]. H.A. Petrovska, T.A. Redko // Method of the optimization dispersive characteristics of the thin-film optical materials. Bulletin Lviv Polytechnic National University, Electronics.-2002.- №455.- P.39-44. (In Ukrainian)

[5]. Ya.M. Bondarchuk. Technology and treatment of the optical materials.-Lviv: Liga-Pres, 2001.-P.242 (In Ukrainian)

ary
Resonance oscillations in quasiperiodic layered structures with lumped inhomogeneities

V. F. Borulko, *Senior Member, IEEE*, D. V. Sidorov, *Student Member, IEEE*
Dnipropetrovsk National University, Dnipropetrovsk, Ukraine

Abstract: Scattering of plane monochromatic waves on quasiperiodic layered structures is considered. Exact solutions for complex amplitudes of reflected and transmitted waves are found by method of transmission matrices. Influence of connection of lumped elements on resonance properties is investigated. Hysteresis behavior is observed for case of Kerr nonlinearity.

Layered Bragg structures are widely observed in nature and applied in engineering. Permittivity or/and permeability of layers are usually supposed to be periodical functions of space coordinate [1-3]. Distortions of periodicity supply additional possibilities for effects of wave scattering [4]. For obtaining desired complex amplitudes of scattered plane waves as function versus incident frequency or angle, spatial distribution of magnitude and phase of quasiperiodic perturbation can be used.

In [5] resonance properties of resonators formed by quasiperiodic Bragg reflectors are investigated by solution of some types of eigenvalue problems. In the first eigenvalue problem the values of complex eigenfrequencies are found for the decaying natural oscillations of the "cold" passive resonators. The other eigenvalue problems correspond to undamped oscillations in "hot" structures containing amplifying elements. In the second eigenvalue problem we must find real frequencies and imaginary part of permittivity or permeability (lasing threshold) at fixed value of real part of permittivity and permeability of amplifying layer [3]. In the third case the generation is achieved by introduction of lumped element with negative resistance into structure. In the fourth eigenvalue problem complex eigen permittivities or permeabilities of amplifying layers are found for fixed frequency of oscillations [6].

Elements of scattering matrices and equivalent complex impedances used for solving the eigenproblems were found by usage of the formalism of the transmission matrix method [1-3]. In this method any layer or lumped element is characterized by the transmission matrix and the matrix of total layered structure can be found as product of matrices of all layers and elements.

Bragg structures without periodicity distortions operate as narrow-band reflectors. Real parts of their complex eigenfrequencies situated outside band of Bragg reflection. Quality factors are rather low. They can be considerably increased by introduction of layer of abnormal length [3] or connection of lumped inhomogeneity [5]. Fig. 1-3 show influence of lumped element, connected in the middle of central layer of Bragg structure, on magnitude of electric field

inside of the structure. Layers of examined structure have following parameters:

$$\varepsilon_{2m-1}=2,\ \varepsilon_{2m}=1,\ \mu_{2m-1}=1,\ \mu_{2m}=1,\ H_{2m-1}=H_{2m}=\lambda_0/4,$$

where ε is permittivity, μ is permeability, and H is electrical thickness of layer. Total number of layers is equal to 21.

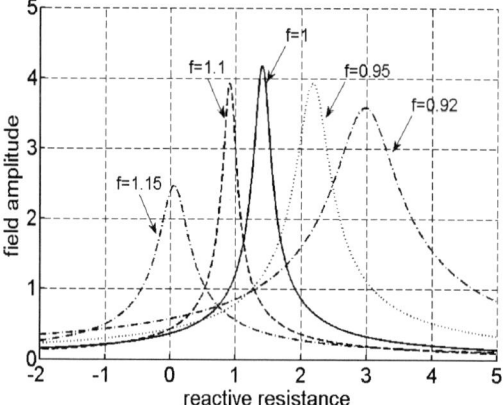

Fig. 1. Field magnitudes as function of reactive impedance of series inhomogeneity connected in the middle of central layer of periodic Bragg structure.

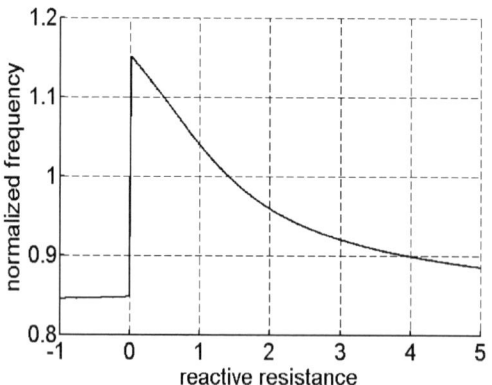

Fig. 2. Frequency of maximum magnitude as function of reactive impedance of series inhomogeneity connected in the middle of central layer of periodic Bragg structure.

CAOL*2010 International Conference on Advanced Optoelectronics & Lasers, 10-14 September, 2010, Sevastopol, Ukraine

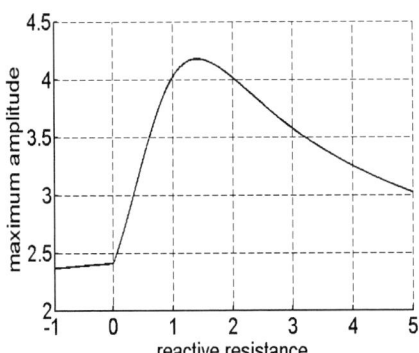

Fig. 3. Maximum amplitude as function of reactive impedance of series inhomogeneity connected in the middle of central layer of periodic Bragg structure.

Quasiperiodic one-dimensional structures can be used in generating devices. "Hot" eigen parameters have not simple correlation with eigenfrequencies of "cold" structure. Positions of real frequencies for unwanted eigen modes and their number for hot eigenproblem have no similarity with complex eigenfrequencies for the same structure. Some of maximum of transmission coefficients does not form poles in domain of real frequency and imaginary part of permittivity or permeability. Amplifying in structure can be described not only by imaginary part of permittivity but also imaginary part of permeability or refractive index [3]. If electric thickness of amplifying layer is equal to half of wave length, all three approach give very close results. But for shorter thickness results become sensible to position of amplifying layer relative to maximum of electrical field (in the first approach) or magnetic field (in the second approach).

Very thin layers with high contrast in permittivity can be approximately described as parallel lumped inhomogeneities. Model of series element is applicable for thin layers with high contrast in permeability.

We have considered influence of small amplification caused by introduction of small negative series resistance or large shunt resistance into resonator. If the series resistance or parallel conductivity is large, absolute value of their reflection coefficients are close to unity and insertion loss or amplification is small, but obtained structure has small proximity to original cold structure. It is important that infinite reflection coefficients are achieved if $Z/Z_0+2=0$ (Z is resistance of active element, Z_0 impedance of waveguiding line, wave impedance of free space in examined situation) for series lumped element and $Z_0/Z+2=0$ for shunt one. The resonances that close to these conditions may not be considered as actual one but as formal resonances. Results considerably depend on point of connection of lumped element. The largest influence is obtained in point of maximal magnetic field for series element and electric field for shunt one. Special example can be constructed in which amplification is absent only on frequency of cold resonance.

Eigen values of negative resistance supplying undamped oscillations have been found by two methods, external and internal. External method is based on two-dimensional searching on plane (Z, f) the poles of the transmission coefficient of resonator with connected lump element. In internal method total structure is divided into two parts by point of connection of lumped element. Then we calculate equivalent complex impedances of each part in point of connection. Total equivalent complex impedance is found by rules of series or parallel resistance respectively for cases of series or parallel connection. Resonance frequencies are found from condition of nullification of imaginary part of total impedance. Factor of resonator quality are defined by real part of conductance or resistance respectively for cases of series or parallel connection. If lumped element is single, numerical results of external and internal methods coincide but internal method is simpler. Advantage of external method is possibility of consideration of multiple lump elements.

Total equivalent complex conductivity of Bragg resonator for series symmetrical connection is shown in Fig. 4. On frequency of Brag reflection the imaginary part of equivalent conductivity is equal to zero and its real part has small value. This corresponds to case of good quality of resonator for "hot application". Fig. 5 shows results of calculation of eigen values of real frequencies and real equivalent resistance for parallel connection in the central layer of resonator with one abnormal layer. There are "unsuccessful" points of connection for manifestation of resonance. Fundamental resonance of cold structure on the frequency of Bragg reflection f_0 does not appear for symmetrical parallel connection and for series connection in the end of central layer. But these points have the best quality for opposite type of connection. Intermediate points implement fundamental resonance for both types of connection but with lower values of "quality factors". Minimization of dependence on point of connection can be achieved by usage of combinational series-parallel lump elements.

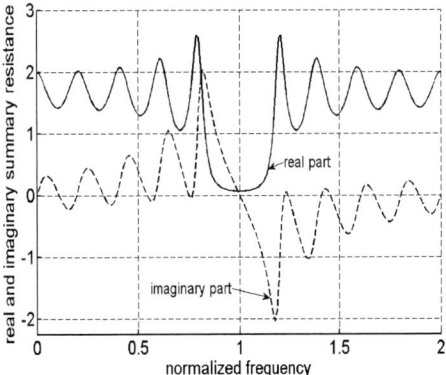

Fig. 4. Real part (the solid line) and imaginary part (the dashed line) of total equivalent complex impedance of Bragg resonator for series symmetrical connection

978-1-4244-7043-3/10 $26.00 © 2010 IEEE   163

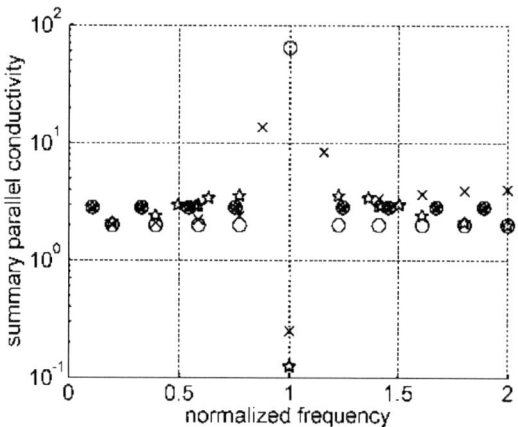

Fig. 5. Eigen values of real frequencies and real part of equivalent conductivities for parallel connection of lumped inhomogeneity in Bragg structure. Results obtained for exactly symmetrical connection ($l/d_{11} = 1/2$, where d_{11} is geometrical thickness, l is distance from beginning of central layer to connection point) are shown by circles. Asymmetric connections are presented by x-es ($l/d_{11} = 3/4$), and by pentagram ($l/d_{11} = 1$).

Taking into account of nonlinearity of lumped element is very important from practical and principal points of view. Model of Kerr nonlinearity [7] was used for investigation of transmission of electromagnetic waves through Bragg structure with parallel nonlinear element with negative conductivity depending on magnitude of electric field in point of connection. Hysteresis behavior is observed for dependence of transmitted magnitude from incident magnitude. This hysteresis is stronger for case when electric length of central layer is half of wavelength (Fig. 6) than for case of quarter of wavelength (Fig. 7).

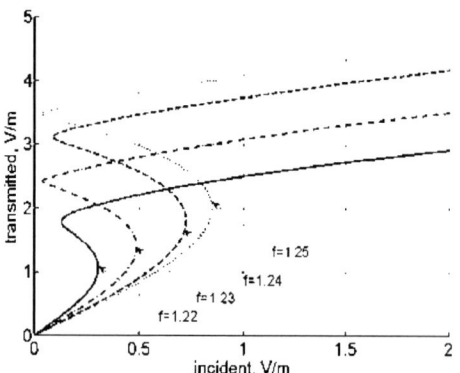

Fig. 6. Magnitude of transmitted wave as function of magnitude of incident wave for Bragg structure with negative nonlinear parallel conductance inside layer of half wavelength

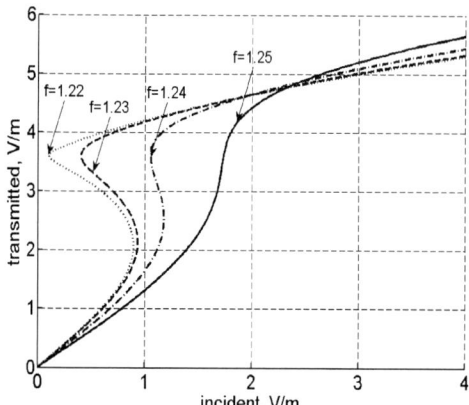

Fig. 7. Magnitude of transmitted wave as function of magnitude of incident wave for Bragg structure with negative nonlinear parallel conductance inside layer of quarter wavelength

Fulfilled investigations have shown that simplified model of lumped inhomogeneities can be useful for describing many resonance phenomena in Bragg structures with distortions of periodicity. It allows analyzing both linear and nonlinear effects.

REFERENCES

[1] C. Elachi, "Waves in active and passive periodic structures: A review," *Proceedings of the IEEE*. vol. 64, No. 12, pp. 1666-1698, 1976.

[2] M. Born, E. Wolf, *Principles of Optics*, Pergamon Press, Oxford, 1975.

[3] V.O. Byelobrov and A.I. Nosich "Mathematical analysis of the lasing eigenvalue problem for the optical modes in a layered dielectric cavity with a quantum well and distributed Bragg reflectors", *Opt. Quant. Electron.*, Vol. 39, pp. 927-937, 2007.

[4] A.G. Barriuso, J.J. Monzon, and L.L. Sanchez-Soto, A. Felipe "Comparing omnidirectional reflection from periodic and quasiperiodic one-dimensional photonic crystals," *Opt. Express*, Vol. 13, No. 11, pp. 3913-3920, 2005.

[5] V.F. Borulko, D.V. Sidorov "Eigen modes of quasiperiodic layered resonators" Proc. of MSMW2010, Kharkov, Ukraine, A-3, 2010

[6] M.S. Agranovich, B.Z. Katsenelenbaum, A.N. Sivov, N.N. Voitovich, "Generalized Method of Eigenoscillations in Diffraction Theory", WILEY-VCH Verlag Berlin GmbH, Berlin, 1999.

[7] M. W. Feise, I. V. Shadrivov and Yu. S. Kivshar. "Tunable transmission and bistability in left-handed bandgap structures". *Phys.*, vol. 14: 0401094-1-3, 2004.

CAOL*2010 International Conference on Advanced Optoelectronics & Lasers, 10-14 September, 2010, Sevastopol, Ukraine

Disc and microdisc resonators with strip-like diffraction gratings

O.I. Belous, A.I. Fisun, *Senior Member, IEEE,* and V.L. Pasynin
A. Usikov Institute of Radio Physics and Electronics, National Academy of Sciences of Ukraine
12, Academician Proskura St., Kharkiv 61085, Ukraine

Abstract: Impact of a diffraction grating on the spectrum of disc and microdisc resonator has been studied using FTDT method. Conditions of the single-mode excitation of whispering gallery mode of disc resonator was estimated.

I. INTRODUCTION

Dielectric disc resonators are gaining a wide acceptance due to a high Q-factor of whispering gallery modes [1]. In recent years the semiconductor, crystal and polymer microdisc resonators are implemented in lasers [2]. At the same time the spectrum of whispering-gallery oscillations is defined as the rather thick grid of equidistant azimuth oscillations of both E- and H-polarizations. In some specific case the thickness of spectrum hampers the use of disc resonators.

The goal of this report is the frequency sparsing of the forced oscillation spectrum of dielectric disc resonators. It has been suggested the strip-like diffraction grating be applied th the mode selection of disc resonato.

II. NUMERICAL SIMULATION

Numerical analysis: To solve the problem on the excitation of oscillations in the quasioptical dielectric resonators we made the following modeling assumptions:
- the numerical experiments were carried out with the use of the finite-difference time-domain (FTDT) method [3];
- consideration is given using the 2-D initial-boundary problem $(\partial / \partial x \equiv 0)$ for *E*-polarized field;
- since the electromagnetic properties of the half-disk resonators are identical to the disk resonators [1], for the sake of convenience we will carry out the computer-assisted simulation for the model being depicted in Fig. 1.

The half disk resonator is placed at the metallic conducting plane. The initial wave is incident upon the resonator through port L_1 with an iris. The reflected wave is detected through waveguide L_1 as well. The transmitted wave is received by port L_2. Both coupling holes are of the same diameter b. The strip-like grating or metallic continuous screen are applied to the side surface of half-disk resonator. The time-domain characteristics convert to the frequency ones through the Fourier transform. The numerical experiment was carried out using the algorithm developed in [4]. In the computer simulation three cases are examined:

- excitation of the resonator without diffraction grating (well established application of a dielectric resonator);
- investigation of resonator with screening;
- study of the resonator spectrum with diffraction grating.

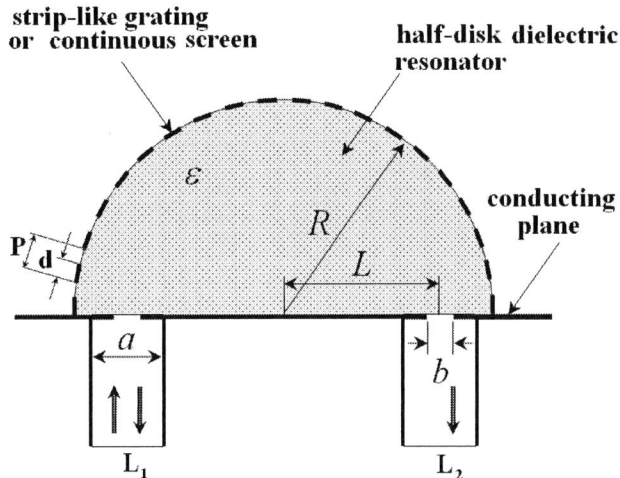

Fig. 1. Two-dimensional model of half-disk dielectric resonator.

Simulation results: In the first and second versions of the numerical experiment the spectrum is of an equidistant nature. The resonance frequencies of forced oscillations are the function of the number of half-waves $\lambda/2$ being fitted perfectly along the lateral surface of the resonator. This result is a matter of common knowledge and lends support to the validity of the method for numerical analysis.

The screening of the resonator leads to the shift of the frequency spectrum which can be attributed to the displacement of the field caustic surface to the resonator centre and, as a consequence, the optical length of resonator becames a somewhat diminished one. In addition, the less intensive resonance responses are observed in the screened resonator. In the present paper the identification of the last mentioned oscillations is of no interest. Figure 2 illustrates the spectrum of the dielectric resonator without the diffraction grating (a) and spectrum of the resonator equipped with the diffraction grating (b).

978-1-4244-7043-3/10 $26.00 © 2010 IEEE

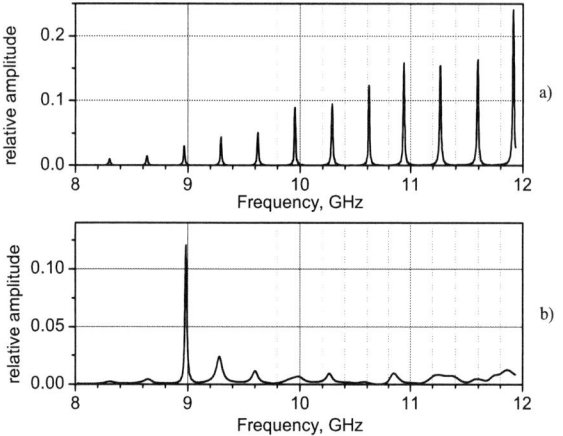

Fig. 2 Simulation results for half-disk dielectric resonator without diffraction grating (a) and equipped by one (b).

As indicated in the picture, the diffraction grating affects the frequency sparseness. The selectiveness is the function of the filling factor of the grating:

$$\theta = d\!\big/\!P$$

where d is the strip width and P is the period of grating. The frequency of maximum response is determined by the length of the resonator lateral surface and period of the grating

$$2\pi R = 2nP$$

In the simulation process it has been found that the mode selection is the most efficient approximately at $\theta = 0.3$ (see Fig.2 b).

III. Conclusion

Using simulation in the FDTD technique it was demonstrated that the strip-like diffraction gratings are the promising candidate for whispering-gallery mode selection.

The essence of the selection process is as follows: The incident wave is transformed to the surface one by a strip-like diffraction grating at the dielectric substrate if the condition $\lambda/2 = P$ is fulfilled. If the diffraction grating applies to the lateral surface of the dielectric resonator, the direction of the surface wave is coincident with the propagation direction of the whispering-gallery mode. If geometrical dimentions of the grating and resonator are chosen in such a way that the integer number of half waves n fits the lateral surface of the resonator, the surface wave serves to excite the whispering-gallery mode with azimuthal number n. The other waves go off into outer space. The results of computer-assisted simulation is supported experimentally [5] and has of practical importance.

References

[1] A.Ya. Kirichenko, Yu.V. Prokopenko, Yu.F. Filippov, and N.T. Cherpak, "Quasi-Optical Solid-State Resonators" // Kyiv. Naukova dumka. 2008. 296 p. (in Russian).

[2] H. Cao et al., "Optically pumped InAs quantum dot microdisc lasers," *Appl. Phys.Lett.*, vol. 76, no. 24, pp. 3519-3521, 2000.

[3] O.I. Belous, O.N. Sukhoruchko, and A.I. Fisun, "Microwave resonator," patent of Ukraine #90831.

[4] Sirenko Yu.K., Ström S., Yashina N.P. Modeling and analysis of transient processes in open resonant structures. New methods and techniques. New york: Springer House. 2007.

[5] O.I. Belous, O.N. Sukhoruchko, and A.I. Fisun, "Excitation of forced modes in the dielectric disk resonator – diffraction grating system," J. Nano-Electron. Phys., vol. 2, no. 1, pp. 89-93, 2010 (in Russian).

CAOL*2010 International Conference on Advanced Optoelectronics & Lasers, 10-14 September, 2010, Sevastopol, Ukraine

Spiral Resonators with Whispering Gallery-like Modes

O. V. Goroshko, R. V. Golovashchenko, V. N. Derkach

Usikov Institute of Radiophysics and Electronics of the National Academy of Sciences of Ukraine

12 Ac. Proskura St., Kharkov, 61085, Ukraine

goroshko_elena@ire.kharkov.ua

Abstract: The results of experimental study of spectral, power and field characteristics of spiral resonators with whispering gallery-like modes (WGM) in the millimeter wave band by means of the computerized measuring stand are presented. Design of tunable filters and systems with directional radiation on the base of such resonators is discussed.

Now much attention is given to study of properties of resonators that partially confine electromagnetic wave by total internal reflection at the dielectric sidewalls and using such structures as versatile building blocks for photonic devices. In particular, microresonators with directional radiation and photonic devices on their basis, such as coupled resonator optical waveguides (CROW) at optical wavelengths [1] is refer to such devices. Meanwhile, alongside with well developed theory of calculation of CROW [2] experimental investigation of such structure in optics is difficult. Investigation of systems of coupled resonators in the millimeter wave band can simulate the behavior of electromagnetic waves in complex systems in the optical range and to develop controlled elements for millimeter waves. Therefore, such resonators cannot be used as devices with the directional radiation. A number of studies proposed to use for this purpose coupled disk, half-disk and spiral resonators [3]. The spiral resonators shape can be described as a disk with the radius linearly depending on the azimuthal angle φ as $r(\varphi) = r_0(1 + \varepsilon\varphi/2\pi)$ where r_0 is the radius at $\varphi = 0$, and ε is the change rate of the radius. The radius mismatch at $\varphi = 2\pi$ gives a notch of width $r_0\varepsilon$, which is typically a small fraction of the radius.

This paper describes a computerized experimental stand for studying spectral characteristics and electromagnetic field distribution in open resonance systems [4], [5]. The results of studying the characteristics of spiral dielectric resonators in the millimeter wave band are presented.

We used the computerized experimental stand for investigation of modal spectrum, Q-factor and resonance field distribution of single, half-disk and spiral resonators. The block-diagram of the stand is presented in Fig. 1. Experimental stand consists of 3D scanner, extra-high frequency (EHF) module, control system (CS) and personal computer (PC). 3D scanner gives a possibility to move an active or passive probe in 3D volume $250 \times 250 \times 250$ mm^3 with minimum step of 0.1 mm and top speed of 5 mm/s. EHF-module consists of EHF

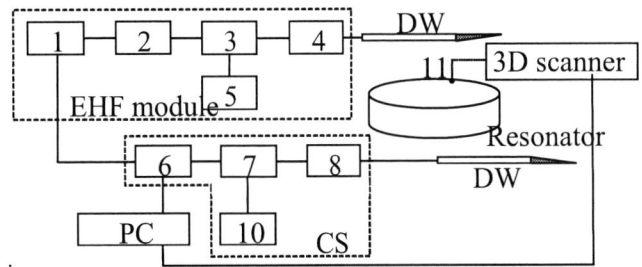

Fig. 1. Block-diagram of the stand. 1 – generator, 2 – ferrite isolator, 3 – modulator, 4 – attenuator, 5 – low-frequency generator, 6 – hardware and software complex, 7 – amplifier, 8 – detector, 9 – PC, 10 – oscilloscope, 11 – probe.

oscillator, waveguide elements and antennas for excitation of resonators under test, detectors and lock-in amplifier.

CS and PC provided tuning of the EHF oscillator in the required frequency band, and controlling of 3D scanner and receiving and acquisition of signals from detectors.

We investigated spiral resonators made of materials with small losses in EHF band (teflon, polystyrene). Frequency band is 50–78 GHz. Radius of spiral is $r_0 = 18$mm. The change rate of radius is $\varepsilon = 4$mm. The modal spectrum of resonators was registered at tuning of frequency in the given frequency band with the minimum step about 100 kHz. Fig. 2 shows an example of the spectrum of the polystyrene spiral resonator.

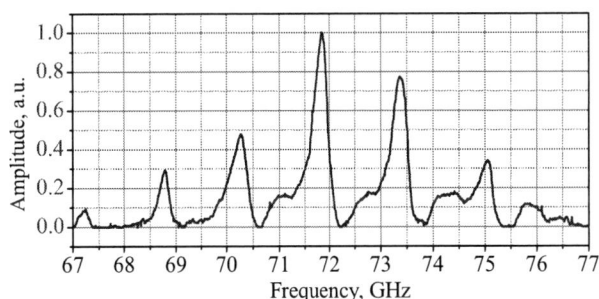

Fig. 2. Spectrum of the polystyrene spiral resonator.

It is clearly seen that in such cavity modal spectrum is formed by two whispering gallery-like modes. They differ by intensity and Q-factor.

978-1-4244-7043-3/10 $26.00 © 2010 IEEE

Electromagnetic field distribution on surfaces of the resonator was investigated by the "small perturbations" technique. In the "transmission" regime the investigated resonator was placed on the adjusting platform between exciting devices. As exciting devices dielectric waveguide (DW) antennas in the form of the dielectric waveguides of rectangular or round section were used. The ends of waveguides were made narrowed as well as were coated with an absorbing material for diminution of wave reflection. The perturbing probe in the form of a sphere in diameter of much less than wavelength was made of an absorbing material and disposed on a tip of a quartz stick in diameter of 0.3 mm. Then a probe was moved linear near to resonator surface by means of the 3D scanner. Thus the resonant transfer ratio (RTR) of the resonator changed depending on location of a probe due to partial absorption of electromagnetic energy by probe. Decrease of amplitude of RTR (dark places on a field distribution map) was registered when the probe was located in a point of maximum of the electrical component of electromagnetic field of the resonant oscillation. The current coordinate and RTR values were recorded by PC. The example of distribution of electrical component of WGM electromagnetic field of spiral resonator made of polystyrene is presented in Fig. 3.

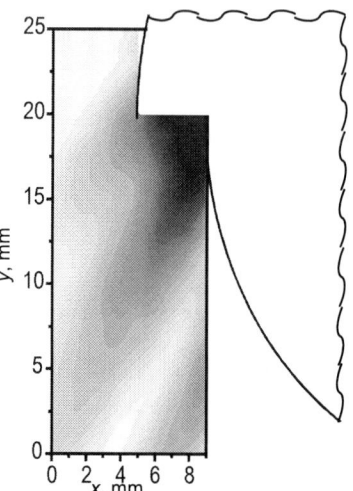

Fig. 4. The distribution of electrical component of the electromagnetic field near the spiral notch at half of the height of the resonator.

The beam of directed radiation about 30^0 to the normal of notch is more intensive and has more narrow directional radiation pattern. Therefore, we believe that such spiral resonators can be used in systems requiring a unidirectional radiation.

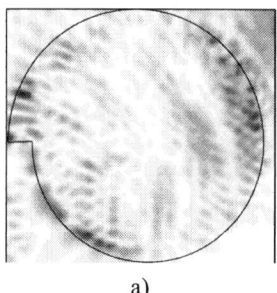 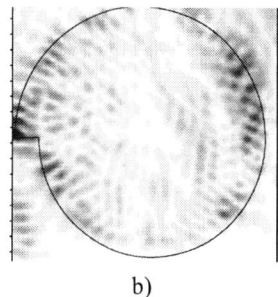

a)   b)

Fig. 3. The distribution of electrical component of electromagnetic field of whispering gallery-like modes on the surface of spiral resonator on 71.34 GHz (a) and 71.86 GHz (b).

Oscillation at 71.86 GHz has a great intensity and Q-factor in comparison with oscillations at a frequency of 71.34 GHz. Due to reflection of electromagnetic waves on the inner corner of the step distribution of the electromagnetic field has different from the whispering gallery modes. The oscillation at a frequency of 71.86 GHz has a high intensity electric field component in the notch of the spiral. Therefore, we assumed the existence of directional radiation from the notch is on this frequency. Distribution of electrical component of the electromagnetic field near a spiral cut in half the height of the resonator is presented in Fig. 4.

We observed two beams of directional radiation near to spiral notch. The angles of directional radiation with respect to the normal of notch were approximately as 30^0 and 50^0.

REFERENCES

[1] S. V. Boriskina, "Coupling of whispering-gallery modes in size-mismatched microdisk photonic molecules," Opt. Lett., vol. 32, no. 11, pp. 1557–1559, June 2007.

[2] S. V. Pishko, P. Sewell, T. M. Benson, and S. V. Boriskina, "Efficient analysis and design of low-loss whispering-gallery-mode coupled resonator optical waveguide bends," J. Lightwave Technol., vol. 25, no. 9, pp. 2487–2494, Sept. 2007.

[3] A.W. Poon, X. Luo, H. Chen, G.E. Fernandes, R.K. Chang, "Microspiral Resonators for Integrated Photonics", Optics & Photonics News, vol. 19, no. 10, pp. 48-53, 2008.

[4] R. V. Golovashchenko, E. V. Goroshko, A. V. Varavin, A. S. Plevako, and V. N. Derkach, "Hardware and software complex for mm-wave spectroscopic research," in Proc. of 16th International Crimean Conference on Microwave and Telecommunication Technology, CriMiCo'06, Sept. 11–15, 2006, pp. 817–818.

[5] V. N. Derkach, R. V. Golovashchenko, and E. V. Goroshko, "Coupled-cavity structures on the base of whispering gallery disk dielectric resonators at millimeter wave band", in Proc. of the 10th Anniv. Intern. Conf. on Transparent Optical Networks, ICTON 2008, Athens, Greece, June 22–26, 2008, vol. 4, pp. 234–237.

CAOL*2010 International Conference on Advanced Optoelectronics & Lasers, 10-14 September, 2010, Sevastopol, Ukraine

Metal waveguide resonator with a tilted mirror

A.V.Volodenko, O.V.Gurin, A.V.Degtyarev, V.A.Maslov, *Member IEEE*,
V.A.Svich, *Member IEEE*, A.N.Topkov, T.F.Ruban
V.N.Karazin Kharkiv National University, Kharkiv, Ukraine

Abstract: A study was made of the influence of misalignment of the plane mirrors of a metal waveguide resonator for the submillimtre laser ($\lambda = 0{,}4326$ mm) on the energy losses, polarization and mode structure of the radiation. It was found that misalignment of the resonator only weakly affected the energy losses, but exerted a strong influence on the mode structure and polarization of the radiation

I. INTRODUCTION

There has been repeated mention in the literature of the high mode selectivity and stability against misalignment of the resonators of waveguide lasers. For example, it was shown in [1] that introduction of additional losses into the resonator (wires), misalignment of the mirrors, and application of an external magnetic field merely altered the output power but did not change the radiation mode structure. The influence of misalignment of the mirrors of a plane-plane resonator on the output power of a waveguide laser was experimentally investigated in [2, 3]. The waveguide resonator was found in [2] to be considerably more stable than an open plane-plane resonator, although the question of the mode structure of the radiation from a misaligned resonator was not considered. The results obtained in [3] provide evidence of a strong influence of misalignment of the plane mirrors on the mode structure of waveguide resonator. All these data refer to the study of the characteristics of the radiation of waveguide lasers of optical and mid-IR ranges. Effect of misalignment of the flat mirrors on the characteristics of the radiation of waveguide quasi-optical resonators (WQOR) submillimetre lasers in the literature is not reviewed. Experimental study of the this influence was the purpose of this work.

II. EXPERIMENTAL SETUP

Figure 1 presents the scheme of the experimental setup for studying the mode spectrum of WQORs, their output radiation intensity distribution and polarization. To obtain symmetric resonance curves and to study the output radiation intensity of the resonator, the resonator was investigated in the "passage" regime [4]. The resonator was formed by hollow circular metal waveguide (*10*) of diameter 13.8 mm and length 600 mm. Mirrors (*9, 11*) represent a two-dimensional nickel grid with a period 100 mm made of strips of width 25 mm and thickness 17 mm. The transmission coefficient of the grid at a wavelength of 0.4326 mm, at which measurements were performed, was 6 %.

All the elements of the resonator were mounted on an IZA-2 measuring line which provided the precision displacement (with a misalignment of no more than $1''$) of output mirror

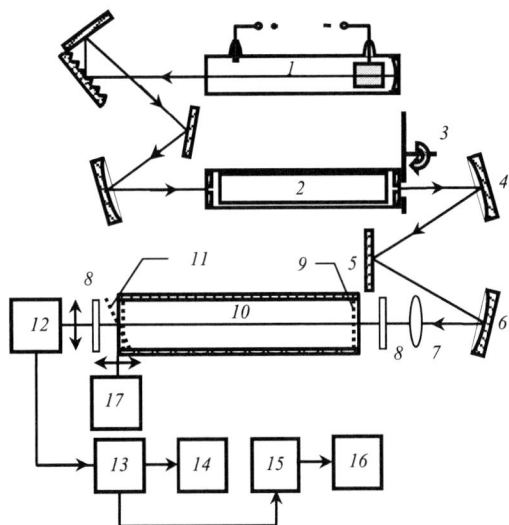

Fig. 1. Scheme of the experimental setup: (*1*) CO_2 laser; (*2*) submillimetre cell; (*3*) mechanical modulator; (*4 − 6*) mirrors; (*7*) Teflon lens; (*8*) attenuator,(*9, 11*) mirrors; (*10*) copper waveguide; (*12*) pyroelectric detector; (*13*) amplifier; (*14*) oscilloscope; (*15*) ADC; (*16*) computer; (*17*) electric drive.

(*11*) along the resonator optical axis with the help of electric drive (*16*). The resonator was excited through semitransparent mirror (*9*) by an optically pumped submillimetre laser consisting of CO_2 pump laser (*1*) and submillimetre cell (*2*). The submillimetre laser operating on formic acid molecule (HCOOH) emitted at 0.4326 mm. The laser radiation was modulated with mechanical modulator (*3*) with a frequency of about 100 Hz. Then using a quasi-matching path consisting of a spherical (*4, 6*) and (*5*) flat mirror the laser beam was introduced into the studied resonator at an angle of 0.01 rad to the resonator axis, that excluded hit of the radiation reflected from the resonator in the laser. Attenuator (*8*) was set to avoid feedback between the resonator and the detector. For the excitation of higher cavity modes in front of him teflon lens (*7*) was placed with a focal length of 30 cm. The radiation passed through the resonator was detected with devices (*12 − 16*).

The measurement method was similar to that described in [4]. The spectrum of resonator eigenmodes was recorded by varying the resonator length with the help of electric drive (*17*). The total energy losses δ per round trip in the resonator were determined from the measured width of the resonance curve. The total energy losses per round trip can be written:

978-1-4244-7043-3/10 $26.00 © 2010 IEEE

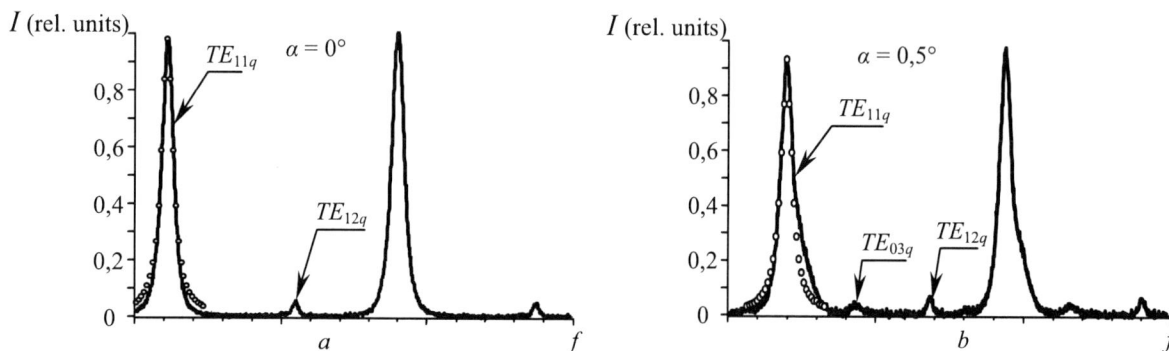

Fig. 2. Tuning characteristics of the resonator at coaxial excitation.

$$\delta = \delta_w + \delta_r + \delta_d + \delta_t,$$

where δ_w are the waveguide losses, δ_r are losses in the reflectors due to the absorption and transmission, δ_d are losses caused by diffraction at the sites of free space, δ_t are losses due to a misalignment of mirrors. Having calibrated mirrors closely to the waveguide it is found δ_0 at the angle of misalignment $\alpha = 0$. Tilting one of the mirrors and measuring the total losses of resonator modes it is possible to find the dependence of its additional losses on the misalignment angle:

$$\delta_t(\alpha) = \delta(\alpha) - \delta_0.$$

The transverse modes were identified by intermode intervals, which were calculated from their phase shifts after a round trip in the resonator, and from the theoretical transverse intensity distributions [5]. The transverse intensity distributions near the output mirror of the resonator were measured by scanning pyroelectric detector (12) with a spatial resolution of 1 mm in a plane perpendicular to the laser beam. The degree of polarization Π was measured by dividing the beam passing through the cavity into two orthogonally polarized components with minimum and maximum power by means of the analyzer. As the analyzer the one-dimensional lattice superimposed on a polyethylene film with polarizing ability of 99 % was used.

III. EXPERIMENTAL RESULTS.
3.1. COAXIAL RESONATOR EXCITATION BY GAUSSIAN BEAM

At coaxial excitation of resonator on its entrance the slightly divergent beam of radiation with a Gaussian intensity profile with a diameter of 1/10 of maximum intensity equal to the diameter of the used waveguide was formed. The characteristic shape of tuning curve of the resonator at the misalignment angle α of the output mirror relative to the optical axis $\alpha = 0$ (a) and $\alpha = 0,5$ (b) is shown in Fig. 2. Two and three TE_{11q}, TE_{12q}, TE_{03q} modes there are in the spectrum of the resonator, respectively. Transverse intensity distribution of the first mode in the first case corresponds to the TE_{11} waveguide mode and the measured degree of polarization is equal to 91 %. Comparison

of calculated and measured intermode intervals suggests that the second observed mode is the TE_{12q} mode. The measured total energy losses per round trip in the resonator for the TE_{11q} mode in this case have made 31%. The error in measuring the total losses in the experiment is equal to ± 2,5%.

When misalignment occurs in the cavity the excitation of higher modes having a smaller degree of polarization is observed and the structure of the field changes in it. This is confirmed in Fig. 3 the dependence of the additional energy losses and the degree of polarization of the TE_{11q} mode as a function of the misalignment angle of output mirror. Thus, when the mirror is misaligned in the plane perpendicular to the direction of the polarization vector of the exciting cavity radiation a weaker dependence of the cavity modes characteristics on the misalignment is observed than the distortions are in a plane parallel to the direction of the polarization vector. It is caused apparently by that in the first case the basic contribution to the formation of the fundamental cavity modes asymmetrical waveguide TE_{1m} modes make, and in the second case – symmetric TE_{0m} modes having smaller losses.

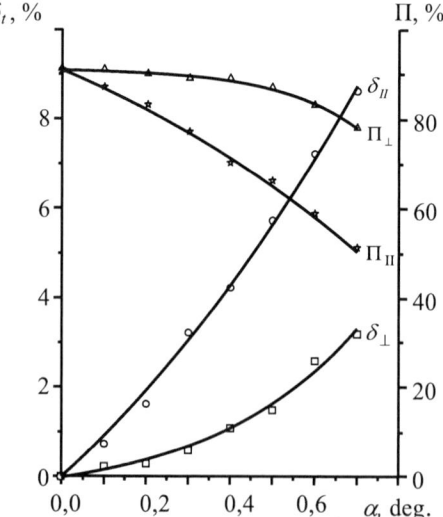

Fig.3. Dependence of the additional energy losses δ_t and the degree of polarization Π for the TE_{11q} mode as a function of tilt α of the mirror.

CAOL*2010 International Conference on Advanced Optoelectronics & Lasers, 10-14 September, 2010, Sevastopol, Ukraine

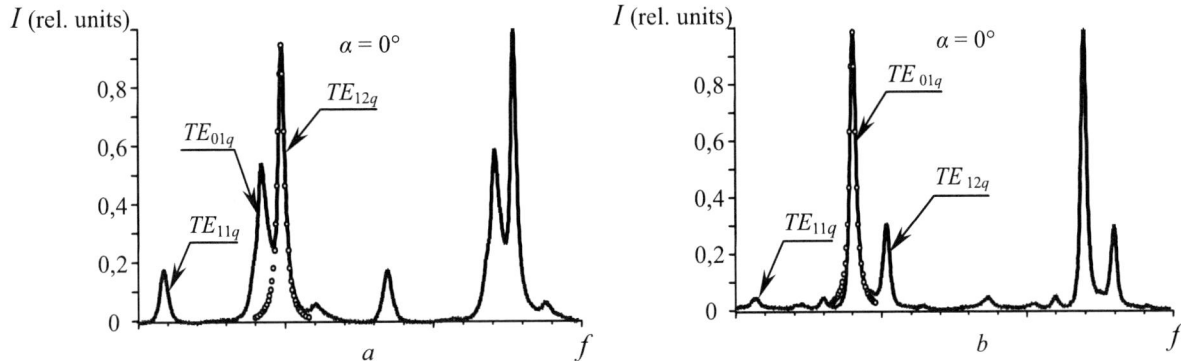

Fig. 4. Tuning characteristics of the resonator in the optimization of the transmission coefficient on the TE_{12q} (a) and TE_{01q} (b) modes.

This is confirmed and more a weak dependence of polarization degree of TE_{11q} mode on the misalignment angle in the second case.

3.2. RESONATOR EXCITATION BY FOCUSED BEAM OF RADIATION

At excitation of the resonator by focused beam of radiation the reproducibility of the spectrum is deteriorating due to the sharp dependence of conditions of the excitation of different modes on the position of the beam waist. Shifting the lens (7) across and along the beam, it was possible to increase the coefficient of transmission of the resonator on any chosen mode.

The characteristic form of tuning characteristics of the resonator by optimization of the transmission coefficient for the TE_{12q} mode is shown in Fig. 4a. Thus in the spectrum three lowest cavity modes were observed. The measured degree of polarization of the TE_{12q} mode is 45 %, that a little below the calculated values for the waveguide TE_{12} mode. It can be

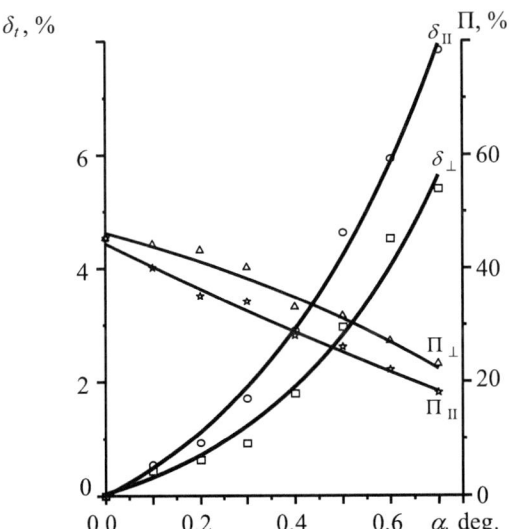

Fig.5. Dependence of the additional energy losses δ_t and the degree of polarization Π for the TE_{12q} mode as a function of tilt α of the mirror.

caused by the near arrangement for the TE_{12q} and TE_{01q} modes in the spectrum of the resonator. The total losses of the TE_{12q} mode at $\alpha = 0$ are equal 18 %. The dependence of the additional energy losses and the degree of polarization for the given mode, caused by misalignment of a mirror, is shown in Fig. 5. Here too, as well as for the TE_{11q} mode, there is a greater dependence of the additional energy losses from tilt of the mirror in a plane parallel to the direction of the polarization vector of the exciting radiation than for misalignment in a plane parallel to the direction of the polarization vector of the exciting radiation. However, for the TE_{12q} mode this difference is expressed more weakly than for the TE_{11q} mode. So, at an angle of misalignment 0.7° the difference of additional losses at various directions for tilted mirrors for TE_{11q} mode is about 6 %, and for the TE_{12q} mode is about 3 %. Similarly the degree of polarization of the radiation which have transited through the resonator changes.

The characteristic form of tuning characteristics of the resonator by optimization of the transmission coefficient for the TE_{01q} mode is shown in Fig. 4b. In the spectrum of the cavity as well as at the tuning on the TE_{12q} mode the TE_{11q}, TE_{12q} and TE_{01q} modes are observed. The measured value of the degree of polarization of the TE_{01q} mode does not exceed 8 %. Some difference of polarization of this mode from circular can be caused non-ideal a used waveguide (an ellipticity, curvature, etc.) or influence of the TE_{12q} mode, which is on the frequency distance from the TE_{01q} mode smaller than the width of the resonance curve at 0,1 level on the intensity of the maximum. The total losses of the TE_{01q} mode at $\alpha = 0$ are equal 17 %. The dependence of the radiation characteristics for this mode, due to tilt of the mirror, is shown in Fig. 6. In this case, the dependence of energy losses on the direction of tilt of the mirror relative to the polarization of the exciting radiation within the limits the measurement error was not observed. This, apparently, it is possible to explain by the absence of a field of this mode at the waveguide walls. Change the degree of polarization also does not depend on a direction of tilt of the mirror. Increasing the degree of polarization of radiation at the angle of tilt of the mirror it is possible to explain by the appearance in the spectrum of the resonator of higher modes, having a greater degree of polarization of radiation.

978-1-4244-7043-3/10 $26.00 © 2010 IEEE   171

CAOL*2010 International Conference on Advanced Optoelectronics & Lasers, 10-14 September, 2010, Sevastopol, Ukraine

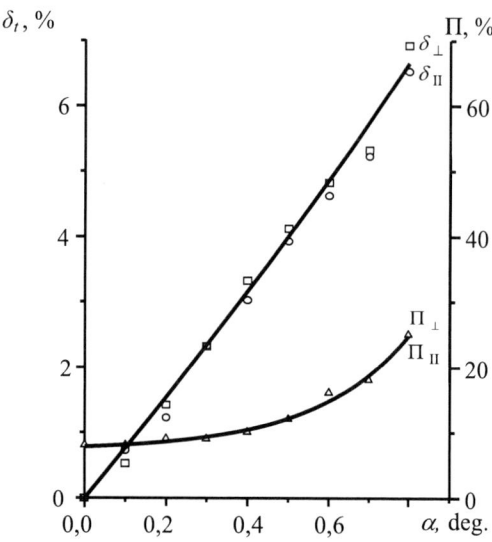

Fig. 6. Dependence of the additional energy losses δ_t and the degree of polarization Π for the TE_{01q} mode as a function of tilt α of the mirror.

IV. CONCLUSION

The effect of the misalignment of plane mirrors of metal waveguide resonator submillimetre laser ($\lambda = 0{,}4326$ mm) on the energy losses, polarization and mode structure of radiation is experimentally investigated. It is shown that misalignment of the resonator in a plane parallel to the plane of polarization of the exciting radiation leads to more rapid growth of energy losses for the TE_{11q}, TE_{12q}, modes than at the misalignment of mirrors in a plane perpendicular to the polarization of the exciting radiation. The energy losses of TE_{01q} modes do not depend on the direction of the plane of misalignment of mirrors relatively plane of polarization of the exciting radiation. The degree of polarization of the TE_{11q} and TE_{12q} resonator modes decreases almost linearly with increasing of a angle of misalignment of mirrors. A large dependence of the degree of polarization of these modes is observed at the misalignment of the mirror in the plane parallel to the direction of the polarization vector of the exciting radiation than at the misalignment in a perpendicular plane. For TE_{01q} mode such difference in the behaviour of the degree of polarization it is not observed.

REFERENCES

[1]   R.E.Jensen, M.S.Tobin, "CO_2 waveguide gas laser", Appl. Phys. Lett., vol. 20, no. 12 , pp. 508-510, 1972.

[2]   B.A.Kuzyakov, V.F.Khor'kov, "Small single-mode quantum generator on carbon dioxide", Sov. J. Comm. Tech. Electron., 1981, vol. 26, no. 3, pp. 610-612, 1981.

[3]   V.V.Antyukhov, A.F.Glova, O.R.Kachurin et al., "Output characteristics of a waveguide CO_2 laser with plane mirrors", Sov. J. Quantum Electron., vol. 15, no. 8, pp. 1063-1065, 1985.

[4]   A.V.Degtyarev, V.A.Svich, V.M.Tkachenko, A.N.Topkov, "Metal waveguide resonators for submillimeter lasers with uniform reflectors", in: Applications of Millimetre and Submillimetre Radiowaves, Kharkov, Institute of Radiophysics and Electronics, National Academy of Sciences of Ukraine, pp. 105-111, 1993.

[5]   Ya.N.Fel'd (ed.) *Spravochnik po volnovodam (Handbook on Waveguides)*, Moscow, Sov.Radio, 1952.

Measurement of linear polarization of laser radiation with two-grating thin-wire bolometer

V.M. Kuzmichov[1], S.V. Pogorelov[2], B.V. Safronov[1], V.P. Balkashin[1], I.A. Priz[1]

[1]V.N. Karazin Kharkiv National University, Quantum Radiophysics Department
4, sq. Svobody, Kharkiv, 61077, Ukraine, ph. +380 57 707 51 57
[2]National Pharmaceutical University, Department of Physics,
53, str. Pushkinskaja, Kharkiv, 61002, Ukraine, ph. +380 57 771 81 59 e-mail: svpog@yahoo.co.uk

Abstract: The method of measurement of direction of linear polarization of intense and wide aperture laser radiation with two grating thin wire bolometer has been improved. Bolometric elements of gratings are arranged no mutually perpendicular directions.

Measurement of linear polarization of intense and wide-aperture laser radiation is an actual problem. The possibility of measurement of linear and elliptic polarization parameters of radiation with three-gratings thin-wire bolometer is shown in the works [1,2]. Linear polarization measuring problem with two-grating thin-wire bolometer is not solved exhaustively in the work [3] as single-valuedness of an angle of polarization direction at mutually perpendicular position of gratings elements is undetermined. The ones angles of radiation polarization relative to gratings are coincide and the others are opposite in direction to one another which give us the same direction of radiation polarization.

The aim of the work is one-valued determination of a directional angle of linear polarization of radiation with two-grating thin-wire bolometer.

An ambiguity of the determination of the radiation polarization angle is solved with arrangement of grating's bolometric elements not mutually perpendicular.

Let angles of bolometric elements of gratings are $\psi_1 = 0$ and $\psi_2 \neq \dfrac{\pi}{2}$. Let us consider pulse mode of laser radiation when duration of the pulse t_p is considerable lesser than time heat constant of bolometric elements τ; bolometer works in the linear mode when its relative increase of resistance under radiation is in proportion to absorbed energy. Under such conditions grating signals U_i are

$$U_i = \frac{\Delta R_i}{R_{i0}} = \eta^E \frac{\overline{E}}{m} K_i, \ i = 1, 2, \qquad (1)$$

where R_{i0} and ΔR_i - are the initial and absolute increase of resistance of i-th grating; $\eta^E = \alpha q^E / c$ - is the transformation coefficient of bolometric elements; α - is temperature coefficient of resistance of bolometer material;

c – is specific heat capacity of bolometer material; q^E - is a factor of absorption efficiency (FAE) of the bolometer for E-polarized radiation when its electric vector is in the line of bolometer element axis; m- is linear mass of bolometric element; $\overline{E}$ - is averaged linear optic energy incident on bolometric elements; k_i - are polarized coefficients of interaction radiation with bolometer elements of gratings which describe the dependence of relative absorption of optic radiation from the directional angle of radiation polarization and are

$$k_i = \cos^2(\varphi - \psi_i) + K_D \sin^2(\varphi - \psi_i), \ i = 1, 2, \qquad (2)$$

where φ - is directional angle of radiation polarization; $K_D = \dfrac{q^H}{q^E}$ - is coefficient of dichroism of bolometric elements; q^H - is FAE of bolometric element for H-polarized radiation when its electric field is perpendicular to bolometer element axis.

The sum of grating signals is

$$U_1 + U_2 = \eta^E \frac{\overline{E}}{m}(k_1 + k_2) \qquad (3)$$

Normalized signals of gratings are evaluated subject to (3)

$$U_{in} = \frac{U_i}{\eta^E \dfrac{\overline{E}}{m}} = k_i = (k_1 + k_2)\frac{U_i}{U_1 + U_2} \qquad (4)$$

and are equal to polarized coefficient of interaction k_i.

Probable two angle of radiation polarization relative to each grating are determined with formulae (2) and (4)

$$\varphi_{i1, i2} = \psi_i \pm \arcsin\left\{\frac{1}{K_D - 1}\left[k_1 + k_2 \frac{U_i}{U_1 + U_2} - 1\right]\right\}^{1/2} \qquad (5)$$

The sum $k_1 + k_2$ for mutually perpendicular bolometric elements of gratings does not depend on the angle φ and is $k_1 + k_2 = K_D + 1$. This sum does not be a constant value in the case of not mutually perpendicular arrangement of bolometric elements of gratings and will depend on the angle φ. The dependence of $k_1 + k_2$ from the angle φ at values of coefficient of dichroism $K_D = 1,8$ and angles $\psi_1 = 0$ and $\psi_2 = 85°$ is shown at Fig. 1. One can see that the sum may vary relative the average value $K_D + 1 = 2.80$ within the range $\pm 2.5\%$.

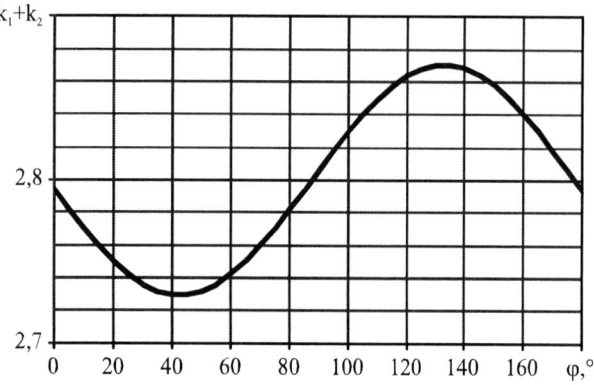

Fig.1. Dependence $k_1 + k_2$ from the angle φ

In the first approximation the sum $k_1 + k_2$ is assumed equal $k_1 + k_2 = K_D + 1$ and measured angles $\varphi'_{i1,i2}$ are estimated with (5). After that the sum $(k_1 + k_2)$ is determined with (2) where angles $\varphi'_{i1,i2}$ have to be chosen close to one another in optic beam section for the both gratings. Values of angles $\varphi_{i1,i2}$ are determined more exactly with the estimated sum $(k_1 + k_2)$ at (5) where the main part of systematic error at the expense of no perpendicularity of bolometric elements of gratings is eliminated.

Directions of mismatched angles $\varphi_{i1,i2}$ within the beam section are distinguished at the angle $(\frac{\pi}{2} - \psi_2)$. Thus, the ambiguity of determination of the angle of radiation polarization direction is solved completely.

Let us consider an example of application of the measuring method. Let $\psi_1 = 0$, $\psi_2 = 85°$, $K_D = 1,8$ and $\varphi = 30°$. According to the formula (2) polarized coefficients of interaction will be $k_1 = 1,2$ and $k_2 = 1,5368$. Grating signals are in proportion to coefficients k_1 and k_2. Let us assume $k_1 + k_2 = K_D + 1 = 2,8$ in the first approximation. Values of angles $\varphi'_{11,12} = \pm 30,47°$ and $\varphi'_{21,22} = 85° \pm 57,76°$ have been evaluated with (5). The close to each other angles are $\varphi'_{11} = 30,47°$ and $\varphi'_{22} = 27,24°$. The sum $k'_1 + k'_2$ has been calculated for the angle φ'_{11} - $k'_1 + k'_2 = 2,7363$; and the more exact values of angles $\varphi_{11,12} = \pm 29,98°$ and $\varphi_{21,22} = 85° \pm 54,98°$ have been calculated with (5). The matched angles are $\varphi^*_{11} = +29,98°$ and $\varphi^*_{22} = +30,02°$. The averaged value of measured angles is

$$\overline{\varphi} = \frac{1}{4}(\varphi^*_{11} + \varphi^*_{22} + \varphi^{**}_{11} + \varphi^{**}_{22}) = 29,995°,$$

which is agreed up to $0,005°$ with the angle of linear polarization of measured radiation.

Thus, the measuring method of linear polarization of laser radiation by two-grating thin-wire bolometer with no mutually perpendicular directions of bolometric elements of gratings has been improved.

REFERENCES

[1] Kuzmichov V.M., Pogorelov S.V., Kuzmichova E.V. Measurement of parameters of elliptic polarization of laser radiation with thin-wire bolometer // Ukrainsky metrologichny journal. – 2009. – No 2. – P.-35-38.

[2] Kuzmichov V.M., Pogorelov S.V., Safronov B.V., Balkashin V.P., Priz I.A., Kohns P. Measurement of linear or elliptic polarization of laser radiation with three-gratings bolometer // Radiofizika i radioastronomiya – 2009. – V.14. - No 2. – P.-214-221.

[3] Kuzmichov V.M., Pogorelov S.V., Safronov B.V., Balkashin V.P., Priz I.A. Two-gratings bolometric gauge of linear polarization of laser radiation // Metrologiya – 2009 – No5, - P.17-22.

CAOL*2010 International Conference on Advanced Optoelectronics & Lasers, 10-14 September, 2010, Sevastopol, Ukraine

Novel semiconductor optical amplifier with tapered active channel

Yu.O. Kostin, A.A. Lobintsov, and S.D. Yakubovich, *Senior Member, IEEE*
SUPERLUM DIODES Ltd. P.O.Box-70, Moscow 119454 Russia
Fax: +7 495 720 5465
E-mail: yakubovich@superlumdiodes.com

Abstract: Tapered semiconductor optical amplifier (SOA) based on SQW improved nanostructure (gain maximum at 840 nm) was used as output power amplifier in MOPA-system with SLD as master oscillator. CW output power of up to 0.5 W ex standard multimode fiber (MMF) was obtained.

INTRODUCTION

A well known approach to the realization of high-power SLD-based light source is MOPA system with SLD as master oscillator. Recently high performance reliable SOA for 820-860 nm spectral range based on improved SQW (GaAl)As/GaAs nanostructure was developed [1]. When used as power amplifier in MOPA-system of the above mentioned type it ensured CW output power level of 50 mW ex single-mode fiber (SMF). The aim of present work was to manufacture tapered SOA based on the same nanostructure and to reach with its usage the output power level of several hundreds mW ex MMF.

EXPERIMENTAL RESULTS

SOA samples studied in present work were based on SQW (GaAl)As/GaAs heterostructure described in [1]. The configuration of its tilted ($\alpha = 7^\circ$) active ridge waveguide is shown on Fig.1a. It contains two sections: the input single-mode section ($W_{SM} = 4.0$ μm, $L_{SM} = 600$ μm) and the output tapered section ($W_{OUT} = 65$ μm, $L_T = 1000$ μm). SOA was tested as chip-on-submount in the prototype of MOPA-system shown schematically on Fig.1b. Spectrally matched SLD-module with output SMF was used as master oscillator. To prevent the suppression of its emission by the back emission of SOA the fiber-optic polarization insensitive isolator OFR Model IO-F-840 was used. Polarization controller OZ Optics Model BB-500-11-850 at SOA input ensured optimal polarization of the input signal. The output signal could be measured in free space or coupled into standard 50 μm MMF.

At first SOA samples were studied as SLDs (without input signal) in single-pass and double-pass operation modes. The corresponding L-C characteristics and spectra at $I_{SOA} = 1000$ mA are shown on Fig.2. Double-pass operation mode was realized using SMF reflector at SOA input. Far field patterns in

single-pass operation mode at different pumping levels are shown on Fig.3. Angle in junction plane is counted off from the output beam axis (23° with respect to normal to crystal facets).

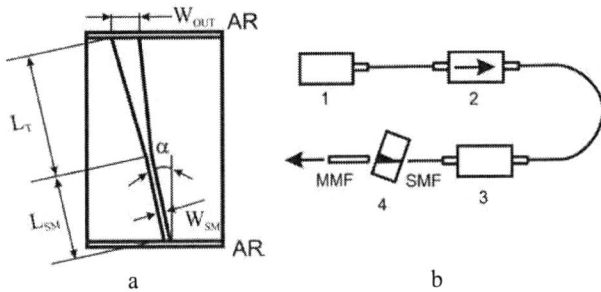

Fig. 1. Configuration of tapered SOA (a) and of the experimental MOPA-system (b): 1 – SLD-module; 2 – optical isolator; 3 – polarization controller; 4 – SOA (chip-on-submount).

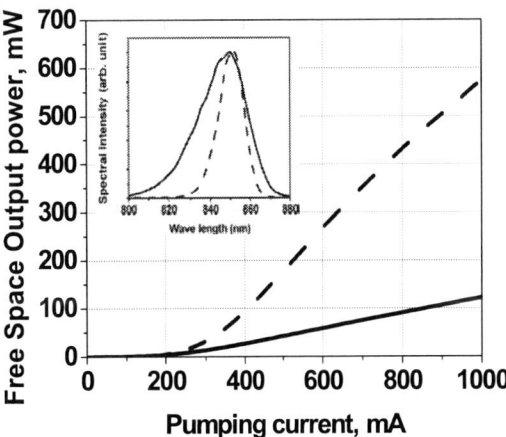

Fig. 2. L-C characteristics and spectra (insert) in single-pass (solid curves) and double-pass (dashed curves) free-running modes.

978-1-4244-7043-3/10 $26.00 © 2010 IEEE   175

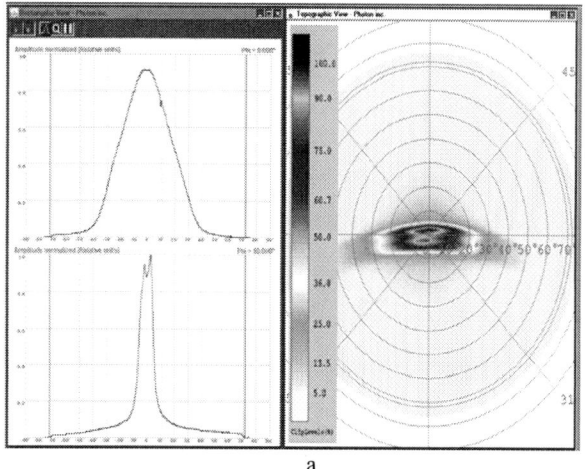

a

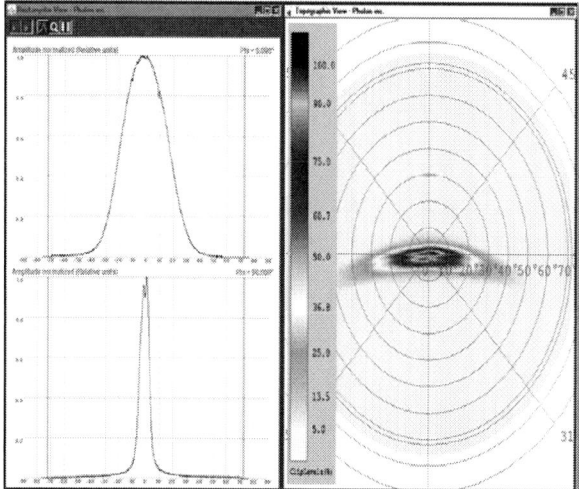

b

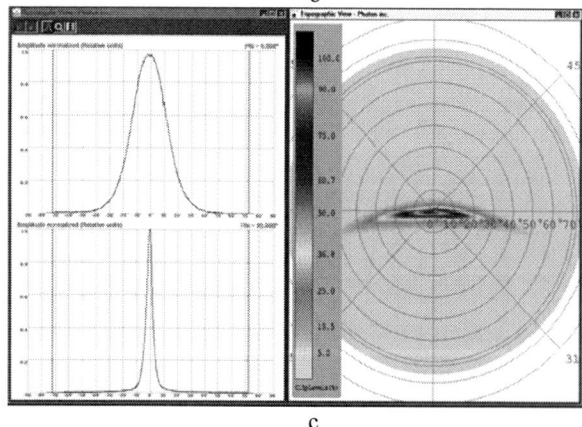

c

Fig.3. Far field patterns in single-pass operation mode
at I_{SOA} = 300 mA (a), I_{SOA} = 700 mA (b) and
I_{SOA} = 1000 mA (c).

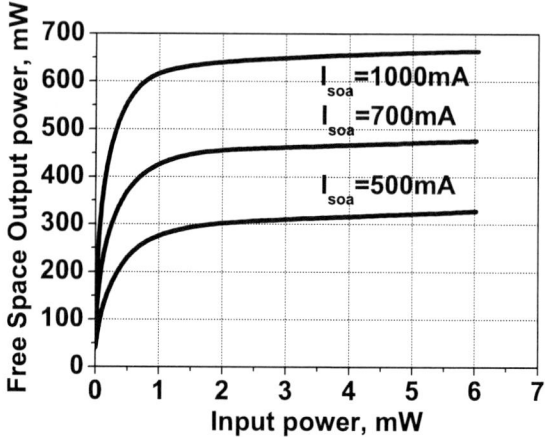

Fig.4 Transmission characteristics of MOPA-system at
different pumping levels.

Transmission characteristics of MOPA-system (CW, 25°C) at different SOA pumping levels are presented on Fig.4. At I_{SOA} = 1000 mA saturated output power of 670 mW was obtained. Emission spectrum at this operation point was similar to that observed in free-running double-pass operation mode (λ = 855 nm, FWHM = 14 nm). The coupling efficiency into 50μm MMF lensed pigtail reached 75% allowing output power ex MMF near 0.5 W. This value is far from the attainable maximum. Single-mode SOAs with output aperture of 4.0 μm based on the same heterostructure [1] possessed CODM threshold of more than 200 mW. It is obvious that using optimized heatsink and thermostabilization system at higher injection currents such tapered SOA in the same MOPA-system may ensure much higher output power. It will be the goal of our further experiments. The reliability of the light source of this kind needs of course separate investigations.

CONCLUSION

The prototype of MOPA-system with SLD as master oscillator and tapered SOA at 850 nm as power amplifier was investigated. CW output power of 0.5 W ex MMF was obtained.

REFERENCES

[1] Lobintsov A.A., Uspensky M.V., Shishkin V.A., Shramenko M.V., Yakubovich S.D., "High-Performance Semiconductor Optical Amplifier of Spectral Band 820 860 nm, *Quantum Electronics*, 40 (4), 305-309 (2010).

Effect of laser beam shape on differential-phase measurements

Yu. V. Pilgun, E. N. Smirnov

National Taras Shevchenko University of Kyiv, Ukraine

Abstract: Accuracy of differential-phase measurements depends on actual shape of laser beams formed by focusing optics. In present work two factors affecting beam profile are considered: aperture limiting and asymmetrical intensity irregularities. Ripples in measured signal are found to be caused by beam sidelobes. Asymmetry of beams leads to direction-dependent measurements.

Differential-phase method permits measuring of differences in optical path with high precision [1, 2]. In the method investigated object is tested with two focused laser beams (fig. 1). After passing the object, beams acquire different phase shift, which can be recovered from photodetector signal. Optical frequencies of probe beams are shifted with acousto-optical device (not shown), and photodetector senses beat frequency. Optical path difference is obtained from phase of this beat frequency signal. This allows detecting variations in optical path with sensitivity of $\lambda/100$ or even better.

To recover the whole profile of the object pair of beams is moved over the object's surface. The acousto-optical device that shifts optical frequency also performs scanning of beams. Phase differences are measured in each point of scanning and combined to obtain full data row.

Distribution of light field at photodetector plane is formed by contribution both from the object features and the beam shape. We expect to measure phase shift caused only by the object. But phase profile of the beam overlaps with the object profile and also affects measurements. If we assume focused

Gaussian beam, it has flat wavefront and no problems are expected. But in real device we rarely have ideal beams. Optical scheme always contains some limiting diaphragms, so diffraction sidelobes can appear in beam profile after focusing on the object. If initial beam has some intensity irregularities, it also could lead to distortion of focused beam.

To investigate possible effects on differential-phase measurement, we considered a simple model of beam distortion. Initial Gaussian beam is limited by aperture and have single intensity downfall (fig. 2). We included downfall because acousto-optical device often introduce such a flaw,

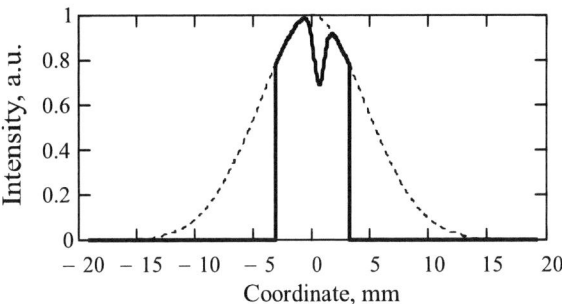

Fig. 2. Beam profile before focusing optics.
(Aperture limited Gaussian with downfall at side)

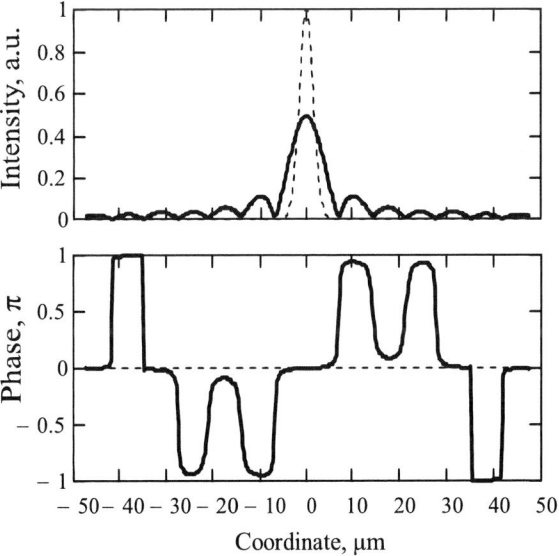

Fig. 3. Beam profile in the object plane.
Every other sidelobe has almost $\pm\pi$ phase shift.

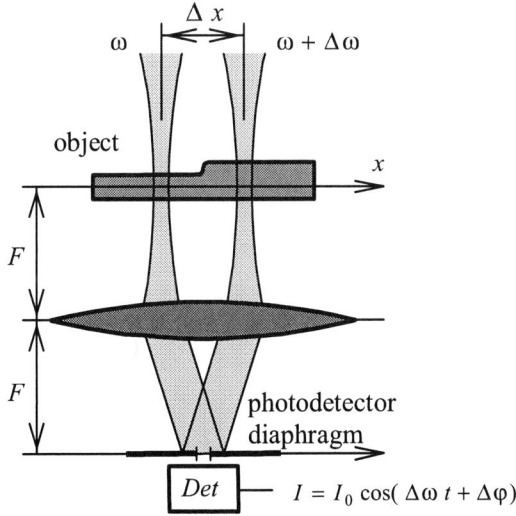

Fig. 1. Scheme of differential-phase measurements.
Pair of beams is scanned over the object.

978-1-4244-7043-3/10 $26.00 © 2010 IEEE

which is related to non-uniformity of sound field in the device.

After focusing on the object (fig. 3) not only intensity profile of the beam is distorted, but also phase profile is affected by introduced irregularities. Moreover, the beam becomes asymmetrical. Due to intensity downfall of initial beam is shifted to the side, resulting beam has asymmetrical phase jumps in sidelobes. Measuring objects with such beams could give inaccurate results.

Our experiments suggest that most irregular results are obtained when investigated object has sharp phase jumps. So we have simulated differential-phase measuring of step-like profile using distorted beams. To estimate introduced errors we performed this simulation also with ideal Gaussian beams (see fig. 4, 5)

Calculations shows that beam sidelobes produce ripples on the measured signal. Influence of sidelobes is not easily seen for objects with relatively slow changes of phase. But when variation of object phase profile is sharp it leads to appearing of pronounced ripples. Even though intensity of side lobes is relatively low, variation of beam phase is significant and this disturbs measured values considerably. Due to method is highly-sensitive to phase, beam with sidelobes feels presence of object's phase change at quite far distances comparing to main lobe size.

Asymmetry of probe beams brings to another effect – it makes measurement procedure direction-dependent. If the object is rotated 180 degrees and scanned in backward direction result will be not just mirrored but essentially different. Asymmetrical beam responses differently to rising and falling steps of the same height. This effect is clearly seen in fig. 5b.

Presented results help to explain origin of ripples and asymmetry which appear in experimental data. Unfortunately it is rather difficult to make side-by-side comparison of calculated and experimental data, because exact shape of beams cannot be easily measured.

REFERENCES

[1] Somekh M.G., Valera M.S., Appel R.K., "Scanning heterodyne confocal differential phase and intensity microscope," *Appl. Opt.*, vol. 34, no. 22, pp. 4857-4868, 1995.

[2] Carlson T.B., Denzer S.M., Greenlee T.R., Groschen R.P., Peterson R.W., Robinson G.M., "Vibration-resistant direct-phase-detecting optical interfero-meters," *Appl. Opt.*, vol. 36, no. 28, pp. 7162-7171, 1997.

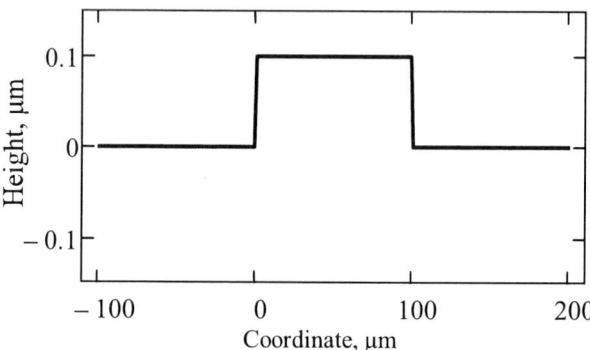

Fig. 4. Phase profile of test object

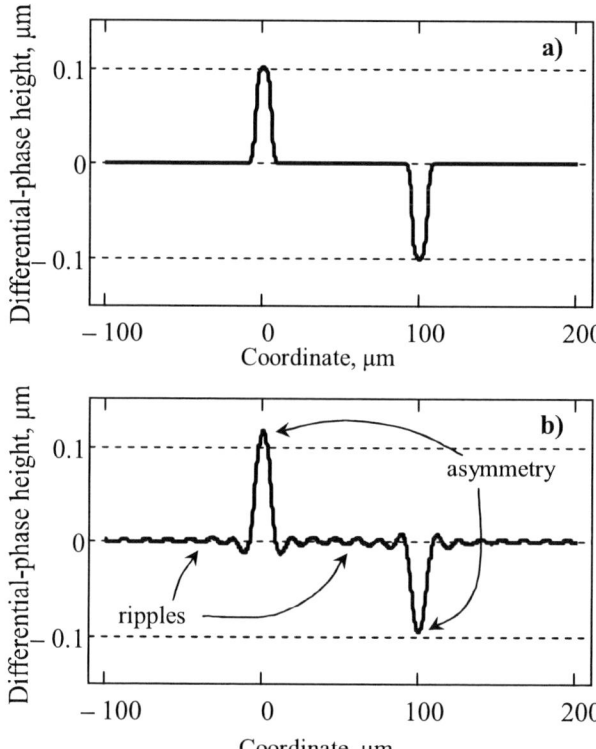

Fig. 5. Simulation of differential-phase measurements:
a) profile measured with pair of ideal Guassian beams;
b) profile measured with distorted probe beams

CAOL*2010 International Conference on Advanced Optoelectronics & Lasers, 10-14 September, 2010, Sevastopol, Ukraine

The relations between the basic characteristics of scattered laser irradiation field and the location objects surface parameters

[1]Arakcheev P.V., [1,2]Semerenko D.A., [1]Maryanina A.S., [1]Nazarov S.I., [1]Buryi E.V.

[1] N.E. Bauman Moscow State Technical University, Moscow, Russia
[2] Moscow State University of Instrument Engineering & Computer Science, Moscow, Russia

Abstract- The system of relations of basic characteristics of laser irradiation field scattered by location object surface was determined. For estimation of parameters of object surface and reconstruction the registration conditions of characteristics of laser irradiation field were established. The requirements to complex of models describing characteristics of the laser irradiation field of the object surface were made.

I. INTRODUCTION

The usage of high-speed data transfer systems, relaying of the radio- and TV- programs, providing a great number of consumers with the navigational data require the intensive development of the near space.

As a result, the number of satellites which provides these tasks grows up. At the same time celestial mechanic principles predetermine a quite short range of orbit inclination and height, inside which it is located the orbits of most Earth satellites. Space debris elements (SDE) orbits are also located in this range. SDE are the parts of rocket stages, space systems' structural elements and uncontrolled satellites with exhausted operational resource. Therefore, despite of the huge volume of near space, the probability of collision of some functional satellite and SDE exceeds the level, which can be considered insignificant.

The information availability about the STE dynamic allows excluding of the possibility of such accidents. The problem is that a lot of nature factors affect on the STE orbit parameters and rate of its action is quite unknown. Besides, STE prior information is absent often. That is why it is difficult to prognosticate the STE space position with high precision.

II. THE MAIN TASKS INFORMATION AND PRINCIPLE OF MEASUREMENTS

For solving tasks of identification of remote small objects are necessary to estimate overall dimensions. To suppose, the object which have a surface Ω and will observed in a left Cartesian coordinates (CC) - $XOYZ$, the axis OZ have guide such as a vector of observation $\mathbf{V}$, point O is coincide with a center of transceiver of location system. Acquisition of overall dimensions x_{max}, y_{max}, z_{max} of observing part of surface Ω_V (fig. 1) and can be use for identity objects or in some cases for determine a position one respect to CC of observation.

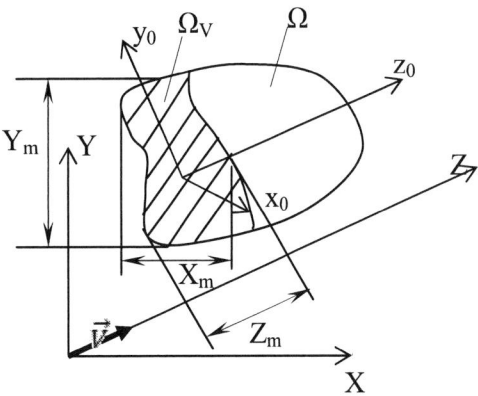

Figure 1. Object position in cartesian coordinates.

In case of position determination of an object the impulses of laser irradiation $P(t)$ are used, if a pulse duration is comparable with propagation time of plane wave in direction of surface Ω, that information of magnitude z_{max} can be taken from analysis of pulse shape of scattering Ω_V and broadband photo detector will register in the vicinity of a point O:

$$I(t,\alpha,\beta) = P(t) \otimes A(t) \otimes g(t,\alpha,\beta) \otimes A(t),$$

where $I(t,\alpha,\beta)$ – intensity of detected irradiation, α and β – angle which show orientation of the object in CC of observation, $A(t)$ – pulse characteristic atmospheric channel, and $g(t,\alpha,\beta)$ – pulse characteristic of scatter of the object. The output signal $U(t)$ of photo detector (photo detector and amplifier were broadband) will be determined by the next equation

$$U(t) = A_0 \cdot I(t,\alpha,\beta)\big|_{\alpha,\beta} \otimes u(t),$$

where A_0 – proportional coefficient, $u(t)$ – pulse characteristic of photo detector.

Designing informative systems of the attributes by usage an integral transformation $U(t)$ allow to solve the task of automotive recognition of N objects with known $\Omega_1, \Omega_2,..., \Omega_N$, where $N \gg 2$ were shown in the works [1-5]. The recognition can be produce in a case when main characteristics of scatter of individual part of surface Ω will not lead to loss of individual properties $U(t)$

978-1-4244-7043-3/10 $26.00 © 2010 IEEE

Determination of the overall dimensions x_{max}, y_{max} is easy to make when an image of the object can be reconstruct. Though, in a case of observation of remote objects no lighting SDE, and this case have meet quite seldom. As source of information of projection Ω to plane XOY is possible to apply parameters of fourth order correlation function (CF) of scattered field irradiation in case when it's difficult to obtain valid object image.

If the field of intensities are homogeneous and randomly that immediate estimation CF can be obtained by calculation of spatial convolve of intensities [6]:

$$\hat{\Gamma}_{S_V}^{(2,2)}(\delta\mathbf{r},\tau) = \frac{1}{S_V}\int_{S_V} I(\mathbf{r},t)\cdot I(\mathbf{r}+\delta\mathbf{r},t+\tau)d\mathbf{r} \quad (1)$$

where S_V – square of flat surface in observation range when intensity of scatter irradiation is registered $I(\mathbf{r},t)$, $\mathbf{r}$ – radius-vector, determining a position of point in a surface limit S_V, t – time. If a distance to object is large ($R^2 \gg S_V$ and $S_V \perp \vec{\mathbf{n}}_z$) and for flat object it is true $\tau \approx 0$. At location of an object axis symmetric a next equation can be used

$$\hat{\Gamma}_{L_V}^{(2,2)}(\delta\mathbf{r},\tau) = \frac{1}{L_V}\int_{L_V} I(\vec{r},t)\cdot I(\vec{r}+\delta\mathbf{r},t+\tau)d\vec{r} \,,$$

where L_V – randomly selected segment on a surface $S_V : (\mathbf{r},\delta\mathbf{r})\in L_V$.

Estimation ρ_0 of position first zero of CF $\Gamma_{1,2}^{(2,2)}(\delta\mathbf{r})$ is connected with angular size of object γ, which allow to determine a maximal magnitude of cross linear size of object $l_{max} = \gamma R$ at unknown distance R to one. Keep in mind a small error of estimation it can be obtained $x_{max} = \gamma_x R$ and $y_{max} = \gamma_y R$ previously find magnitude γ_x and γ_y [2] for two types of orientation L_V: 1) $L_V \parallel \mathbf{n}_x$; 2) $L_V \parallel \mathbf{n}_y$ at location of an object without axis symmetry. Note, the system of estimation γ, obtained at different orientations of L_V, it makes that possible to reconstruct a contour L_Ω of projection Ω_V on to plane XOY. The family of counted estimations of complex 2nd order correlation function allows reconstructing the surface contour of the observed object, which image could not be obtained by the optical watches facilities. Nevertheless, in general case the reconstructed counter is not show feature of surface of location object. Rather, accuracy reconstruction L_Ω can be obtained as a result of usage estimation of CF of 6th order of counter of intensities field. In the general case this function can be found from equation

$$\hat{\Gamma}_{(1,2,3)}^{(3,3)}(\delta\vec{r}_1,\delta\vec{r}_2,\tau_1,\tau_2) = \frac{1}{S_V}\int_{S_V} I(\vec{r},t)I(\vec{r}+\delta\vec{r}_1,t+\tau_1)I(\vec{r}+\delta\vec{r}_2,t+\tau_2)d\vec{r}.$$

The application of CF characteristics of field scatter irradiation function to detect parameters Ω_V we suppose that registration of the irradiation it will permitted by photo detectors system. The detectors are placed along axis OX with constant step Δx of coordinate systems. A calculated cumulant of 4th and 6th order CFs by usage of equation for differential phase characteristics

$$d\varphi(k\cdot\Delta x) = \varphi[(k+1)\Delta x] - \varphi(k\cdot\Delta x),$$

it possible to build phase function $\varphi(k\cdot\Delta x)$, reciprocal intensities functions $|\dot{J}(k\cdot\Delta x)|$, module which is determined through cumulant of 6th order CF $K_{2,3}^{(2,2)}(k\cdot\Delta x)$:

$$|\dot{J}(k\cdot\Delta x)| = \sqrt{K_{2,3}^{(2,2)}(k\cdot\Delta x)}\,,$$

$$\varphi(k\cdot\Delta x) = \sum_{m=1}^{k-1} d\varphi(m\Delta x) + k\cdot\varphi(\Delta x).$$

Then, we can evaluate a intensities function of scattered irradiation in area of objects $\hat{I}_\Omega(x,y)$ if amplitudes and phase of reciprocal intensities function are known

$$I_\Omega(x,y) = \frac{(\lambda R)^2}{\zeta}\int_{-\infty}^{\infty}\int \dot{J}(x_1,y_1;0,0)e^{j\psi}\exp\left[j\frac{2\pi}{\lambda R}(x_1 x + y_1 y)\right]dx_1 dy_1,$$

which is inverse two dimensional Fourier transform multiplication reciprocal intensities function $\dot{J}(x_1,y_1;0,0)$ and phase multiplier $e^{j\psi}$. If R is large, we can assume $e^{j\psi} = 1$. The results of surface reconstruction text objects are shown in a fig. 2.

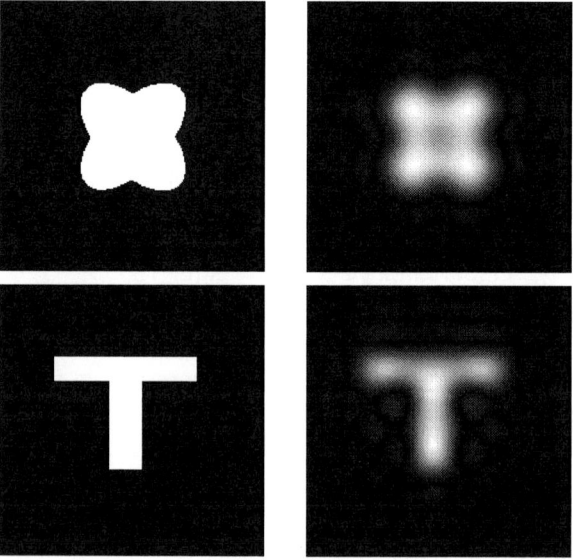

Figure 2. The results of calculation $\left(\hat{I}_\Omega\right)_{kl}$ for different objects.

It's necessary to consider a system of estimation of flow properties of the primary photo electrons (photo counters) obtained by usage of photo electronic multiplier. In optical range, of wavelength the detection is quantum mechanical process. Probability of registration n photoelectrons during time of observation T and in case an irradiation photocathode with constant intensity can be described by Puasson's law

$$P(n,T,t) = \frac{\langle n \rangle^n}{n!} \exp\left(-\langle n \rangle\right),$$

where $\langle n \rangle = \eta \langle W \rangle$ – average amount photo electrons registering in a range T, with time t, η – quantum efficiency photocathode, $\langle W \rangle$ – average magnitude of integral intensity. In case, when $W(t) \neq const$, it should be used Mandel's formula:

$$P(n,T,t) = \frac{1}{n!}\left\langle (\eta W)^n \exp(-\eta W) \right\rangle = \frac{1}{n!}\int_0^\infty (\eta W)^n \exp(-\eta W) P(W) dW,$$

where $P(W)$ – distribution of intensity integral W. If field of irradiation have M degree of freedom (DOF) and for every ones is true ratios $\langle n \rangle_M = \langle n \rangle / M$, that

$$P(n,T,t) = \frac{\Gamma(n+M)}{n!\Gamma(M)}\left(1+\frac{M}{\langle n \rangle}\right)^{-n}\left(1+\frac{\langle n \rangle}{M}\right)^{-M},$$

where $\Gamma(\bullet)$ – gamma - function, that is negative binominal distribution photo counters. DOF of M field scattered irradiation is equal to product of magnitudes polarization, temporal and spatial DOF. For single-mode pulse laser irradiating an object is executed approximate equality between coherence time of irradiation and pulse duration has linear polarization. The DOF of registering field of irradiation is that equal to magnitude of lattice DOF which in a first approximation is determined by square ratio of coherence irradiation S_c in area of receive aperture to square of this aperture S_B: $M = S_c / S_B$.

If scatter of irradiation by surface of object is conform to the Lambert's law that $S_c \sim \lambda^2 R^2 / S_\Theta$, where S_Θ – projection square Ω_V to plane XOY, λ – irradiation wavelength, R – distance to the object. Thereby, object S_Θ is estimated at known λ, R, S_B and distribution magnitude M is detected by curve view from experimental data. Thus, without usage of optical methods the object sizes can be determined.

III. CONCLUSION

In case when optical methods are not applicable for forming estimation of geometrical characteristics of a surface of remote small object that these parameters are evaluated by analysis irradiation scatter field of object. Fast detector can detect an object length in direction of viewing line at single position of registration timing characteristics (properties).

Single position registration of photo detector of scatter light in mode of "count photons" is allow to estimate projection square of irradiated object surface part to plane which is perpendicular to viewing line by way determined statistical characteristics of flow initial photo counters. Multi position registration of count intensities field allow to determine the properties of 4th and 6th order CF. These properties are help to get estimation of cross dimensional sizes of object and to reconstruct image of a surface.

ACKNOWLEDGEMENT

This work was supported by grant of Russian Board on Education and Science nk-530p/3.

REFERENCES

[1] Appazov R.F. Marks in hurt and memory - Simferopol.: Dolia, 2001. - 415 p.

[2] Roy A. Orbital Motion. – Bristol, Adam Hilger, 1988. – 532 p.

[3] Aksenov E.P. Theory of motion of artificial satellite of Earth – Moscow, Nauka, 1977. – 360 p.

[4] Zuev V.E., Zuev V.V. Actual problems of optics atmospheric, in 9 parts, p. 8. Optical remote probing of atmospheric, Saint – Petersburg, Gydrometeoizdat, 1992. – 232 p.

[5] Buryi E.V. Synthesis of an object recognition system based on the profile of the envelope of a laser pulse in pulsed lidars // Quantum electron, 1998, 28 (5). Pp. 458-462.

[6] E V Buryi, A L Mitrofanov, "Estimate of the fourth-order coherence function by a spatial convolution method and its potential applications in laser information systems", Quantum electron, 1996, 26 (5), Pp. 449–453.

[7] Buryi E.V., Rozdestvin V.N., Prospects for the development of methods of obtaining information about objects in the form of laser radar systems and methods of its use, Vestnik MGTU im. Baumana. Priborostroenie. – 1998.

[8] Buryi E.V., Kosygin A.A. Estimation of the angular dimensions of objects and reconstruction of their shapes from the parameters of the fourth-order radiation correlation function // Quantum electron, 2004, 34 (10). - Pp. 979-982.

[9] Buryi E.V., Smirnova Yu. L. Effect of the quantum nature of detecting low-intensity radiation on the distance measurement error in pulsed laser ranging // Quantum electron, 2004, 34(12). Pp. 1147-1150.

CAOL*2010 International Conference on Advanced Optoelectronics & Lasers, 10-14 September, 2010, Sevastopol, Ukraine

Diagnostics of laser radiance penetration into material by multi-channel pyrometer

V.S. Golubev, A.V. Dubrov Yu.N. Zavalov, V.D. Dubrov, N.G. Dubrovin
ILIT RAS, 1 Svyatoozerskaya St., 140700 Shatura, Moscow Region

Abstract: Investigations of the melt removing in the gasjet-assisted CO_2 laser cutting by pyrometer have been performed for steel plate 3, 6 mm and 10 mm thick are reported. The measurements of local brightness and brightness temperature were conducted for different values of cutting speed and assisted gas pressure.

The technology of cutting sheet materials with gas lasers is currently widely spread. However, when this technological procedure is automated, upgrading of its quality and reliability presents an urgent problem. In [1-2] it has been suggested that pulsations of brightness of thermal emission from the zone of laser beam action on metal should be measured with the aim of laser cutting control, and the correlation between pulsations of emitted light brightness and averaged surface roughness has been revealed. Papers [2-3] display a relationship between the pulsation spectrum of metal melt heat emission and the striation frequency of the cut side edge. A system of laser cutting process monitoring that uses the data on radiation pulsation over three spectral bands has been suggested in [4], which permitted the brightness temperature pulsations to be measured. Paper [5] discusses an optical system representing an array of photo sensors to exert monitoring of the cavity processes in laser welding. It is also advised that optical fibers should deliver the luminous flux from the radiation zone to the photo sensors. Paper [6] describes the application of high speed photography to follow the dynamics of melt in the cutting area. We have previously carried out the comparison studies on the pulsation spectrum of integrated luminosity of emission from the laser cutting zone, the spectrum of brightness temperature under two-color pyrometry, and the spectrum of the side edge roughness, resulting from laser cutting [7]. This paper reports the results of measuring the pulsations of local brightness temperature in the zone of laser beam effect on metal by the use of a four-channel two-color pyrometer.

The experiments make use of a «Trumatic 2530» machine equipped with a TLF 1500 CO_2 laser produced by Trumpf GmbH company, Germany.
Cutting of a low-carbon steel sheet 10 mm thick in oxygen as an assist gas was made with "optimal" cutting parameters: the nozzle of $\varnothing 1.4$ mm was located at 0.8 mm from the sheet upper surface; the lens focus of 127 mm was 0.5 mm above the sheet surface. In the experiments on cutting different specimens the cutting speed was varied from 16 to 24 mm/s, the recommended speed being 20 mm/s. Oxygen was used as an assist gas, its pressure being varied from 0.3 to 0.8 bar.

In another series of experiments cutting of a low-carbon steel sheet of 3 or 6 mm thick in oxygen as an assist gas was made with "optimal" cutting parameters: the nozzle of $\varnothing 1$ mm was located at 0.3 mm from the sheet upper surface; the lens focus of 127 mm was 0.3 mm below the sheet surface. In the experiments on cutting different specimens the cutting speed was varied for 6 mm thick plate from 25 to 40 mm/s, the

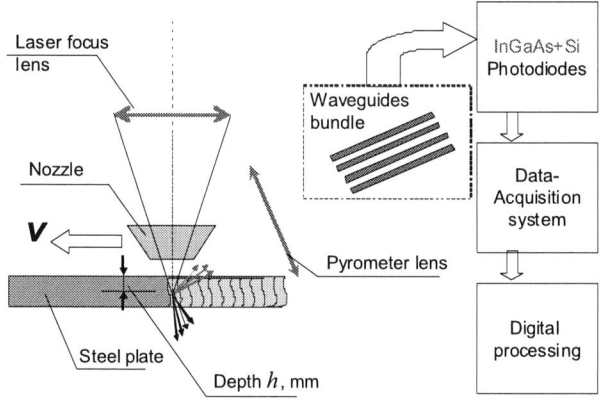

Fig. 1.Experimental set-up

recommended speed being 33 mm/s, and from 40 to 60 mm/s for 3 mm thick of metal plate. The pressure of oxygen employed as an assist gas was varied from 1 to 6 bar.

Fig. 1 depicts the experimental set-up. A portion of light emitted by a region of cut zone heated by laser radiation is converged by an optical lens and illuminates the end faces of four optical fibers located at an angle of ~30° to the plane of the sheet as shown in Fig. 1. The light is delivered by an optical fiber (core diameter 62 μm) to the photodiodes the photocurrent of which is amplified, and then it is measured by the 14-bit data acquisition and processing system and stored with the access time of 250 μs per channel of measurement. The arrays of K1713-05 photodiodes (Hamamatsu company) are used. This type of detector incorporates an infrared transmitting Si photodiode mounted over an InGaAs PIN photodiode along the same optical axis. In this case, the local part of luminescence falling at the end face of fiber is heated. In view of strong enough temperature dependence ($\sim T^4$) of the heated surface luminosity, the most heated regions of the cut zone are supposed to exert primary effect on the signal of the photo sensor. The illumination of the photodiode depends not only on the temperature in the heated region, but on other

CAOL*2010 International Conference on Advanced Optoelectronics & Lasers, 10-14 September, 2010, Sevastopol, Ukraine

factors as well, among them the area of melted material luminosity, the distance from luminescence source (sources), the heated surface luminosity pattern, and the angular sensitivity of optical sensors. The employment of a lens in the pyrometer allows the spatial selectivity of measurements to be improved. The measurement of brightness temperature by the two-color pyrometry method reduces the influence of intervene factors on the source brightness.

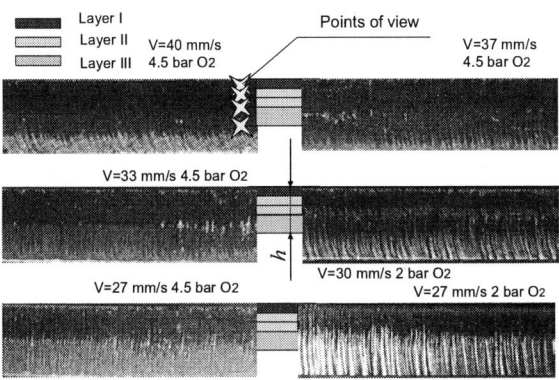

Fig. 2. The samples of 6 mm thick

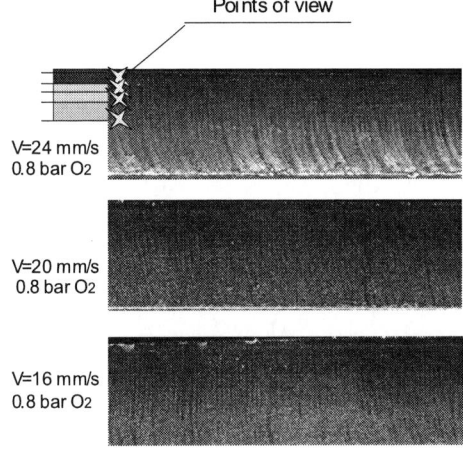

Fig. 3. The samples of 10 mm thick

The local brightness temperature has been taken for different values of luminosity region depth in the channel:

-in the samples of 3 mm thickness the depth was 0.8, 1.6, and 2.4 mm;

-in the samples of 6 mm thickness the depth was 0.8, 1.6, 2.4 and 4 mm;

-in the samples of 10 mm thickness the depth was near 0.8 mm, 1.6 and 3.2 mm. Figs 2 and 3 present the photographs of the samples cut with laser and the points of local brightness temperature observation. The photos show different layers formed by the joint action of laser radiation and melt flowing under assist gas blow [8]. Temperature T* behavior measured on different depth of penetration of laser beam into the mild steel of 6 mm thikness is shown on fig. 4-5.

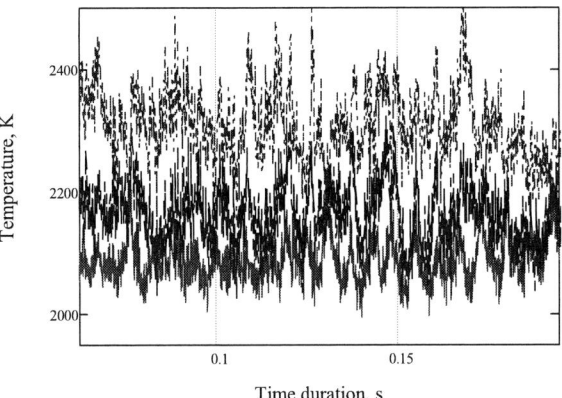

Fig. 4. Temperature T* behavior, mild steel of 6 mm thikness. The cutting velocity is 30 mm/s, pressure of oxygen is 4.5 bar. Temperature increases with depth: 0.8, 2.4, (top line) 4 mm.

The obtained early, [7], results show that relying upon the measurements of the brightness temperature average values and fluctuation amplitude, it is not possible to solve the so-called inverse problem, e.g. to find out the cutting speed from the data of statistical sampling. So the development of algorithms for control and monitoring of the cutting parameters will require the measurement of additional parameters.

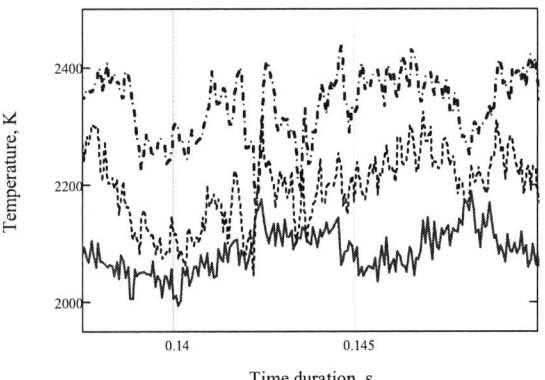

Fig. 5.The same as fig. 4 in another scale.

For this purpose, the measurements of time fluctuations of local illuminance of the photodiode ends were performed, and with the use of a specific program these data permitted generating the time spectra of brightness temperature pulsations. The procedure of data processing has been earlier described in [7]. Averaging of pulsation power spectra makes possible enhancing the relative amplitude of resonance oscillations in the fluctuation spectrum against the background of random oscillations. The N=32 or 64 realizations were averaged.

978-1-4244-7043-3/10 $26.00 © 2010 IEEE   183

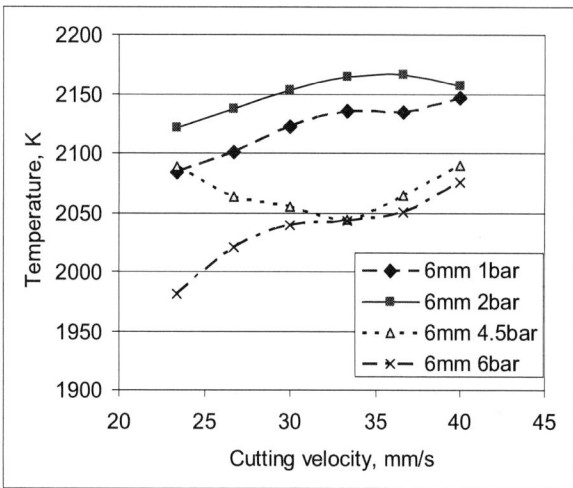

Fig. 6. Averaged brightness temperature T*.

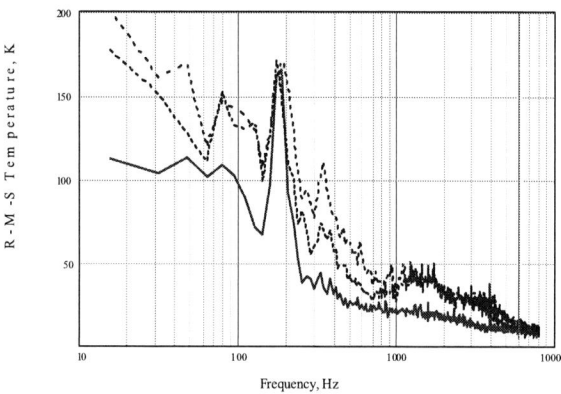

Fig. 7. The noise spectra for data fig. 4-5.

Fig. 6 illustrates the dependence of the brightness temperature T* averaged values at various depths in the cut on the speed of beam motion with respect to the sample of 10 mm thickness at the pressures of 4.5 bar (a) and 2 bar (b).

Figs. 7-8 display the corresponding spectra dependences of the root-mean-square oscillation amplitude $\sigma(T^*)$. The similar dependences have been obtained in laser cutting of 10 mm thick samples.

The dynamics of melt motion in gas-jet assisted CO2 laser fusion cutting has been studied with the use of a multi-channel pyrometer employing an optical fiber in the optical unit. The results of measuring the local brightness of melt luminosity and the brightness temperature in laser cutting of low-carbon steel sheet of 3 mm, 6 mm and 10 mm thickness are presented depending on the velocity of beam movement and the assist gas pressure. The spatial resolution around 100 µm for measuring local brightness temperature has been attained. It has been shown that the changes of assist gas pressure and velocity lead

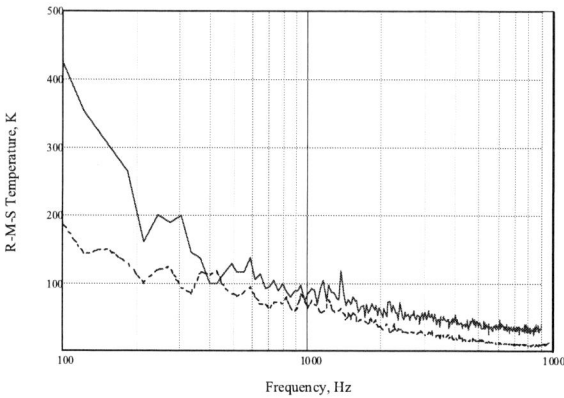

Fig. 8. The noise spectra for sample 3 mm thickness, 0.8 (top line), and 2.4 mm depth, speed 55 mm/s, 1 bar of O_2.

to a variation in the pulsation spectrum of local brightness temperature.

REFERENCES

[1] Hansmann M., Decker I., Ruge J. "On-line control of the laser cutting process by monitoring the shower of the sparks", Power Beam Technology, Brighton, England, 10-12 Sept. 1986, p.440-445.

[2] Decker I., Heyn H., Martinen D. et al., "Process monitoring in laser beam cutting on its way to industrial application", Proceed. SPIE, v.3097, pp. 29-37 (1997).

[3] Sforza P., Santecesaria V., "Analytical dependence of the roughness of the cut edge on the experimental parameters and process monitoring", Proceed. SPIE, v.2207, pp. 836-847 (1994).

[4] Sforza P., de Blasiis D., Lombardo V. "Three –module sensor for CO2-laser welding and cutting processes", Proceed. SPIE, v.3097, pp. 97-107 (1997).

[5] Engel T., Kane M., Fontaine J., "Temporal ccharacterization of plasma cw high-power CO2 laser-material interaction: contribution to the welding process control", Proceed. SPIE, v.3097, pp. 727-734 (1997).

[6] Takashi Fushimi, Hideyuki Horisawa, Shigeru Yamaguchi et al., "A fundamental studying of laser cutting using a high speed photography" Proceed. SPIE, v.2888, pp. 90-95 (2000).

[7] Dubrov V.D., Dubrovin N.G., Zavalov Yu.N. et al. "The spectrum of pulsation of temperature of melting kern in gas-assisted laser cutting", , In: Technical digest of the International Conference on Coherent and Nonlinear Optics and the Lasers, Applications, and Technologies conference (ICONO/LAT2005), May 11-15, 2005, St.Peterburg, p.69.

[8] V.S. Kovalenko, V.V. Romanenko, L.M. Oleschuk, Wasteless Processes of Laser Cutting, Tehnika, Kiev, 1987 (in Russian).

CAOL*2010 International Conference on Advanced Optoelectronics & Lasers, 10-14 September, 2010, Sevastopol, Ukraine

Optical measurements of ferroelastics $(Cs_3Bi_2I_9)$: new phenomena and optical devices

F.V. Motsnyi

V.Lashkarev Institute of Semiconductor Physics, National Academy of Sciences of Ukraine,
Pr.Nauky 41, 03028 Kyiv, Ukraine
National Academy for Statistics, Accounting and Audit, State Committee for Statistics of Ukraine,
Pidgirna str. 1, 04107 Kyiv, Ukraine,   E-mail: fv_motsnyi@isp.kiev.ua

Abstract: The review report deals with the experimental and theoretical studies of new phenomena registered (the nontraditional temperature shift $E_g(T)$, the change of exciton-phonon interaction and the giant thermodynamical optical effect).The results are discussed on the basis of exciton-phonon interaction and the ferroelastic phase transition. The optical active devices are proposed.

Ferroelastics is a peculiar new class of crystalline solids in which the spontaneous strain of a crystal lattice is appeared relatively of initial one as a consequence of structure phase transition from more symmetrical (paraelastic) phase into less symmetrical (ferroelastic) one. At cooling through the Curie point T_c (without any external action) the ferroelastic domains appear in the ferroelastic crystals in such a manner that the total strain of a sample (taking into account the stress sing in each domain) is zero. Such crystal partition into domains corresponds to the minimum of the strain energy of a crystal. One of ferroelastics is the $Cs_3Bi_2I_9$ layered semiconductor having high anisotropy of chemical bonding: a strong ion-covalent bonding in the separate layer sandwich and a weak Van der Waals interaction between the neighbouring sandwiches.

In this article, the powerful optical methods [1-6] are used for the manifestation of new physical phenomena and the following problems are solved towards this purpose:

- the domain structure observations of $Cs_3Bi_2I_9$ are performed using the polarization microscope at heating and cooling
- the Raman spectra of $Cs_3Bi_2I_9$ are studied in heating regime over temperature range 5-300 K (involving the ferroelastic phase transition point T_c = 220 K)
- the exciton reflection spectra of $Cs_3Bi_2I_9$ are investigated at temperatures 4.2-300 K with light polarization $\mathbf{E}\perp\mathbf{c}$
- the separation of complicated contours of optical spectra into components is considered on the basis of Van Hove singularities.

The typical Raman spectra of $Cs_3Bi_2I_9$ layered crystals in the low frequency region at 300, 220, 80 and 5 K are presented in Fig.1. These spectra are rich. Temperature lowering down to T_c = 220 K make the Raman spectrum clear. At T ≤ 80 K the Raman spectra are modified essentially: the doublet lines appear and the intensities of the lines change (Fig. 1, Fig. 2). The interpretation of the spectra are given in table. The detection of the low-frequency doublets in the region of

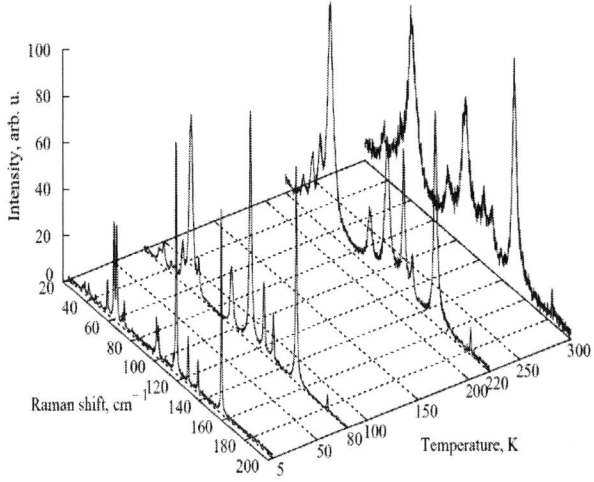

Fig. 1. The Raman spectra of $Cs_3Bi_2I_9$ taken in unpolarized light at temperatures 5, 80, 220 and 300 K. λ_{exc} = 632.8 nm.

transition vibrations due to Davydov splitting the lines 37.9, 45.0, 61.4 and 68.3 cm^{-1} (caused by the dynamical interaction between the $[Bi_2I_9]^{3-}$ molecular ions occupying two non-equivalent position in the crystal lattice), or splitting of the degenerate vibrations of E symmetry (Bethe splitting) (the line 97.4 cm^{-1}) due to the removal of degeneracy under spontaneous strain that appears when a high-symmetry hexagonal crystal lattice becomes low-symmetry monoclinic by passing through the phase transition point T_c = 220 K [4,6].

The ferroelastic phase transition at T_c = 220 K belongs to those of the first order but close to the second one [1,6]. The soft mode and the frequency softening are not observed [4].

The temperature shift of $E_g(T)$ of $Cs_3Bi_2I_9$ at $\mathbf{E}\perp\mathbf{c}$ is described very well by Varshni formula:

$$E_g(T) = E_g(0) - \frac{\alpha \cdot T^2}{T + \theta}, \qquad (1)$$

where $E_g(0)$ = 2.857 eV, α = 7·10^{-4} eV·K^{-1}, Θ = 258.654 K. The energy position of $E_g(T)$ does not change at T < 45 K

978-1-4244-7043-3/10 $26.00 © 2010 IEEE   185

CAOL*2010 International Conference on Advanced Optoelectronics & Lasers, 10-14 September, 2010, Sevastopol, Ukraine

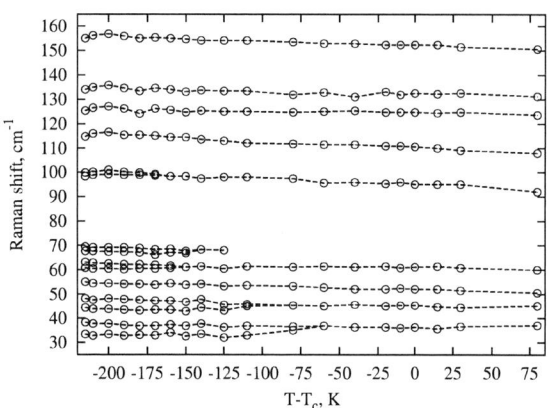

Fig. 2. Temperature dependences of the position of low-frequency lines 34.4, 38.3, 44.3, 48.2, 55.1, 60.8, 63.2, 67.7, 69.5, 98.3, 99.8, 114.8, 125.3, 134.0 and 155.0 cm^{-1} in the Raman spectra of $Cs_3Bi_2I_9$ taken in unpolarized light. λ_{exc} = 632.8 nm.

Table. Line frequencies in the Raman spectra of $Cs_3Bi_2I_9$ and $Rb_3Bi_2I_9$ and their ratios. λ_{exc} = 632.8 nm. T=300K.

$\nu_{Cs_3Bi_2I_9}$, cm^{-1}	$\nu_{Rb_3Bi_2I_9}$, cm^{-1}	$\nu_{Cs_3Bi_2I_9}/\nu_{Rb_3Bi_2I_9}$	Interpretation of Raman spectrum of $Cs_3Bi_2I_9$
-	24	-	
37	37	1	
45	45	1	Vibrations of $[Bi_2I_9]^{3-}$ as a whole
52	52	1	
60	62	0.97	
-	67	-	
-	76	-	
92	-	-	Vibrations of $[BiI_6]^{3-}$–$[BiI_6]^{3-}$ inside of bioctahedron
108	-	-	
124	-	-	
131	131	1	Vibrations of Bi–I bonds inside of $[BiI_6]^{3-}$ octahedron
151	145	1.04	

and shift to the long-wavelength side with temperature up to 300 K without any anomalies at T_c = 220 K [3,6]. Such behavior of Eg(T) is not peculiar for layered semiconductors but is typical for the well-studies ones. The exciton binding energy Ry is 279 meV [2,3].

The experimental temperature dependence of the half-width H(T) of the exciton band (n=1) of $Cs_3Bi_2I_9$ (open circles) is depicted in Fig. 3. Its fitting on the basis of Toyozawa theory (solid line) for a weak (5 K ≤ T ≤ 150 K) (a) and a strong (150 K ≤ T ≤ 220 K) (b) exciton-phonon interaction with different phonon frequencies and empirical linear extrapolation (H(T) = k(T − 225K) + H$_o$, where k = 0.758 meV·K^{-1} and H$_o$ =116.3 meV) (225 K ≤ T ≤ 300 K) (c). The best correlation between experiment and theory exist for

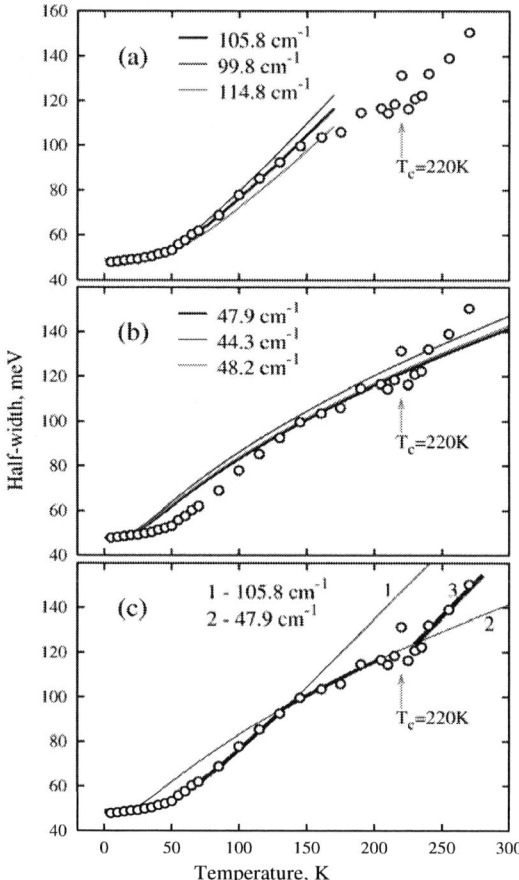

Fig. 3. The modeling of temperature behavior of the half-width H(T) of the exciton absorption band (n=1) of $Cs_3Bi_2I_9$.

effective phonons with frequencies 114.8 and 105.8 cm^{-1}, respectively. The change of exciton-phonon interaction from a weak to a strong in the same sample is registered as temperature increases [3,6]. The region 183-221 K corresponds to a heterophase structure where ferroelastic and paraelastic phases coexist [1,3,6]. So, the $Cs_3Bi_2I_9$ is layered semiconductor in the hexagonal phase and behave like traditional one in the monoclinic phase. Temperature T* = 150 K is an temperature below which a crystal loses the nature of layered substance.

The oscillations observed in the $Cs_3Bi_2I_9$ reflection spectra (Fig. 4) with the wavelength are caused by oscillations of the refractive index in time (n = n$_0$ + α·η), where η is the order parameter (the spontaneous strain) and n$_0$ is the refractive index in high temperature (hexagonal) phase in which η=0) [5]. In non-study–state the order parameter η is described by the Landau-Khalatnikov's equation:

$$\frac{d\eta}{dt} = -\Gamma \frac{d\Phi(\eta)}{d\eta} \ . \qquad (2)$$

CAOL*2010 International Conference on Advanced Optoelectronics & Lasers, 10-14 September, 2010, Sevastopol, Ukraine

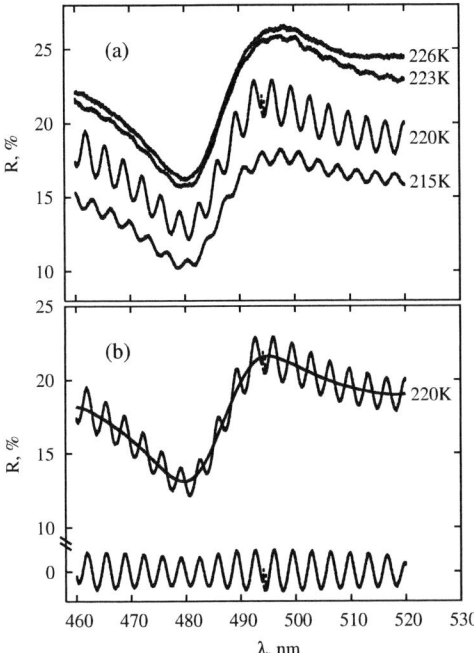

Fig. 4. The oscillations at the background of exciton band of $Cs_3Bi_2I_9$ (a) and the exciton reflection spectrum with oscillations and the modulation component (b). T_c= 220 K

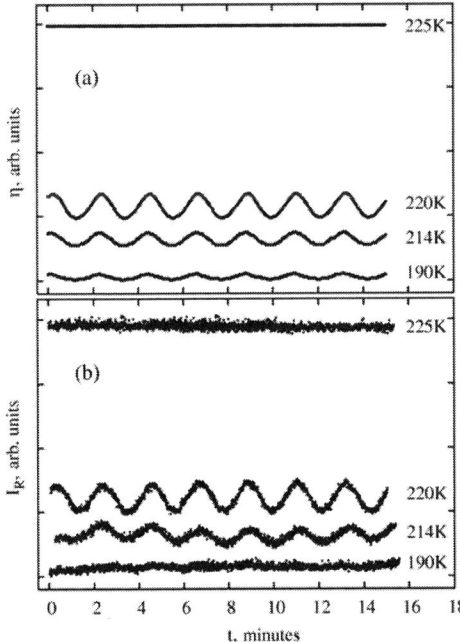

Fig. 5. The temperature dependence: (a) the order parameter $\eta(t)$ from equation (2) ($a = 1$, $b = -1$ and $c = 10$) and (b) the intensity $I_R(t)$ of the light reflected from a sample.

Here $\Phi(\eta) = (a\frac{\eta2}{2} + b\frac{\eta4}{4} + c\frac{\eta6}{6})$ is the free energy, Γ is a coefficient. In studied system the temperature periodically depends on the time. Therefore in formula (2) we take $(T + \Delta_r \cdot Cos\ \omega t)$ instead of T were Δ_r and ω is the amplitude and frequency of a source of the temperature vibrations, respectively. The results obtained testify the correction of the model proposed.

Thus, a new giant thermodynamical optical effect [5] near the ferroelastic phase transition point was found and explained on the base of a model that takes into account the temperature dependence of the refractive index through the order parameter of the crystal.

This phenomenon gives possibilities to apply the powerful optical methods to study the phase transitions at different influences, to use it for the modulation and amplification of output light, the manufacturing of high precise temperature sensors and optical bistabile cells, and also as non-destructive express optical method for the appearance of phase transitions.

REFERENCES

[1] I.Girnyk, O.Krupych, I.Martunyuk-Lototska, F.Motsnyi and R.Vlokh."Phase coexistence in $Cs_3Bi_2I_9$ ferroelastics: optical, dilatation and ultrasonic velocity studies", *Ukr. J. Phys. Opt,* Vol, 4, no. 4, pp. 165-169, 2003.

[2] V.G.Dorogan, M.P.Lisitsa, F.V.Motsnyi and O.M.Smolanka "Self-coordinated method for separation of optical function spectra into components", *Ukr. Phys. J.,* vol. 48, no. 2, pp.181-185, 2003.

[3] V.F.Machulin, F.V.Motsnyi, E.Yu.Peresh, O.M.Smolanka and G.S.Svechnikov, "Effect of temperature variation on shift and broadening of exciton band in $Cs_3Bi_2I_9$ layered crystals", *Fiz. Nizkikh Temperatur,* vol. 30, no. 18, pp.1283-1286, 2004.

[4] F.V.Motsnyi, M.V.Vuychik, O.M.Smolanka and E.Yu.Peresh "Phase transition in $Cs_3Bi_2I_9$ ferroelastic: investigation by Raman scattering technique", *Functional Materials,* vol.13, no. 3, pp.1-6, 2005.

[5] F.V.Motsnyi, O.M.Smolanka, V.I.Sugakov and E.Yu.Peresh "Giant thermodynamical optical effect near ferroelastic phase transition point in $Cs_3Bi_2I_9$ layered crystal", *Solid Stat. Communs,* vol. 137, pp. 221-224, 2006.

[6] F.V.Motsnyi, O.M.Smolanka, E.Yu.Peresh, S.V.Virko and V.G.Dorogan "Change of exciton-phonon interaction in layered ferroelastic crystals ($Cs_3Bi_2I_9$)", *Physica: Condensed Matter,* vol. B 403, pp. 2838-2841, 2008.

978-1-4244-7043-3/10 $26.00 © 2010 IEEE

CAOL*2010 International Conference on Advanced Optoelectronics & Lasers, 10-14 September, 2010, Sevastopol, Ukraine

Optical interferometry and data analysis of laser-produced plasmas

A.A. Aliverdiev[1], D. Batani[2], R. Dezulian[2], T. Vinci[2,3], A. Benuzzi-Mounaix[3], M.Koenig[3], and V. Malka[4]

[1] Institute of Physics DSC RAS, 367003, ul. Yaragskogo, 94, Makhachkala, Russia
[2] Dipartimento di Fisica "G. Occhialini", Università di Milano-Bicocca, Milano, Italia
[3] Laboratoire pour l'Utilisation des Lasers Intenses, Ecole Polytechnique, Palaiseau, France
[4] Laboratoire d'Optique Appliquée, Ecole Polytechnique, Palaiseau, France

Abstract: We present our experimental investigations of the temporal evolution of plasmas produced by high power laser irradiation of various types of target materials. We obtained and analysed "high-quality" time-resolved streak-camera interferometric data on the evolution of the plasma electronic profile, which can directly be compared to analytical models and numerical simulations. For aluminium targets, the agreement with 1D simulations performed with the hydrocode MULTI is excellent, at least for large times ($t \geq 400$ ps).

Our experiment was realized at the LULI laboratory using 2 beams from the Nd:glass high power laser system converted to 2ω and together delivering a typical intensity of 10^{14} W/cm^2 on target. The temporal profile is approximately trapezoidal with rise and fall time of 150 ps and a flat top duration of 600 ps, the spot diameter produced by the lens coupled to the PZP had a full width at half maximum (FWHM) of 400 μm with central flat top region of 200 μm diameter. A probe beam (Nd:YAG laser converted to 2ω) was coupled to Mach-Zehnder interferometer and to a streak-camera with ps resolution. The diagnostic system allowed the evolution of the plasma density profile to be measured as a function of time.

Fig. 1a shows a typical streak camera interferogram. Our data are only 1D in space, in direction perpendicular to the target surface. Therefore, it is not possible to perform the Abel inversion, procedure, which usually associated to the analysis of interferometric data. Let's notice that the size of spot (FWHM of 400 μm) is comparable to the distance from the target, therefore plasma expansion can be considered 1D at least at the first order. [1,2]

The plasma-induced phase shift in a point x of the interferometric picture is

$$\Delta\phi(x,t) = \frac{2\pi}{\lambda}\left(L - \int_{-L/2}^{L/2} n(x,y,t)\,dy\right),$$

where λ the laser wavelength, n the plasma refraction index, L the effective plasma length.

Assuming the usual expression for n in 1D approximation

$$\Delta\phi(x,t) = \frac{2\pi L}{\lambda}\left(1 - \sqrt{1 - \frac{n_e(x,t)}{n_c}}\right),$$

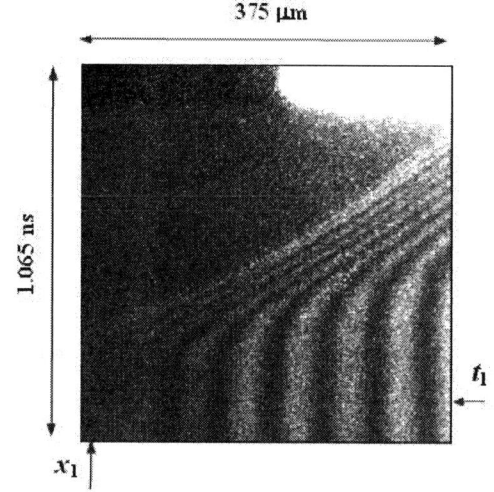

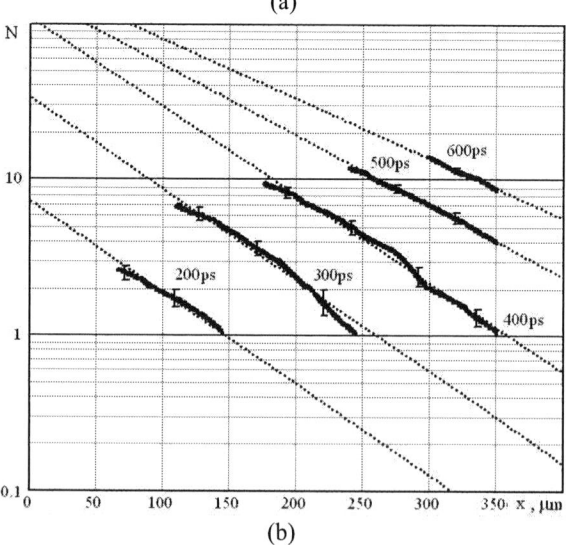

Fig. 1. (a) the typical interferogram recorded with the streak camera (Al target, shot #5); (b) experimental profiles of π-phase shifts N. [1]

978-1-4244-7043-3/10 $26.00 © 2010 IEEE

CAOL*2010 International Conference on Advanced Optoelectronics & Lasers, 10-14 September, 2010, Sevastopol, Ukraine

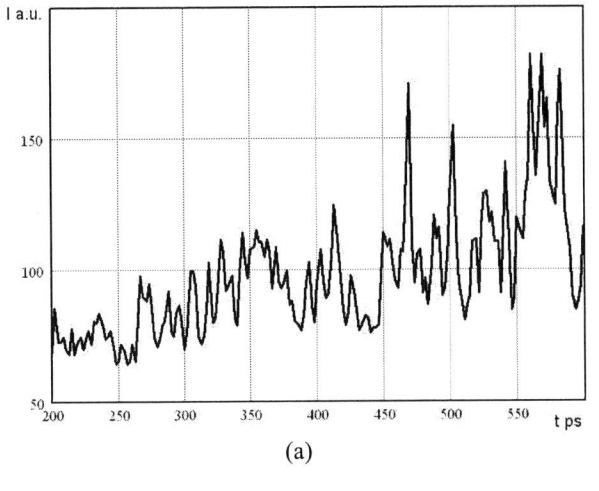

(a)

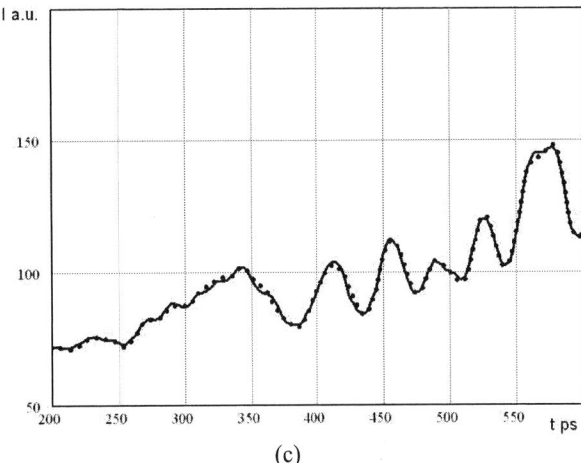

(c)

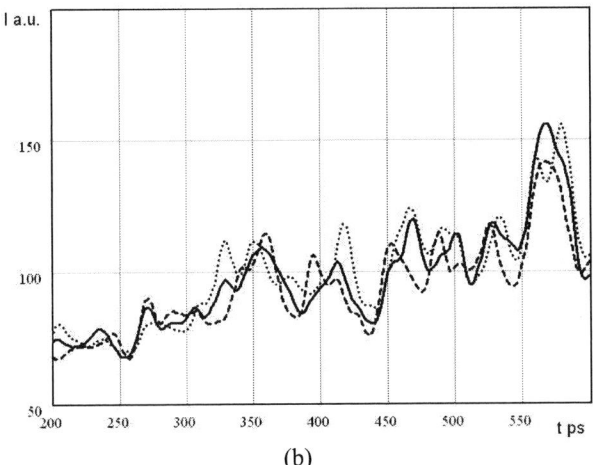

(b)

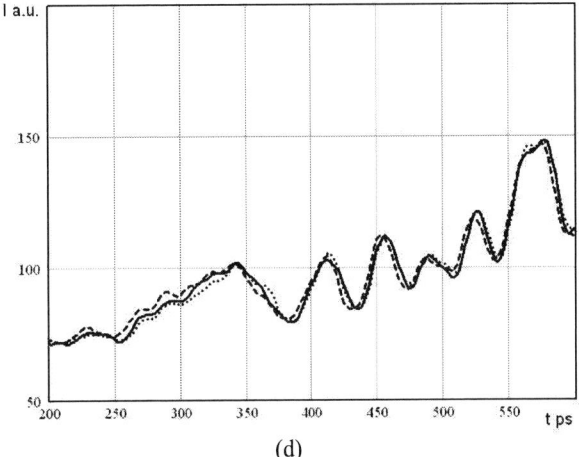

(d)

Fig. 2: (a) time-dependence of intensity of a streak-camera image (Al, shot #5, for the 3rd minimum from the right); (b) the Gauss-filter application for the line on fig. 2a (solid line) and for the neighbour intensity lines ± 1 pixel.

Fig. 2: (c) the same but with the prior time-shift averaging in the spatial interval ±10 pixels (7.3 μm) with characteristic velocity 3.5 10^7 cm/sec; (d) time-shift averaging in the spatial interval ±7.3 μm (10 pixels) with different characteristic velocities: 4.6 10^7 cm/sec (dotted line) and 3.5 10^7 cm/sec (solid line). Finally a time Gauss-filter has been applied after the averaging.

where n_c the electron density, $n_c = \dfrac{4\pi^2 c^2 \varepsilon_0 m_e}{\lambda^2 e^2}$ the critical density. For $n_e \ll n_c$ we obtain: $n_e = \dfrac{n_c \lambda}{L} N$, where $N = \Delta\phi/\pi$ the number of π-phase shifts. A typical dependence N vs. x for a set of times is presented in fig. 1b.

For the precise evaluation of N we have applied a technique of averaging with time shifts [3]. Indeed, how we can see from fig. 2, the time behaviour of intensity from a streak-camera image for a fixed x is affected by a strong noise, which can result both in neglecting or shifting of a real extremes (maxima and minima of interferograms) and in the appearance of false extremes.

False extremes can usually be removed by traditional means (e.g., by applying Gauss-filter), but the shift of real extremes remains a problem (fig. 2b). Averaging over a small range of values of x also doesn't give good results because it smoothes the picture.

In alternative, averaging with time shifts with a characteristic velocity on the order of the real velocity gives both a good resolution and a good contrast in the obtained densitometry. Fig. 2c shows the stability of this procedure for a small variance of spatial co-ordinate, and fig. 2d demonstrates the stability for a quite large variance of characteristic velocity.

Finally the precision of π-phase shifts in time-spatial square had an order of 1-2 pixels. The dependences $N_t(x)$ are well interpolated by straight (dotted) lines, i.e. by exponential

978-1-4244-7043-3/10 $26.00 © 2010 IEEE

189

profiles. By shifting the origin of the co-ordinates to the found from the strength lines intersection ($x_0=-130\mu m$, $t_0=100ps$), we can write:

$$\ln(N(x,t)) = b_0 - b_1 \cdot x \,.$$

Starting from our experiment results for $N(x,t)$ and using the least square method, we find the time dependences of b_0 and b_1. The exponential profile of $N_e(x)$ is in a good agreement with theoretical 1D self-similar models of isothermal plasma expansion according to which

$$n_e(x,t) = n_c \exp(- x/(c_s \cdot t))$$

where $c_s(cm/s) = 9.79 \cdot 10^5 \sqrt{\gamma Z^* T_e/\mu}$ the adiabatic sound velocity, γ the adiabatic constant ($\gamma=1$ if electrons are isothermal), μ the atomic number, Z^* the ionisation degree, T_e (eV) the electron temperature, which in classical models (where energy is absorbed at the critical surface in the plasma and transported inward by electronic conduction) is typically given by $T_e(eV)=10^{-6}(I_L\lambda^2)^{2/3}$ where I_L (W/cm^2) the laser beam intensity, and $\lambda(\mu m)$ the wavelength.

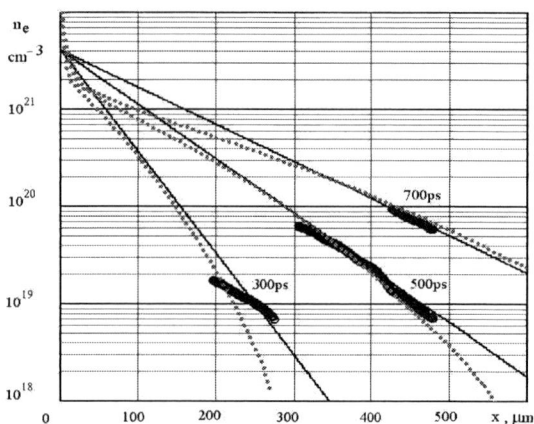

Fig. 3: The dependence of electron density (Al target, shot #5): experimental data (rings), calculation from the 1D self-similar model (solid straight lines), results of 1D MULTI simulation (grey dot lines). (Al, shot #5)

Finally we have found a brilliant accordance of our experimental data and with 1D self-similar model both with 1D (both 2D [2]) MULTI simulations [4] for aluminium targets (fig. 3). The situation is quite different for gold targets, where the radiation transport influence is significant. Nevertheless the perfect correspondence of plasma length determined from b_0 and the sound velocity – from b_1 and good gives us a good repeatability of experiments (results for other shots are reported) gives us a base about good conclusion for our methodic.

The work was partially supported by SPIE, ESF (SILMI, 2783), INTAS (06-1000014-5638), Cariplo Foundation - Landau Network - Centro Volta, and RFBR (09-01-96508)

REFERENCES

[1] Aliverdiev A., Batani D., Dezulian R., Vinci T., Benuzzi-Mounaix A., Koenig M., and Malka V. "Coronal hydrodynamics of laser-produced plasmas", Phys. Rev. E.2008, vol. 78, 046404.

[2] Aliverdiev A., Batani D., Dezulian R., Vinci T., Benuzzi-Mounaix A., Koenig M., and Malka V. "Hydrodynamics of laser-produced plasma corona by optical interferometry", Plasma Phys. Control Fusion. 2008, vol. 50, 105013

[3] Aliverdiev A.A. "Integral-Geometric Methods for the Time-Resolved Optical Diagnostic", *Optical Memory & Neural Networks (Information Optics)*, 2006, vol. 15, No. 2, pp. 97-104.

[4] Ramis R., and Meyer-ter-Vehn J., "MULTI 2D - A Computer Code for Two-dimensional Radiation Hydrodynamics," MPQ Report 174. Garching, Germany: Max-Planck-Institut für Quantenoptik, 1992

CAOL*2010 International Conference on Advanced Optoelectronics & Lasers, 10-14 September, 2010, Sevastopol, Ukraine

Measuring surface distribution of narrowband radiation wavelength by colorimetric method

A.V.Kraiski[1], T.V.Mironova[1], T.T.Sultanov[1], V.A.Postnikov[2]

[1] P.N. Lebedev Physical Institute RAS, Moscow, Russia

[2] Institute of Physico-chemical medicine, Moscow, Russia

Abstract: We propose a method to measure the narrowband radiation wavelength when having a digital picture of the emitting surface. The accuracy of the wavelength determination is better then 1 nm. The method was tested with pictures of the Hg spectrum and with the continuous spectrum of the filament lamp. The degree of homogeneity of the holographic sensor swelling was studied by the method in the stationary case as well as in the non-stationary one.

The problem emerged since one needs to determine the swelling degree distribution along the surface of holographic sensors [1-4]. Such a sensor is a Denisyuk hologram, i.e. a layered periodic structure. They are used to analyze components of solutions. In the present work the colorimetric method based on the usual digital camera [5, 6] was developed and used for these measurements.

When working with holographic sensors, there is a problem of measuring the homogeneity of response of sensor properties over its surface. An inhomogeneity can emerge due either to inhomogeneities of the object under consideration, or to those of the sensor properties. The hologram thickness is few tens of micrometers. Its reflection spectrum has the spectral width 5-20 nm. Because of it, one suffices to use the response from two color channels.

The main point of the method is as follows. In every pixel of the image the radiation is measured in not less than two channels with different spectral sensitivity. In the working range, the ratio of the spectral sensitivities of at least two channels is monotonous and from the ratio of the channel signals one can uniquely determine the average wavelength of the narrowband radiation. If the wavelength is out of the working range, one needs the third additional channel in order to determine the wavelength uniquely.

As the main working camera we used the 5-megapixel camera Sony F717. We have not managed to find its spectral characteristics (as well as those of many other cameras) in the existing literature, hence, we made our own studies (fig.1). Because of the constructive defect of the camera (similarly to many others), the working range turns out to be restricted by the two regions: 570-605 nm and 450-535 nm. The spectral response of the camera was studied and gauged with the incandescent lamp. We constructed the characteristic function that allows one via color value of the pixel to determine the wavelength of the radiation acting at this pixel. The approach is tested for the images of the incandescent lamp spectrum obtained in the spectrograph with diffraction grating, and for

the images of the holographic sensor at the stationary state (580 nm) and at the transition process of dehydration of the sensitive layer in the alcohol solution. The initial reflection of the sensor was in the red part of the spectrum out of the working range.

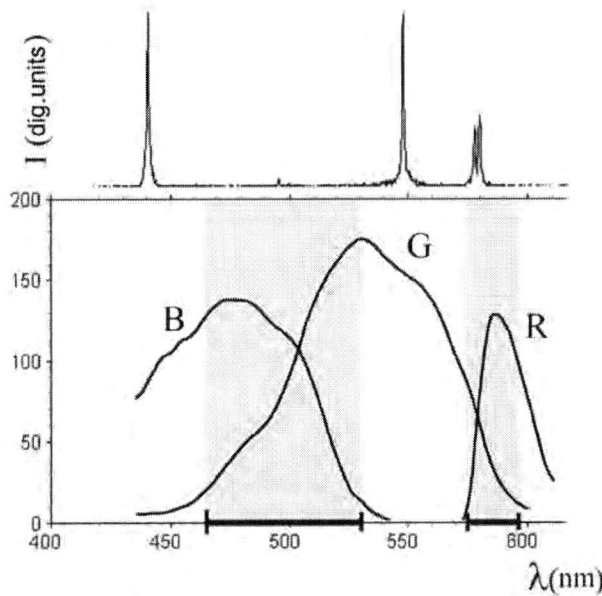

Fig.1. Above – the mercury lamp spectrum, below – the signals of color sensors of the camera in the photo of the spectrum of an incandescent lamp. Ranges of color sensitivity (working ranges) are marked [6] (is reprinted by permission of Turpion Ltd).

For the part of the lamp spectrum in the yellow working range of the camera the approach gives practically the linear dependence of the wavelength on the coordinate and the correct restoring of the wavelengths of the yellow mercury spectrum doublet presented at the image. The restored wavelength is stable in the perpendicular direction.

At some moment of the transition process the reflection wavelength decreases up to 580 nm at the periphery but changes smoothly in the direction to the central part of the hologram up to 598 – 599 nm. In addition to the low frequency changes one can see small scale fluctuations of the wavelength. The standard deviation (SD) from the average over the

978-1-4244-7043-3/10 $26.00 © 2010 IEEE   191

window increases with the window size from 0.5 nm for the 4-pixels window up to 1.8 nm for the 2500-pixels window (S on fig.3). The data for any fixed size window is the average received by the continuous scanning of the 500x500 pixels region by the window of this size. The rate of the SD growth changes weakly with increase of the window size. It seems to be connected with the large scale of the heterogeneities.

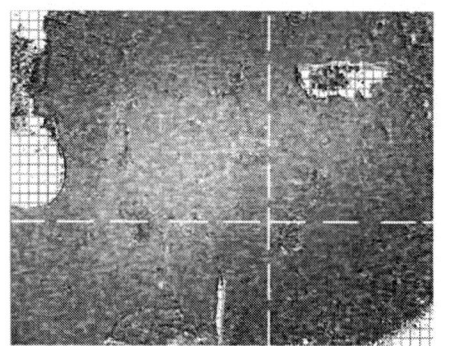

a)

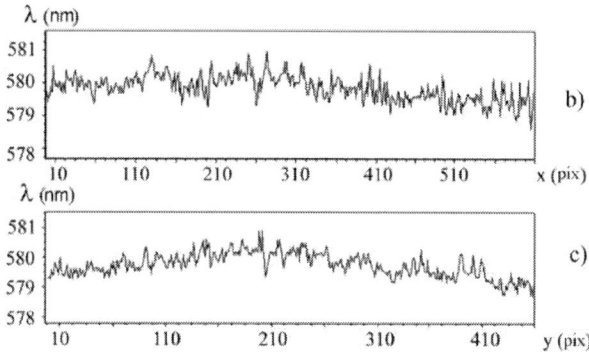

b)

c)

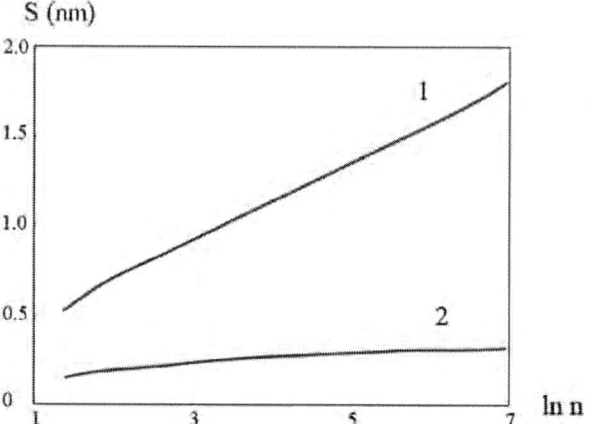

Fig.3. The dependence of calculated wavelength noise S on the size of the window. 1 is the curve for transition process, 2 is the curve for stationary process [6] (is reprinted by permission of Turpion Ltd).

The work is partly supported by the grant within the RAS Presidium program of fundamental research "Fundamental Science for Medicine".

Fig.2. Map of wavelength distribution over the image in the final stationary station. Grid areas mark the points out of operation range. Luminosity of grey points is proportional to the calculated wavelength (a).
The horizontal (b) and the vertical (c) sections of the wavelength map. Lines on the photo indicate sections. [6] (is reprinted by permission of Turpion Ltd).

After the transition to the stationary state, the whole hologram reflects in the spectral range of 1-1.5 nm width (fig2). There are no large scale heterogeneities in the hologram. The small scale fluctuations significantly decrease. SD changes from 0.16 for the smallest averaging window to 0.32 for the largest one. The rate of the SD growth decreases steeply with increase of the window size. This is an evidence of the high homogeneity of the hologram in the stationary state. Comparing these characteristics for the stationary and non-stationary states, one gets an evidence of heterogeneous transition processes.

Thus, the colorimetric method is suitable for study the processes inhomogeneous in space and the hologram quality.

REFERENCES

[1] Marshall A.J. et al.,2006, Analyte-responsive holograms for (bio)chemical analysis, J.Phys.Condens.Matter.,18, S619-626; USA patent №5989923 от 23.11.1999)

[2] V.A.Postnikov, A.V.Kraiskii, T.T.Sultanov, V.E.Tikhonov. Hydrogel holographic sensors sensitive to an acid media. XVIII Intenational scool-seminar "Spectroscopy of molecules and crystals" 20.09-28.09.2007, Beregove, Crimea, Ukraine, Abstracts p.261.

[3] Kraiski A.V., Postnikov V.A., Sultanov T.T., Tikhonov V.E. Al'man. Klin. Med., 17 (2), 108 -111 (2008).

[4] A.V. Kraiski, V.A. Postnikov, T.T. Sultanov, A.V. Khamidulin. Holographic sensors for diagnostics of solution components. Quantum Electronics 40 (2) p.178 - 182 (2010)

[5] A.V. Kraiski, T.V.Mironova, V.A. Postnikov, T.T. Sultanov, V.I.Sergienko, V.E. Tikhonov. Sposob izmernia dliny volny uzkopolosnogo svetovogo izluchenia colorimetricheskim sposobom. Zaiavka na poluchenie patenta RF na izobretenie ot 21.05.2008 reg №2008119917

[6] A.V. Kraiski, T.V.Mironova, T.T. Sultanov. Colorimetric method for measuring wavelength distribution of narrow-band light radiation with digital camera. Quantum Electronics 40 (7) p.652 - 658 (2010)

CAOL*2010 International Conference on Advanced Optoelectronics & Lasers, 10-14 September, 2010, Sevastopol, Ukraine

Basing of the television-picture generator application for distance monitoring of turbulent atmosphere

G.N. Dolya[1], A.N. Katunin[2], A.N. Bulay[2], E.S. Chudovskaja[1]
[1] V. Karazin Kharkov National University, Kharkov, Ukraine
[2] I. Kogedub Kharkov University of Aircraft, Kharkov, Ukraine

Abstract: Basing of television-picture generator application for remote monitoring of turbulent atmosphere, which ground on scatterogram local maximum angular position analysis realized. Showed that environmental turbulence number increase lead to corresponding measurand fluctuations degree changes, which determined possibility of television-picture generators application in inflammation detection systems.

Introduction

Environmental control active methods permanent improvement determined by wide range of problems, which solved with using of laser optoelectronic monitoring system [1-3]. Main problems which are solved successfully: laser distance atmosphere gas analysis; laser distance monitoring of atmosphere aerosol pollution; laser distance monitoring of marine surface oil pollutions and distance monitoring of shelf zone turbidity. Furthermore, laser distance monitoring methods are used for turbulent atmosphere operation factors measurement and for ignitions detections [2].

Technical realization of laser optoelectronic monitoring system for ignitions detections permit to put transmitter and receiver at one part of laser beam propagation path through investigated material volume, and to put light-reflector (the mirror at simple case) at other part [4]. Using of light-reflecting coating (LRC), which represented a set of tiny balloons or tiny prisms, as a light-reflector allow increasing greatly range of such systems. Moreover, radiation scattered at LRS, have characteristic structure in the form of a set chaotic based spots (speckles), in the receiver plane [5].

Inflammations detection laser systems are based on reflected laser emission fluctuation intensity measurement when it propagates through turbulent layer, produced by source of inflammation. Such measurement, generally, accompanied by errors, caused by external residual radiation, internal photodetector noises, etc. At the same time, environmental turbulence causes fluctuations of refractive index at radio path, and as a result, leads to distortion of radiation scatterogram form when it reflected by LRC.

On basis of this effect it's possible to build up turbulent distortion sensors, grounds on back-scattered radiation angular fluctuations registrations by television receiver, which allow analyzing bivariate distribution of laser emission intensity.

Aim of work to base possibility of television-picture generator application for remote monitoring of turbulent atmosphere.

Possibility of the television-picture generator application for distance monitoring of turbulent atmosphere

Using of LRC, which represented aggregate of tiny balloons or tiny prisms, allow increasing range of laser systems, as well as sensed signal parameters measurement accuracy. The greatest efficiency provide for using of LRC on basis of tiny prisms, which size is about of 100 mcm. Such object surface fragment is showed at fig. 1, and diffraction pattern of laser emission scattered by LRC – at fig. 2.

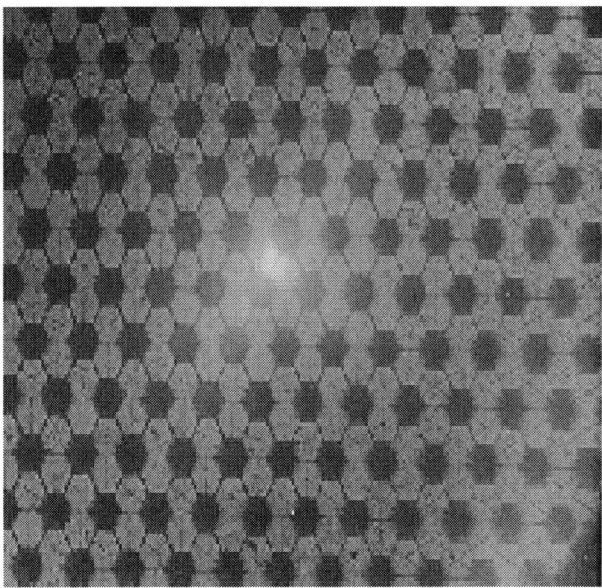

Fig.1. - Surface fragment of LRC on basis of tiny prisms.

LRC apparatus function described by expression $I = F(U) f(NW)$ [6], where function $F(U)$ correspond to diffraction pattern at elementary light-reflector, and function $f(NW)$ determined by LRC spatial grates parameters. Presence of probe beam phase distortions lead to amplitude-phase distribution distortions of wave, which diffract at LRC spatial grates, and this lead to changes of scattered by LRC radiations diagram. Also diagram changed and in case of back propagation through turbulent atmosphere layer. Distortions

978-1-4244-7043-3/10 $26.00 © 2010 IEEE

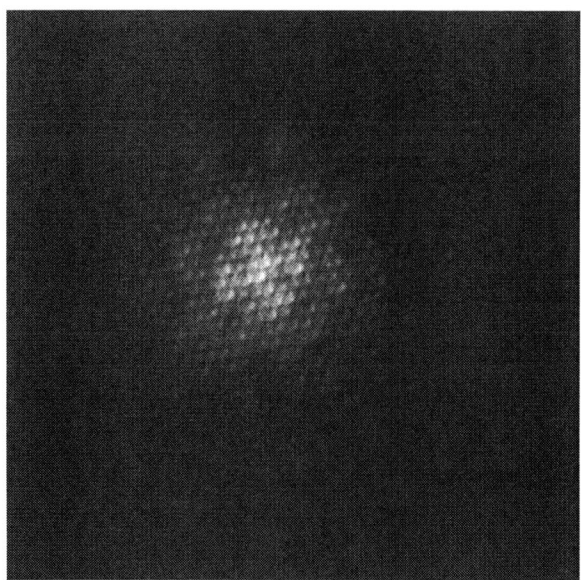

Fig.2. - Diffraction pattern of laser emission scattered by LRC.

are expressed in scatterogram local maximums angular fluctuations, and this may be registered by photosensitive matrices area of charged-coupled device (CCD). At simple case turbulent atmosphere maybe presented in the form of thin phase screen, describable by significances of two parameters – spatial correlation distance of heterogeneities (phase distortions)

$$\rho = \left(C_n^2 k^2 L\right)^{-3/5} \qquad (1)$$

and dispersion of light wave phase fluctuation when it propagate through fragment of the phase screen

$$\sigma_\varphi^2 = C_n^2 k^2 L L_0^{5/3}, \qquad (2)$$

where L – path length, k – wavenumber, L_0 – external scale of turbulence, C_n^2 – fluctuation structural constant of the refractive index.

Correspondingly, these parameters are connected by the ratio

$$\rho = \frac{L_0}{\sigma^{6/5}}. \qquad (3)$$

The processes of focused laser emission propagation through phase transparency, reflection and diffraction on LRC surface analysis realized by computer simulation. Here taken into account LRC surface texture (Fig. 1), and regularities (1), (2) and (3). At first, in the absence of the turbulence at laser beam radio path, amplitude-phase distribution of laser emission diffracted intensity was calculated and one of the scatterogram local maximums were measured. Then, the next value of phase incursion random quantity, distributed by normal law, and again calculated shift quantity of local maximum etc. Calculation results, for different values of parameter L_0 are presented at fig. 3, where on abscissa axis laid values of phase incursion mean-square distance (MSD), and on ordinate axis – scatterogram maximum MSD shift values.

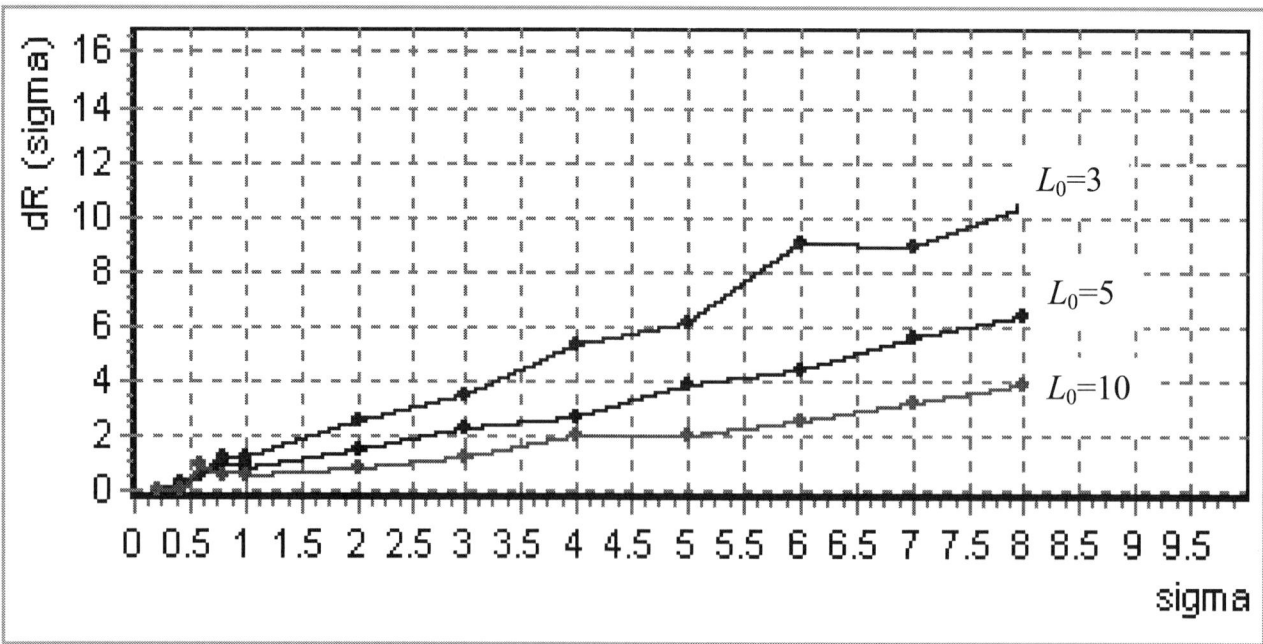

Fig. 3. – Dependence of scatterogram maximum MSD shift value from phase incursion MSD for different values of external scale of turbulence L_0.

Presented results show, that increase level of turbulent distortion cause monotone increase of local maximum position fluctuations dispersion and that is the base for detection method of air turbulence change of state at beam radio path, for example, for the purpose of early inflammations detection.

Conclusions

In the work, basing of possibility of television-picture generator application for remote monitoring of turbulent atmosphere, which ground on scatterogram local maximum angular position analysis realized.

Computer simulation results argue, that phase fluctuations of laser emission at radio path lead to appearance of local maximums fluctuations in back-scattered radiation.

Showed that environmental turbulence number increase lead to corresponding measurand fluctuations degree changes, which determined possibility of television-picture generators application for distance monitoring of turbulent atmosphere for the purpose of early inflammations detection

REFERENCES

[1] V.E. Zuev. Spreading of the laser radiation in atmosphere, 1981.

[2] Dolya G.N., Katunin A.N., Mazanov V.M. Comparative analysis of possibilities of the interferometry methods for observation of the phase heterogeneities in the air environment, Eastern European magazine of the advanced technologies, 2006, Issue 2(20), pp. 61–64.

[3] N. D. Ustinov, I. N. Matveev, V. V. Protopopov. Processing Methods of the Optical Fields in Laser Locations, 1983.

[4] F.I. Sharovar Methods of early detection of the combustions. 1988.

[5] Jones R., Whikes K. Holographic and speckle interferometry: Translation from Engish, 1986.

[6] R. Ditchbern, Physical Optic, 1965.

Experimental Investigations of Influence of Additionally Induced Polarization of the Ground State of Atoms on Multiphoton Transitions

I. I. Bondar', V. V. Suran

Department of Physics, Uzhgorod National University, Voloshin str., 54, Uzhgorod, 88000 Ukraine

Abstract - **Multiphoton transitions in the *Ba* atom are experimentally studied in the presence of strong nonresonant radiation that additionally polarizes atoms in the ground state. It is found that the additionally induced polarization of Ba atoms leads to an increase in the probability of multiphoton transitions from this state.**

I. INTRODUCTION

Recently [1], we have found that an additional strong perturbation of the spectrum of bound states of *Ba* atoms during their multiphoton excitation leads to an increase in the probability of this excitation. This fact indicates that the probability of multiphoton excitation of atomic states depends on the magnitude of an additional polarization of atoms in these states. The polarization of an atom in the state n under the action of optical radiation is defined as $P_n(\omega) = \alpha_n(\omega)\varepsilon$, where α_n is the dynamic polarizability of the atom in the state n, ω is the frequency of the radiation, and ε is the strength of the electric field created by the radiation. It is interesting to clarify how the additional polarization of atoms in initial states affects the probability of multiphoton transitions. This was the aim of this work.

We experimentally studied the effect of an additionally induced polarization of the ground state of the *Ba* atom ($6s^2\ ^1S_0$) on four-photon transitions to the $6p^2\ ^1S_0$ and $6p^2\ ^3P_0$ states. These four-photon transitions were excited using radiation from a color-center laser whose frequency (ω_1) can be varied. The radiation frequencies that correspond to the four-photon transitions between the unperturbed ground state $6s^2\ ^1S_0$ and the excited $6p^2\ ^1S_0$ and $6p^2\ ^3P_0$ states are, respectively, $\omega_0 = 8593$ cm^{-1} and $\omega\ _0 = 8624$ cm^{-1}. Note that these four-photon transitions were experimentally studied in a number of our previous works (see, e.g., [2–3]).

As a source of strong nonresonant radiation that creates an additional polarization of *Ba* atoms in the ground state, we used a dye laser. The characteristics of the radiation of the dye laser were selected such that the probability of ionization of the *Ba* atom caused by this radiation would be small, while the polarization of the atom in the ground state caused by this laser would be considerably greater than the polarization caused by the radiation of the color-center laser. In connection with this, the radiation frequencies of the dye laser (ω_2) were chosen such that they would correspond to direct multiphoton ionization of the *Ba* atom from the ground state; this ensured a small

probability of the ionization of the atom. To obtain a strong polarization of *Ba* atoms in the ground state, the radiation frequencies of the dye laser should be such that the absolute values of the dynamic polarizabilities of this state (α_0) corresponding to these frequencies would be large. To this end, the frequencies of the dye laser were chosen in such a way that, on the one hand, they would be close to the frequency that corresponds to a one-photon transition from the ground $6s^2\ ^1S_0$ state to the first resonant state $6s6p\ ^1P_1^\circ$ ($\omega_{hm} = 18060$ cm^{-1}), and, on the other hand, they would sufficiently differ from this frequency in order to avoid the effect of mixing of the ground and resonant states. Based on this, the radiation frequency ω_2 of the dye laser was either 17375 or 17735 cm^{-1}.

We measured the yield of Ba^+ ions formed as a result of the action of the laser radiation on the beam of *Ba* atoms. Note that, to ionize the *Ba* atom from the ground state by the radiation of the color-center laser, absorption of five quanta is necessary, whereas, for the ionization by the dye-laser radiation, three quanta should be absorbed. At the same time, to ionize the *Ba* atom in the $6p^2\ ^1S_0$ or $6p^2\ ^3P_0$ state, the atom should absorb one radiation quantum of either the color-center or the dye laser.

In our experiments, we studied the joint action of the color-center and dye-laser radiations on *Ba* atoms. The radiation frequency of the color-center laser was varied in the interval $\omega_1 = 8380$–8650 cm^{-1}, into which the frequencies ω_0 and ω_0 fall. The radiation frequency of the dye laser was fixed. Both laser radiation beams were superimposed in space and were focused into the beam of *Ba* atoms. The two radiations were linearly polarized. The electric vectors of these radiations were parallel. The results obtained in these experiments were compared with the results obtained under the conditions where the radiations of the color-center and dye lasers separately acted upon *Ba* atoms. The electric field strength created in the focal region by the radiation of the color-center laser was $\varepsilon_1 = 2 \times 10^6$ V/cm and that created by the dye laser was $\varepsilon_2 = 5 \times 10^4$ V/cm.

II. EXPERIMENTAL RESULTS

The results obtained in our experiments are shown in Fig. 1. As is seen from this figure, when the color-center laser radiation acts alone on *Ba* atoms, the yield of Ba^+ ions exhibits a resonance maximum, whose shape is asymmetric and which is shifted with respect to the frequency $\omega_0 = 8593$ cm^{-1}. Based on our results from [3–5], this maximum can be attributed to

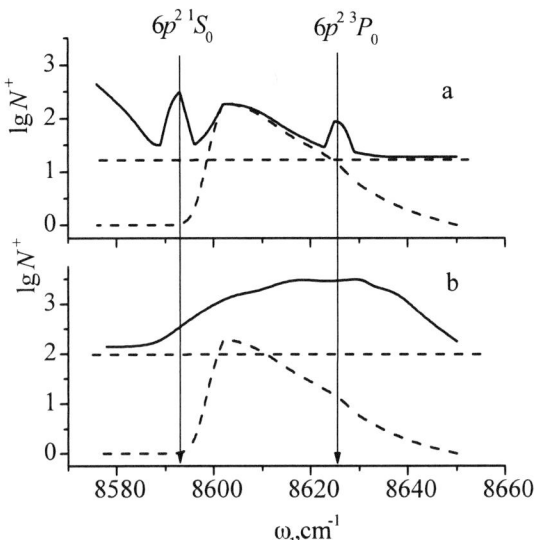

Fig. 1. Dependence of the yield of Ba^+ ions (N^+) on the radiation frequency ω_1 of a color-center laser measured under conditions when this radiation acts alone on Ba atoms (dashed curves) and simultaneously with the dye-laser radiation (solid curves). The dashed horizontal lines indicate the average yield of Ba^+ ions obtained when only the dye-laser radiation acts on Ba atoms. The vertical arrows indicate the radiation frequencies of the color-center laser corresponding to the four-photon excitation of the unperturbed $6p^2\,^1S_0$ and $6p^2\,^3P_0$ states ($\omega_0 = 8593$ cm^{-1} and $\omega_0 = 8624$ cm^{-1}). The data were obtained for two values of the radiation frequency of the dye laser: $\omega_2 =$ (a) 17375 and (b) 17735 cm^{-1}.

a four-photon excitation and ionization of the $6p^2\,^1S_0$ state, which is strongly perturbed by the dynamic Stark effect. In this case, the strong perturbation of the $6p^2\,^1S_0$ state is determined by the fact that the radiation frequencies of the color-center laser are close to frequencies that correspond to one-photon transitions from this state to states with lower energies. As a result of this, the absolute value of the dynamic polarizability α_n of the $6p^2\,^1S_0$ state is large.

Consider now the results obtained upon simultaneous action of the radiations of the dye and color-center lasers. Fig. 1 shows that, in this case, the yield of Ba^+ ions also exhibits a resonance structure. The yield of Ba^+ ions corresponding to this structure is generally greater than the yield of these ions caused by the separate action of either of the two lasers, as well as than the sum of these separate yields.

An analysis shows that this resonance structure cannot result from a multiphoton excitation of states of the Ba atom by resonant processes that involve quanta of the two lasers used, i.e., Raman excitation processes and excitation processes where quanta of both radiations are summed.

As follows from Fig. 1, the character of the above resonance structure depends on the radiation frequency of the dye laser. Thus, if this frequency is 17375 cm^{-1} (Fig. 1a), the yield of Ba^+ ions shows clearly pronounced maxima near the frequencies $\omega_0 = 8593$ cm^{-1} and $\omega_0 = 8624$ cm^{-1}, which correspond to the four-photon excitation of the unperturbed $6p^2\,^1S_0$ and $6p^2\,^3P_0$ states.

When the radiation of the color-center laser acts alone on Ba atoms, there are no such maxima in the yield of Ba^+ ions. The absence of the maximum in the case where the unperturbed $6p^2$ $^1S_0$ state is excited only by the radiation of the color-center laser results from the fact that an efficient four-photon excitation of this state is possible only if the intensity of the color-center laser radiation is high enough as to strongly perturb this

The occurrence of the maximum near the frequency $\omega_0 = 8593$ cm^{-1} in the case where the radiations of the color-center and dye lasers simultaneously act on Ba atoms indicates that the effective four-photon excitation of the $6p^2\,^1S_0$ state in the presence of the dye-laser radiation occurs at considerably lower intensities of the color-center laser than in the case where this laser acts alone on Ba atoms. That is, the presence of the radiation of the dye laser during the four-photon excitation of the $6p^2\,^1S_0$ state by the radiation from the color-center laser significantly increases the probability of this process.

The absence of the maximum corresponding to the excitation of the $6p^2\,^3P_0$ unperturbed state under the action of only the radiation of the color-center laser is explained by the fact that the probability of the corresponding four-photon transition is low. When the radiations of the two lasers act simultaneously, the maximum caused by the four-photon excitation of the $6p^2$ $^3P_0$ state only slightly differs in amplitude from the maximum caused by the four-photon excitation of the $6p^2\,^1S_0$ state. That is, the presence of the radiation of the dye laser during the four-photon excitation of the $6p^2\,^3P_0$ state also increases the probability of this process.

The radiation frequency of the dye laser (17375 cm^{-1}) significantly differs from frequencies of transitions from the $6p^2\,^1S_0$ and $6p^2\,^3P_0$ states to other states and, at the same time, it is close to the frequency of the one-photon transition from the ground state to the $6s6p\,^1P_1^o$ resonance state. It follows from this that the action of the radiation of the dye laser in the course of the four- photon excitation of the $6p^2\,^1S_0$ and $6p^2\,^3P_0$ states leads to an increase in the polarization of the Ba atom in the ground state and does not lead to an increase in its polarization in these excited states. In this case, the polarization of the Ba atom in the $6p^2\,^1S_0$ and $6p^2\,^3P_0$ states is determined by the radiation of the color-center laser, since its frequency is close to frequencies that correspond to one-photon transitions from this state to states with lower energies.

Now, let us consider the results that were obtained when the frequency ω_2 of the dye-laser radiation was chosen to be 17735 cm^{-1} (Fig. 1b). In this case, the resonance structure that appears under the simultaneous action of the radiations of the two lasers substantially differs from the above structure that also occurs as a result of the simultaneous action of the two laser radiations, but with the frequency ω_2 of the dye laser being 17375 cm^{-1}. Now, the resonance structure is observed as a broad resonance maximum, which differs in frequency from the maximum that appears under the action of the radiation of the color-center laser alone. Furthermore, the yield of Ba^+ ions that results from simultaneous irradiation by the color-center and dye lasers considerably exceeds that obtained under irradiation by the color-center laser alone. This resonance maximum is attributed to the four-photon excitation of the $6p^2$ $^1S_0$ state from the ground state strongly perturbed by the dye-laser radiation. This is evidenced by the following facts.

The resonance maximum in the vicinity of the frequency ω_0 = 8593 cm^{-1}, which corresponds to the four-photon transition between the unperturbed ground state and the $6p^2$ $^1S_0$ state, is absent. The detuning $\delta = \omega_{nm} - \omega_2 = 325$ cm^{-1} is in this case smaller than when the radiation frequency of the dye laser was 17375 cm^{-1} ($\delta = 685$ cm^{-1}). Therefore, the dynamic polarizability α_0 of the ground state of the Ba atom is in this case substantially greater than in the case considered above. The corresponding Stark shift ($\Delta E = -\alpha_0\varepsilon^2/4$) of the ground state should also be considerably greater than in the case where the radiation frequency of the dye laser was 17375 cm^{-1}, and can significantly exceed the effective radiation linewidth of the color-center laser. As a result of this, the resonant transition between the ground state and the $6p^2$ $^1S_0$ resonant state should manifest itself in the yield of Ba^+ ions not as a symmetric maximum in the vicinity of the frequency ω_0 = 8593 cm^{-1} but rather as a broad asymmetric maximum, which is shifted toward higher frequencies with respect to ω_0, since the shift of the ground state under the action of the radiation of the dye laser is positive. It is this pattern that is observed in our investigation.

To confirm the explanation given above, we measured the Stark shift of the ground state of the Ba atom caused by the radiation of the dye laser at frequencies that are close to ω_2 = 17375 and 17735 cm^{-1}. The magnitude of the shift was determined by the width of resonance maxima in the dependence of the yield of Ba^+ ions on the radiation frequency of the dye laser. The radiation at a frequency close to the frequencies ω_2 indicated above can cause two-photon excitation of bound states of Ba atoms. In this case, radiation frequencies strongly differ from frequencies corresponding to one-photon transitions from these states to other states with lower energies. Therefore, their perturbation by the radiation of the dye laser is insignificant. As far as the ground state is concerned, since the radiation frequencies of the dye laser are close to the frequency of the one-photon transition from this state to the first resonant state (ω_{nm} = 18060 cm^{-1}), its perturbation should be considerable. That is, the shape of corresponding resonance maxima in the yield of Ba^+ ions will be mainly determined by the perturbation of the ground state of the Ba atom.

For these experiments, we chose the two-photon excitation of the $6p^2$ $^3P_0$ and $6s7d$ $^3D_2$ triplet states. The frequencies corresponding to two-photon transitions to these unperturbed states (17247 and 17881 cm^{-1}) are close to the frequencies ω_2 indicated above. The results of these measurements are presented in Fig. 2.

As is seen from Fig. 2, the width of the resonance maximum caused by the two-photon excitation of the $6p^2$ $^3P_0$ state hardly exceeds the radiation linewidth of the dye laser ($\Delta\omega \approx 3$ cm^{-1}), whereas the width of the maximum caused by the excitation of the $6s7d$ $^3D_2$ state considerably exceeds the linewidth of this radiation. These facts indicate that, indeed, the ground state of

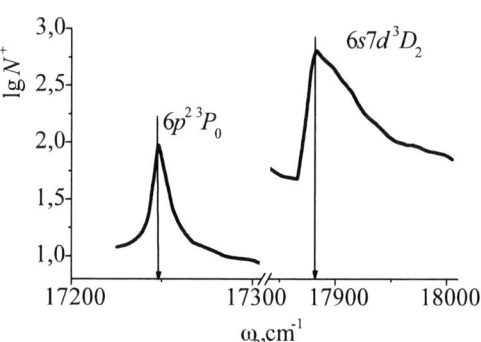

Fig. 2. Dependence of the yield of Ba^+ ions (N^+) on the radiation frequency ω_2 of a dye laser. The vertical arrows indicate the radiation frequencies of the dye laser corresponding to the two-photon excitation of the unperturbed $6p^2$ $^3P_0$ and $6s7d$ $^3D_2$ states (17247 and 17881 cm^{-1}).

the Ba atom is much more perturbed by the radiation of the dye laser in the vicinity of the frequency ω_2 = 17735 cm^{-1} than in the vicinity of the frequency ω_2 = 17375 cm^{-1}. This confirms the above identification of the maximum in the yield of Ba^+ ions shown in Fig. 1, which was obtained under the simultaneous action of the radiations of the color-center and dye lasers, as caused by the four-photon transition from the perturbed $6s^2$ $^1S_0$ ground state to the unperturbed $6p^2$ $^1S_0$ state.

The absence of the clearly pronounced maximum corresponding to the four-photon excitation of the $6p^2$ $^3P_0$ state is explained by the fact that the probability of this process is in this case smaller than the probability of the four-photon excitation of the $6p^2$ $^1S_0$ state.

It is obvious that, when the radiation frequency ω_2 of the dye laser is 17735 cm^{-1}, the additional polarization of the Ba atom in the ground state is considerably greater than when this frequency is 17375 cm^{-1}. Fig. 1 shows that the yield of Ba^+ ions formed as a result of the ionization of Ba atoms via the four-photon resonance with the $6p^2$ $^1S_0$ state in the presence of the dye-laser radiation with the frequency ω_2 = 17735 cm^{-1} is also much greater than when this frequency is 17375 cm^{-1}. That is, the additional polarization of Ba atoms in the initial state in the course of the four-photon excitation of the $6p^2$ $^1S_0$ state leads to a higher probability of this process.

Therefore, our results show that the creation of an additional polarization of atoms in initial states during multiphoton transitions leads to an increase in the probability of these transitions.

REFERENCES

[1] I. I. Bondar', V. V. Suran. Linear Dichroism in the Two-Photon Transition between the Perturbed $6s^2$ $^1S_0$ and $5d7s$ $^3D_2$ States of a Ba Atom in the Presence of Additional Strong Nonresonant Radiation. *Opt. Spektrosc.* vol. 102, no.1, pp. 43-48, 2007.

[2] I. I. Bondar', V. V. Suran. Influence of the Stark effect on multiphoton ionization of atoms when the dynamic polarizability depends strongly on the laser frequency. *JETP*, vol. 86, no.2, pp.276-283, 1998.

[3] I. I. Bondar', V. V. Suran, Study of the Stark Effect Induced in the Ba Atom by the Dye Laser Radiation. *Opt. Spektrosc.* vol. 85, no.3, pp. 327-232, 1998.

Influence of semiconductor photoreceiver spectral responsivity at different temperature on optical measurements

A. V. Polyakov, M. A. Ksenofontov

Institute of Applied Physics Problems of Belarussian State University, Minsk, Belarus

Abstract: The analytical expressions allowing to calculate spectral responsivity of semiconductor receivers of optical radiation with an error no more 5 % in all spectral range are developed. Criteria of a choice of radiation wavelength and accuracy of radiation power stabilization of the injection laser in view of temperature dependence of semiconductor photodetector spectral responsivity have been received.

Optical measurement plays an important role in a variety of scientists, medical, and industrial application. For last decades the direction using optical radiation and unique properties of optical environments for registration of various physical influences was generated. The successes in the field of semiconductor radiation sources, photoreceivers, optical fiber with small attenuation have resulted in occurrence and rough development optoelectronic measuring systems (OMS) on the basis of fiber-optic sensor (FOS). Now process of creation OMS begins with development of mathematical model of FOS functioning with the subsequent computer modeling of FOS work. It allows to estimate, how selected constructive decision corresponds to the given specifications. At this stage it is very important to use adequate and exact mathematical models.

The transformation function of FOS represents complex multistage dependence of registered value on an output of FOS X from external influence F_{in}:

$$X = P(I_{IL}) f_{FD} \left\{ y_j \left[x_i \left(F_m (F_{in}) \right) \right] \right\} S(\lambda) k , \quad (1)$$

where $P(I_{IL})$ is optical radiation power created by a radiation source at course on it a current I_{IL}; f_{FD} is photodiod converter function; $S(\lambda)$ is integrated spectral responsivity of the photoreceiver; k is losses index of optical radiation power at $F_{in} = 0$.

One of major parameters on which quality of the sensor is estimated is the size of maximum allowed temperature error:

$$\delta_\theta = \frac{\left| X_\theta - X_{\theta_0} \right|}{X_{\theta_0}} \cdot 100\% . \quad (2)$$

As the fiber-optic channel is enough thermostable, the basic contribution to a temperature error will bring transmitter-receiver block. In this case for an estimation of a temperature error of FOS the finding of the analytical expressions describing temperature dependence of parameters of injection lasers (IL) and semiconductor photodetectors is necessary.

One of basic FOS regular errors is dependence of semiconductor photodiode (FD) spectral responsivity $S(\lambda,\theta)$ on wavelength radiation λ and temperature θ. This error is most essential at the $\lambda = 1,55$ μm, appropriate to the optical fiber (OF) minimal losses and a maximum of spectral sensitivity for Ge- and InGaAs-FD. For definition $S(\lambda)$ in most cases use empirical curves at any fixed parameters received with the help of expensive certification equipment or a reference material. The basic lack of such approach is its complexity caused besides attraction of certain means, calculation n-order approximation polynomial (depending on desirable accuracy), and also impossibility to look after with its help continuous dynamics of function $S(\lambda)$ change depending on temperature.

In the given work the approach based on reception of analytical expressions for functions $S(\lambda)$ is offered proceeding from physical laws underlying functioning of the semiconductor photodiode. As a sample the PIN photodiode of a series J16 and C30617, produced by "EG&G Optoelectronics" was chosen. These detectors are high-quality germanium and InGaAs photodiodes designed for the 0,8 to 1,8 μm wavelength ranges. The FD responsivity is determined by a ratio:

$$S(\lambda) = \frac{e\lambda}{hc} T_{tr} \exp(-\alpha x) \left(1 - \exp\left[-\alpha L_{fd} \right] \right), \quad (3)$$

where e is electron charge; h is Planck constant; c is light speed, T_{tr} is transmission coefficient of radiation, α is absorption coefficient, L_{fd} is thickness of the depleted layer of FD material, x is depth location from a surface of the depleted layer.

As for Ge under action of falling near IR range radiation most probable are the indirect transitions with emitting of Raman phonon, the expression for interzoned absorption coefficient looks like [1]:

$$\alpha(\theta) = N^* \sigma_b \left(\frac{\lambda_c(\theta)}{\lambda} \right)^2 \left(1 - \frac{\lambda}{\lambda_c(\theta)} \right)^2 \Bigg/ \left[1 - \exp\left(-\frac{E_f}{k\theta} \right) \right], \quad (4)$$

where $N^* = N_A N_D / (N_A + N_D)$ is reduced impurity concentration, N_A is acceptor concentration, N_D is donor concentration; $\lambda_c = 1,24 / [\Delta E(\theta) + E_f]$ is long-wave infrared border of radiation absorption, ΔE is width of the forbidden zone, $\Delta E = 0,66$ eV at $\theta = 300$ K; E_f–energy of Raman phonon, $E_f = 0,037$ eV; σ_b is boundary value of photoionization cross-sectional $\sigma_b \approx 1,25 \cdot 10^{-18} / \lambda_c^2$ sm^2.

The width of layer L_{fd} in FD with sharp asymmetrical p-n – junction does not depend on radiation wavelength and is determined by a doping level the semiconductor, its relative dielectric permittivity, concentration of own carriers, voltage of return displacement and temperature. The carried out researches in an interval of temperatures from -30 °C up to $+70$ °C have shown, that with growth of temperature L_{fd} is increased practically linearly and this increase is very insignificant. Relative change of width of the impoverished layer at increase of temperature at one degree makes $2,7 \cdot 10^{-5}$ 1/K. Therefore in all practically important temperature range L_{fd} can be counted not dependent on temperature.

In the Fig. 1 is submitted calculate with the help of the formulas (3) – (4) functions $S(\lambda)$ for Ge-FD (solid curve) and InGaAs-FD (dotted line). At accounts the following meanings were used for Ge-FD: $L_{fd} = 1,2 \cdot 10^{-4}$ sm, $x = 10^{-7}$ sm, $N_A = 2 \cdot 10^{16}$ sm^{-3}, $N_D = 10^{18}$ sm^{-3}, $\theta = 293$ K, $T_{tr} = 0,6...0,65$ and for InGaAs-FD: $L_{fd} = 3,2 \cdot 10^{-4}$ sm, $x = 3 \cdot 10^{-7}$ sm, $T_{tr} = 0,8$. The border of FD responsivity in the range of the large wavelength is determined in basic of the width of material forbidden zone, and the fall of responsivity in the range of short waves is connected to absorption of light near to a surface and increase of losses at the expense of superficial recombination of the photoexited carriers. In the Fig. 1 are marked off experimental magnitudes received at research germanium PIN photodiodes of a series J16 (daggers) and InGaAs PIN photodiodes of a series C30617 (triangles). As it is visible from a Fig. 1, experimental and calculated values differ no more than on 5 % in all a spectral range. Than the received analytical approximation is much accurately then known bell-shaped or triangular approximation [2]. A Ge-photodiode generates a current across the p-n or p-i-n junction when photons of sufficient energy are absorbed within the active region. As the forbidden zone width of the semiconductor depends on temperature $\Delta E = \Delta E_0 (\theta = 0\,\mathrm{K}) - \delta_E \theta$ (for Ge $\Delta E_0 = 0,75$ eV, $\delta_E = 4 \cdot 10^{-4}$ eV/K), therefore according to (3) – (4) responsivity is a function of wavelength and detector temperature.

Temperature changes have little effect on the detector responsivity at wavelengths below the peak, but can be important at the longer wavelengths (Fig. 2). For example, at $1,2$ μm the change in response of a room temperature detector is less than $0,1$ % per °C, while at $1,7$ μm the change is approximately $1,5$ % per °C (Fig. 2). Uniformity of response within the active region of a room-temperature detector is typically better than $\pm 0,5$ % at $1,55$ μm.

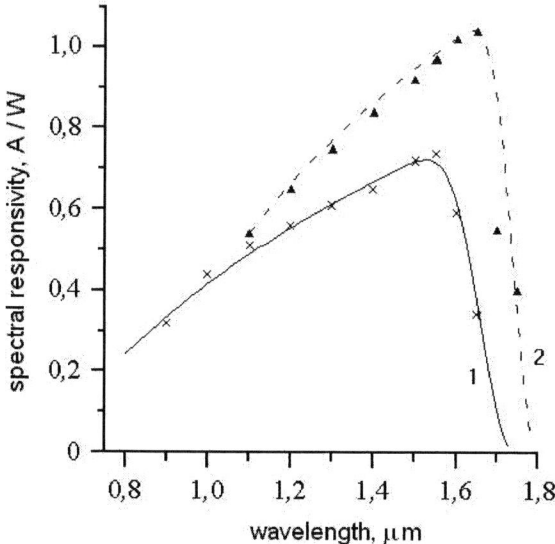

1 – Ge-FD, 2 – InGaAs-FD

Fig. 1. Spectral responsivity of PIN photodiodes

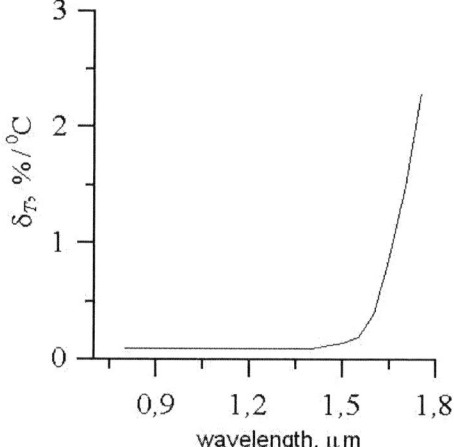

Fig. 2. Temperature coefficient of responsivity at 25ºC

As follows from (1), at value definition of a temperature error it is necessary to take into account that fact, that change of temperature influences not only spectral responsivity FD, but also radiation power and wavelength of IL generation. Temperature dependence of radiation wavelength is expressed as follows:

$$\lambda(\theta) = \lambda_0(\theta_0) + \delta_\lambda(\theta - \theta_0). \qquad (5)$$

Power of IL radiation at temperature change can be calculated under the formula:

$$P_{IL} = \eta_{IL} \frac{hc}{e\lambda} \left(I - I_{th}(\theta_0) \exp\left[(\theta - \theta_0)/\theta_{IL} \right] \right), \qquad (6)$$

where $I_{th}(\theta_0)$ is threshold current at reference temperature θ_0; θ_{IL} is characteristic temperature; η_{IL} is an external quantum output of IL. For InGaAsP/InP –IL θ_{IL}=50...70 K, $\delta_\lambda = 0{,}09$ nm/K for $\lambda_0 = 1{,}55$ µm and $\delta_\lambda = 0{,}11$ nm/K for $\lambda_0 = 1{,}3$ µm.

On Fig. 3 are presented calculated according to equations (1) – (4) in view of (5) – (6) dependences of a temperature error in a range of temperatures from 0 °C up to plus 50 °C for two wavelengths. As initial the temperature at which calibration of the device was carried out got out, equal is usual 20–25 °C, and corresponding initial wavelengths were equal λ_0=1,3 µm and λ_0=1,55 µm. From the lead calculations follows, that the value of a temperature error can reach 50-80 %. Lower temperature error for λ_0=1,55 µm explains as follows. As according to (5), with decrease of temperature the threshold current of IL generation decreases. At a constant pump current it leads to increase in radiation power. Simultaneously with this at reduction of temperature the maximal value $S(\lambda)$ decreases and displaced in short-wave area. Thus, reduction of spectral responsivity of a photodetector in neighborhood 1,55 µm partially compensates increase in radiation power of the injection laser. For wave length 1,3 µm spectral sensitivity of the photodiode practically does not change at change of temperature and in this case the temperature error is completely determined by change of IL radiation power.

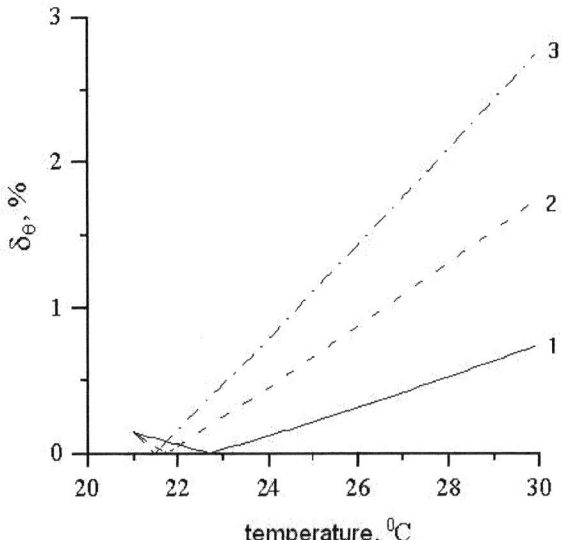

$1- \delta_P$=2 %; $2- \delta_P$=3 %; $3- \delta_P$=4 %
Fig. 4. Additional temperature error at various accuracy of stabilization of the injection laser radiation power

The offered mathematical model for account $S(\lambda, T)$ is fair not only for Ge-FD, but also for others indirect zone semiconductor photoreceivers consideration of the appropriate parameters describing a material of the semiconductor. It is due to note, that for other types of semiconductor FD it is necessary to take into account, that near to edges of fundamental absorption $\alpha \sim (h\nu - E)^\gamma$, where $\gamma = 1/2$ for the solved direct transitions, $\gamma = 3/2$ for the forbidden direct transitions $\gamma = 2$ for indirect transitions.

The developed analytical model allows to estimate for a design stage a temperature error of any optoelectronic information-measuring systems as which except for fiber-optic sensors can be optical pulse reflectometer, optical measuring equipment of radiation power, measuring equipment of losses in a fiber optical path, etc.

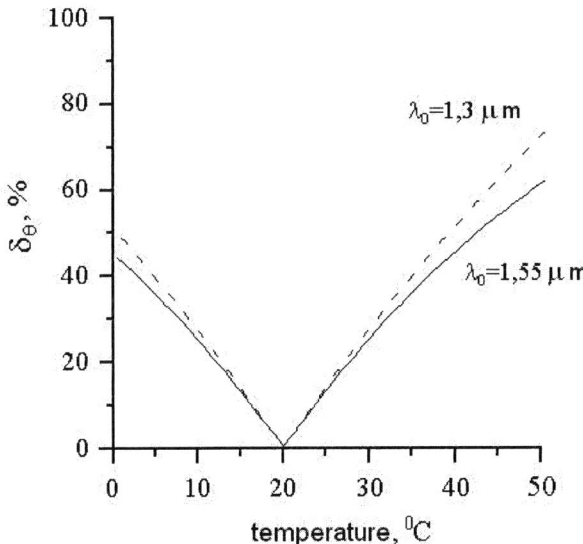

Fig. 3. Temperature error of the FOS

From the lead calculations follows, that the basic contribution to a temperature error makes change of IL radiation power. Therefore it is necessary to apply additional measures on stabilization of laser radiation power. On Fig. 4 calculations of a FOS temperature error for λ_0=1,55 µm are presented provided that IL radiation power is stabilized with relative error δ_P=2 % (a curve 1), δ_P=3 % (a curve 2) and δ_P=4 % (a curve 3). Changes of radiation power can arise both due to temperature fluctuations, and due to fluctuations of a IL pump current.

REFERENCES

[1] A. V. Polyakov, M. A. Ksenofontov, "Temperature error of fibre-optical measuring systems", *J. of Engineering Physics and Thermophysics*, vol. 80, no. 4, pp. 138–142, 2007.

[2] F. O. Huck, R. E. Davis, "Computational modeling for multispectral sensor design" *Proc. SPIE*, vol. 278, pp. 23–39, 1981.

CAOL*2010 International Conference on Advanced Optoelectronics & Lasers, 10-14 September, 2010, Sevastopol, Ukraine

Research of bolometer parameters of the trellised receiver of laser radiation

Kokody N.G., Pak A.O., Safronov B.V., Balkashin V.P., Priz I.A., Perepechai M.P.
V.Karazin Kharkiv National University, Chair of the Quantum Radiophysics
Svobody Sq.4, Kharkiv, 61077, Ukraine

Abstract: The experimental research of thermalphysic properties of thin metal bolometers (platinum and nickel) in a range between room temperature and the temperature close to melting is carried out. The factor of heat exchange with external environment, the efficiency factor of radiation absorption in a visible range, the time constant of bolometer cooling are measured. The polynomial approximation of the data received is carried out.

The trellised receivers are used for measurement of the spatial-power, polarization and temporary characteristics of laser radiation. They contain several grids that consist of thin metal wires. The laser radiation heats up these wires and they change their resistance. After measuring resistance of each wire and processing the received data on special algorithms, it is possible to receive the information about the distribution of intensity in a beam of radiation, the character of polarization, the form of a pulse of radiation and other characteristics.

The signals from the receiver depend on such parameters of bolometer as length and diameter, density, thermal capacity and conduction, temperature factor of resistance, factor of heat exchange with external environment, factor of efficiency of radiation absorption.

All these parameters depend on temperature. Such dependences need to be taken into account for signals processing because bolometers can be heated up by laser radiation to several hundred degrees. The dependences of density, thermal capacity, conduction, resistance on temperature are represented in the reference books [1, 2]. The factor of heat exchange and the factor of efficiency of radiation absorption depend on the concrete properties of bolometers. The authors investigated the dependence of these parameters on temperature.

The factor of heat exchange. Bolometer is heated up by an electrical current. The heating temperature is determined by the value of resistance. The capacity dispersed in bolometer is measured. It is equal to the capacity that is reverberated to the external environment $\alpha_p l(T - T_0)$, here l is the length of bolometer, T is its temperature, T_0 is the temperature of external environment, α_p is the linear factor of heat exchange.

Factor of heat exchange can be calculated under the formula

$$\alpha_p = \frac{i^2 R}{l(T - T_0)}, \qquad (1)$$

where i is a current through bolometer, R is its resistance.

The measurements were performed for platinum bolometers with diameters 10 and 20 microns and for nickel bolometer with diameter 40 microns. The results are shown in Fig. 1. The factor of heat exchange weakly depends on a diameter of a wire. But nevertheless such dependence exists — thick wires have stronger heat exchange, than thin ones. The results of the measurements are well coordinated with the results received in the work [3].

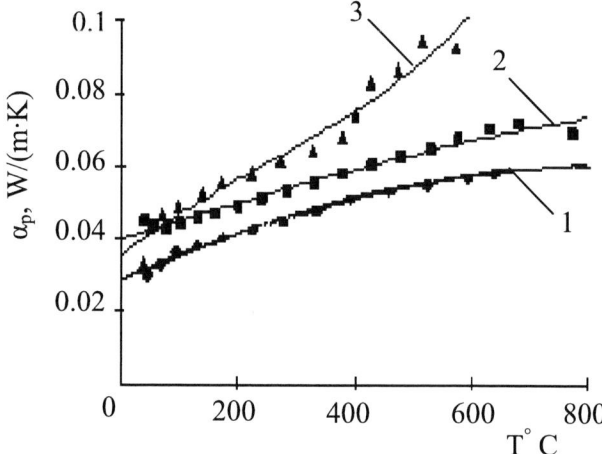

Fig. 1. Dependence of the factor of heat exchange on temperature. 1) Pt – 10 μm, 2) Pt – 20 μm, 3) Ni – 40 μm

The factor of heat exchange has two components — convective streams of air and thermal radiation.
Capacity of thermal radiation:

$$P_{rad} = \sigma S(T^4 - T_0^4),$$

where σ is the Stefan-Boltzmann constant, S is a radiating surface of bolometer. The factor of radiation is accepted to be equal to unit.

The calculations made according to the formula mentioned above show that the contribution of thermal radiation is increased with the increase of bolometer diameter. At temperature $600°$ C for bolometer with a diameter 10 microns this contribution is equal to 3 %, for 20 microns it is 5 %, for 40 microns — 7 %.

The basic contribution to heat exchange is given by air convective streams.

978-1-4244-7043-3/10 $26.00 © 2010 IEEE   202

The efficiency factor of radiation absorption. In the experiment the dependence on temperature of a signal from bolometer during measurement of energy of an optical pulse was investigated. The heating of bolometer was small — no more than 3 % from initial temperature. Therefore the processes can be described by the linear equations.

The increase of bolometer temperature during the absorption of energy E is equal

$$\Delta T = \frac{Q_{abs} E}{mc},$$

where $\Delta T = T_{max} - T$, T is the maximal temperature, T is the initial temperature, Q_{abs} is the factor of efficiency of absorption, m is the mass of bolometer, c is the specific heat.

Hence

$$Q_{abs} = \frac{mc(T_{max} - T)}{E}. \qquad (2)$$

The temperature of bolometer was determined by the value of signals U and U_{max}:

$$U = iR = iR_0(1 + \alpha_R T),$$
$$U_{max} = iR_{max} = iR_0(1 + \alpha_R T_{max}), \qquad (3)$$

where i is a current, R is its resistance, α_R is the temperature coefficient of resistance [2]:

$$\alpha_R = 0{,}00393 - 5{,}31 \cdot 10^{-7} \cdot T \quad \text{K}^{-1} \text{ — for Pt,}$$

$$\alpha_R = 0{,}0156 - 5{,}78 \cdot 10^{-6} \cdot T \quad \text{K}^{-1} \text{ — for Ni.}$$

From the formulas (3) follows that

$$T_{max} - T = \frac{\Delta U}{iR_0 \alpha_R},$$

where $\Delta U = U_{max} - U$, R_0 is the resistance at 0° C.

After this transformation we receive:

$$Q_{abs} = \frac{mc(T)}{E} \left[\frac{1}{\alpha_R(T)} + T \right] \frac{\Delta U}{U}. \qquad (4)$$

Energy of radiation E from a pulse lamp that got to bolometer, and the mass of bolometer are not known precisely. Therefore, the formula (4) only gives the temperature dependence Q_{abs} but can not give absolute values.

Therefore, under the known formulas [4] the values of Q_{abs} for platinum and nickel cylinders were calculated for the values of a complex refractive index for a visible range. These values are used for plotting the diagrams. The dependences are shown in Fig. 2.

The diagrams for platinum bolometers are almost equal. On the diagram for nickel bolometer the maximum is seen at $T \approx 280^\circ$ C. It is located near a Curie point ($T_c = 350^\circ$ C) but does not coincide with it because the curve is influenced by summands $1/\alpha_R$ and T.

From works [3, 5] it follows that the value Q_{abs} increases with increasing temperature. The difference of the dependence that is received here is explained by the fact that measurements and calculations in works [3, 5] cover the infra-red range while results of the given research cover the visible range. The complex refractive index varies according to the temperature in the visible and the infra-red ranges.

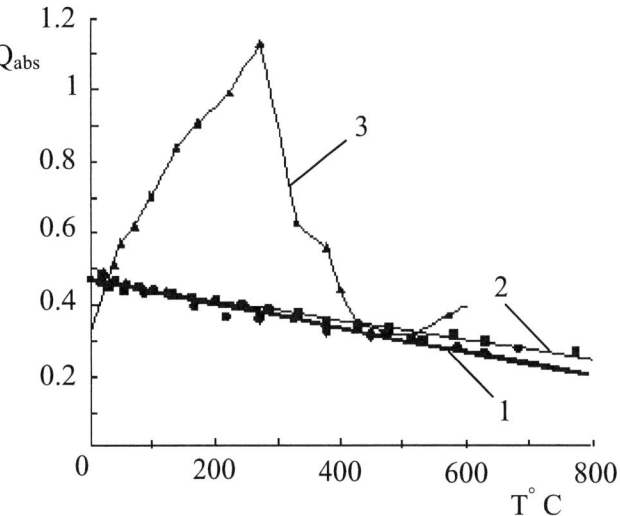

Fig. 2. The dependence of the factor of efficiency of absorption on temperature. 1) Pt – 10 μm, 2) Pt – 20 μm, 3) Ni – 40 μm

The thermal time constant defines lag effect of bolometer as a receiver of radiation. It was determined during processing a course of cooling bolometer after heating by an optical pulse. It is described by dependence $\exp(-t/\tau)$, where τ is the thermal time constant.

The diagrams of dependence are shown in Fig. 3. τ decreases with increasing temperature. This is a result of increasing heat exchange.

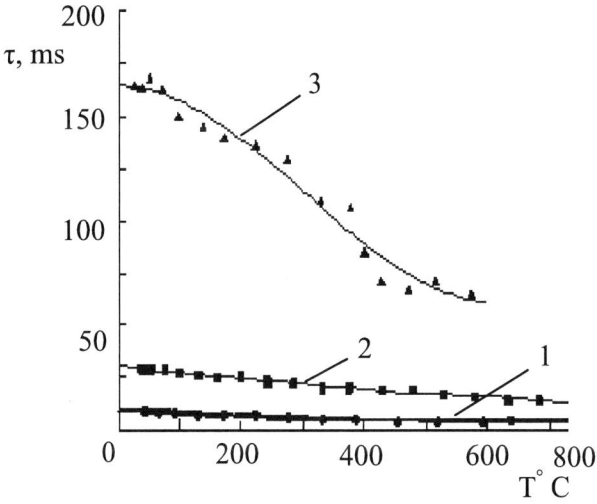

Fig. 3. The dependence of the thermal time constant on temperature. 1) Pt – 10 μm, 2) Pt – 20 μm, 3) Ni – 40 μm

The measurement of the form of a pulse laser radiation. The received dependences $Q_{abs}(T), \alpha_p(T), \tau(T)$ were used in the algorithm of processing signals from bolometer during the measurement of the form of a pulse of laser radiation.

The source of radiation in the experiment was the neodymium pulse laser working in a mode of free generation. The energy of a pulse is equal to 15 J, the diameter of a beam of radiation — 15 mm, the width of a pulse of radiation about 1 ms.

The receiver of radiation is the grid with the size 25 × 25 mm^2. It contains 16 platinum bolometers with a diameter of 10 microns located on the distance of 1.5 mm from one another. The electrical vector of radiation is directed parallel to the axis of bolometers.

The average temperature of bolometers that are located in the area of the maximal intensity, achieves 400° C, maximal — 1500° C. The parameters of bolometers vary strongly at this temperature.

The purpose of experiment is the measurement of the form of a pulse of radiation of the laser. The thermal time constant of bolometer is about 10 ms, therefore, the signal from bolometer can not repeat the form of a pulse of radiation. To solve this problem is possible with a help of computer signal processing based on the use of the heat balance equation:

$$mc\frac{dT(t)}{dt} + \alpha_p lT(t) = Q_{abs}P(t),$$

where T is the temperature of bolometer, m is its mass, l is the length, c is the specific heat, α is the linear thermal conductivity factor, Q_{abs} is the efficiency factor of radiation absorption of bolometer.

This equation assumes the identical temperature in the bulk of bolometer. This assumption can be accepted, because the diameter of bolometer is small — from 10 up to 100 microns. The time of the heat distribution for such distances in metal is tens microseconds. This time determines the minimal value of pulse duration which can be measured.

This equation can be transformed into the form:

$$P(t) = \frac{mc}{Q_{abs}}\left[\frac{dT(t)}{dt} + \frac{1}{\tau}T(t)\right], \qquad (5)$$

where $\tau = \dfrac{mc}{\alpha_p l}$ is the thermal time constant of bolometer.

Having measured the function of temperature change of bolometer in time variation and having processed a signal from the receiver under the formula (5), it is possible to restore the form of a pulse of radiation — the function $P(t)$. It can also be done when the thermal time constant τ is more than the width of a pulse that is to measure the form of short pulses of radiation with the receiver with large inertia.

In these calculations it is necessary to take into account the dependence of parameters of bolometer on temperature. For platinum bolometer with a diameter of 10 microns:

$$c(T) = 132 + 0{,}0250 \cdot T + 1{,}35 \cdot 10^{-6} \cdot T^2 \ \text{J/(kg·K)},$$

$$\tau(T) = 9{,}95 - 0{,}0180 \cdot T + 1{,}39 \cdot 10^{-5} \cdot T^2 \ \text{ms}, \quad (6)$$

$$Q_{abs}(T) = 0{,}474 - 3{,}352 \cdot 10^{-4} \cdot T.$$

The data on the dependence are taken from the literature [2].

In a Fig. 4 thick line 1 shows the form of a radiation pulse of the laser measured by the photo diode, thin line 2 — the results of the restoration of the pulse form of radiation with the use of the formula (5). The curves well coincide. The shaped line 3 shows the results of signal processing according to the formula (5) without the dependence of bolometer parameters from temperature. This curve differs strongly from curve 1.

Thus, one can see that during the measurements of the characteristics of powerful laser radiation it is necessary to take into consideration these dependences.

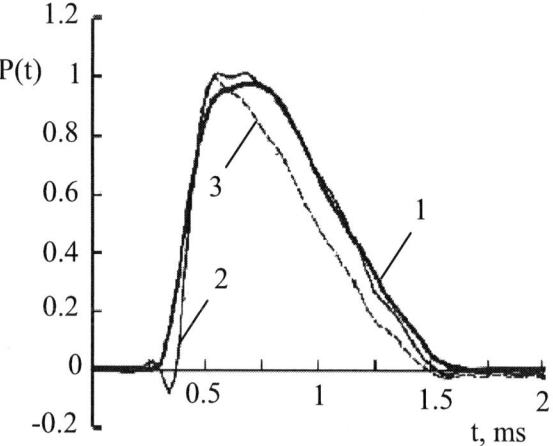

Fig. 4. A pulse of radiation of the laser

REFERENCES

[1] Table of physical quantities. The directory. Edited by academician Kikoin I.K. — M.: Atomizdat, p. 1008, 1976 [in Russian].

[2] Zinoviev V.E. Thermalphysic properties of metals at high temperatures. — M.: Metallurgy, p. 384, 1989 [in Russian].

[3] Perepechai M.P. Research of interaction of intensive infra-red radiation with metal cylinders with the purpose of creation through passage measurer of capacity of laser beams. Candidarte's thesises Kharkiv, p.182, 1979 [in Russian].

[4] Kerker M The scattering of light and other electromagnetic radiation. — N.Y., London: Academic Press, p.666, 1969.

[5] Kuzmichov V.M., Pogorelov S.V. Transformation of energy of a pulse of the laser with platinum thin-wire bolometer // Ukrainian metrological journal, vol. 2, p. 42 – 47, 2003 [in Russian].

Measurement of intensity distribution of radiation in a laser beam by shear mechanism

N.G. Kokodiy, Li Zhenhua
Kharkov National University, Kharkov, Ukraine

Abstract: The method of measuring of the intensity distribution by knife-edge is proposed. The algorithm and program for handling of the signal from receiver are developed. By using the MATLAB to carry out the simulation analysis, the method is proved that it can accurately recover the complicated energy distribution.

INTRODUCTION

Distributing of intensity in space is important characteristic of laser radiation. Matrix receivers are used often in visible range. Measuring transformers can be photodiodes or thermal transformers. In first case devices can be used for registration of both continuous and pulse radiation, but they work in a spectral range, wrap-round only visible and near infra-red to the spectrum area, and usually counted on measuring of parameters of narrow low-powered beams of radiation. In second case a spectral range is considerably wider, and measuring can be worked with beams of high-powered, but only with lasers, workings in the single mode or rarely followings impulses. For measuring of parameters of radiation of continuous lasers this method is useless, because distributing of temperature on the surface of matrix does not repeat distributing of intensity in the beam of radiation. At large power of radiation receivers cannot be put in a direct beam, and it is needed to use coupler. It complicates a device and increases his error.

Bolometric grates, made out of thin metallic wire can measure the parameters of powerful radiation, located on the way of distribution of laser beam [1]. Resistance of separate wire is proportional to its temperature that intensities of radiation are in this place.

Bolometric grates possess many dignities. They do not distort the beam of radiation, have a small losses, can be used in beams with the large of cross-sectional. The lacks of them are fragility of construction, and instability of parameters, caused by gradual oxidization of wires, and the change of absorption of radiation, difficulty of the use in the narrow beams of radiation.

The grid receivers of radiation are produced by a firm Prometec (Germany). They can measure distributing of intensity of radiation and cross-section in laser beams by a diameter from 20 to 60 mm at power from 100 to 10000 W in visible, near and middle infra-red range of spectrum.

The method of measuring of diameter of Gaussian laser beam is described in [2]. A knife moves across a beam. Dependence of the reflected or passing power of radiation from

position of knife is measured. Defining it, it is possible to find the diameter of beam of radiation.

The lack of such method of measuring is that he supposes the Gaussian law of distributing of intensity in the section of beam of radiation. It is just for lasers, workings in the mode of TEM_{00}, for example, for many He-Ne lasers. But CO_2 lasers, as a rule, have more difficult distributing of intensity in a beam. Then described method of measuring for them is unacceptable.

THEORY AND EXPERIMENT

The arbitrary distributing of intensity can be measured, using a few knives, crossings a beam in different directions. In the device of firm Beam Master (Germany) 3 knives are used at the analysis of beams, are close to Gaussian, and 7 knives in the case of substantial ungaussian beams. A device is used for the analysis of focal area of beam. Distinguish ability of device is 0.1 мкм. Photo receivers are used on the basis of Si or $InGaAs$, which work in a spectral range from 400 to 1800 nm. Signals from them are processed by a computer, and results are shown on a display in a digital and graphic form.

It is shown in this work, that it is possible to measure distributing of intensity of any kind by the special location of one knife of radiation receiver.

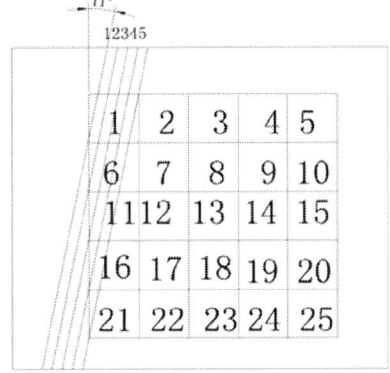

Fig. 1. Geometry of problem

Geometry of task is shown on Fig. 1. An entrance window of device is a square by the sizes of 25 x 25 mm. Its area is divided to 25 elementary squares the sizes of 5 x 5 mm. In the beginning the window is closed by a knife, which then moves along the axis x. The edge of knife is locating by a angle

$\varphi = 11.31°$ to the axis y ($\tan(\varphi) = 1/5$) and then knife moves along the axis of x with the step of 1 mm. On the picture are shown a few first positions of knife .They are marked to the numbers from 1 to 5. Power of radiation, which is getting through a window in each of these positions, is measured.

It is considered that radiation intensity in elementary square is the same. Then a few first correlations look so:

1) $0.1\ I_1\ S = P_1$
2) $0.3\ I_1\ S + 0.1\ I_6\ S = P_2$
3) $0.5\ I_1\ S + 0.3\ I_6\ S + 0.1\ I_{11}\ S = P_3$ (1)
4) $0.7\ I_1\ S + 0.5\ I_6 + 0.3\ I_{11} + 0.1\ I_{16}\ S = P_4$
5) $0.9\ I_1\ S + 0.7\ I_6\ S + 0.5\ I_{11}\ S + 0.3\ I_{16}\ S + 0.1\ I_{21}\ S = P_5$

I_1, I_2, I_3,… are intensity of radiations in elementary squares 1, 2, 3…, S is area of elementary squares 1, 2, 3... The system has 30 equations. The vector of right parts is measured in the experiment. The number of unknowns is 25. Redundancy of system is instrumental in weakening of influence of errors of measuring of powers P_1, P_2, P_3 ... in the final result.

For verification of method a numeral experiment was worked. The function of distributing of intensity of radiation is:

$$I(x,y) = \exp\left[-\frac{(x-x_0)^2 + (y-y_0)^2}{r_0^2} \right]$$

where $x_0 = 10$ mm, $y_0 = 15$ mm are coordinates of center of beam, $r_0 = 5$ mm is his mean-square diameter. It was calculated value of intensities I_1, I_2, I_3, ... I_{25} in the centers of elementary squares.

Powers of $P_1\ P_2\ P_3$, … P_{30} were calculated for the equation system (1), part of which is shown higher. , They are imitating signals from photo-receiver, measured in the experiment. These results were used as experimental data at the solving of reverse problem — reconstruction of function distributing of radiation intensity by the solving of the system (1).

Results of decision function I(x,y) are shown in Fig.2b, on Fig.2a the initial function of I0(x,y), obvious that functions I0(x,y) and I (x,y) are conformed well between itself

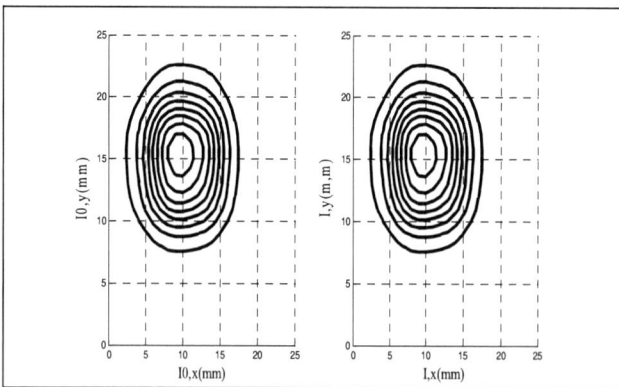

Fig. 2. The initial function *I0(x,y)*, and results of reconstruction of the function *I(x,y)*

It was experimental verification of method. The source of radiation was incandescent lamp with the reflector, which is allowing focusing its light. On the window of receiver headed the beam of light with diameter 20 mm. The receiver of radiation is photodiode, placed in a cylindrical scattering cavity with an entrance window 25 x 25 mm.

The knife moved across a light beam as shown on Fig.1. The signals from the photo-receiver were used for the solving of the system of equations (1) as its right part. The results of calculations are shown on Fig. 3. We can see the maximum of function I(x,y) and area of high intensity by a diameter near 20 mm, how it was in the experiment. But a quality of the map of radiation intensity distributing is bad. Function *I(x,y)* is strongly cut up, especially on the edges of window.

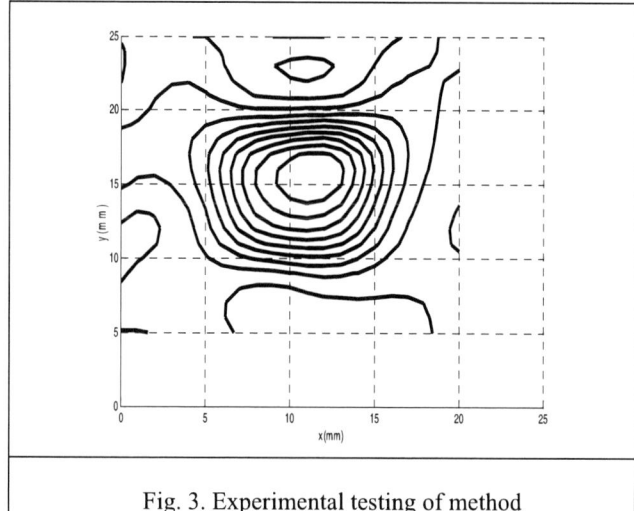

Fig. 3. Experimental testing of method

The reason of it is strong influence of errors of measuring of signals from photodiode to the results of calculations. This task, like most reverse tasks is an incorrect problem.

A numeral experiment was made for evaluated by influence of errors of measuring of signals from a photo-receiver to the result of calculation of function *I(x,y)*. For this purpose in the values of vector of P (the right part of the system of equation (1)) were entered random additions, which is set in specified limits, and calculated the error of determination of function *I(x,y)*.

Calculation of error of determination of function now is not specified by standards. There are a few formulas for this purpose [3,4]. We were used the formula [4]:

$$\varepsilon_{out} = \sqrt{\frac{\sum_{m=1}^{M}\sum_{n=1}^{N}\left(I_{m,n} - I0_{m,n}\right)^2}{\sum_{m=1}^{M}\sum_{n=1}^{N}\left(I0_{m,n}\right)^2}}$$

There are *I0* and *I* are initial and reconstructing functions of radiation intensity distributing, *M* and *N* are number of points along the axes of x and y, in which these functions were calculated.

The error of measuring of receiver signal was determined on a formula [4]:

$$\varepsilon_{in} = \sqrt{\frac{\sum_{q=1}^{Q}\left(P_q - P0_q\right)^2}{\sum_{q=1}^{Q}\left(P0_q\right)^2}}$$

where $P0$ is ideal signal from a receiver, P is signal with errors, $Q = 30$ is the number of points ,in which a signal was measured.

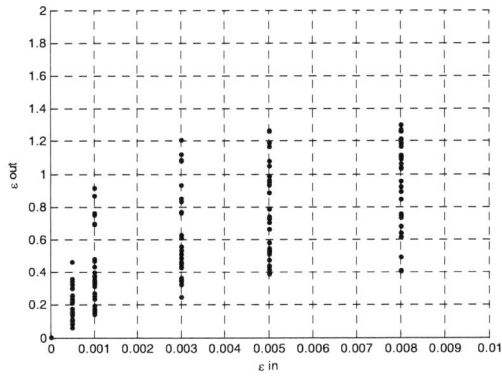

Fig. 4. Influence of the errors in the signal of receiver on result

The results of tests are shown on Fig. 4. We can see that errors in the signal of receiver are influencing on result very much — error of result in 100 times more than error is in the signal from receiver.

There are different methods of diminishing of influence of these errors. There are smoothing of signal by different algorithms, regression methods, Tikhonovs method of regularization of solve and others. The choice of the best method is an object of further researches.

CONCLUSION

In this work was shown principle possibility of measuring of function of distributing of intensity in the beam of radiation in the device with one moving knife.

REFERENCES

[1] Kokodii N.G., Kokodii D.N. Measurement of the distribution of intensity in a beam of thermal radiation. *High Temperature*, 2007, vol. 45, No. 2, pp. 255-260.

[2] Khosman J.M., Garetz B.A. Measurement of a gaussian laser beam diameter through the direct inversion of knife-edge data. *Applied Optics*, 1983, Vol. 22, No. 21, pp. 3406-3410.

[3] Fienup J.R. Invariant error metrics for image reconstruction // *Applied optics*, 1997, Vol. 36, No. 32, pp. 8352-8357.

[4] Tikhonov A.N., Arsenin V.Y. Methods for solving incorrect problems. M.: Science, 1986. - 286 pp. (in Russian)

[5] G.T.Herman. Image Reconstruction from projections. Fundamentals of reconstructive tomography. – M: 1983. - 349 pp.

CAOL*2010 International Conference on Advanced Optoelectronics & Lasers, 10-14 September, 2010, Sevastopol, Ukraine

Holographic Shack-Hartmann wavefront sensor

Dmytro V. Podanchuk[*], Volodymyr P. Dan'ko, Myhailo M. Kotov
National Taras Shevchenko University of Kyiv.
64, Volodymyrska st., Kyiv, 01033, Ukraine
[*]Corresponding author. E-mail address: pdv@univ.kiev.ua

Abstract: Principles of constructing the holographic wavefront sensors are considered. Method of production of the two-channel holographic lenslet arrays with compensation of specified aberrations and its parameters are presented. The possibility of adaptation of the holographic sensor for measuring the wave fronts with different curvature is shown.

The Shack-Hartmann wavefront sensors are widely used in adaptive optics, metrology and some medical applications. The basic elements of the sensor are lenslet array and photodetector situated in lenslet focus. The principle of sensors work consists in measuring the local slopes of the tested wave front relatively to the reference plane wave front. The constructive simplicity and possibility of simultaneous measurements of local slopes for a whole aperture in two orthogonal directions are regarded as its advantages. But the common Shack-Hartmann wavefront sensor has essential limitations concerned with contradiction between angular sensitivity and measurement range: when the focal length of the lenslet array is reduced then the measurement range expands, but contrariwise the angular sensitivity reduces.

One of the means to resolve this problem is the usage of the holographic lenslet arrays in the Shack-Hartmann wavefront sensor. It allows improving the functionality of the method and increasing the aberration measurement maximum with simultaneous retention of the needed angular sensitivity. There are several methods of making the holographic wavefront sensor. One of them consists in usage of the holographic lenslet array with two space-separated foci and the algorithm of measurement results co-processing [1]. In this case the nonlinear features of the holographic recording material can be used for dual focus lenslet array production [2].

Another design of holographic sensor, when a few wave fronts with specified shape are recorded into the optical memory of the lenslet array, is considered in the work. It allows shifting the measurement rage of the sensor in appropriate way that is impossible when common refractive lenslet arrays are used. In that way the sensors with several space-division channels for simultaneous correction and analysis of the tested wave front can be produced.

The laboratory setup created for experiment make possible the holographic lenslet array recording with precompensation and the distorted wavefront testing. The optical scheme (Fig.1) consists of two channels. The reference channel is used for holographic lenslet array registration. The object channel could be used on the stage of recording of the

holographic lenslet array with aberration precompensation and in the case of wavefront testing by the recorded lenslet arrays.

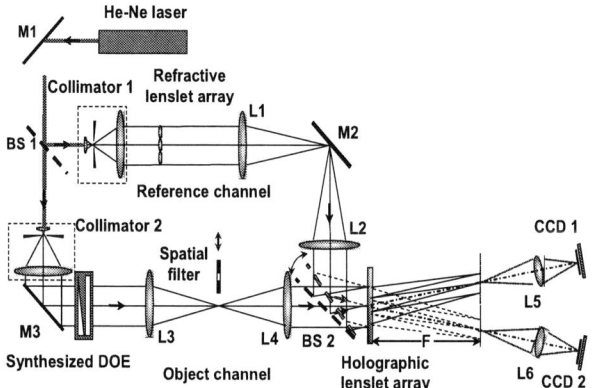

Fig.1. Optical setup of two-channel holographic wavefront sensor

For creating the Reference channel the part of laser beam is reflected by the beam splitter BS1 and extended to the diameter of 12mm by Collimator 1. The reference channel contains the reference Refractive lenslet array which is projected to the hologram plane by the telescopic system (objectives L1, L2), mirror M2 and beam splitter BS2. Refractive lenslet array has 20×20 lenslets with 24 mm focal length and lenslet size of 0.4×0.4 mm. Beam splitter BS2 is constructed movable. It makes possible a recording of two holographic lenslet arrays in one aperture of holographic plate. These lenslet arrays are used in two measuring channels consist of lens L5 and CCD1 or lens L6 and CCD2 respectively.

The laser beam passed through BS1 is used for creating the Object channel. This beam is expanded by Collimator 2 and by the help of mirror M3 illuminate the synthesized diffraction optical element (DOE) with compensation of carrier spatial frequency or tested object. DOE or object plane is translated to the holographic plane by the telescopic system (objectives L3, L4). In the telescopic system L3, L3 the Spatial filter is inserted to cleanup beam from many diffractive orders of DOE, this filter separates necessary order of diffraction. This channel allows a recording of holographic lenslet arrays with precompensation of aberrations, which are inserted by Synthesized DOE, or for creating the tested wavefront.

978-1-4244-7043-3/10 $26.00 © 2010 IEEE

Synthesized diffractive optical element is binary Leith–Upatnieks hologram. Hologram is the result of interference of plane wave and wave with aberrations.

$$I^*(x,y) = I_0 \cdot \left| e^{iw(x,y)} + e^{i\frac{2\pi}{\lambda}x\sin\phi} \right|^2, \qquad (1)$$

where I_0 is beam intensity; φ is angle between zero-order and the 1st order of reconstruction; $w(x,y)$ is arbitrary wavefront aberration defined as superposition of the standard aberrations with certain weight coefficients [3]:

$$w(x,y) = \sum_{n,m} \alpha_n^m w_n^m(x,y), \qquad (2)$$

where w_n^m is a Zernike polynomial, α_n^m is a corresponding weight coefficient.

To compensate a slope of diffracted wave the DOE was used with the compensator represented by the optical wedge.

For example, Figs.2, 3 show interference patterns for synthesized (a) and reconstructed from DOE (b) wave fronts with reference plane wave.

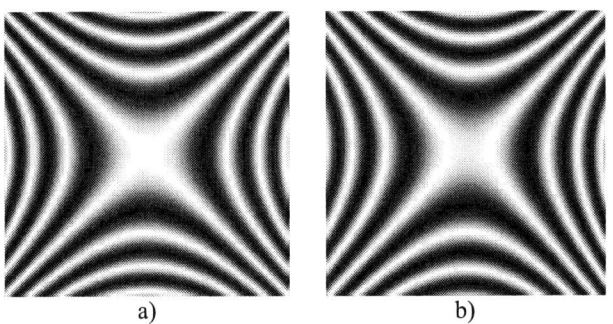

a)   b)

Fig. 2. Computed (a) and reconstructed (b) interference patterns for wave front with astigmatism. Set Zernike's coefficient equals 20, the reconstructed one equals 19.2.

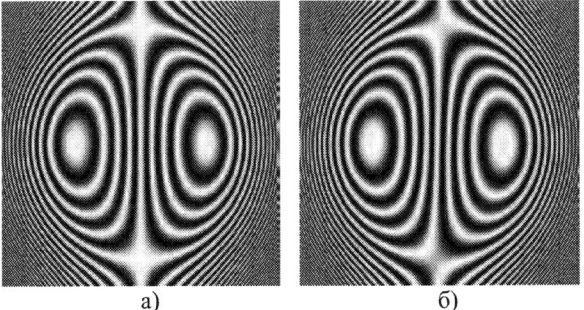

a)   б)

Fig. 3. Computed (a) and reconstructed (b) interference patterns for wave front with coma. Set Zernike's coefficient equals 40, the reconstructed one equals 40.6.

The synthesized wave fronts have aberrations of astigmatism with coefficient $\alpha_2^2 = 20$ (Fig.2) and coma with

$\alpha_3^1 = 40$ (Fig.3) that corresponds to the maximal phase incursion along the aperture edges of 3.18 λ and 6.36 λ respectively. Wave fronts reconstructed from the DOE are measured by the Shack-Hartmann sensor with refractive lenslet array. Coefficients estimated from wavefront reconstruction are used for plotting the interference patterns shown on the Fig.2 (b), Fig.3 (b). Reconstructed and synthesized aberration coefficients differ in a value of 4% at the most; the difference is concerned with DOE's mounting misalignment.

The setup described above made possible to make two-channel holographic lenslet arrays with correction of arbitrary given aberration for both channels. Actually, aberration value depends on limiting spatial frequency of the synthesized DOE only. The hologram recording is realized on holographic plates PFG-01(Slavich, Russia). It should be mentioned that numbers of channels for an array can be increased to four by certain setup improving. Then, of course, the diffractive efficiency will be decreased for each channel, but we suppose that the holographic lenslet array produced in this way is competitive in comparison with the refractive one because its functionality and inexpensive production.

Principle of aberration correction with help of holographic wavefront sensor is described below by the example of spectacles astigmatic lens testing with usage the ordinary lenslet array and the holographic defocus precompensated array. The experiment is carried out in the following way. At the first, the DOE is removed from setup and holographic lenslet array is recorded with the help of plane object beam, as the diffractive analogue of classic refractive array. The tested lens is placed instead the DOE. During the measurements, the holographic lenslet array illuminated by the tested wave, so the reconstructed wave forms the hartmannogram on the surface of a photodetector CCD (768×576, the effective pixel size is 15.5 μm). The estimation of lens parameters gives the values of defocus and astigmatism coefficients equal to $\alpha_2^0 = 88.0$ and $\sqrt{\left(\alpha_2^{-2}\right)^2 + \left(\alpha_2^2\right)^2} = 18.7$ respectively.

Table 1. Estimated values of defocus and astigmatism coefficients for astigmatic spectacles lens

angle of lens rotation	defocus α_2^0	astigmatism $\sqrt{\left(\alpha_2^{-2}\right)^2 + \left(\alpha_2^2\right)^2}$
ordinary lenslet array		
30°	88.0	18.7
75°	87.5	17.5
120°	87.4	17.2
165°	87.6	18.4
holographic array with defocus precompensation		
30°	-1.1	18.4
75°	-0.8	17.1
120°	-0.5	17.2
165°	-0.6	18.0

CAOL*2010 International Conference on Advanced Optoelectronics & Lasers, 10-14 September, 2010, Sevastopol, Ukraine

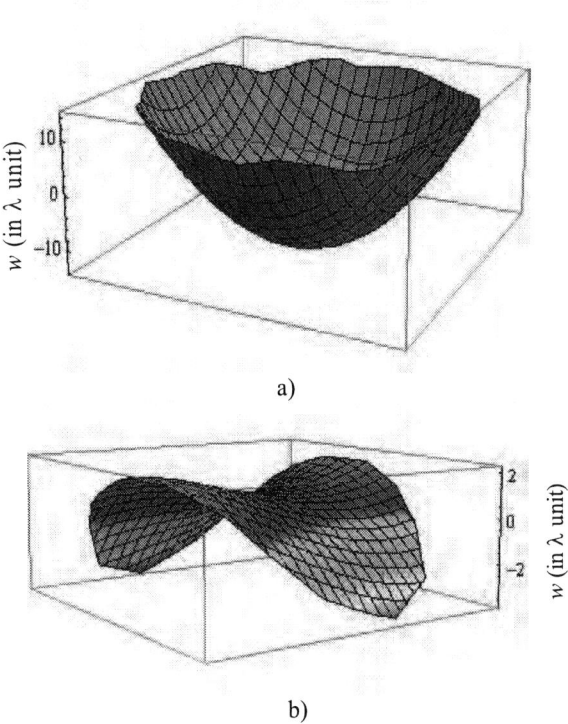

a)

b)

Fig. 4. Reconstruction of the wave front produced by the astigmatic lens: a) the ordinary lenslet array; b) the holographic array with defocus precompensation

At the next stage the DOE generating the spherical object way with appropriate defocus measured above is produced. Then with help of this DOE the holographic lenslet array with defocus precompensation is recorded. This array is used for measuring the wave front behind the astigmatic lens. Now the corresponding coefficients have values of −1.1 and 18.4. All measured coefficients are given in the Table 1. Thus the values of astigmatism coefficient measured with the different array are close (the difference is 2% at the most) despite the repeated tested lens mounting to the optical setup.

Fig. 4 shows the wave front behind the astigmatic lens reconstructed with help of the ordinary lenslet array (a) and

the holographic array with defocus precompensation (b). In the case (a) astigmatism is disappeared against the defocus, contrariwise the case (b) allows to visually estimate the wavefront aberration.

One more advantage of this approach to wavefront sensor designing is the possibility of making the adaptive measurement channel based on reprogrammable holographic lenslet array [4]. The peculiarity of the method is sequential comparison of the tested time-varying optical wave with the specified wave that are synthesized and recorded in holographic memory of the array. The sensor works in the following way. At the first stage wavefront analysis is carried out with the holographic lenslet array that is the analog of the common refractive array. When the sensor measurement range is achieved, wave front with a specified shape is synthesized, the next holographic lenslet array is recorded and measurements are continued. In such a way the sensor adaptation for the wavefront time variations is realized.

The realization of this approach in practice needs the usage, for example, electrically controlled space-time light modulators for making the diffraction optical elements and reversible recording medium for holographic lenslet array.

REFERENCES

[1] Jung-Young Son, Dmytro V. Podanchuk, Volodymyr P. Dan'ko, Kae-Dal Kwak. "Shack-Hartmann wavefront sensor with holographic memory". Optical Engineering, 42(11), p. 3389-3398, 2003.

[2] Dmytro V. Podanchuk, Volodymyr P. Dan'ko, Myhailo M. Kotov, Jung-Young Son, Yong-Jin Choi. "Extended-rang Shack-Hartmann wavefront sensor with nonlinear holographic lenslet array". Optical Engineering, 45(5), 053605, 2006.

[3] Born M, Wolf E. Principles of Optics, Pergamon Press, Oxford, 1983.

[4] D.V. Podanchuk, V.N. Kurashov, V.P. Dan'ko, M.M. Kotov, N.S. Sutyagina. "Shack-hartmann wavefront sensor with holographic lenslet array for the aberration measurements in a speckle field". Semiconductor Physics, Quantum Electronics & Optoelectronics, v. 11, № 1. p. 29-33, 2008.

978-1-4244-7043-3/10 $26.00 © 2010 IEEE

The commercially available version of Shack-Hartmann wavefront sensor

P.N. Romanov, V.E. Zavalova, A.V. Kudryashov, A.L. Rukosuev

"Adaptive Optics NightN" Ltd., Russian Federation

Abstract: The development of low-cost Shack-Hartmann wavefront sensor (SHWS), based on USB2-interface of CMOS camera, is presented in this report. The description of main principle of SHWS and example of wavefront measurements are given. A comparative characteristics and spheres of sensors applications, based on USB-2 and IEEE-1394 types interfaces, are shown below. But certainly there are advantages and disadvantages of each type of these sensors. That is why each type has more suitable spheres of application. This is the subject of our discussions.

I. INTRODUCTION

Shack-Hartmann method is widely used for phase measurement of wavefront investigation alongside with interferometer and other methods. Shack-Hartmann wavefront sensors have sufficiently simple scheme, therefore they are relatively easy adjusted, less sensitive to wavelength variations of measuring radiation and mechanical vibrations. Besides, this type of the sensors has capability to work with the pulsed radiation. This resulted in the fact that SHWS became the most popular wavefront sensor in practical and scientific problems. Some of them are applied in laser technology, micro-technology, laser ophthalmology, astronomical devices, optical communications and etc. Wavefront sensor is used for the measurements and testing of wave front aberrations in adaptive optical systems too. But the desire for making the device as the mass production device at detailed studying of this question encounters difficulties. And difficulties have basic character. The given work is devoted this question.

II. PRINCIPLE OF MEASUREMENTS BY SHACK-HARTMANN WAVEFRONT SENSOR

The principle of sensor is well known and based on measurements of local slopes of the wavefront (wf) df $/dn$ (n-vector normal to wf) between reference front (as usually flat) and current wf. Local slopes are proportional to focal spots displacements (shifts) - S. A two-dimensional array of discrete micro-lenses array (lenslet) divides the incoming wavefront into a number of parts. Each sub-aperture provides a separate focus on the detector of the CCD camera, fig.1. The position of each spot is displaced by local wavefront aberrations. Wavefront reconstruction is performed according to modal estimation and uses least-squares solution. The solution gives us Zernike coefficients a_k of Zernike polynomial expansion

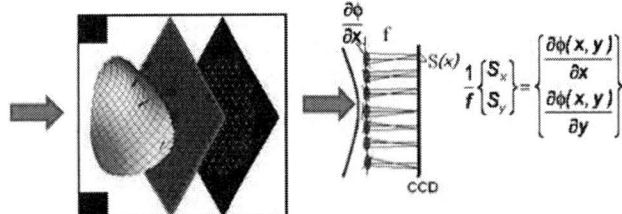

Fig.1. Illustration of Shack-Hartmann principle.

Z_k (x,y) of wavefront. They represent types of wavefront aberrations (focus, astigmatisms, comas, ets).

Software is provided [1] for capturing, displaying, saving the wave fronts data and representation of wavefront analysis data.

It has the following features:

Wavefront Reconstruction for Circular & Rectangular pupil.

Both phase & Intensity measurements of Wavefront.

Wavefront Parameter Display:

P – V Error; RMS wavefront error;

Tilt, focus error, astigmatism, coma, spherical aberration and higher order aberration in terms of Zernike and Seidel coefficients;

Fringe representation; 2D & 3D wavefront image representation; Far- feild laser beam intensity in 2D & 3D (PSF).

Laser beam characterisation parameters:

Measurement of laser beam diameter, beam waist & location;

Measurement of Beam Divergence;

Measurement of Laser Beam Quality, M^2.

Hardware is consists of CCD (or CMOS) camera with microlens array build-in holder to camera and beam resizing optical system, for example telescope (on request).

III. COMPARISON BETWEEN SENSORS, BASED ON USB-2 AND FIREWIRE CAMERAS

The comparison includes three typical directions. These are the comparison of USB-2 and IEEE-1394 type's interfaces [2], the main features and spheres of applications. They are represented on the next tables 1-3.

TABLE 1. Difference between USB-2 and firewire interface.

Items	USB 2.0	FireWire IEEE-1394-a
Data Rate	480Mbits/s (60 MB/s)	400Mbits/s (50 MB/s)
Sustained Data Rate (READ)	33MB/s	38MB/s
Sustained Data Rate (WRITE)	27MB/s	35MB/s
Architecture	Master (host)-Slave(client) configuration requires overhead on the host (master) side in order to maintain the transfer of data between host and client (slave), hence reducing overall sustained data rates. If communication between two clients is needed, then the data rates are reduced even further	Peer-to-Peer connections between devices, allowing a camcorder to transfer data "computer-free."
Connection with personal computers	Cable with Maximum length - 5 m	Cable with Maximum length - 4.5 m
Designed for	Convenience	Speed data transfer in hard drivers for external video devices
The main Advantage	• Standard interface of personal computers and good for driving low-power devices, low-bandwidth devices like keyboards, mice, memory sticks and etc • Low-cost, mass production	Sustained data rates perfect for video or media devices
The main Disadvantage	• It only supplies 3 W of power to connected devices. High-performance cameras equipped with a USB interface requires external power, namely, USB connector be plugged directly into a motherboard's USB port, or into a powered hub. • There is some delay between data transfer request and real start of data transfer	Not usually standard on personal computers and doesn't submitted to the faceplate
Swap (switching on)	hot-swappable	

TABLE 2. Main features of SHW sensors.

Sensors, based on USB 2.0	Sensors, based on IEEE-1394
Simplicity in realization	
Wavefront reconstruction, intensity determination and beam quality estimation of laser radiation simultaneously	
Wide spectral range of wave-front measurements, including non-coherent radiation	
Wide dynamic range	
Low sensitivity to mechanical noises and vibrations and easy built-in any optical scheme	
Non-reliable operation in real-time because of probability of missing some data	Possibility to measure both CW and pulse radiation and single shots in real-time
We can be sure in accuracy of measurements at analysis of saved frames	High accuracy of measurements is the same both for live video and for saved frames
Low-cost	Fast measurements

TABLE 3. Spheres of applications.

Sensors, based on USB 2.0	Sensors, based on IEEE-1394
-	Can be used both as device for wavefront measurements and as element of Adaptive Optical System to correct for wavefront aberrations of laser radiation
-	Scientific research centers, namely, modern scientific Centers for laser fusion
-	Accessories for real-time measurements of the output beam for laser precise technologies
-	Medical and diagnostics centers
Educational and research laboratories	
Accessories for industrial laser system for testing and alignment of optical elements during preventive work	

CONCLUSION

Analysis of advantages and disadvantages of sensors, based on USB-2 cameras, (table1-3) and also the experience of such type sensors use, allows concluding:

It is device is very simple, especially is perfect device to show Shack-Hartmann principle demonstration in some cases of applications.

REFERENCES

[1] A.G. Alecsandrov, V.E. Zavalova, A.V. Kudryashov, A.L. Rukosuev, P.N. Romanov, V.V. Samarkin, Yu.V. Sheldakova. *Quantum Electronics*, "Shack-Hartmann wavefront sensor for measuring the parameters of high power pulsed solid-state lasers" v.40, no4, pp. 321-326,

[2] USB-WARE. (2005), 'FireWire – USB Comparison', [Online], USB-WARE. Available from <http://www.usb-ware.com/fi rewire-vs-usb.htm> [December 20, 2005] 11. P. 831-833.

High intensity laser beam wavefront diagnostics and correction at the Advanced Laser Light Source facility

A. Alexandrov[3], S. Fourmaux[1], J. C. Kieffer[1], A. Kudryashov[2], F. Martin[1], T. Ozaki[1], S. Payeur[1]

[1]Institut National de la Recherche Scientifique, Université du Québec, Varennes, Québec, Canada
[2]Moscow State Open University, Moscow, Russia
[3] Active Optics NightN Ltd., Moscow, Russia

Abstract: The Advanced Laser Light Source (ALLS) is a Canadian facility consisting of several high intensity laser systems based on Ti:Sapphire technology. These lasers deliver 25-35 fs laser pulse time duration, from 1 mJ up to 5 J energy per pulse, high repetition rate from 10 Hz to 5 kHz, and up to 200 TW of instantaneous power or 50 W of average power. The aim of this facility is, first to produce laser based radiation source ranging from THz up to the hard x-ray and acceleration of high energy particles; second to develop and optimize such laser based sources and study their potential applications in field such as biomedical imaging, time resolved x-ray spectroscopy or molecular imaging. We will present the laser beams wavefront diagnostics and the correction methods by means of deformable mirrors used at ALLS.

I. INTRODUCTION

Recent progress in femtosecond laser technology involving both chirped pulse amplification (CPA) technique and Ti:Sapphire amplification crystals have provided access to laboratory scale high intensity laser systems. These laser systems can now produce high peak intensities (10^{15}-10^{20} W/cm^2) at high average power (above 10 W). They can be used to develop high average flux sources of particles or radiation which are of interest for applications outside the fields of fundamental research [1].

This growing interest in laser matter interaction has opened the way to new applications, which include bright x-ray sources [2-4], high energy particles acceleration [5-7] or nuclear activation [8, 9]. However, higher peak intensities and average power, necessary to increase source brightness and energy or even to reach new laser-matter interaction regimes, are required to go beyond proof of principle and successfully implement novel applications.

It is highly efficient to optimize the focusing of the laser beam to reach maximum intensity for a given energy and minimum pulse duration. Obtaining the minimum focal spot diameter requires correcting the wavefront of the laser beam and any aberrations associated with the focusing optics. Adaptive optics systems using deformable mirrors have already been used for laser beam wavefront correction on several high power laser systems with both low and high repetition rate [10, 11].

Wavefront characterizations are performed at ALLS using Shack-Hartmann wavefront sensors. Beam correction is achieved with deformable mirrors inserted under vacuum after

the gratings compressor. The deformable mirrors we use to correct optical aberration are bimorph mirrors composed of piezoelectric-ceramics electrodes controlled by high voltages [12].

II. EXPERIMENTAL TOOLS

A. Laser systems

The Advanced Laser Light Source is a Canadian founded laser facility which includes three commercial prototype laser systems. These are all Ti:Sapphire CPA based laser systems operating at 800 nm. The specifications of each of the ALLS laser system are described in table 1. Laser systems 2A and 2B share a common front end including the oscillator, the regenerative amplification stage operated at 100 Hz repetition rate and a 4 pass amplification stage also operated at 100 Hz. The facility has open access for external users, and each year, ALLS annual report provides an overview of the work in progress at the facility.

	Pulse duration after compression	Pulse energy after compression	Repetition rate	Instantaneous power after compression	Average power
System 1A	25 fs	5 mJ	1 kHz	0.2 TW	5 W
System 2A 10 Hz output	35 fs	250 mJ	10 Hz	8 TW	2.5 W
System 2B 100 Hz output	35 fs	80 mJ	100 Hz	2.3 TW	8 W
System 3	25 fs	5 J	10 Hz	200 TW	50 W

Table 1. Summary of ALLS laser systems characteristics.

B. Wavefront sensors

Shack Hartmann wavefront sensors are used on each laser system with adapted telescope to measure wavefront distortions. The Shack-Hartmann wavefront sensor is composed of a two dimensional array of micro lenses apertures that divides the wavefront surface into a number of sub-beams (approximately 300 sub-apertures). Each micro lens (14 mm focal length) aperture provides a separate focus on the detector of the camera. The local tilt shift relative to the reference focus position of the wavefront allows us to reconstruct the wavefront phase map of the beam. To measure the correct displacement produced at each micro lens focus, the shortest local radius of curvature of the pulse wavefront should remains large enough to keep the local micro lens focus displacement small compare to the lenslet pitch (250 μm) and allow an easy correspondence between the lenslet focus positions produced by the reference and distorted wavefront.

C. Deformable mirrors

To correct for the beam aberration, a closed-loop retroaction algorithm is used to find and apply the phase conjugate of the beam wavefront measured by the Shack Hartmann sensor on a deformable mirror surface. This operation is performed in real time and takes a few seconds. Deformable mirrors are operational both on laser system 2A at the 10 Hz output and on laser system 3.

On laser system 2A compression is achieved under vacuum and the deformable mirror is used after the grating compressor. The deformable mirror is a bimorph mirror with 60 mm clear aperture and 55 mm active aperture. It is composed of 31 piezoelectric-ceramics electrodes controlled by high voltages. An imaging system allows to control that the beam is correctly centered inside the deformable mirror. The beam wavefront at the deformable mirror position is transposed onto the wavefront sensor via a leak of a mirror located just after the deformable mirror. An imaging system is then used to observe the beam focal spot at the target center position. To take into account within the correction the aberration introduced by the focusing optics, the beam wavefront at the off axis parabola mirror position can also be transposed onto the wavefront sensor via another optical system.

On the laser system 3, compression is achieved under vacuum and the deformable mirror is located after the grating compressor. The deformable mirror is a bimorph mirror with 120 mm clear aperture and 100 mm active aperture. It is composed of 48 piezoelectric-ceramics electrodes controlled by high voltages. An imaging system allows to control that the beam is correctly centered inside the deformable mirror. The beam wavefront at the deformable mirror position is transposed onto the wavefront sensor via a leak at the back of a mirror located after the deformable mirror. Another leak is used to monitor at all time the far field of the beam at the focus of a 1 m focal length lens to be able to check for the correct operation of the correction loop. The beam focal spot can also be imaged at the target center position using an imaging system that can be moved inside the laser beam during the experiments.

III. APPLICATIONS OF ADAPTIVE OPTICS AT ALLS

A. Optimization of the laser beam wavefront

The laser beam wavefront can be degraded through its propagation in the amplification stages of the laser system. To optimize the achieved intensity at the focal spot of a high quality off axis parabola with the 200 TW ALLS laser system (25 fs, 5 J, 10 Hz) we did characterize both the beam wavefront at a mirror leak located before the focusing optics (RMS value 0.471λ without correction – Fig. 1a - and RMS value 0.063λ after correction - Fig. 1c) and the focal spot size

via a high magnification ×50 imaging system (beam diameter 8.6 x 14.8 μm² at 1/e² before correction – Fig. 1b - and 7 x 7.8 μm² at the 1/e² of the laser beam peak intensity after correction – Fig. 1d). The improvement is clearly observed on the adjacent figures. The resultant calculated intensity is in excess of 10^{20}W/cm² [3].

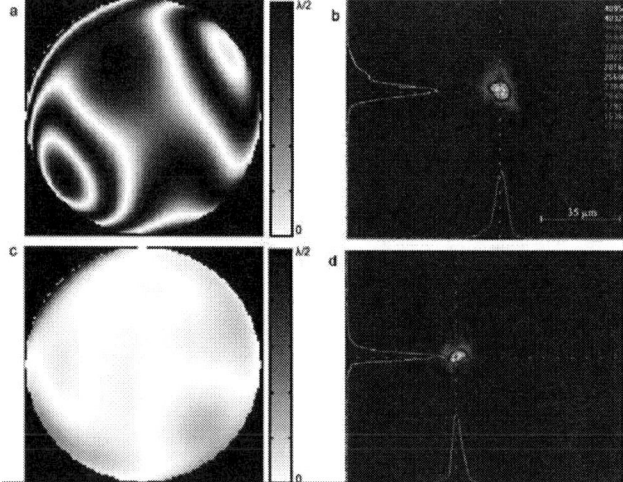

Fig. 1. ALLS laser system 3, a - beam wavefront before correction, b - focal spot with a f/3 off axis parabola before correction, c - beam wavefront after correction, d - focal spot with a f/3 off axis parabola after correction.

B. Optimization of the focusing when a highly aberrated optics is used

The second application is optimization of the focusing when a highly abberated optics is used. The ALLS laser system 2A (8 TW, 35 fs, 250 mJ, 10 Hz) wavefront RMS nominal value is below 0.1λ as the aberration produced by its propagation in the amplification stages of the laser system is low. Thus, beam wavefront correction is not required. The use of a poor quality f/4 off axis parabola degrades the beam focusing which exhibit strong astigmatism. The beam wavefront measurement is made after the focusing optics using a high magnification imaging system composed by a combination of a microscope objective followed by a lens in order to measure the laser beam wavefront at the position of the off axis parabola. Fig. 2 shows the beam focal spot observed at the focus of the laser beam before (Fig. 2a) and after correction (Fig. 2b). After correction the focal spot size is close to the diffraction limit.

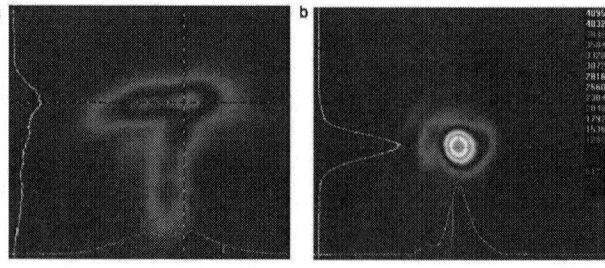

Fig. 2. ALLS 10 TW laser system, a - focal spot before correction, b - after correction.

This demonstrates the potential of beam wavefront correction to allow the use of poor quality and inexpensive off axis parabola or the use of spherical mirror in off axis geometry to focus the laser beam.

C. Potential for beam wavefront correction due to thermal loading

The ALLS laser system 2B (2 TW, 35 fs, 100 mJ, 100 Hz) has a thermal load greater than 10 W on the compression gratings located under vacuum. When the laser pulse focal spot is measured at the target center chamber position under vacuum with 100 mJ at the compressor entrance we clearly observe a focal spot distortion due to the thermal loading on the compressor gratings. Figure 3 show laser pulse focal spot pictures observed for different time delay following insertion of the laser beam in the experimental set-up: from 5 s up to 130 s [13].

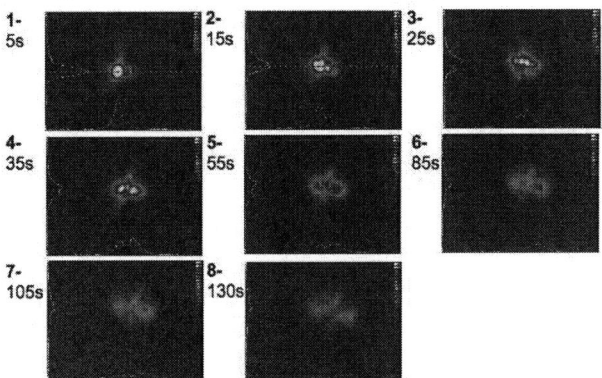

Fig. 3. Focal spot measured at the target center chamber position under vacuum for 400 nm (a KDP crystal is used to produce second harmonic laser beam and increase the laser pulse contrast), 110 mJ at the compressor entrance and 15 mJ incident at the target position. Each picture on this figure is taken for a different delay following insertion in the experimental set-up and indicated below the picture number for a range from 5 up to 130 s.

Characterization of this thermal loading has been achieved by wavefront measurements at a mirror leak before the f/3 off axis parabola focusing optics. The main aberration resulting from the thermal loading is defocusing up to 70 mJ energy inserted at the compressor entrance which correspond to a density energy on the grating of 2 W/cm². A simple method to correct for the effect of the thermal loading is to move the focusing optics in order to retrieve the focus. However, above this energy the aberration doesn't essentially consist of defocusing and so it is difficult to compensate it only by moving the focusing optics. A deformable mirror could be one solution to solve this thermal loading issue [14].

ACKNOWLEDGMENT

We acknowledge strong support from Jean Phillipe Moreau on the ALLS laser system 2A. The ALLS facility has been founded by the Canadian Foundation for Innovation (CFI). This work is founded by NSERC, the Canada Research Chair program and by Ministère de l'éducation du Québec.

REFERENCES

[1] T. Ozaki, J. C. Kieffer, R. Toth, S. Fourmaux and H. Bandulet, "Experimental prospects at the Canadian advanced laser light source facility," *Laser and Particle Beams*, vol. 24, 101, 2006.

[2] A. Rousse, K. Ta Phuoc, R. Shah, et al., "Production of a keV X-Ray Beam from Synchrotron Radiation in Relativistic Laser-Plasma Interaction," *Phys. Rev. Lett.*, vol. 93, 135005, 2004.

[3] S. Sebban, T. Mocek, D. Ros, et al., "Demonstration of a Ni-Like Kr Optical-Field-Ionization Collisional Soft X-Ray Laser at 32.8 nm," *Phys. Rev. Lett., vol.* 89, 253901 2002.

[4] L. M. Chen, P. Forget, S. Fourmaux, et al., "Study of hard x-ray emission from intense femtosecond Ti:Sapphire laser-solid target interactions," *Phys. of Plasmas,* vol. 11, 4439, 2004.

[5] S. P. D. Mangles, C. D. Murphy, et al., "Monoenergetic beams of relativistic electrons from intense laser-plasma interactions," *Nature,* vol. 431, 535, 2004.

[6] C. G. R. Geddes, Cs. Toth, J. Van Tilborg, et al., "High-quality electron beams from a laser wakefield accelerator using plasma-channel guiding," *Nature,* vol. 431, 538, 2004.

[7] J. Faure, Y. Glinec, A. Pukhov, et al., "A laser-plasma accelerator producing monoenergetic electron beams," *Nature,* vol. 431, 541, 2004.

[8] S. Fritzler, V. Malka, G. Grillon, et al., "Proton beams generated with high-intensity lasers: Applications to medical isotope production," *Appl. Phys. Lett.*, vol. 83, 3039, 2003.

[9] J. Magill, H. Schwoerer, F. Ewald, et al., "Laser transmutation of iodine-129," *Appl. Phys.* B, vol. 77, 387, 2003.

[10] F. Druon, G. Chériaux, J. Faure, et al., "Wave-front correction of femtosecond terawatt lasers by deformable mirrors," *Opt. Lett.*, vol. 23, 1043, 1998.

[11] B. Wattelier, J, Fuchs, J. P. Zou, et al., "High-power short pulse laser repetition rate improvement by adaptive wave front correction," *Rev. Sci. Instrum.,* vol. 75, 5186, 2004.

[12] A. V. Kudryashov and V. V. Samarkin, *Proc. 2nd Int. Workshop on Adaptive Optics for industry and Medicine, Durham, England, Ed. G. Love*, World Scientific, 193, 2000.

[13] S. Fourmaux, C. Serbanescu, L. Lecherbourg et al., "Investigation of the thermally induced laser beam distortion associated with vacuum compressor gratings in high energy and high average power femtosecond laser systems," *Optics Express*, vol. 17, 178, 2009.

[14] S. Fourmaux, C. Serbanescu, L. Lecherbourg et al., "Characterization of the laser beam distortion due to the thermal load on high average power femtosecond laser systems," *Proc. Photonics North 2009, Ed. R. Vallée, Proc. SPIE*, vol. 7386, 2009.

Optical investigation of InGaAsP/InP double heterostructure wafers

V. Rakovics

Hungarian Academy of Sciences, Research Institute for Technical Physics and Materials Science, Budapest, Hungary

Abstract: Complex and informative luminescent spectra have been obtained by using visible sources for excitation of InGaAsP/InP double heterostructure diode wafers. The thin contact layer transforms the high energy exciting light to lower energy photons which can excite the active layer, although the InP confining layers are not transparent for primary exciting photon.

The GaInAsP/InP alloy system is widely used for fabricating optical devices, because GaInAsP can be epitaxially grown lattice matched on InP substrates over a wide range of bandgaps from 1.34 eV (920 nm) to 0.74 eV (1680 nm) at room temperature. [1, 2] GaInAsP/InP is an ideal material system for the fabrication of double heterostructure devices. As InP has higher bandgap than the lattice-matched GaInAsP active layer, absorption losses inside the devices structure can be minimized. Liquid phase epitaxy (LPE) were used for development of the device structures, as the composition of the active layer can be relatively easily adjust by weighing appropriate amount of the materials into the growing melts. By development of the precise experimental phase diagram, LPE growth of new GaInAsP device structures became fast and reliable fabrication method. [3-6]

Photoluminescence is widely employed as a non-destructive characterization technique for epitaxial device structures, yielding measurements that may be correlated with final device performance. In order to obtain useful correlations, however, a number of experimental factors must be considered. We discuss herein the effect of probe source wavelength on the results of experiments. InGaAsP/InP double heterostructure diode wafers (Fig. 1.) grown by liquid phase epitaxy (LPE) were characterized by photoluminescence (PL) [3, 7] and infrared transmission measurements[8].

Layer structure and growth conditions of the investigated diodes were described in our previous works [5, 6,]. Double heterostructure InGaAsP/InP diodes were grown in a computer-controlled LPE apparatus equipped with a multibin slider boat. Single-phase melts with relatively high supersaturations were used for all growth experiments. Four-layer structures were grown on (100)-oriented InP substrate consisting of an n-InP buffer layer (3–4 µm), an undoped GaInAsP active layer (1–2 µm), a p-InP confinement layer (6–10 µm), and a p+ InGaAsP contact layer (0.5–1 µm). In order to get efficient diodes the contact layer and the confining layers must be transparent to the radiation emitted by the active layer. 1120nm contact layer was used in short wavelength diodes and 1180nm contact layer for long wavelength emitting diodes.

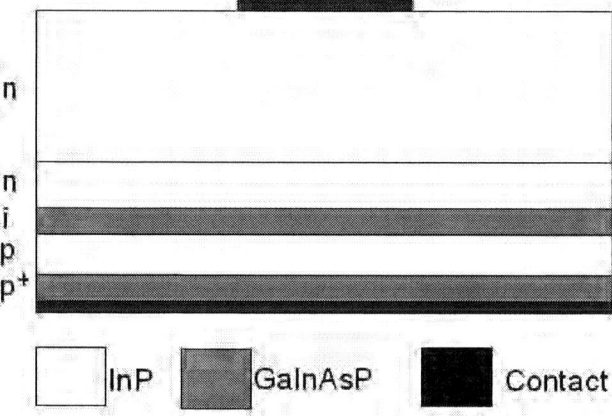

Fig. 1. Schematic cross-section of InGaAsP/InP double heterostructure diode.

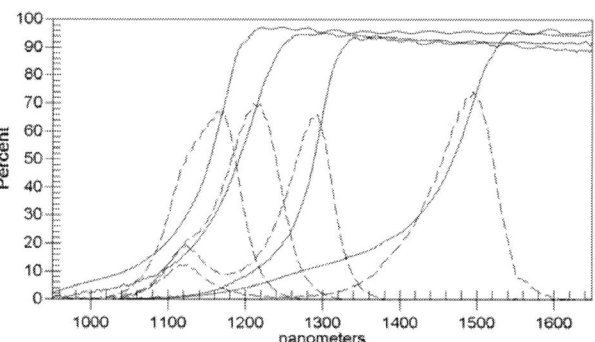

Fig. 2. IR transmission and photoluminescence spectra of 1180 nm, 1220 nm, 1390 nm ang1500 nm LED wafers. Dashed line shows the luminescent spectra at 1120 nm excitation wavelength.

PL and IR transmission spectra have been measured by fiber optic diode array spectrometer. Tungsten halogen lamp source was used for transmission measurements. Fig. 2. shows that information obtained from luminescence and infrared transmission are compatible regarding active layer compositions, but in the case of thick active layers the thickness and compositions of quaternary contact layers remain hidden in transmission measurements, as the active layer is not transparent for higher energy photons.

The optical band gap of direct band gap semiconductors can be obtained from their optical absorption spectrum using $\alpha = A(h\nu - E_g)^{1/2}/h\nu$, where α is the absorption coefficient, A is a constant, $h\nu$ is the energy of the absorbed light, and E_g is the band gap [3].

PL peaks corresponding to 1180 nm contact layer are also not appearing in the spectra. As it is shown complex and informative luminescent spectra have been obtained by using visible sources for excitation of InGaAsP/InP double heterostructure diode wafers. The thin contact layer transforms the high energy exciting light to lower energy photons which can excite the active layer, although the InP confining layers are not transparent for primary exciting photon. The peaks corresponding to the active and the contact layers are clearly seen, but the intensity of the peak are smaller.

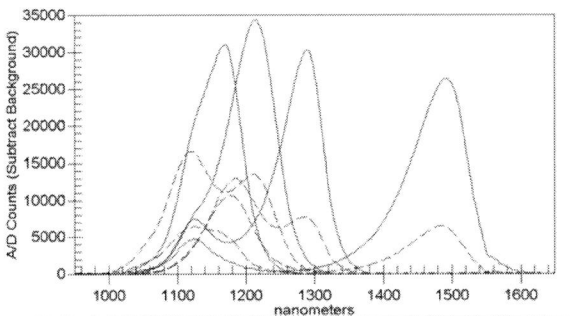

Fig. 3. PL spectra of InGaAsP/InP heterostructure wafers. Solid lines show the spectra with 1120 nm excitation and dashed lines show with 525 nm excitation wavelengths.

Intensity ratio of the corresponding active and contact layers is influenced by their thicknesses, and is also affected by the excitation wavelength. Fig. 4 shows the PL spectra of 1510 nm LED wafer at three different excitation wavelengths.

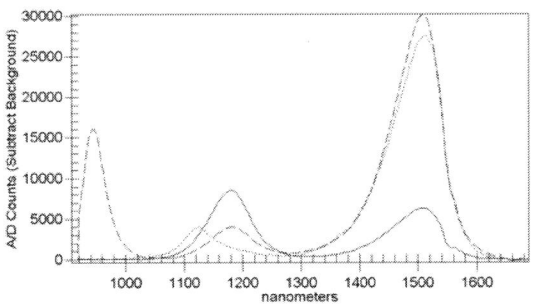

Fig. 4. PL spectra of InGaAsP/InP double heterostructure LEDs with 1510 nm active and 1180 nm contact layers. Excitation wavelengths: 1120 nm dotted line, 940 nm dashed line, 525 nm solid line.

The two short wavelength peaks are only side effect of the scattering excitation light from the back side of the wafer, but both characteristic peaks are visible at visible and near infrared

excitation. Strong characteristic peaks were obtained by application of 940 nm excitation wavelength.

Fig. 5 shows PL spectra of 1180, 1220, 1290 and 1490 nm LED wafers under green LED excitation

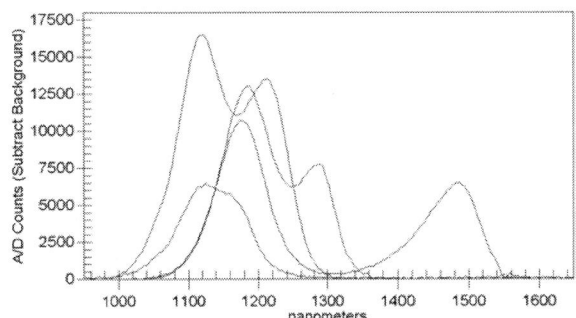

Fig. 5. PL spectra of 1180, 1220, 1290 and 1490 nm LED wafers under green LED excitation.

High power green LED souces with excellent characteristics concerning optical power and beam quality are commercially available. They have excellent stability and spectral purity. Their IR radiation is negligible, so they do not disturb the infrared luminescence of InGaAsP layers. PL peaks of the active layers measured under green and IR excitation are seen in Fig. 6.

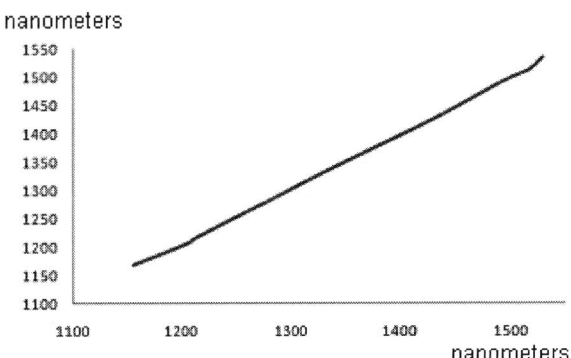

Fig. 6. Comparison of PL peaks obtained with two different excitation wavelengths (525 and 940 nm).

The peak wavelengths obtained by the evaluation of the PL spectra measured with two different excitation sources are practically the same values. As we pointed out above, the intensity of the luminescent peaks are very sensitive to layer thicknesses and doping levels, PL mapping of the double heterostructure wafers with green LED excitation can effectively reveal growth imperfections. In Fig. 7. PL spectra of the same wafer can be seen after thinning the contact layer from 0.8 µm to 0.5 and 0.4 µm. The PL peak of the contact layer decrease and the peak of the active layer slightly increase with contact layer thickness reduction.

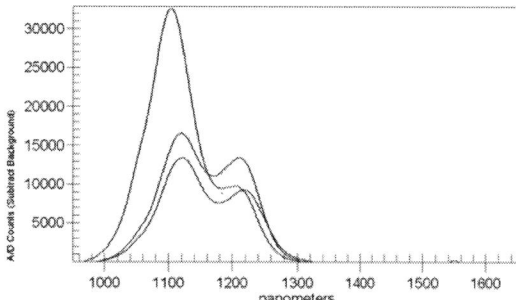

Fig. 7. PL spectra of 1220 nm InGaAsP/InP double heterostructure diode wafer after thinning the contact layer from 0.8µm to 0.5 and 0.4 µm. Excitation wavelength is 525 nm.

In summary PL intensity spectra are influenced by absorption and energy transfer between the quaternary layers. The peak positions are precisely indicating the composition of the lattice-matched quaternary layers, but for determination of the layer thicknesses an additional measurement, such as optical transmission, is necessary. After calibration of PL peak intensity ratios with quaternary layer thicknesses, the PL measurement is suitable for quality control of InGaAsP/InP double heterostructure wafers.

REFERENCES

[1] P. A. Houston, Rewiew Growth and characterization of InGaAsP lattice-matched to InP," *Journal of Materials Science* 16, (1981) 2935-2961.

[2] S.Adachi, *Physical Properties of III-V Semiconductor Compounds*, John Wiley and Sons, 1992.

[3] R. E. Nahory, M. A. Pollak, W. D. Johnston and R. L. Barns, "Band gap versus composition and demon- stration of Vegard's law for Ini-xGaAsyPl-y lattice matched to InP," *Appl. Phys. Lett.* 33 (1978) 659-661.

[4] K. Nakajima, A. Yamaguchi, K. Akita and T. Kotani, "Composition dependence of the band gaps of In1-xGasAsl-yPy quaternary solids lattice-matched on InP substrates," *J. Appl. Phys.* 49, (1978) 5944-5950.

[5] V. Rakovics, J. Balázs, S Püspöki, and C. Frigeri, "Influence of LPE growth conditions on the electroluminescence properties of InP/InGaAs(P) infrared emitting diodes" *Mater. Sci. Eng.* B 80,(2001) 18-22.

[6] Rakovics V, Püspöki S, Balázs J, Réti I, Frigeri C, "Spectral characteristics of InP/InGaAsP Infrared Emitting Diodes grown by LPE", *Materials Science and Engineering* B91-92, (2002) 491-494.

[7] D. Z. Garbuzov," Photoluminescence studies of InGaAsP double heterojunctions", *Czechoslovak Journal of Physics*, 34 (1984) 365-373.

[8] V. Rakovics, J. Balázs, I. Réti, S. Püspöki and Z. Lábadi, "Near-Infrared Transmission Measurements on InGaAsP/InP LED Wafers", *Phys. Stat. Sol.* (C) 0, No3, (2003) 956-960.

Determination of active region overheating temperature of GaN-based light emitting diodes promising for laser media pumping

E. V. Lutsenko, V. N. Pavlovskii, A. V. Danilchyk, M. V. Rzheutski, A. G. Vainilovich, V. Z. Zubialevich, A. V. Muravitskaya, and G. P. Yablonskii

Stepanov Institute of Physics NAS of Belarus, Minsk, Belarus

Phone: +375-17-2949025, Fax: +375-17-2840879, e-mail: v.pavlovskii@ifanbel.bas-net.by

Abstract: Overheating temperature of active region of LEDs made of Bridgelux chips was determined using long-wavelength shift of the EL spectra as well as the voltage reduction at constant current. It was shown that direct liquid active cooling of LED chip reduces drastically its thermal resistance that is promising for pumping of laser active media.

Pumping of laser active media by emission of ultrabright light emitting diodes (LEDs) requires high efficiency, operational stability and long life time of LEDs. Possibilities of efficiency enhancement and durability prolongation of bright InGaN based LEDs are not exhaust to date. There exist already Nd and polymer lasers pumped by LED emission [1, 2]. Further increase of the LED efficiency can lead to an extension of nomenclature of laser active media pumped by their emission. Life time of LEDs working at high current values and consuming electric power more than 1 W depends strongly on active region overheating temperature during their operation. The lower is the overheating temperature the larger may be the injection current and the longer is LED life time at nominal current. In this connection, correct assessment and control of this temperature takes up especial importance. One of the ways of the LED technology development is an improvement of heat removal from the diode active region during its operation and thus a reduction of its overheating temperature. A number of techniques may be used to move downward LED operation temperature. One of the ways to reduce this temperature consists in using LED chip active direct cooling, for example, by liquid flow washing the chip. The aim of this work was determination of the overheating temperature of LED chip active region using long-wavelength shift value of their EL spectra as well as voltage reduction at constant current during operation and lowering this temperature by means of direct active cooling of the LED chip by, for example, deionized water washing flow.

Usually metal radiators of various designs serve as passive heat sink devices. At that heating may occur not only of the LED itself but heating of adjacent radiator region during LED operation. At environment temperature instability, this cooling type leads to instability of the LED optical characteristics inadmissible at their use in photometric technique, scientific research etc. Stabilization of LED chip temperature for the account of active liquid cooling may improve LED output parameters (optical power, spectral composition stability) and prolongate its life time.

To estimate active region overheating temperature by operational constant current (350 mA), heating of LED (made of Bridgelux BXCA4545 chip) by an external source was carried out firstly at pulsed current of the same value with low duty cycle. Low duty cycle (D=1 %) of the short pulses of the current (τ_p=1 μs) ensures only insignificant current heating of the LED chip in this case. At that, EL spectrum position for each LED temperature nearly equal to the active region temperature was determined [3]. Thus a temperature shift value of the EL spectrum relative to its position at environment temperature in dependence on the active region temperature was found. Then the nominal operational direct current of constant value was applied to the diode at room temperature and EL spectrum position of LED was measured in steady state regime. Further, difference between EL spectra positions at pulsed current without external heating and at constant operational current was compared with EL spectrum shift at external heating. The active region temperature at constant current was believed equal to that at pulsed current with external heating for which the EL spectrum maximum shift was the same. Such measurements allow as well following time evolution of the active region overheating temperature immediately after LED activation if one has a possibility to register quickly EL spectra. It should be noted that it is fundamentally important to measure the spectral shifts at equal pulsed and constant currents in order to exclude influence of carrier concentration effects on accuracy of overheating temperature determination.

Fig. 1 presents EL spectra of the blue Bridgelux LED at different temperatures. The EL spectrum is smooth, its FWHM is equal to 22 nm and intensity maximum is positioned at 467 nm at 20°C. Clearly visible long-wavelength shift of the EL spectrum occurs at temperature change from 20 to 40°C at constant operational current.

Fig. 2 shows a dependence of blue EL band on temperature of Bridgelux LED heated by an external source at pulsed current of nominal operational value with low duty cycle. At such excitation, active region temperature can be believed to be

CAOL*2010 International Conference on Advanced Optoelectronics & Lasers, 10-14 September, 2010, Sevastopol, Ukraine

constant and equal to environment temperature. Thus, at such injection level, current heating of LED active region (corresponding to long-wavelength shift of the EL spectrum) is negligibly small [4-7] and the observed experimentally long-wavelength shift of the EL band is caused mainly by a narrowing of InGaN layers band gap caused by the external heating of LED.

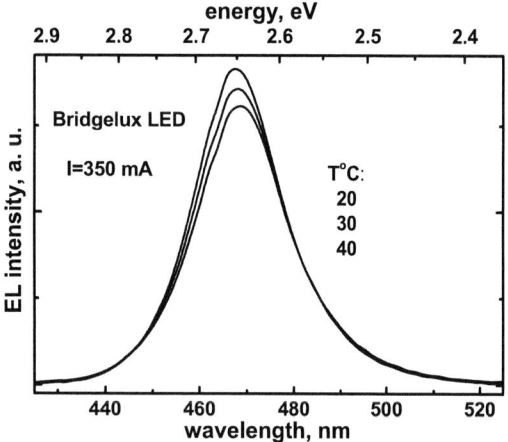

Fig. 1. EL spectra of blue Bridgelux LED at T = 20–40°C under pulsed current I=350 mA.

It is seen from the figure 2 that heating by 20°C corresponds to alteration of the wavelength by 0.73 nm to the long-wavelength side for the given LED. For short temperature interval such as 20°C, EL spectrum peak position may be approximated for this diode by a linear fit shown in Fig. 2 by solid line. Thus coefficient of temperature shift equal to 27.4°C/nm was obtained for the investigated chip. Such calibration can be used as a reference for determination of LED active region temperature at nominal currents causing its heating.

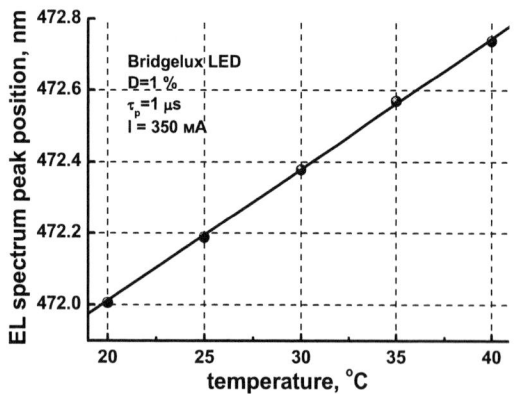

Fig. 2. EL spectrum maximum position of blue Bridgelux LED in dependence on temperature at pulsed current ($\tau_p=1$ µs, D=1 %, I=350 mA) at heating by an external source.

If one has a possibility to measure EL spectra quickly including measurements at early stages after LED activation than there appears an opportunity to follow overheating temperature kinetics with good time resolution. The spectrum should shift to the long-wavelength side in the course of time for the account of active region overheating increase by large current. Fig. 3 shows EL spectrum peak position versus time at passive cooling of the LED chip (triangles) and active cooling by water (circles) of the laboratory LED at 0, 30 and 60 s after LED activation. Direct active cooling of the LED chip surface was ensured by water flowing through the chamber between chip surface and dome-shaped lens glued to LED housing at lens perimeter. Very large difference of the EL spectrum position change was seen for two kinds of cooling during first minute after LED chip activation.

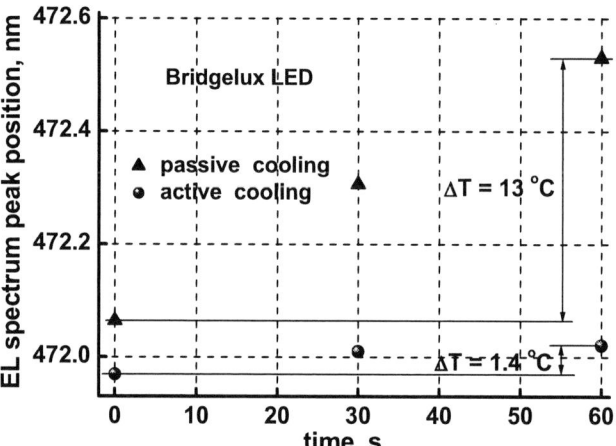

Fig. 3. EL spectrum peak position of LED chip versus time at nominal current I=350 mA at passive (triangles) and active (circles) cooling of the chip.

The active region overheating temperature reaches the value of 13°C at passive cooling whereas it is equal only to 1.4°C at active cooling. A stabilization of the active region temperature is seen in Fig. 3 at active cooling during the first minute. This confirms that water flow ensures enough efficient heat removal from the light emitting active region and stabilizes quickly its temperature and EL spectrum. This means also that thermal resistance of the LED chip actively cooled through the surface is lower than that at its conventional passive cooling.

The second method of the overheating temperature evaluation consists in measuring of applied to LED voltage at constant operation nominal current value. At that temperature calibration can be done using LED heating by an external source at pulsed (not heating) current of the same value with low duty cycle.

Fig. 4 shows a dependence of voltage of Bridgelux LED on its temperature at pulsed ($\tau_p=1$ µs) current I=350 mA with duty cycle of 1/100. Well visible decrease of the voltage is observed at LED temperature alteration by a value of about 50°C. This dependence can be fitted by a line and it gives a coefficient of

978-1-4244-7043-3/10 $26.00 © 2010 IEEE   220

4.7 mV/K for this LED. It should be noted that it is important for correct measurements to use pulsed current and constant current of the same value as it was mentioned earlier.

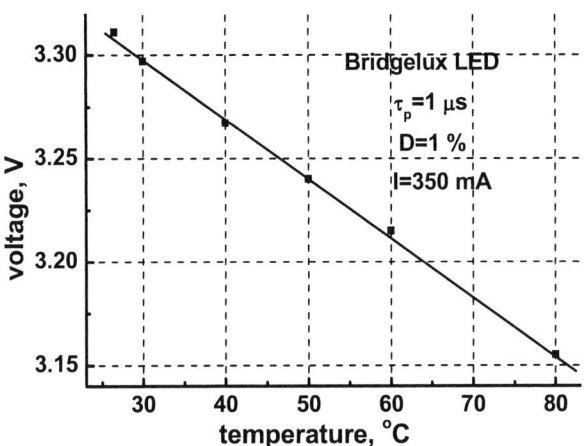

Fig. 4. Voltage of Bridgelux LED in dependence on its temperature at constant pulsed current I=350 mA.

Fig. 5 presents change of voltage at direct current I=350 mA through the active region with time during first minute after activation at passive and active cooling of the chip. Permanent voltage decrease is observed at passive heat removal evidencing substantial active region temperature rise. Much lower voltage drop is observed only during first 10 seconds after activation at active water cooling of the chip after that the voltage and correspondingly the overheating temperature stabilize. The overheating temperature for actively and passively cooled chip amounted the values of 1.7 and 11°C correspondingly which are in reasonable agreement with the values obtained using method of EL spectrum shift.

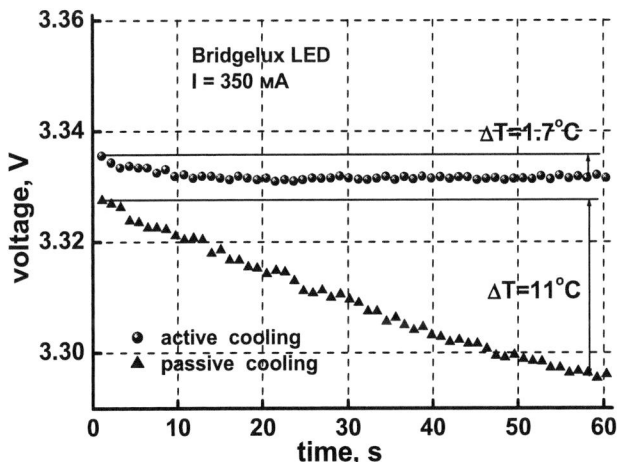

Fig. 5. Change of voltage at constant current through the active region I=350 mA with time during first minute after activation at passive and active cooling of the LED chip.

Because thermal resistance of LED actively cooled by water washing flow is much lower than that at a conventional cooling, LEDs with actively cooled chips are promising for creation of large light fluxes necessary for pumping of laser active media. Despite large operational LED currents necessary for laser active media pumping, low thermal resistance of LED will allow to ensure long LED life time as well as lower quantum efficiency drop caused by overheating. We believe that LED chip arrays with direct chip liquid cooling will allow reaching large light fluxes necessary for laser active media pumping and they are promising also for replacement of traditional pumping sources.

REFERENCES

[1] Self-contained LED-pumped single-crystal Nd: YAG fiber laser / J. Stone, C. A. Burrus // Fiber and Integrated Optics, Vol. 2, Iss. 1, 1979 , p. 19-46.

[2] Hybrid optoelectronics: A polymer laser pumped by a nitride light-emitting diode / Y. Yang, G. A. Turnbull, and I. D. W. Samuel // Appl. Phys. Lett. 92, Issue 16, 2008, P. 163306-163306-3.

[3] Growth, Stimulated Emission, Photo- and Electroluminescence of InGaN/GaN EL-Test Heterostructures / E. V. Lutsenko [et al.] // Phys. stat. sol. (c).-2002. - No. 1, P. 272-275.

[4] Zhao, L. X. Degradation of GaN-based quantum well light-emitting diodes / L. X. Zhao, E. J. Thrush, and C. J. Humphreys // Journ. Appl. Phys. - 2008, -Vol. 103, P. 024501-024501-6.

[5] Vaitonis, Z. Measurement of the junction temperature in high-power light-emitting diodes from the high-energy wing of the electroluminescence band / Z. Vaitonis, P. Vitta, and A. Žukauskas // Journ. Appl. Phys. - 2008,-Vol. 103, P. 093110-093110-7.

[6] Tsai, P. C. Lifetime Tests and Junction-Temperature Measurement of InGaN Light-Emitting Diodes Using Patterned Sapphire Substrates / P. C. Tsai, R. W. Chuang, and Y. K. Su // Journ. of Lightwave Technology. - 2007. -Vol. 25, No. 2, P. 591-596.

[7] Liao, M. P. Carrier Concentration and Junction Temperature Dependencies of Illumination Efficiency of GaN Power Light-Emitting Diodes / M. P. Liao // Mater Quantum Electronics and Laser Science Conference (QELS): Abstracts of International conference. Baltimore, Maryland, USA, Poster Session I (JTuA) May 6, 2007./ Baltimore, Maryland, 2007.

Locating the Waist of a Gaussian Laser Beam

K. I. Muntean

National Scientific Centre "Institute of Metrology", Kharkov, Ukraine

Abstract: The interferometric method of locating the waist position of a gaussian laser beam is considered. The implementation of the method by a plane-parallel plate and a Michelson interferometer is described. Methodological and instrumental errors of the method are estimated.

High-precision laser interferometry, optical communications, location, range finding and other scientific and applied laser applications are in need of measurement of the laser beam spatial parameters.

From the Gaussian beam theory it follows, that for a complete description of the spatial geometry of laser beam in the optical system in the general case, one must set the radiation wavelength and two spatial parameters of the original beam. Theoretically it is permissible to use any spatial parameters, but in practice it is convenient to measure the angle of divergence of the beam of radiation and the coordinate of the waist plane.

It is known more than twenty various methods of measuring the angle of divergence [1-4]. Just the such number of methods indicates that none of them is not very good. Almost all of this methods are modified method of laser beam sections, or modified method of the focal spot, or modified method of recording of the directivity diagram, which are recommended to use the legitimate standards [5, 6].

The least error of measurement, including small angle of divergence, which represent greatest practical interest, is provided by method of recording of the directivity diagram. For its realization a good access to the beam waist is necessary, but it is impossible in practice in most cases.

Measuring a coordinate of the waist plane is less popular, although this option in some cases have a crucial role. It is known method of calculating the coordinate of the waist plane in a laser cavity by the formula [7]

$$l = L(R_2 - d) / (R_1 + R_2 - 2L)$$

where L, R_1 and R_2 - cavity length and radii of curvature of the laser mirrors, respectively. However this method is not suitable for determining the coordinate of the waist plane of the laser beam, which is converted by any optical element.

It is known method of determining the coordinates of the waist plane of the laser beam concerning the device which forms this laser beam [8]. According to this method the divergence angle θ and the diameter D of the laser beam is measured, and the distance l from the plane of measuring up to the waist plane is calculated by the formula

$$l = \left(D^2 - \lambda^2 / \pi^2 \theta^2\right)^{1/2} / 2\theta$$

where λ - the wavelength of the laser radiation. The drawback of this method is the relatively large error of measurement of the waist plane coordinate.

Also it is known methods of determining the coordinate of the waist plane by recording diameters beam of the radiation. [9] or by recording parameters of the diffraction pattern [10].

In accordance with [9], the beam-waist radius, and hence the corresponding diameter of a Gaussian laser beam is located and measured electronically. This is accomplished by periodically interrupting the laser beam in at least two parallel planes separated by a preestablished distance to obtain pulsive signals representative of the total power of the beam. The pulsive signals are differentiated to obtain signals which are proportional to the reciprocal of the beam radii at the planes of interruption. The beam-waist radius of the Gaussian laser beam of known wavelength is a function of the ratio of the amplitudes of the differentiated signals representative of the beam radii at the planes at which the beam is being interrupted.

In [10] a method for the determination of the positions of focal planes and optical axes of converging laser beams is developed on the basis of an analysis of calculated and experimentally observed diffraction patterns. The method requires measurement of the total amount of light bypassing an opaque strip inserted in the focal plane of a laser beam.

These methods are of a general lack of - they require a good access to a physically existing waist, so they have limited applicability.

In [11 - 15] an interferometric method for determining the coordinates of the beam waist of laser radiation is offered, which allows to realize the measurement without actual access to the beam waist. According to this method the original laser beam is split into information and reference coherent beams, which form in the surrounding space non-local two-beam interference pattern. This pattern is similar to the spatial interference pattern of two coherent point sources, and this is a family of nested two-sheeted hyperboloids of revolution with foci at the location of the imaginary waists of laser beams. This interference field at any point in space depends only on the wavelength, the distance between virtual source and the distance from the origin. Therefore knowing the wavelength, the distance between virtual sources and the period of the interference pattern in a recording plane we can calculate the distance from a recording plane up to the origin, i.e. define the waist coordinate. Thus the problem of measuring the spatial parameters of the laser beam, which the plane of the waist is not available or does not exist physically, is reduced to measuring the period of the interference pattern.

The analysis of the known systems of the formation of two interfering beams from the same source shows that the system consisting two parallel planes, which creates two imaginary sources, is applicable only. In this case the distance between imaginary sources is determined only by the parameters of the system formation. Any other system (Fresnel mirrors and Fresnel biprism, Bié split lens, Lloyd's mirror, etc.) form the imaginary sources, the distance between which depends also on the unknown distance from the waist plane of the initial laser beam up to the system of the formation. Technical implementation of the system of two parallel planes can be transparent plane-parallel plate or a Michelson interferometer set up in the zero order. In general case the distance $2C$ between imaginary waists of two interfering beams is equal

$$2C = (2d \cos\alpha)/(n^2 - \sin^2\alpha)^{1/2} \qquad (1)$$

where d - the thickness of the plate; n - refractive index of the plate material; α - angle of incidence of laser beam on the plate.

From (1) the distance between the imaginary waists of interfering beams of radiation, i.e. the focus position of the family two-sheeted hyperboloids, is depended on the angle of incidence of the laser beam on the plane-parallel plate, and hence interference pattern of two Gaussian beams, formed by a system of two parallel planes, is not a complete analogue of the classical interference pattern of two real point sources. When the interference pattern is observed in an arbitrary recording plane, family of the interference fringes can be found as the intersection of the recording plane with the family of hyperboloids. The shape, width and direction of the interference fringes depend only on the position of the registration plane.

Let's consider the cases of observation the interference pattern in practice. We put the origin in the middle between imaginary waists and spend the z-axis through the centers of the two waists. In all cases the observation plane is at a distance $l >> 2C$ from the origin. In principle the observation plane take the next positions:

Let's consider the cases of observation the interference pattern in practice. We put the origin in the middle between imaginary waists and spend the z-axis through the centers of the two waists. In all cases the observation plane is at a distance $l >> 2C$ from the origin. In principle the observation plane take the next positions:

- the observation plane is parallel to the z-axis. Interference pattern is observed as a family of almost equidistant interference fringes, which with some accuracy can be considered as direct;
- the observation plane is located at some angle to the z-axis. This case is realized most often in practice in the measuring of spatial parameters of the laser beam. The interference pattern is observed as a family of curved spaced interference fringes;
- the observation plane perpendicular to the z-axis. The interference pattern is observed as a family of concentric interference rings.

These three cases exhaust the entire set of observed interference patterns. In forming an interference pattern with a plane-parallel plate one can get all three cases of observation, changing the angle of incidence of the radiation beam on the plate. In forming the interference pattern using Michelson interferometer, which is tuned to the zero order, we sold the third case of observation.

The width h of the interference fringes depends on the distance l between the recording plane and the origin. At any point (x, y) the recording plane come two beams, which have an optical path difference

$$\Delta l = \left[(x+C)^2 + y^2\right]^{1/2} - \left[(x-C)^2 + y^2\right]^{1/2} = 2C - m\lambda$$

where m - the order of interference. Reduce the equation to canonical form we obtain

$$x^2 \frac{4}{(2C-m\lambda)^2} - y^2 \frac{4}{4Cm\lambda - m^2\lambda^2} = 1 \qquad (2)$$

and add the equation of the recording plane passing through the point (x, y)

$$x\cos(\alpha-\varphi) + y\sin(\alpha-\varphi) = l \qquad (3)$$

where φ - the angle of incidence of the radiation beam on the plane of registration.

Solving (1) - (3) on l, we obtain

$$l = \frac{hd\sin 2\alpha \cos^2\alpha\left[\cos(\alpha-\varphi) + tg\alpha\sin(\alpha-\varphi)\right]^2}{\lambda\left(n^2 - \sin^2\alpha\right)^{1/2}}$$

where h - the width of the interference fringes. If the plane of registration is positioned perpendicular to interfering beams, that can almost always be done, one have then $\varphi = 0$ and the calculated formula takes more simpler form

$$l = h\,d\sin 2\alpha\ /\ \lambda(n^2 - \sin^2\alpha)^{1/2} \qquad (4)$$

If we are using a Michelson interferometer in place of the period of the interference fringes in the recording plane is more convenient to measure the diameter of concentric rings. The formula in this case acquires the form

$$l = (d(D_{n-1}^2 - D_n^2)\ /\ \lambda n)^{1/2}$$

where d - the thickness of equivalent air plate Michelson interferometer; D_n and D_{n+1} - diameters of any two adjacent interference rings, respectively.

In the calculation formula (4) there is only one variable, namely the period of interference fringes. All other quantities in (4), for each specific implementation of the interferometric method are constant, the absolute value of which is determined in the calibration of a particular measuring device.

To estimate the error of the interferometric method differentiate (4) and obtain the relative error with some excess

$$\frac{\delta l}{l} \approx \left[\left(\frac{\delta h}{h}\right)^2 + \left(\frac{\delta d}{d}\right)^2 + \delta\alpha^2 + \left(\frac{\delta\lambda}{\lambda}\right)^2 + \delta n^2\right]^{1/2}$$

In order of magnitude the particular components of the general error are equal

$$\frac{\delta h}{h} \approx 0,001;\ \frac{\delta d}{d} \approx 0,0001;\ \delta\alpha \approx 0,0001;$$

$$\frac{\delta\lambda}{\lambda} \approx 0,00001;\ \delta n \approx 0,0001$$

hence

$$\frac{\delta l}{l} \approx 0,001$$

Thus, the dominant error of the interferometric method of determining the coordinate of the beam waist is the error of measurement of the interference fringe period.

The method is based on properties of the two-beam

interference. Implementation of the method using a Michelson interferometer is fully consistent with the two-beam interference criteria. At the same time, the implementation of the method using plane-parallel plate not strictly meet these criteria. From a fundamental point of view, plane-parallel plate generates a number of reflected beams of decreasing intensity. In some cases, additional beams can be neglected. For example, a glass plate without reflecting coating in the visible spectrum forms the upper beams of the total intensity of less than 2%. If the intensity of high beams is large enough, then two-beam interference turn into a multi-beam. This case is realized in the formation of the interference pattern using a Fabry-Perot interferometer. Information about the coordinate of the laser beam waist in the multi-beam interference pattern, in our opinion, is lost.

Laboratory testing of the interferometric method was carried out using He-Ne laser with "the plane - sphere" cavity, regular plane-parallel plate on the optical bench of the OCK-2 and regular collimator. Plate thickness is 20.75 mm, the angle of beam incidence on the plate varied in the range 0° - 45°, the distance between the recording plane and the plane of the beam waist at work without the collimator was equal to 75 m. The typical form of obtaining interference patterns at work with the collimator is shown in Fig. 1.

Tests have confirmed the suitability of the present method.

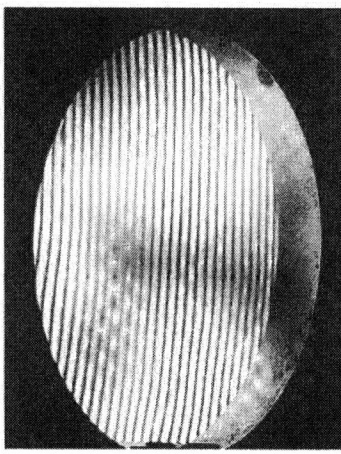

Fig. 1. Typical interferogram for He-Ne laser (632,8 nm, multimode operation, reduced)

The above interferometric method of measuring the coordinates of the laser beam waist was also implemented in the development of a set of equipment for measuring the amplitude-frequency and spatial parameters of a beam of high-power single-frequency CO_2 - lasers with λ = 10,6 μm [16].

REFERENCES

[1] Хирд Г. *Измерение лазерных параметров.* М., Мир,1970.

[2] Высокосов Е.П, Кубарев А.В, Морозов Б.И. и др. Методы измерения расходимости излучения оптических квантовых генераторов. *Измерительная техника, 1973, № 5, с. 32*

[3] Славнов С.Г. Определение угла расходимости излучения ОКГ и методы его контроля. *Измерительная техника, 1976, № 4, с. 45*

[4] Кузьмичев В.М., Погорелов В.И. Измерение диаметра фокального пятна лазерного излучения тонкопроволочным болометром. *Український метрологічний журнал.1999 № 4*

[5] ГОСТ 26089 - 84. Методы измерения диаметра пучка и энергетической расходимости лазерного излучения.

[6] ISO 11146-1 : 2005 "Lasers and laser-related equipment. Test methods for laser beam widths, divergence angles and beam propagation ratios – Part 1: Stigmatic and simple astigmatic beams".

[7] Kogelnik H., Li T. Laser beams and resonators. *Appl Optics, 1966, v. 5, № 10, p. 1550 – 1567.*

[8] Ищенко Е.Ф., Климков Ю.М., *Оптические квантовые генераторы.* М., "Сов. Радио", 1968 г.

[9] Arnaud Y. A. Apparatus for locating and measuring the beamwaist radius of a Gaussian laser beam. *US patent № 3617755, G 02f 1/28.*

[10] Винокуров Г.И., Горбунов В.А., Дятлов В.Д. и др. Метод определения положения фокальной плоскости сходящихся лазерных пучков. *Квантовая электроника, 1976, т. 5, № 3.*

[11] Murty M.V.R.K. The use of a single plane parallel plate as a lateral shearing interferometer with a visible gas laser source, *Appl. Opt., Apr 1964, vol. 3, No 4, p. 531.*

[12] Miyamoto T., Yasuura K. Measurement of the Beam Parameters of a Laser Beam and Its Diffraction Field, Using a Hologram. *Appl Optics, 1971, v. 10, № 1.*

[13] Мунтян К.И, Зимокосов Г.А. Способ определения координаты перетяжки пучка излучения ОКГ. *А.с. СССР № 550917 от 31.10.75 г.*

[14] Coranson R.W., Tart F.D. Infrared Laser Vergence Meter. *"Electro-Optics/Laser 77 Conf and Expo. Proc. Tech Program, Anaheim, Calif, 1977 ", p. 287.*

[15] Тихонов Е.А, Киселев О.В. Интерферометрический метод измерения угловой расходимости лазерных пучков. *Оптический журнал, 2007, т. 74, № 4, с. 42.*

[16] Бабенко К.И., Ефимов Г.В., Зимокосов Г.А. и др. Комплекс аппаратуры для измерения амплитудно-частотных и пространственных характеристик излучения одномодового оптического квантового генератора с λ = 10,6 мкм. *Исследования в области квантовой электроники. Ленинград, 1976, вып. 203 (263), с. 32 - 48.*

CAOL*2010 International Conference on Advanced Optoelectronics & Lasers, 10-14 September, 2010, Sevastopol, Ukraine

High power laser beam position stabilization system by means of adaptive optics.

M. E. Dryagin, A.G. Alexandrov, P. N. Romanov, A. L. Rukosuev, V. V. Samarkin
"Active Optics NightN" Ltd., 115407, Sudostroitelnaya 18 bld. 5, Moscow, Russia, tel. (499) 618-81-71

Abstract: Considered stabilization system serves for correcting the position of the laser beam in space and represents a closed circuit, consisting of the following basic elements: adaptive optical system, presented by two tip-tilt mirrors; the position detector, based on two photodiode sensors; controller.

In powerful gas-discharge lasers there is a problem changing the position of the beam in space, and also bunch tremblings. That in optical system bimorph deformable mirror as much as possible effectively corrected wave front, applies laser beam position stabilization system.

The system works on following algorithm: a laser beam, being reflected from two tip-tilt mirrors, falls on a beam splitter. A beam part, being reflected, falls on bimorph deformable mirror, and the part passes through a beam splitter and is detected by two quadrant sensors. If we expand the part of the scheme, presented in the figure as «Tilt Sensor Unit», we will see the laser beam as a line passing through two quadrant sensors. The controller reads out indications from sensors and, according to internal algorithm, controlling tip-tilt mirrors, aligns the position of the laser beam, aspiring to that it passed through the centers of quadrant photodiodes.

Positioning accuracy tip-tilt mirrors is 0.2 µrad. The system operates with a frequency of 300 Hz in real time.

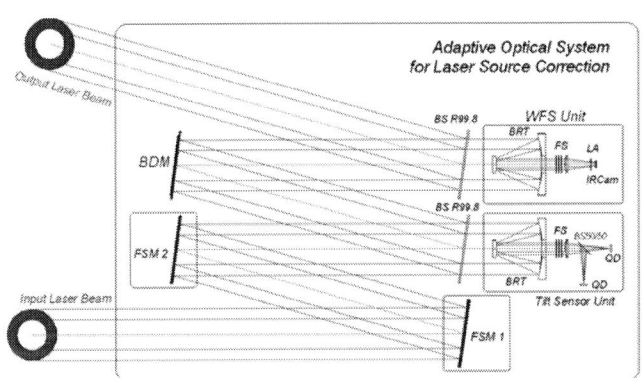

The figure shows the functional scheme of an arrangement of elements in stabilization system:

- FSM1, FSM2 – fast steering mirrors.
- BS – beam splitters.
- Tilt Sensor Unit.
    - BRT – beam reducing telescope.
    - FS – filter set.
    - QD – position detectors, based on two quadrant photodiodes.
- BDM – bimorph deformable mirror.
- WFS Unit – wave front sensor.
    - BRT – beam reducing telescope.
    - FS – filter set.
    - IRCam – infra red camera.

978-1-4244-7043-3/10 $26.00 © 2010 IEEE

CAOL*2010 International Conference on Advanced Optoelectronics & Lasers, 10-14 September, 2010, Sevastopol, Ukraine

Enhanced interferometric technique for non-destructive characterization of crystalline optical materials: automated express refractive index measurements

I. D. Karbovnyk[1], N. A. Andrushchak[2], Ya.V. Bobitskii[2]

[1]Ivan Franko National University of Lviv, Lviv, Ukraine

[2]Lviv Polytechnic National University, Lviv, Ukraine

Abstract: We report an enhanced experimental set-up that allows fast and accurate determination of refractive indexes in optical materials. The set-up is equipped with the high-precision sample positioning system and sensitive detector for the registration of interference fringes shift. Additionally, an approach that can be utilized in order to eliminate the error caused by non-parallel sample edges is discussed. The results of the refraction index measurements in lithium niobate are presented.

Refractive index for a given wavelength is a crucial parameter for optical media. Although, there exists a variety of methods that allow determining the refractive index of the given material [1, 2, 3], iterferometric-turning technique [4] has several advantages in compare to other methods. One of the main arguments in favor of iterferometric-turning measurements scheme is the possibility of fast, non-destructive and accurate analysis of the particular crystalline sample that is prepared for further use in different applications.

In this report we describe the improved implementation of the interferometric- turning technique. The essential goal of the improved design was to achieve higher precision of the rotation angle and eliminate the errors related to sample geometry.

Block-diagram of the experimental automated set-up for the measurements of refractive indices is shown in Fig. 1. Light beam from the laser 1 is split in two by the semitransparent prism 2. One of the beams then travels through the reference arm of the interferometer and is reflected from the mirror 3. The other one passes through the polarizer 5 and the sample 11. Then, after being reflected from the mirror 4, it travels back again through the sample. Eventually, both beams interfere at the photodetector 7 (the lens 6 is used for focusing purposes). Computer 10 is employed for the communication with the analog-to-digital converter module 9.1 and step-motor control unit 9.2 that drives the rotational table 8. The measurement idea is the following. Sample is rotated from the initial position ($\varphi=0$) by the angle φ. Considering the sample thickness as d, the laser wavelength as λ (in our case λ=633 nm) and the number of interference maxima for which the fringe is shifted [5] one obtains the refractive index n [4]:

$$n = \frac{\sin^2 \varphi + (1-\cos\varphi - K\lambda/2d)^2}{2(1-\cos\varphi - K\lambda/2d)}. \quad (1)$$

When the refractive index of the medium, in which the sample is placed, is taken into account, the working formula is [6]:

$$n = \frac{\sin^2 \varphi \cdot n_c^2 + [(1-\cos\varphi) \cdot n_c - K\lambda/2d]^2}{2[(1-\cos\varphi) \cdot n_c - K\lambda/2d]} \quad (2)$$

As it is clear from the expressions (1) and (2), the accuracy of the refractive index determination is directly connected to the precision of the rotation angle. Therefore, one of the primary goals during the set-up modernization was to design high-precision computer-controlled sample positioning system.

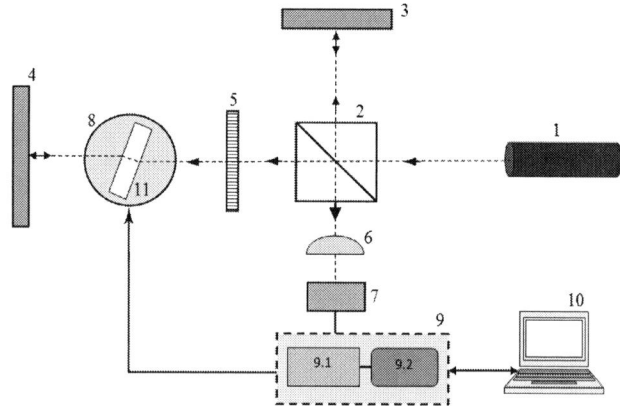

Fig. 1. The principle of the refractive index measurements.

For this purpose we have selected 17HD6403-02 hybride bipolar stepper motor [7] from Moons Industries. The motor can be run in three modes: full step, half step and microstep. The latter mode provides the minimum direct rotation angle of $\delta\varphi$=1.8°/16. In order to realize the microstep operation mode we used a couple of Allegro Microsystems A3955 integral circuit drivers. Each IC ensures pulse width current modulation in one of the motor's windings. Motor supply voltage can be

978-1-4244-7043-3/10 $26.00 © 2010 IEEE

varied from 10 to 50 V, while maximum current in the winding is limited to 1.5A in order to reduce the excess heating of the drivers. Stepper motor drivers were connected to the personal computer via data register of the Centronix parallel interface [8,9].

For the automatic registration of the interference fringes FD-7k photodiode in a photoconductive mode [10] was exploited. Signal from the photodetector amplifier circuit output is transferred to the input of 12-bit serial ADC (Burr-Brown ADC7841). ADC is interfaced to the Centronix interface control register through SPI digital lines.

ADC and stepper control electronic modules are designed as a single unit consisting of three printed circuit boards: power supply board, ADC board and stepper driver board. Stabilized +5 V voltage is used to supply digital circuits and to provide a reference for the ADC.

Sample positioning, rotation angle fixation and recording as well as photodetector signal intensity measurements are fully automated. Experiment control software for Windows has been developed in Borland Delphi environment. It is worth mentioning here as well, that the software includes the program algorithm of the initial sample position detection.

Since the accuracy of the refraction index determination depends also on the quality of the sample preparation (first of all, the experiment accuracy is very sensitive to parallelism of the sample opposite edges) we have carried out the theoretical analysis, deriving the formula which allow to eliminate the error connected with non-ideal sample geometry. The detailed analysis is to be published elsewhere, so here we would like to restrict ourselves to the final expression only, which reads

$$n = \sqrt{\left[\frac{2\cos\varphi - A}{2 - A/(tg\alpha \cdot \sin\varphi)}\right]^2 + \sin^2\varphi}, \qquad (3)$$

where α is the angle determining sample non-parallelism and

$$A = \frac{(K_1 - K_2)\lambda}{2d}$$

(K_1 and K_2 values are determined experimentally, rotating the sample by the angle φ clockwise and counterclockwise, respectively).

The described experimental set-up was approved in the measurements of refractive indices of different crystalline materials, including uniaxial lithium niobate. In particular, we have refined the refractive indices of ordinary (n_o) and extraordinary (n_e) beams for this crystal. It was found that at T=293 K n_o=2.2868 and n_e=2.2032; the error in both cases is estimated to be less than 0.03%. These values are fairly close to previously reported room temperature data for pure LiNbO$_3$ crystals [11, 12].

Summarizing, we have improved the automated technique for fast, precise and non-destructive determination of the refraction index of optical crystals, upgrading hardware and software for accurate sample stage positioning and deriving

expressions allowing to eliminate errors related to sample edges geometry.

REFERENCES

[1] F. L. McCrackin, E. Passaglia, R. R. Stromberg, H. Steinberg, "Measurement of the Thickness and Refractive Index of Very Thin Films and the Optical Properties of Surfaces by Ellipsometry", Journal of Research of the National Bureau of Standards A: Physics and Chemistry, 1963, vol. 67A, No. 4, pp. 363-377.

[2] G. Smith, "Liquid immersion method for the measurement of the refractive index of a simple lens", Appl. Opt., 1982, vol. 21, pp. 755-757.

[3] J. McAndrew, "Differential Dispersion Measurement of Refraction Index", American Mineralogist, 1972, vol. 57, pp. 231-236.

[4] A. S. Andrushchak, B.V. Tybinka, I.P. Ostrovskij, W. Schranz, A.V. Kityk, "Automated interferometric technique for express analysis of the refractive indices in isotropic and anisotropic optical materials", Optics & Lasers in Engineering, 2008, vol. 46, No. 2, pp. 162-167.

[5] Yu. N. Shestopalov, "Methods of counting and interpolating interference fringes when measuring lengths and angles", Measurement Techniques, 1972, vol. 15, No. 2, 212-215.

[6] A.S. Andrushchak, B.V. Tybinka, N.A. Andrushchak, S.S. Dumych. Ukraine Patent #39155, MPK G01N 21/01 from 10.02.2009, Issue 3.

[7] P. Crnosija, B. Kuzmanovic, S. Ajdukovic, "Micro-computer implementation of optimal algorithms for closed-loop control of hybrid stepper motor drives", IEEE Trans. on Industrial Electronics, 2000, vol. 47, No. 6, pp. 1319- 1325.

[8] A.V. Sorokin, "Instrument-to-PC Interfacing Using an Enhanced Parallel Port", Instruments and Experimental Techniques, 2002, vol. 45, pp. 87–91.

[9] P. Thimmaiah, K. R. Rao, E. R. Gopal "External device control using IBM PC's Centronics printer port", J. Instrum. Soc. India, 2004, vol. 32, No. 4, pp. 239-247.

[10] J. G. Graeme, "Photodiode Amplifiers", 1995, New York, McGraw-Hill, 253 p.

[11] D. S. Smith, H.D. Riccius, R. P. Edwin, "Refractive indices of lithium niobate", Optics Communications, 1976, vol. 17, pp. 332-335.

[12] R. S. Weis, T. K. Gaylord, "Lithium niobate: Summary of physical properties and crystal structure", Applied Physics A: Materials Science & Processing, 1985, vol. 37, No. 4, pp. 191-203.

Photophysical, photochemical and lasing properties of dipyrrolylmethene complexes in solutions and solid matrices

R.Kuznetsova[1], L.Samsonova[1], T.Kopylova[1], G.Mayer[1], O.Sikorskaya[1], Yu.Aksenova[1], N.Zulina[1], E.Antina[2], S.Yutanova[2], L.Antina[2], T.Pavich[3], K.Solovyov[3], S.Arabei[4]

[1]Tomsk State University, Tomsk, Russia
[2]Institute of Solution Chemistry RAS, Ivanovo, Russia
[3]B.I.Stepanov Institute of Physics, NAS of Belarus, Minsk, Belarus
[4]Belarusian State Agrarian and Technical University, Minsk, Belarus

Abstract: The spectral-luminescent properties of dipyrrolylmethene derivatives in different solutions and solid silicate matrices doped with these compounds were measured, and their photophysical (γ_{fl}, k_{rad}, τ_{fl}, γ_{phos}, τ_{phos}), lasing (efficiency, λ_{las}), photochemical ($\varphi_{ph}^{355,532}$), limiting (CL) characteristics were determined. The possibilities of applications are discussed.

INTRODUCTION:

The research of BF_2-complexes of dipyrrolylmethene and linear *bis*-(dipyrrolylmethene) metal complexes is intensified because the possibility of using them in new technologies (laser media, limiters, OLEDs, etc.) has shown up. The study of the correlation of the complexes structure and solvatation ability with their properties is necessary for targeted selection of specific compounds. This is the aim of the present study.

MATERIALS AND METHODS:

Two boron fluoride complexes (BODIPY1, BODIPY2), which differ by substitution at the *meso*-position in b (Fig1), and four double-helical complexes of Zn(II) [Zn_2L_2] with *bis*-(dipyrrolylmethene) derivatives (Fig1, c – f) were synthesized

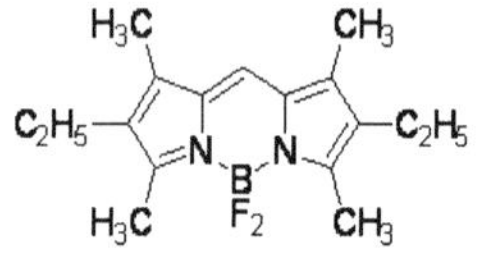

Fig.1. Structures of dipyrrolylmethe derivatives: **a** – BODIPY1, **b** – BODIPY2, [Zn_2L_2] complexes: **c** - **I** – L: R_1=R_2=Me, R_3=H; **d** - **II** – L: R_1=Et, R_2=Me, R_3=H ; **e** - **III** – L : R_1=Et, R_2=Me, R_3=Ph ; **f** - **IV** – L : R_1= Me, R_2= Et, R_3=H.

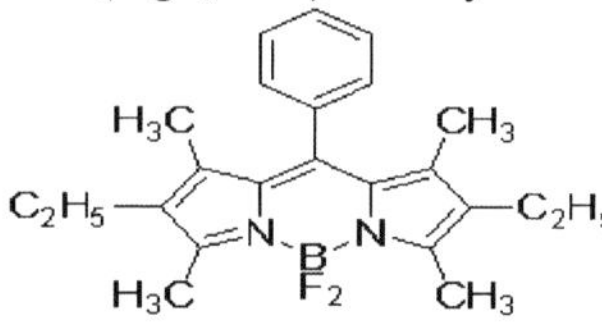

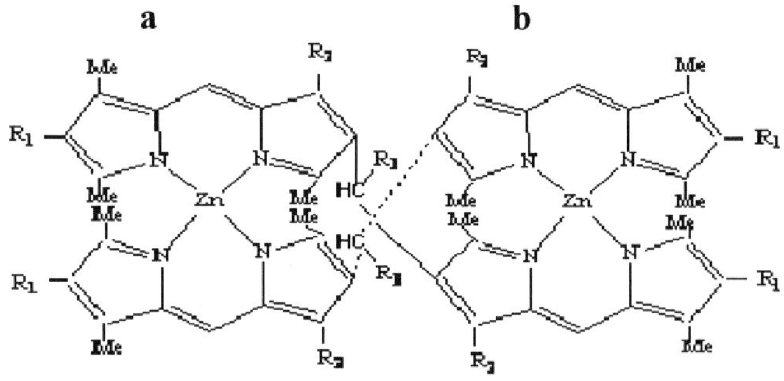

by methods described in [1, 2]. The structure of complexes is confirmed by elemental, NMR, IR, X-ray, chromatography analysis. The solvents were cyclohexene (c/h), chloroform ($CHCl_3$), ethylacetate, ethanol, acetonitryl, dimethylsulfoxide (DMSO). Solid-state matrices doped with dyes were synthesized by sol-gel methods [3, 4]. The complexes were introduced from ethanol solution after hydrolysis stage.

Absorption, fluorescence, excitation fluorescence spectra, fluorescence quantum yields of solutions and solid elements with compounds were measured using spectrometer CM2203 (SOLAR). Spectra and lifetime of long-lived (ms) emission of the frozen solutions (77K) were measured with the use of spectrofluorimeter Cary Eclipse and cryostat Optistat DN (Oxford). Lasing, limiting and photochemical characteristics were studied with 2nd and 3rd harmonics of Nd:YAG laser and laser spectrometer SOLAR. Radiation lifetimes were estimated by methods described in [5, 6]. Quantum yields of phototransformations under laser excitation were determined by method [7].

RESULTS AND DISCUSSION:

1. BODIPYs.

1.1. Spectral-luminescent properties. The range of spectral maxima for BODIPY1 and BODIPY2 in different solvents is narrow: 522-532 nm for S_0-S_1 absorption and 538-548 nm for fluorescence. S_0-S_2 band at $\cong$370 nm has intensity smaller by a factor of 1/10 as compared to S_0-S_1 for all solvents. These results agree with the literature data for other BODIPYs [8], however fluorescence efficiency is substantially higher. Radiation constants and fluorescence quantum yields of studied BODIPY1 and BODIPY2 under S_0-S_1 excitation vary only slightly depending on complex structure and solvent (γ_{fl}=0.8-1). On the other hand, fluorescence quantum yields of these compounds are decreased by factors of 2/3 – 1/2 under S_0-S_2 band excitation (λ_{ex}=350 nm) as compared to S_0-S_1 excitation (λ_{ex}=490 nm). These results presuppose enhancement of nonradiative and photochemical processes from S_0-S_2 state. Spectroscopic features – spectral shifts and intensity depending on solvents and substitutions indicate to charge-transfer character of S_0-S_1 transition from metal to ligand (MLCT). On the basis of spectral-luminescent properties the satisfactory linear dependence of Stokes shifts on the universal interactions function for all solvents except

978-1-4244-7043-3/10 $26.00 © 2010 IEEE

DMSO has been found. It points to specific character of the interactions of DMSO with BODIPY1 and BODIPY2 in the ground and excited states which is in agreement with the results on photostability, that will be considered below.

1.2. *Solid silicate matrices:* The intensity of S_0-S_1 band is decreased and a new band is formed at 484 nm in solid porous silicate matrix doped with BODIPY1 (Fig.2, inset). This absorption corresponds to BODIPY1 complex in protonated form since in acid media BODIPY1 has the same spectrum (Fig.2). pK_a of this process is 0.5. This form is formed by the interaction of a nitrogen atom of BODIPY1 with the silanol groups of the matrix $\equiv$Si-OH. The same spectrum is inherent to the $BF_3 \cdot HL$ form where coordinating p-element B and proton compete. The role of media in competing requires clarifying. This form das not fluoresce as follows from study of emission and excitation spectra. BODIPY2 is not protonated in silica matrix because its basicity is lower. However, BODIPY2 with phenyl substituent evidently aggregates in the pores of matrix which decreases fluorescence yield of BODIPY in silica matrix too.

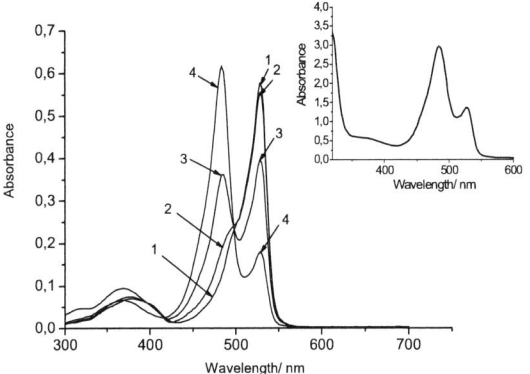

Fig.2. Absorption spectra of BODIPY1 in ethanol – 1 and with additions of HCl from 0.1 to 10 % – 2 – 4 and in silica matrix (inset).

1.3. *Lasing properties.* There are commercial BODIPYs (PM567, PM597, etc.), known active media for tunable lasers. Lasing efficiency of BODIPY1 and BODIPY2 exceed that of R6G and PM567 under 15 MW/cm^2 power excitation (Table).

Table. Lasing properties

Compounds	Conc., mM	Effic., %	λ_{las}, nm
BODIPY1	0.5	59	553+557
BODIPY2	0.5	52	551
R6G	0.1	49	561
BODIPYPM567	0.5	51	556+560
BODIPYPM597	0.1	52	574+579

The more neutral matrices would be selected for solid laser on the basis of BODIPY1 and BODIPY2: for example organic+nonorganic [4, 10], or composite [10, 11].

1.4. *Photostability of BODIPY.* There are literature results of higher resources of laser media on the basis of PM547, PM567 as compared to R6G [9,10]. However in these papers

quantitative characteristics are absent. We measured quantum yields of BODIPY2 transformation in ethanol and DMSO under conditions of spontaneous emission. Phototransformation efficiency in ethanol under UV irradiation (355 nm) is 37 times higher compared to 532 nm irradiation (φ_{phot}^{355}=2.6 10^{-3}, φ_{phot}^{532}=7 10^{-5}). It is in agreement with the decrease in fluorescence efficiency at UV excitation. Phototransformation quantum yield in DMSO is 7 times higher compared to ethanol solution which agrees with specific interactions in this solvent.

2. [Zn₂L₂] complexes of *bis*-(dipyrrolylmethene).

2.1. *Spectral-fluorescent characteristics.*[Zn_2L_2] complexes of *bis-*(dipyrrolylmethene) consist of 4 dipyrrolylmethene units (Fig.1 c-f) and, at first sight, have spectral properties similar to those of BODIPYs: S_0-S_1 absorption band at 526-530 nm and fluorescence band at 544-548 nm in ethanol and cyclohexene solutions. The S_0-S_2 band is at 476-478 nm, as distinct from BODIPYs. However the efficiency of fluorescence of all compounds (c-g) in ethanol solutions is lower by a factor of 1/100 as compared with cyclohexene solutions where quantum yields of fluorescence distinguish 1.

2.2. *Lasing characteristics:* These unusual photophysical properties make it possible to observe lasing of [Zn_2L_2] (I) complex in cyclohexene under excitation by 2nd harmonic of Nd:YAG laser without optimizations (532 nm, 28 MW/cm^2) with efficiency 1.2% on λ_{las}=553.5 nm. The information about lasing properties for similar [Zn_2L_2] complexes of *bis*-dipyrrolylmethenes is absent in the literature.

2.3. *Limiting characteristics:* As to ethanol solutions, it is shown that these [Zn_2L_2] complexes limit the power laser radiation: the transmission of ethanol solutions of [Zn_2L_2](c, d) at 355 nm (W=200 MW/cm^2) is decreased by factors of 1/2-1/3 as compared to linear absorption because the absorbance from excited states is higher than that from ground state.

2.4. *Photochemical characteristics:* Quantum yield of phototransformations under visible radiation (532 nm, W=50 MW/cm^2) for [Zn_2L_2](II) was determined: it amounts to 1.6 10^{-4} at ethanol and 2 10^{-5} at cyclohexene. The photoproduct in both solvents is the same with absorption maximum in the 504-510 nm region.

2.5. *Silicate matrices*: The introduction of [Zn_2L_2](I) in pores of silicate matrices modifies spectral characteristics to protonated nitrogen form (λ_{So-S1}=504nm). The competition of solvation shell formation and coordinating interaction with metals which forms spectral properties will be studied later.

2.6. *Emission characteristics of the frozen solutions.* Fluorescence band of frozen ethanol solution (77K) of [Zn_2L_2] II (Fig.1–d) has shifted to the blue by ~6 nm, and essential increase (by 100 times) of intensity is observed (Fig.3). This effect does not take place in cyclohexene solution of [Zn_2L_2] (II). It means that in ethanol temperature quenching of [Zn_2L_2] fluorescence is observed as consequence of specific intermolecular interactions in the excited state. It is necessary to research this mechanism further.

The measurement of long-lived emission has shown that there are several bands in both solvents (Fig.3), and lifetimes for these bands are different: 750 nm (τ=26 ms), 660 nm (τ=2.7 ms), 546-548 nm (τ=1.3 ms).

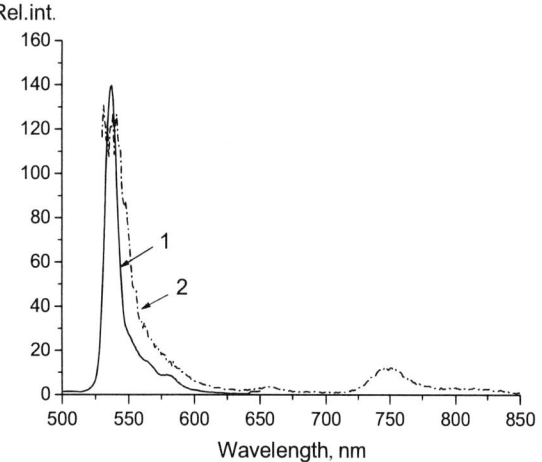

Fig.3. Typical fluorescence – 1 and long-lived emission – 2 in frozen ethanol (77K) of [Zn_2L_2] complex. λ_{ex}=500 nm, (intensity of 1 is multiplied by 0.02)

We suppose that phosphorescence of [Zn_2L_2] complex manifests itself at 750 nm because excitation spectra of this emission coincides with absorption spectra of [Zn_2L_2] complex. The more intensive band in this spectrum coinciding with ordinary fast fluorescence (540 nm) is delayed fluorescence. Long-lived emission at 660 nm may belong to intermediate complex of excited [Zn_2L_2] with solvent or dissolved oxygen from which delayed fluorescence is formed. This is in agreement with lifetimes of 540 and 660 nm bands.

Long-lived delayed fluorescence of smaller by ~1/10 intensity exists at room temperature (300 K), however other bands (750 and 660 nm) are quenched in liquid solutions. This effect may be studied for using in electrophosphorescence [12].

CONCLUSION

We have demonstrated spectral-luminescent, lasing, limiting, photochemical properties of new dipyrrolylmethene derivatives which may be used as active media for tunable lasers and limiters of power radiation. There are perspectives of using these compounds as emission layers in electrophosphorescence, optical sensors for acids, oxygen and sulfur containing compounds. Further thorough study of these applications is necessary.

ACKNOWLEDGMENTS: This work is supported by Federal Target Programm "Research and Research-Teaching Staff of Innovative Russia" 2009-2013, tasks 1.1.№ 02.740.11.0444, 02.740.11.0253 and 1.2.2. No. P565.

REFERENCES

[1] Antina E.V., Beresin M.B., Semeikin A.S., Dudina N.A., Yutanova S.L., Guseva G.B. "BF$_2$ complexes of 2,2'- dipyrrolylmethene and biladiene-a,c: synthesis and spectral properties" XII Young Scientists Conf. Organic Chem., Ivanovo, p.251-253, 2009.

[2] Dudina N., Antina L., Guseva G., Beresin M., Vyugin A. "Synthesis structure and spectral characteristics of Zn(II) helicate with *bis*-(2,4,7,8,9-pentamethyldipyrrolyl-methene-3-yl)methane" XII Young Scientists Conf. Organic Chem., Ivanovo, p.248-250, 2009.

[3] Shaposhnikov A., Kuznetsova R., Kopylova T., Mayer G., Telminov E., Pavich T., Arabei S. "Spectral-luminescent and lasing characteristics of laser dyes in silicate gel-matrices and thin gel-films" Quant. Electron., vol.34, no.8, p.715-721, 2004.

[4] R.Kuznetsova, Yu.Manekina, E.Telminov, G.Mayer, T.Pavich, S.Arabei, K.Solovyov "Spectroscopic, lasing and photochemical characteristics of organic molecules in silicate gel-matrices" Opt. Atmosph. and Ocean, vol.19, no.7, p.653-660, 2006.

[5] Terenin A.N. Photonics of dyes molecules. Nauka 1967.

[6] Qin W., Barauh M., Van der Auweraer M., De Schryver F.C., Boens N. „Photophysical properties of boron dipyrromethene analogues in solutions" J.Phys.Chem.A, vol.109, p.7371-7384, 2005.

[7] R.Kuznetsova, G.Mayer, T.Kopylova., V.Svetlichny, E.Telminov, D.Filinov "Phototransformations of organic compounds under power laser excitation under conditions of nonlinear absorption" High Energy Chem., vol.36, no.5, p.375-380, 2002.

[8] QinW., LeenV., RohandT., W.Dehaen, D.Dedecker, Van der Auweraer M., K.Robey, L.Meervelt, D.Beljonne, B.Averbeke, J.Clifford, K.Driesen, K.Binnemans, Boens N. "Synthesis, Spectroscopy, Crystal Structure, Electrochemistry, Quantum chemical and Molecular Dynamics Calculations of 3-Anilin Difluoroboron Dipyrromethene Dyes" J. Phys. Chem. A, vol.113, p.439 -447, 2009.

[9] F.Arbeloa, T.Arbeloa, I.Arbeloa, I.Garcia-Moreno, A.Costela, R.Sastre, F.Amat-Guerri "Photophysical and lasing properties of dipyrromethene BF$_2$ dyes in liquid solution" Chim.Phys.Lett, vol.299, p.315-321, 1999.

[10] A.Costela, I.Garcia-Moreno, Gomez, F.Amat-Guerri, M.Liras, R.Sastre "Efficient and high photostable solid-state dye lasers based on modified matrices of poly(methyl methacrylate)" Appl. Phys., vol.B76, p.365-369, 2003.

[11] R.Kuznetsova, G.Mayer, Yu.Manekina., V.Svetlichny, E.Telminov, T.Pavich, S.Arabei, K.Solovyov "Spectroscopic and lasing properties of photoexcited organic luminophores introduced into composite gel-systems" Opt. and Spectroscopy, vol.102, no.2, p.234-244, 2007.

[12] R.Evans, P.Douglas, Ch.Winscom. "Coordination complexes exhibiting room temperature phosphores-cence: Evaluation of their suitability as triplet emitters in organic light emitting diodes" Coord. Chem. Reviews, vol.250, p.2093-2126, 2006.

Measurements and Modeling of Optical Distortions Relaxation in High Power Nd:Glass Lasers

V.E. Zavalova, A.V. Kudryashov, A.L. Rukosuev

"Adaptive Optics NightN" Ltd., Moscow

Abstract: Evolution of wavefront aberrations in high-power solid-state laser (Nd:glass laser) was measured and analyzed. The measurements were performed by Shack-Hartmann wavefront (SHW) sensor. The technical opportunities of such type sensors both device for measurements and element of Adaptive Optical Systems to correct for wavefront aberrations are demonstrated in this report. Analysis of damping distortions shows that its damping has thermal nature and can be described well by non-stationary equation of diffusion at pumping and cooling of Nd:glass-slab. Theoretical estimations of aberrations dynamics is in good agreement with the experimental data.

The study of the wavefront aberrations for high-power solid-state laser in shot-to-shot regime is an urgent problem, because the main purpose of such lasers is to improve beam focusable of beam on the target and to increase pulse repetition rate. The result of some experimental studies carried out with a Shack-Hartmann wavefront sensor were represented in [1-2]. The main advantages of this type sensor include: the wide spectral range of operation for both cw and pulsed radiation; measurements even non-coherent radiation; in-sensitivity to vibrations, easy embedding in any optical schema and relatively low cost.

Typical optical schemas of high-power Nd:glass lasers have a lot of optical elements. Beam goes from an oscillator, passes through preamplifiers, amplifiers and focuses on the target. But it was defined that multiply transmission amplifiers are the main sources of wavefront aberrations. The geometry of the slab, direction of pumping and cooling in relation to the coordinates axes are shown on the Fig. 1. Two flash lamps on each side and reflector provide uniform optical pumping of the slab and heat removes along both sides by cooling flow. Laser radiation outputs from lateral facet.

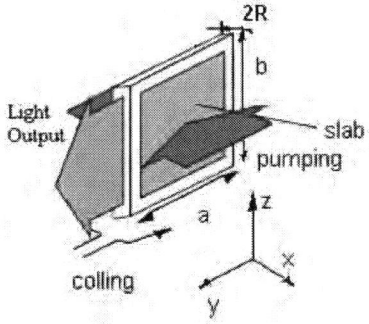

Fig.1. The view of the slab.

As the result, optical path of the beam acquires both statically aberrations, caused by presence a lot of optical elements and dynamical aberrations, caused by pumping. The example of behavior of Dynamics aberrations after pumping, measured by means SHW sensor, is represented on the Fig.2, static aberrations are subtracted.

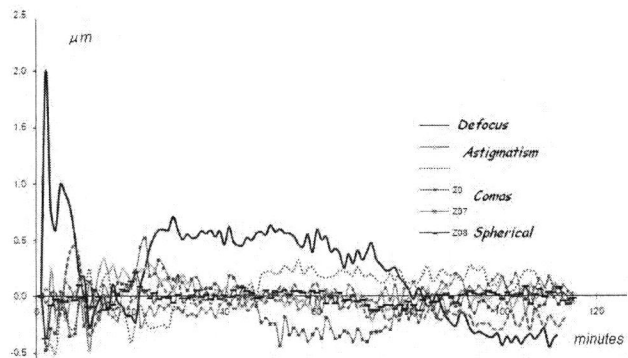

Fig.2. The example of relaxation of dynamics aberrations after pumping.

Long time relaxation is explained by thermal effects. The wavefront becomes predominantly aberrated by thermal lens effects and thermal birefringence astigmatism. Correlation between wavefront aberrations and temperature distributions follows from dependence of refraction index on temperature changes.

The thermal model of cooling of active element of laser radiation amplification – Nd:glass slab after shot is based on the solution of unsteady thermal diffusion (conduction) equation for think plate, because 2R<<a,b; initial temperature distribution TH(x,0) (and boundary conditions taken account that the temperature of environment Tc and heat transfer through the plate surface are given [3]. Solution for dimensionless temperature is obtained in a series. Convergent of series depends on cooling, material properties and spatial scale. And in general solution could be represented as $\theta(x,\tau) = F(F_0, x/R, B_i)$, where dimensionless number B_i characterizes the intensity of heat transfer and dimensionless number F_0 is dimensionless time.

On the Fig.3 we can compare result of measurement of dynamic aberrations evolution and results of estimations performed with use mathematics modeling at the same conditions.

978-1-4244-7043-3/10 $26.00 © 2010 IEEE

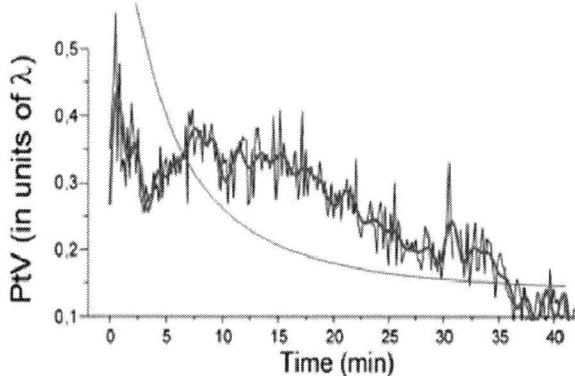

Fig.3. Evolution of the peak to valley (PtV) of the summary wavefront aberrations after a full-energy shot, measured by Shack-Hartman wavefront sensor. Blue line represents the analytical solution of unstationary thermal diffusion (conduction) equation, performed at the same conditions, like at measurements.

Thus, measured wave-front dynamics is correspondent to the temperature evolution in the regular mode of cooling. The increment in the exponential function allows estimating the time interval when steady state or thermal equilibrium is become. Overall thermal stabilization, when dynamic aberrations become comparable with the accuracy of measurements, requests a lot of time period, may be about hour. Because in real laser systems if we try to increase the shot repetition rate above the dissipation time of the thermal load, cumulative thermal load appears and laser beam focusability worsens.

That is why Closed-Loop Adaptive Optical System (CLAOS) should be used in high power lasers, because only it able to correct for not only static but dynamical aberrations for each shot. But the result of correction depends on algorithm of CLAOS. We need in not only good compensation of single shot, but hold good focusable from one shot to another shot. It is not trivial, because aberrations are accumulated by cumulative thermal load effect.

In conclusion we can say that,

- Non-stationary model of active elements cooling gives a good agreement with the results of wavefront dynamics measurements.
- Model allows estimating residual aberrations before the next shot. It can be useful for matching between cooling regime and shots period repetition.
- Estimating of residual aberrations before the next shot can be very useful for pre-correction algorithm for Closed-Loop Adaptive Optical System to control for the corrector surface in real-time.

REFERENCES

[1] A.G. Alecsandrov, V.E. Zavalova, A.V. Kudryashov, A.L. Rukosuev, P.N. Romanov, V.V. Samarkin, Yu.V. Sheldakova. *Quantum Electronics*, "Snack-Hartmann wavefront sensor for measuring the parameters of high-power pulsed solid-state lasers " v.40, no4, pp. 321-326, 2010.

[2] B. Wattellier, J. Fuchs, J. P. Zou, K. Abdeli, H. Pépin, and C. Haefner. "Repetition rate increase and diffraction-limited focal spots for a nonthermal-equilibrium 100-TW Nd:glass laser chain by use of adaptive optics" *Optics Letters*, v.29, no 21, 2004.

[3] V.P. Isachenko, V.A. Osipova, A.S. Sukomel. *Teploperedacha,–* Moscow: Energoizdat. 1981.

Application of $Sm_2Ti_2O_7$ in Technology of Mirrors for He-Ne Laser

V. F. Zinchenko[1], V. I. Maksimenko[2], V. P. Sobol'[2], L. V. Sadkovska[1], Ye. V. Timukhin[1]

[1]A. V. Bogatsky Physico-Chemical Institute of NAS of Ukraine, Odessa, 65080, Ukraine

[2]State Enterprise for Special Instrument Making "Arsenal", Kyiv, 02010 Ukraine

Abstract- **The technological regimes of production of the reflective coat with low energy losses for He-Ne laser mirrors have been worked-off on the base of environmentally safe film-forming material (FFM) $Sm_2Ti_2O_7$ developed for high-refractive layers and convenient FFM SiO_2. The experimental specimens of coat have been produced using the optimized technology. They are characterized by scattering factor σ = 0.010-0.015 %; absorption index κ = 0.001-0.005 %. The optical characteristics of the obtained specimens have been studied.**

I. INTRODUCTION

He-Ne lasers with radiation wavelength λ = 632.8 nm have high stability of radiation frequency at low output power (ranging from several μW to tens mW). Owing to this fact they are effectively used in measuring devices (laser rangers, interferometers, metrology etc.). The mirrors of such lasers are characterized by high values of reflectance ρ, which are close to 1 (ρ>0.99 (99 %)). This parameter, being slightly deviated from 1 (from 100%), could be intractably measured by direct methods. That is why the quality of He-Ne laser mirrors are commonly characterized by mean of other parameters, namely, by transmittance τ, scattering factor σ, absorption index κ and total optical losses $\alpha = \tau + \sigma + \kappa$.

The required parameters of He-Ne laser mirrors are realized only in a case of using of multilayered dielectric interference coats deposited on substrates of optical materials polished to the highest grade of optical accuracy. Such multilayered reflective coats are usually prepared through the alternated vacuum deposition of quarter-wave (λ/4) layers of FFMs with different refractive indices: FFM with higher (H – high) refraction index – n_H, another FFM with lower (L – low) refractive index n_L. Here, the maximal value of reflectance ρ (or the lowest value of transmittance, τ) at given wavelength λ is realized for the odd number of layers N (the first and the last layers should be the high-refractive ones) if the optical thickness of each layer (l_{opt}) is equal to λ/4 ($l_{opt} = nd = \lambda$/4, where n is the refractive index and d is the geometric thickness of layer).

One could obtain the coats possessing the values of ρ and τ close to ideal parameters by using FFM with the values of refractive index, which essentially differ from one another (high values of n_{LH}). The typical values of transmittance τ of the output (signal) mirrors of the commercial He-Ne lasers range from 0.01 % for mini-lasers to 1 % for big lasers (the length of body and resonator ~1 m). In order to reduce the level of total losses of resonator α for other (dead) laser mirrors it is also necessary to reach as low values of τ as possible. However, the reducing of σ and κ parameters to the level lower than 0.01 – 0.001% is also significant, because these parameters make contribution to the total losses of optical energy of the coat. That is why the creation of refractive coats for the mirrors of He-Ne lasers, which scattering factor (σ) and absorption index (κ) do not exceed 0.015% is an urgent problem.

II. TECHNICAL APPROACH

The reflective coat, which consists of separate layers deposited from the developed high-refractive FFM $Sm_2Ti_2O_7$ [1] and low-refractive convenient FFM SiO_2 were constructed by computer simulation method. According to the experimental data for the single-films model the refractive index of $Sm_2Ti_2O_7$ FFM at the wavelength of He-Ne laser λ = 632.8 nm is equal to n_H=2.2 and the refractive index for SiO_2 (silicon dioxide) is equal to n_L = 1.45-1.455. This FFM forms amorphous films at its deposition by electron-beam evaporation. This is the necessary condition to obtain the coat with low scattering [1, 2]. $Sm_2Ti_2O_7$ (samarium dititanate) has been selected among the series of studied multi-component oxides because of it manufacturability at coat deposition, the most optimal optical characteristics, σ, κ of the single layers and because of its compatibility with SiO_2 (by technological and thermo-mechanical properties). The optimization of the technological regimes of coat deposition including the estimation of the quality of produced model and experimental mirrors specimens has been performed by the values of light scattering factor σ, transmittance τ, index of total optical losses α and absorption index κ at the reflection of laser radiation. These parameters have been measured directly at the working wavelength λ = 632.8 nm. He-Ne laser in this case was the source of radiation. Visual evaluation of the reflective coat by optical microscope has been used as an additional mean at the analysis of experimental data. Fig. 1 presents the maintenance for the measurements of scattering factor σ and transmittance, τ. The presented maintenance provides the accuracy at σ and τ measurements not worse than 10^{-3} % and its self-noise consisted of electrical and optical noises is equal to ~$2 \cdot 10^{-4}$ %. This maintenance at σ measurements was calibrated against the standard disperser of lusterless ("opal") glass with σ = 96%, which was mounted instead of the studied mirror 9. This maintenance at τ measurements was calibrated against the standard attenuator of neutral glass with attenuation index of 10^3 ($\tau = 0.1\%$), which was mounted instead of the studied mirror 5.

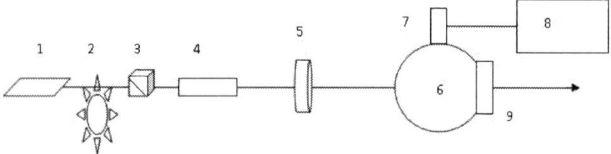

Fig. 1. The maintenance for the measurement of optical losses of laser mirrors: 1 – laser, 2 – modulator, 3 – polarizer, 4 – block of the restrictive diaphragms, 5 – position of mirror at σ measurements, 6 – photometry sphere, 7 – photo-receiver, 8 – unit of detection, 9 – position of mirror holder at photometry sphere at σ measurements.

The method of the phase shift of modulation [3] of the intensity of external laser ray, which passes through the resonator with the studied mirror, was used in a modified mode. The essence of this modification consists in an automated sustaining the frequency of radiation of the external probing laser in a resonance with the frequency of the dead resonator containing the studied mirror. The sensitivity of this method was increased two-fold as compare to that reported in [3] due to performed improvement, reproducibility of the measured values of α was equal to $(2 – 3)\cdot10^{-3}$ %. Meanwhile, this method is too laborious because the procedure requires the alignment of the measuring resonator matching with the external laser at each measurement. That is why; this method is commonly used as calibrating method and randomly to check the data obtained by other methods. The measurements of total optical losses α were performed by the method of threshold excitation current of generation. All express measurements of α values have been performed using the method of threshold excitation current of generation in an active laser I_l. One of the mirrors in a resonator of such laser is removable, so, either the standard or the studied mirror may be placed in this position.

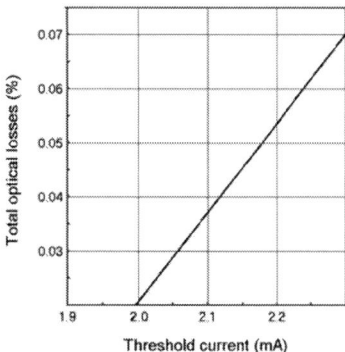

Fig. 2. – Calibration plot of $\alpha = f(I_l)$ dependence.

In this case no matching of the measuring and external lasers is required. In a result the alignment is simplified.

Dependence $\alpha = f(I_l)$ (Fig. 2) for the laser used in the method of measuring the threshold excitation current of generation has been calibrated by means of standard mirrors, the total losses α for which were measured by the method of the phase shift of modulation. This dependence was found to be linear with the slope of 0.25 % / mA at the values of I_l ranging within 2-4 mA. The reproducibility of I_l measuring at certain fixed position of the studied mirror was shown to be not worse than 0.01 mA that corresponds to the accuracy of α determination $2.5\cdot10^{-3}$ %.

III. EXPERIMENTAL RESULTS

The model coats of the mirrors were deposited on KU-1 quartz glass substrates ($\varnothing$ 22 mm, 4 mm thick) preliminary treated by deep grinding and polishing. Such treatment is the indispensable condition to provide high service characteristics of the reflective coat and mirrors in general.

The construction of the developed reflective coat involves 28 layers (14 couples of $Sm_2Ti_2O_7$ - SiO_2 layers) each $\lambda/4$ thick (except for the last one). The last SiO_2 layer is of duplicated thickness ($\lambda/2$) and is optically neutral for the rating wavelength λ (i.e. does not influencing the values of ρ and τ). This layer plays the protective function providing mechanical durability and stability of the coat. Geometric thickness of quarter-wave layers of the studied coat are equal to $d \cong 71.9$ nm for $Sm_2Ti_2O_7$ and $d \cong 109$ nm for SiO_2 at $\lambda = 632.8$ nm. According to the data of X-ray studies the film layers forming such coats are X-ray-amorphous. The activity on the working-off the technological regimes of coat deposition included the refinement of the factors influencing the index of light - scattering factor σ of the laser mirrors. The aim of this activity was to minimize σ and to reduce its value down to $\leq 0.015\%$ and also to minimize the absorption index κ. Our earliest teleological investigations of the mirrors' service characteristics in dependence on technological regimes of its production reveal the main influence of substrate temperature T_s on the absorption index of the coat (κ) (Table). So, the data presented in Table show the reducing of absorption index of the refractive coat from $\kappa = 0.2\%$ at $T_s = 300°C$ to $\kappa \leq 0.01\%$ at $T_s = 350°C$ when the oxygen pressure in the chamber $p(O_2) \geq 4\cdot10^{-3}$.

TABLE
CORRELATION BETWEEN SERVICE CHARACTERISTICS OF MODEL COATS AND TECHNOLOGICAL REGIMES OF ITS DEPOSITION

Substrate temperature, T_s, °C	Oxygen pressure, $p(O_2)$, Pa	Scattering factor, σ, %	Total energy losses, α, %	Absorption index, κ, %
250	$6\cdot10^{-3}$	0.015-0.03	0.32-0.35	0.30-0.34
300	$6\cdot10^{-3}$	0.017-0.025	0.15-0.2	0.14-0.18
	$4\cdot10^{-3}$	0.016-0.027	0.16-0.2	0.13-0.18
350	$6\cdot10^{-3}$	0.017-0.028	0.016-0.028	0.001-0.011
	$4\cdot10^{-3}$	0.015-0.023	0.018-0.025	0.002-0.01
	$2\cdot10^{-3}$	0.018-0.025	0.028-0.038	0.013-0.02

All preliminary studies have been performed at deposition duration $t_H \cong 7 – 10$ min for high-refractive $Sm_2Ti_2O_7$ layers (the rate of deposition was equal to $10 - 7$ nm/min) and $t_L \cong 2$ min for low-refractive SiO_2 layers (the rate of deposition was equal to $\cong 50$ nm/min). The studies of the influence of rate of the layer deposition on the service characteristics of the mirrors have been performed through the investigation of a set of mirror model specimens with the coats deposited at $p(O_2) = 4 \cdot 10^{-3}$ Pa and $T_s = 360°C$ (i.e. under parameters optimized during the experiments described earlier).

At the first stage of investigation the duration of deposition of $Sm_2Ti_2O_7$ layers was varied within 2 to 20 min (this corresponds to the rate of deposition ranging from 36 nm/min to 3.6 nm/min), the rate of deposition of SiO_2 layers being constant. Some increasing of σ is observed at higher deposition rate ($t_H < 5$ min). This is caused presumably by more sharp and intensive heating of FFM and the nearest units of evaporator that leads to the enhanced dusting due to cracking of FFM sediments at these units [4]. At $t_H > 7$ min the increasing of σ is the result of increasing of duration of the evaporated FFM entrainment with the flow and fixation of dust micro-particles in the coat [4]. It should be also noted that some increasing of κ is observed at $t_H < 5$ min because of enhancement of the oxide dissociation and onset the oxygen deficiency in the coat due to more intensive FFM heating at higher rates of deposition.

The similar dependence of σ and κ on the duration of deposition of one SiO_2 layer at $t_H \cong 5$ min displays the increasing of σ at high rate of deposition ($t_L \leq 1$ min). No σ increasing occurs at t_L lower than 5 min. Such increasing may be expected at the subsequent increasing of t_L. Unfortunately, it is impossible to deposit SiO_2 layers at lower rate because of the peculiarities of its evaporation. The absorption index (κ) of the reflective coat does not depend on t_L.

A batch of the experimental specimens of the mirrors for He-Ne lasers (Photo 1) has been produced at the final stage by the optimized technology using the created FFM. The parallel-sided circles of melted KU-1 quartz (Ø22 mm and 4 mm thick) were used as substrates. Both sides of these circles were preliminary treated by deep grinding and polishing.

Photo 1. Experimental specimens of the mirrors for He–Ne laser with the multilayered coats based on SiO_2 and $Sm_2Ti_2O_7$ FFMs.

The refractive coats were deposited by electron-beam evaporation of the environmentally safe FFM: SiO_2 and the developed FFM $Sm_2Ti_2O_7$. As it was stated earlier the construction of the created refractive coat consists of 28 layers (14 pairs of $Sm_2Ti_2O_7$ - SiO_2 layers) each $\lambda/4$ thick (except for last SiO_2 layer, which is of double thickness ($\lambda/2$), optically neutral to the nominal wavelength λ and does not influence ρ and τ values). The geometrical thicknesses of the quarter-wave layers of the created coat are equal to: $d \cong 71.9$ nm for $Sm_2Ti_2O_7$, and $d \cong 109$ nm for SiO_2. The created coats were found to be X-ray-amorphous according to the data of X-ray studies. As it was experimentally found the refractive multilayered coats of the mirrors for He-Ne lasers produced by the optimized technology are characterized at the wavelength λ = 632.8 nm by the following optical parameters: scattering factor σ = 0,010-0,015 %; absorption index κ = 0,001-0,005 %.

IV. CONCLUSIONS

• The construction of multilayered refractive coat has been simulated on the base of the developed environmentally safe $Sm_2Ti_2O_7$ FFM and SiO_2 oxide. The data on optical characteristics of the single-layered films deposited from the appropriate FFM have been taken into account upon simulation of the coat construction.

• The technological regimes of multilayered refractive coat production have been optimized: substrate temperature at deposition t_S = 350–370°C; content (pressure) of oxygen in a vacuum chamber at deposition of high-refractive layers $p(O_2) \geq 4 \cdot 10^{-3}$Pa; duration and rate of $Sm_2Ti_2O_7$ layers deposition $t_H \cong 5$ min.

• The experimental specimens of the mirrors with low level of energy losses for He-Ne lasers have been produced. Optical characteristics of the mirrors were determined to be as follows: scattering factor σ = 0.010-0.015 %; absorption index κ = 0.001-0.005 %.

ACKNOWLEDGMENT

The Scientific-Technological Center of Ukraine (STCU) (under project No 1356) has supported this study.

REFERENCES

[1] V. F. Zinchenko, G. I. Kocherba, V. P. Sobol', O. V. Mozkova, V. Ya. Markiv, and N. M. Belayavina, "Peculiarities of structure and optical & operational properties of thin-film coatings based on oxides, fluorides and chalcogenides of metals", *Physics and Chemistry of Solid State*, vol. 11, pp. 204-210, 2010 (in Ukrainian).

[2] O. D. Vol'pyan and P. P. Yakovlev, "The influence of heat treatment on optical properties of Ta_2O_5 films", *Opticheskij zhurnal*, vol. 69, pp 29–31, 2002 (in Russian).

[3] J. M. Herbelin and J. A. McKay, "Development of laser mirrors of very high reflectivity using the cavity-attenuated phase-shift method", *Appl. Opt.*, vol. 20, pp. 3341-3344, 1981.

[4] D. T. Wei, "Ion beam interference coating for ultralow optical loss", *Appl. Opt.*, vol. 28, pp. 2813-2816, 1989.

Modeling of XeCl excilamps, taking into account of process of halogen regeneration

S. S. Anufrik, A. P. Volodenkov, K. F. Znosko

Yanka Kupala State University of Grodno, 22, Ozheshko Street, 230023, Grodno, Belarus,

Fax: +375 (152) 73-19-10; e-mail: a.volodenkov@grsu.by

Abstract: Kinetic model for XeCl*-molecules active medium of an excilamps, taking into account of the halide donors (HCl, Cl$_2$) regeneration process is presented. Conditions being discussed, when XeCl*-molecules formation and the regeneration dynamics of the halide donors are possible to consider separately.

SIMULATION METHOD

Modeling of electro-discharge XeCl excilamps is sufficiently difficult physical and mathematical problem. The model should take into account and describe processes which occur both in the active environment, and in system of excitation of the volume discharge. Generally the computer model includes the following modules and databases.

1) The module of the solving of Boltzmann equation for the electron energy distribution function (EEDF) [1]. This module on composition of a mixture, on value of a degree of ionization and set E/N (E - intensity of an electric field in an interelectrode gap; N - full concentration of particles) allows to find EEDF and accordingly to define rates of plasma-chemical reactions with participation of electrons, and also to define electron mobility.

Rate factors of reactions with participation of electrons were obtained by averaging on EEDF expressions of next type.

$$k = \left\langle \sigma(\varepsilon) \cdot \sqrt{2\varepsilon/m} \right\rangle \qquad (1a)$$

For the solving ready program Bolsig + [2], which automatically calculates rate factors of reactions, is used. Essential advantage of this program is the account of electron - electron collisions and presence of a separate file with a database on cross-sections of reactions with electron participation depending on their energy (Siglo.sec). This file contains sections for 15 gas component: N$_2$, O$_2$, H$_2$, Cl$_2$, F$_2$, HCl, CF$_4$, SiH$_4$, CH$_4$, SF$_6$, He, Ne, Ar, Kr, Xe. As the file is written in a text format it allows the user to fill up easily independently a database on cross-sections by additional component, for example, to take into account influence on EEDF presence in mix XeCl*, Xe$_2$*, Xe$_2$$^+$ and so on.

2) The module of the solving of system of the equations of plasma-chemical reactions [3]. Models of XeCl-excilamp (halogenide HCl, Cl$_2$) are investigated and analyzed. Two channels of formation of XeCl*-molecules were taken into account: harpoon reaction and three-body recombination.

On the base of these models the program module for solving of system of the equations of plasma-chemical reactions in electro-discharge XeCl-excilamp was developed.

This module allows to find the time dependence of concentrations of electrons, ions, atoms and molecules in various energy levels in plasma. The module allows to determine various local characteristics of plasma. In simplest case the next plasma-chemical reactions must be taken into account.

Ionization and excitation

$Xe + e \rightarrow Xe+ + e + e; (ki)$;

$Xe + e \rightarrow Xe* + e; (ke1)$;

$Xe* + e \rightarrow Xe+ + e + e; (ks1)$;

$Xe^* + Xe^* \rightarrow Xe^+ + e + Xe (k_{Pen})$

Vibration excitation of HCl

$HCl(Vi) + e \rightarrow HCl(Vj) + e (kij)$;

Dissociative attachment

$HCl(Vi) + e \rightarrow Cl^- + H; (kia,)$;

Production of XeCl-molecules

$Xe* + HCl(v) \rightarrow XeCl* + H (k_{harp})$;

$Xe+ + Cl^- + M \rightarrow XeCl* + M; (\beta_{rec})$

Quenching of XeCl- molecules

$XeCl* + e \rightarrow Xe + Cl + e; (kq)$;

$XeCl* + N \rightarrow Xe + Cl + N; (\tau q)$

Emission of XeCl- molecules

$XeCl* \rightarrow Xe + Cl + h\nu; (\tau_{sp})$;

The power density of radiation of the excilamps (the power, obtained from unit of volume) is determined by next expression.

$$P = \frac{|XeCl^*| \cdot h\nu}{\tau_{sp}} \qquad (1b)$$

[XeCl*] is concentration of excimer molecules; τ_{sp} - time constant of spontaneous emission.

3) The module of the solving of the equations of an electric circuit [4]. This module describes work of system of excitation of the volume discharge in active medium. On the total resistance of plasma this module allows to define time dependence of E/P, formed by excitation system in active medium.

PROCESS OF REGENERATION OF HCL MOLECULES

Process of regeneration of HCl molecules is taken into account as follows. Molecule HCl is formed owing to three-partial association [5].

$$H+Cl+Ne=HCl+Ne \ (kr1=1,1 \ 10^{-39} \ cm^6/s) \qquad (2)$$

Then, regeneration can occur on the following channels [6].

$$H+Cl_2 =HCl+ Cl \ (kr2=1,45 \ 10^{-10} \ exp(-590/T) \ cm^3/s) \ (3)$$

$H_2+Cl=HCl+H$ (kr3=3,5 10^{-11} exp(-2290/T) cm^6/s)   (4)

Value T is equal to temperature of translational motion.

The molecule of hydrogen is formed on the following reaction [7].

$$2H+Ne=H_2+Ne \text{ (kH}_2=3 \ 10^{-33} \text{ cm}^6\text{/s)}\qquad(5)$$

The molecule of chlorine is formed on the following reaction [8].

$$2Cl+Ne=Cl_2+Ne \text{ (kCl}_2=5 \ 10^{-33} \text{ cm}^6\text{/s)}\qquad(6)$$

In brackets rate factors, being taken from works [5-8], are resulted. On the basis (2-6) the following system of the kinetic equations has been written (Ne is used as buffer gas):

$$\frac{d[HCl]}{dt}=kr1\cdot[Ne]\cdot[H]\cdot[Cl]+kr2\cdot[H]\cdot[Cl_2]+kr3\cdot[H_2]\cdot[Cl]$$

$$\frac{d[Cl_2]}{dt}=kCl_2\cdot[Ne]\cdot[Cl]\cdot[Cl]-kr2\cdot[H]\cdot[Cl_2]$$

$$\frac{d[H_2]}{dt}=kH_2\cdot[Ne]\cdot[H]\cdot[H]-kr3\cdot[H_2]\cdot[Cl]\qquad(7a)$$

$$\frac{d[Cl]}{dt}=-kr1[Ne]\cdot[H]\cdot[Cl]+kr2\cdot[H]\cdot[Cl_2]-kr3[H_2]\cdot[Cl]-2kCl_2[Ne]\cdot[Cl]^2$$

$$\frac{d[H]}{dt}=-kr1[Ne]\cdot[H]\cdot[Cl]-kr2[H]\cdot[Cl_2]+kr3[H_2]\cdot[Cl]-2kH_2[Ne]\cdot[H]^2$$

In system of the equations (7a) the following designations are used: [H], [Cl], [H$_2$], [Cl$_2$], [HCl] - concentration of corresponding atoms and molecules. The system of the equations (7) was solved numerically.

We have used the next initial condition:

$$[HCl](0)=2,2\cdot10^{16} \ 1/cm^3; \ [Cl_2](0)=0,1\cdot10^{16} \ 1/cm^3;$$

$$[H_2](0)=0,1\cdot10^{16} \ 1/cm^3;\qquad(7b)$$

$$[Cl](0)=4,2\cdot10^{16} \ 1/cm^3;[H](0)=4,2\cdot10^{16} \ 1/cm^3.$$

We regard 50% of HCl molecules to be dissociated at initial moment. At Fig.1 kinetics of HCl regeneration is represented for mixture, which is typical for XeCl-excilamp: 2 Torr HCl; 30 Torr Xe; 308 Torr Ne.

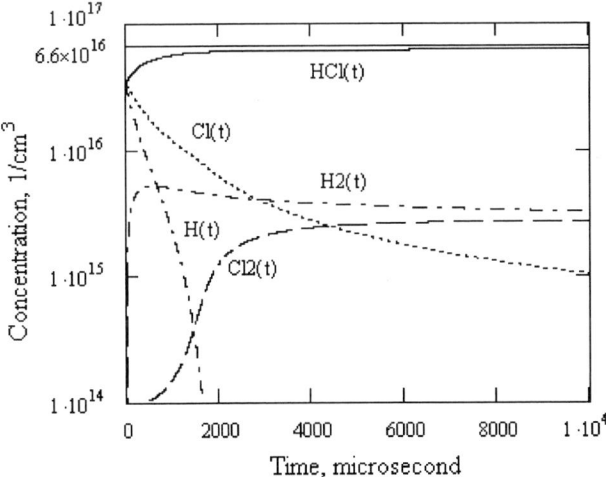

Fig. 1. Kinetics of HCl regeneration for excilamp mixture. Characteristic constant of regeneration time is about 2000 microseconds for our mixture.

At Fig.2 kinetics of HCl regeneration is represented for mixture, which is typical for XeCl-excilamp: 2 Torr HCl; 30 Torr Xe; 102,7 Torr Ne.

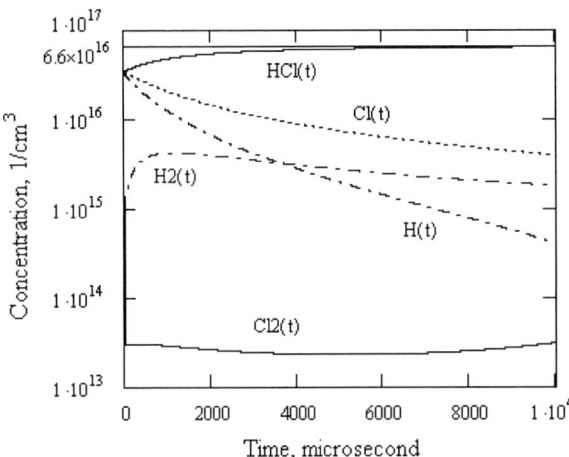

Fig. 2. Kinetics of HCl regeneration for excilamp mixture.

Characteristic constant of regeneration time depends on the general pressure of a mixture and is about 6000 microseconds for this mixture. The established level of HCl molecules concentration depends on pressure and is equal 6,163 10^{16} 1/cm^3 for conditions of Fig. 1 and 6,404 10^{16} 1/cm^3 for conditions of Fig 2. Addition of H$_2$ molecules (1 Torr) in excilamp mixture increases steady-state concentration of HCl molecules up to 6,798 10^{16} 1/cm^3. At Fig.3 kinetics of HCl regeneration is represented for mixture: 1 Torr H$_2$; 2 Torr HCl; 30 Torr Xe; 308 Torr Ne.

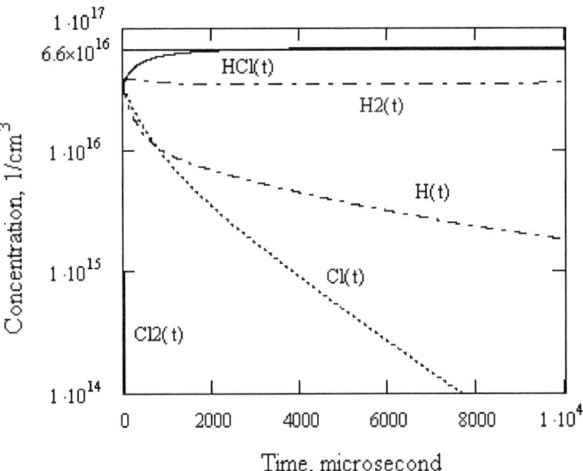

Fig. 3. Kinetics of HCl regeneration for excilamp mixture with H$_2$ addition.

At fig.4 kinetics of HCl regeneration is represented for mixture, which is typical for XeCl-laser: 2 Torr HCl; 30 Torr Xe; 3008 Torr Ne. Characteristic constant of regeneration time is about 200 microseconds for laser mixture. The established level of HCl molecules concentration is equal $5{,}617 \cdot 10^{16}$ $1/cm^3$ for conditions of Fig. 4.

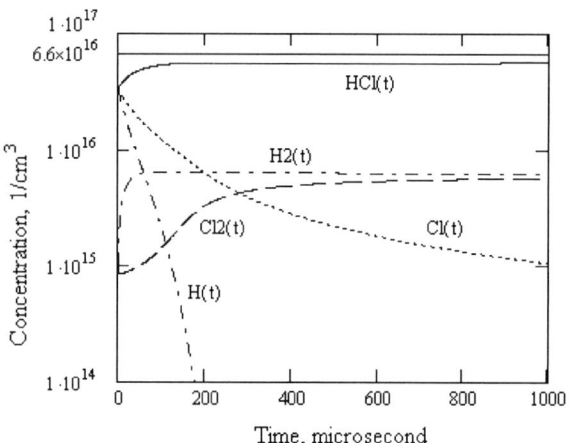

Fig. 4. Kinetics of HCl regeneration for laser mixture.

PROCESS OF REGENERATION OF CL_2 MOLECULES

Process of regeneration of Cl_2 molecules is taken into account as follows. Molecule Cl_2 is formed owing to three-partial association by reaction (6).

On the basis (6) the following system of the kinetic equations has been written (Ne is used as buffer gas):

$$\frac{d[Cl_2]}{dt} = kCl_2 \cdot [Ne] \cdot [Cl] \cdot [Cl], \quad \frac{d[Cl]}{dt} = -2kCl_2 \cdot [Ne] \cdot [Cl] \cdot [Cl] \quad (8)$$

The decision of the equations system (8), provided that during the initial moment of time $t=0$ concentration of atoms of chlorine is equal $[Cl]$ (0), has the following kind.

$$[Cl](t) = \frac{[Cl](0)}{1 + 2 \cdot kCl_2 \cdot [Ne] \cdot [Cl](0) \cdot t} \quad (9a)$$

$$[Cl_2](t) = [Cl_2](0) + \frac{[Cl](0)}{2} \cdot \left[1 - \frac{1}{1 + 2 \cdot kCl_2 \cdot [Ne] \cdot [Cl](0) \cdot t}\right] \quad (9b)$$

The equations (8) have a constant of time equal to the following value.

$$\tau_{ac} = \frac{1}{2 \cdot kCl_2 \cdot [Ne] \cdot [Cl](0)} \quad (10)$$

The physical sense of a constant of time consists that it is numerically equal to a time interval during which initial concentration of atoms of chlorine will fall in 2 times owing to their association in molecules of chlorine. Constant of time of process of regeneration of Cl_2 influences not only a rate of reaction (6), but also dissociation degree of molecules at the initial moment of time. We regard 50% of Cl_2 molecules to be dissociated at initial moment. At Fig. 5 dependence of process of regeneration of molecules of chlorine on time is represented for mixture, which is typical for XeCl-excilamp: 2 Torr Cl_2; 5 Torr Xe; 100 Torr Ne. Characteristic constant of regeneration time is about 460 microseconds for our mixture.

In the conclusion we shall note, that the account of regeneration can be used for definition of output characteristics in a pulse-periodic mode.

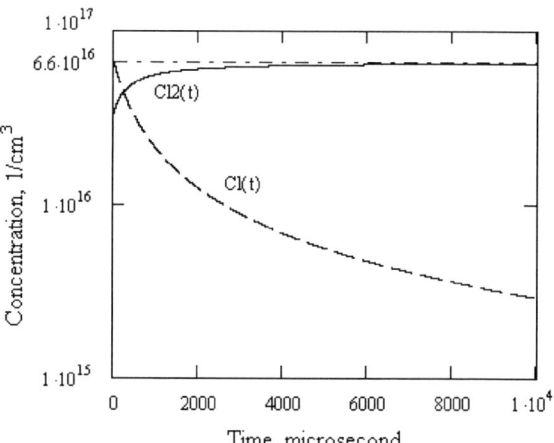

Fig. 5. Kinetics of Cl_2 regeneration for excilamp mixture.

REFERENCES

[1] G. J. M. Hagelaar, L. C. Pitchford. "Solving the Boltzmann equation to obtain electron transport coefficients and rate coefficients for fluid models," *Plasma Sources Sci. Technol.*, vol.14, no. 1, pp.1-12, 2005.

[2] http://www.codiciel.fr/plateforme/plasma/bolsig/bolsig.php.

[3] A. N. Panchenko, A. S. Poliakevich, A. A. Sosnin, V. F. Tarasenko. "Glow discharge in excilamps of low pressure," *Proceedings of institutes of higher education. Physics.* vol. 42, no. 6, pp. 50-66, 1999.

[4] S. S. Anufrik, V. O. Shkleinik, A. P. Volodenkov, K. F. Znosko. "XeCl-excilamps computer modeling," *Proceedings of the VII symposium of Belarus and Serbia on physics and diagnostics of laboratory and astrophysical plasmas (PDP`2008), September 22-26, 2008, Minsk, Belarus*, pp. 118-121, 2008.

[5] R. Bruzzese. "Comparison between experimental and theoretical results on discharge constriction in XeCl (asterisk) lasers," *Nuovo Cimento, Lettere, Serie 2.*, vol. 40, pp. 45-52, 1984.

[6] R. T. Watson. "Rate Constants for Reactions of ClOx of Atmospheric Interes," *J. Phys. Chem. Ref. Data.*, vol. 6, no. 3, pp. 871-917, 1977.

[7] A. V. Phels. "Cross sections and swarm coefficients for H+, H2+, H, H2, and H- in H2 for energies from 0,1eV to 10 keV," *J. Phys. Chem. Ref. Data.*, vol. 19, no. 3, pp. 653-675, 1990.

[8] D. L. Donohoue, D. Bauer, A. J. Hynes. "Temperature and pressure dependent rate coefficients for the reaction of Hg with Cl and the reaction of Cl with Cl: A pulsed laser photolysis-pulsed laser induced fluorescence study," *J. Phys. Chem. A.*, vol. 109, pp. 7732 – 7741, 2005.

XeCl-excilamp with the capacitance discharge

S. S. Anufrik, A. P. Volodenkov, K. F. Znosko

Yanka Kupala State University of Grodno, 22, Ozheshko Street, 230023, Grodno, Belarus,
Fax: +375 (152) 73-19-10; e-mail: a.volodenkov@grsu.by

Abstract: Average power of radiation made 0.6 W (frequency of pulses repetition F=50 Hz), and efficiency of 3 %. Excilamp brightness was 30 mW/cm^2 approximately. With 1cm^3 of active medium power radiation 74 mW/cm^3 was obtained, that exceeds radiation power with 1cm^3 for glow discharge excilamps at low pressure.

EXPERIMENTAL SETUP

The results of experimental research on XeCl-excilamps using capacitance discharge are presented. Advantage of this excilamp consists in simplicity of its design.

The excilamp construction is submitted in a Fig. 1. The excilamp electrodes (1) were situated outside of active volume. The nipple (2) was used for gas mixture pumping-out. Diameter of a quartz tube (3) made 16 mm, the walls thickness was 1 mm and irradiator length was ~ 200 mm. The lamp cell has a total volume of 0.03 l and the inetrelectrode volume between electrodes (1) varied from 0,0075 l up to 0.015 l.

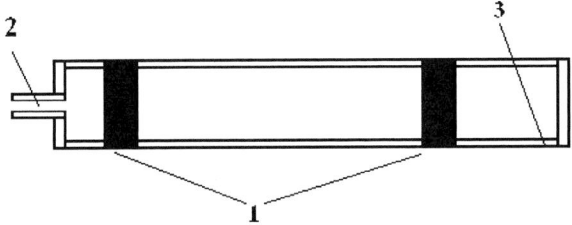

Fig. 1. The design of excilamp with capacitance discharge...

Excitation of the capacitance discharge was carried out by two-polar pulse groups with duration ~ 0,3 microseconds. For excitation the modified power supply [Fig. 2] of nitrogen laser ЛГИ-21 was used [1, 2].

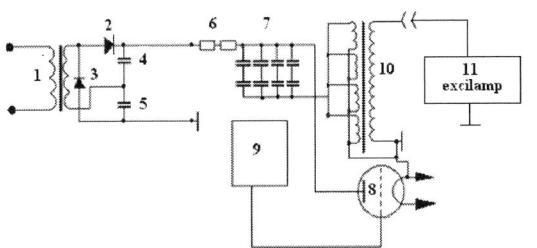

Fig. 2. The simplified power supply circuit of an excitation system of excilamp.

At Fig. 2 the next designations are used: 1- high-voltage transformer; 2, 3 – diodes; 4,5 – capacitors; 6 – charging resistance; 7 – storage capacity; 8 – thyratron; 9 -block of management of the thyratron, 10 – high-voltage cable transformer; 11 – excilamp.

Mains supply voltage (220 V) is connected to primary winding of high-voltage transformer (1). High voltage (5kV) from secondary winding of high-voltage transformer (1) is supplied to voltage-doubling circuit on diodes (2, 3) and capacitors (4, 5). The high voltage (10 kV) from out of voltage-doubling circuit is applied to the battery of storage capacitors (7) by means of charging resistance (6). The thyratron (8) is fired by pulses of controlled frequency from pulsed oscillator (9). Storage capacity (7) is discharged through primary winding of high-voltage cable transformer (10). The high voltage (~35 kV) from secondary winding of high-voltage cable transformer (10) is supplied to excilamps electrodes (11).

EXPERIMENTAL INVESTIGATION AND OPTIMIZATION OF EMISSION CHARACTERISTICS

Emission characteristics at interelectrode distance 10 cm are investigated. The dependence of relative value of average power of radiation from HCl pressure is submitted in Fig.3. The optimal value of partial pressure of HCl is equal to 2,5 Torr.

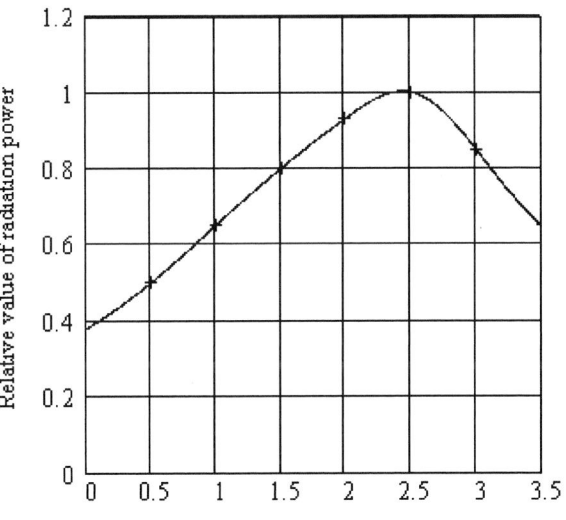

Fig. 3. The dependence of relative value of average power of radiation from HCl pressure (total pressure P=60 Torr; partial pressure of Xe: P_{Xe}=20 Torr).

The dependence of relative value of average power of radiation from Xe pressure is submitted in Fig.4. The optimal value of partial pressure of Xe is equal to 20 Torr.

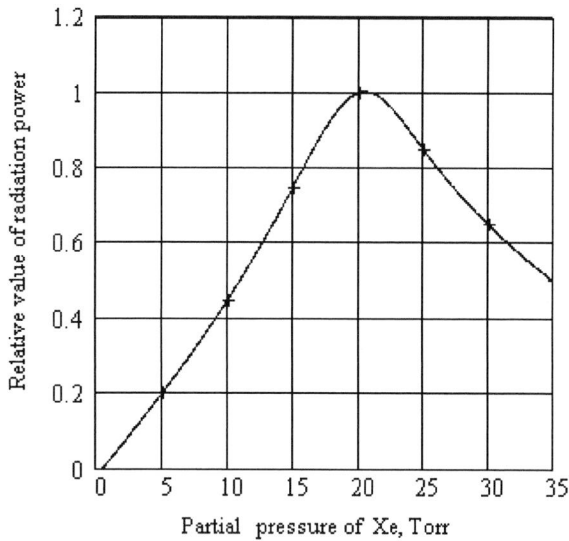

Fig. 4 The dependence of relative value of average power of radiation from Xe pressure (total pressure P=60 Torr; partial pressure of HCl: P_{HCl}=2,5 Torr).

The dependence of relative value of average power of radiation from total pressure of mixture is submitted in Fig.5

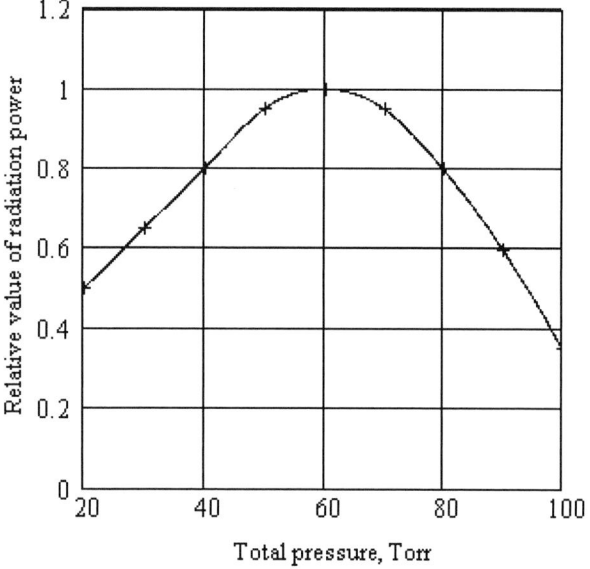

Fig. 5 The dependence of relative value of average power of radiation from total pressure of gas mixture (partial pressure of HCl: P_{HCl}=2,5 Torr; partial pressure of Xe: P_{Xe}=20 Torr) at interelectrode distance 10 cm.

On basis of realized experiments the optimal intermixture containing is found to be equal to: 2,5 Torr HCl, 20 Torr Xe,

37,5 Torr He. Average radiation power made ~ 0.6 W (pulses frequency F=50 Hz), and efficiency of ~3 %. Excilamp brightness was 15 mW/cm^2. With 1cm^3 of the active medium power of radiation 37 mW/cm^3 was obtained. This value is comparable to power of radiation with 1см^3 of active medium for excilamp of the glow discharge of low pressure [3].

Research of output emission characteristics of excilamp is executed at interelectrode distance 5 cm. Optimization of composition of a gas mix has been carried out. The optimal intermixture containing is found to be equal to: 2,5 Torr HCl, 20 Torr Xe, 77,5 Torr He.

The dependence of relative value of average power of radiation from total pressure of mixture is submitted in Fig.6.

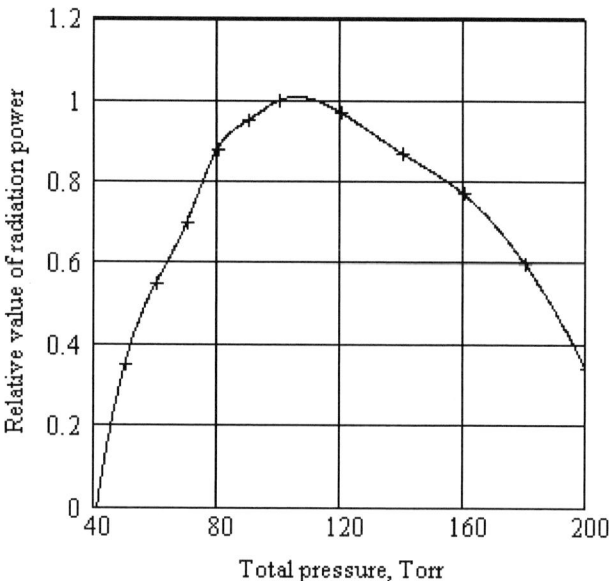

Fig. 6 The dependence of relative value of average power of radiation from total pressure of gas mixture (partial pressure of HCl: P_{HCl}=2,5 Torr; partial pressure of Xe: P_{Xe}=20 Torr) at interelectrode distance 5 cm.

Optimal the intermixture containing: 2.5 Torr HCl, 20 Torr Xe, 77.5 Torr He. Average radiation power made ~ 0.6 W (pulses frequency F=50 Hz), and efficiency of ~3 %. Excilamp brightness was 30 mW/cm^2. With 1cm^3 of the active medium power of radiation 74 mW/cm^3 was obtained. This value exceeds power of radiation with 1см^3 of active medium for excilamps of the glow discharge of low pressure [3].

It is necessary to note, that at interelectrode distance of 5 cm, the discharge were filled with all volume of an excilamp. At calculation of power of radiation we took into account only an emission between electrodes which were on distance of 5 cm from each other. Therefore the real efficiency of excilamp is higher at least in 2-3 times.

Besides for optimum composition of a mix dependence of relative value of power of radiation on value of amplitude of voltage pulse has been investigated. Dependence carries identical character as for interelectrode distances of 5 cm, and 10 cm. This dependence is submitted in Fig. 7. Dependence of relative value of power of radiation on frequency of repetition of pulses has been investigated and this dependence is submitted in figure 8.

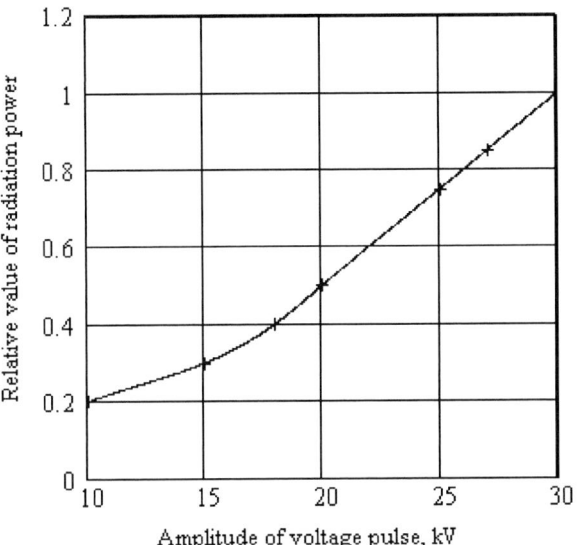

Fig. 7 The dependence of relative value of average power of radiation from value of amplitude of voltage pulse.

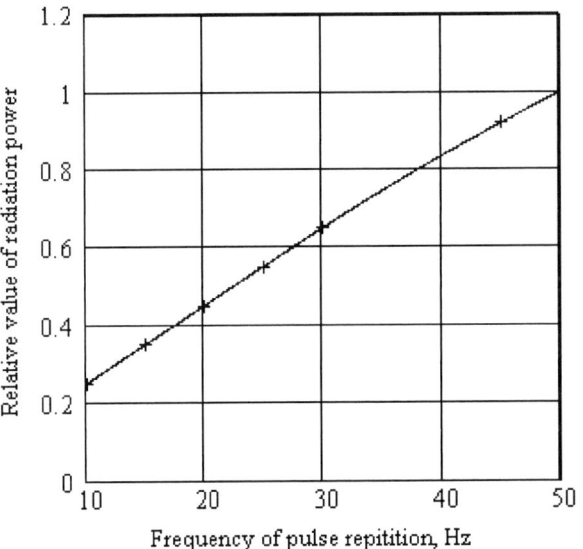

Fig. 8 The dependence of relative value of average power of radiation from frequency of repetition of pulses.

CONCLUSIONS

Emission characteristics at interelectrode distance 10 cm are next.

Optimal the intermixture containing: 2.5 Torr HC1, 20 Torr Xe, 37.5 Torr He. Average radiation power made ~ 0.6 W (pulses frequency F=50 Hz), and efficiency of ~3 %. Excilamp

brightness was 15 mW/cm^2. With 1cm^3 of the active medium power of radiation was 37 mW/cm^3.

Emission characteristics at interelectrode distance 5 cm are.

Optimal the intermixture containing: 2.5 Torr HC1, 20 Torr Xe, 77.5 Torr He. Average radiation power made ~ 0.6 W (pulses frequency F=50 Hz), and efficiency of ~3 %. Excilamp brightness was 30 mW/cm^2. With 1cm^3 of the active medium power of radiation was 74 mW/cm^3.

The excilamp has very simple design and very simple excitation system also. We have developed the technique of XeCl-excilamps computer modeling [4-6]. The results, being obtained as a result of application of this technique, are in a good agreement with experimental results, represented in this work.

In summary we shall note, that the basic result of the research executed on sources of not coherent radiation, consists in development of a design and practical creation of XeCl-excilamps with excitation by the capacitance discharge [7].

REFERENCES

[1] S. S. Anufrik, A. P. Volodenkov, K. F. Znosko. "The simple excilamp on basis of barrier discharge." *Proceedings of 5th International Conference on Plasma Physics and Plasma Technolodgy (PPPT 5), September 18-22, 2006, Minsk, Belarus,* vol. 2, pp. 851-854, 2006.

[2] S. S. Anufrik, A. P. Volodenkov, K. F. Znosko. "The some types of XeCl-excilamps." *Proceedings of SPIE,* vol. 7009, pp. 70090Q-70090Q9, 2008.

[3] A. N. Panchenko, A. S. Poliakevich, A. A. Sosnin, V. F. Tarasenko. "Glow discharge in excilamps of low pressure," *Proceedings of institutes of higher education. Physics.* vol. 42, no. 6, pp. 50-66, 1999.

[4] S. S. Anufrik, V. O. Shkleinik, A. P. Volodenkov, K. F. Znosko. "Technique of XeCl-excilamps computer modeling." *LFNM'2008, September 29 – October 4, 2008 Alushta, Crimea, Ukraine,* pp.48-50, 2008.

[5] S. S. Anufrik, V. O. Shkleinik, A. P. Volodenkov, K. F. Znosko. "XeCl-excilamps computer modeling," *Proceedings of the VII symposium of Belarus and Serbia on physics and diagnostics of laboratory and astrophysical plasmas (PDP`2008), September 22-26, 2008, Minsk, Belarus,* pp. 118-121, 2008.

[6] S. S. Anufrik, A. P. Volodenkov, K. F. Znosko, N. V. Mihuta. "Modeling of electrodischarge XeCl-excilamps," *Digest of scientific works "Laser and optical-electronics technique",* Issue 11 pp.122-129, 2008.

[7] S. S. Anufrik, A. P. Volodenkov, K. F. Znosko, M. M. Salehudinov. "XeCl-excilamps of capacitance discharge," *Digest of scientific works "Laser and optical-electronics technique",* Issue 11 pp.140-147, 2008.

Point of care fiber optical sensor for non-invasive multi parameter monitoring of blood and human tissue biochemistry

V. A. Saetchnikov[1], E. A. Tcheriavskaia[1], G. Schweiger[2]

[1]Belarusian State University, Minsk, Belarus
[2]Ruhr - Universitaet Bochum, Bochum, Germany

Abstract- Compact fiber optical and thermal sensor for non invasive measurement of blood biochemistry including glucose, hemoglobin and it's derivatives concentrations is developed as a prototype of the point-of-care diagnostic devices for cardiologic, tumour and diabetic patients. Integrated platform for data acquisition, data processing and communication to remote networks has been developed on the pocket PC. Method of non-invasive monitoring of human blood biochemistry based on spatially localized NIR diffuse scattering spectroscopy and metabolic heat measurements has been developed. Multivariate statistical analysis and cluster analysis was applied to convert sensor pickup signals into physicochemical variables.

Most proposed methods for glucose determinations are based on monitoring extremely weak near-infrared glucose absorption spectral features or use the effect of glucose on the tissue scattering coefficient, which is a nonspecific effect. There are several problems to get a proper accuracy of measured data (unexplained biological noise, chance correlation with other biological or instrumental events, data overfitting, lack of appropriate control experiments etc). More promised approach is to consider the physiologic response of the body to changes in glucose concentration.

We developed a method, involves the measurement of physiologic indices related to metabolic heat generation and local oxygen supply, which correspond to the glucose concentration in the local blood supply. The principles underlying the method are unique in that measurements are performed on indices associated with the oxidative metabolism of glucose. The metabolic oxidation of glucose in the human body provides most of the necessary energy for cellular activities. Body heat generated by glucose oxidation is based on the subtle balance of capillary glucose and oxygen supply to the cells. Hence, blood glucose can be estimated by measuring the body heat and the oxygen supply. The method was derived from the observation that the homeostatic circadian rhythm of the human body is dependent on the interrelationship between metabolic heat, local oxygen supply, and glucose concentration. The variables in this model consist of heat generated, hemoglobin (Hb) concentration, oxy hemoglobin (HbO_2) concentration, and blood flow rate. So both thermal and optical sensors are used to measure these values simultaneously in an individual and, through statistical manipulations quantitatively investigate a

blood biochemistry, for example, the concentration of blood glucose.

The method was derived from the observation that the homeostatic circadian rhythm of the human body is dependent on the interrelationship between metabolic heat, local oxygen supply, and glucose concentration. The variables in this model consist of heat generated, hemoglobin (Hb) concentration, oxy hemoglobin (HbO_2) concentration, and blood flow rate. All variables should be estimated after compensation for the appropriate environmental conditions.

We used noninvasive thermal and optical sensors on different parts of cutaneous tissue of an individual (fingertip, arm, ear lobe etc.) to measure thermal generation, blood flow rate, Hb and HbO_2 concentration. Investigations were concentrated on working through methodology and experimental realization of diffuse optical reflective spectroscopic measurements combined with thermal emission and diffuse scattering.

The measuring device consists of a sensor pickup and electronics integrated with a computer. The sensor pickup contains of various radiation sources, heat and optical detectors, which interact with the human body when different parts of cutaneous tissue of an individual (fingertip, arm, ear lobe etc.) is placed on it or monitored. The sensor pickup measures the following physicochemical indices in an individual: thermal generation, heat balance, blood flow rate, Hb and HbO_2 concentrations as well as several environmental conditions. The various sensors are used for proper enable correction for effects of different endogenous interferences.

The heat generated by glucose oxidation in the human body is measured by the principle of heat diffusion which represents the sum of evaporative and non evaporative heat loss. Optical measurements of the device are based on the principles of diffuse reflectance because skin absorption of various wavelengths is determined by the total absorptive and scattering properties of the epidermis and the dermis.

The calibration and glucose measurement processes were performed independently. Methods of multivariate statistical analysis were applied to convert various signals from the sensor pickup into physicochemical variables. Multivariate statistical analysis involving the variables from sensor signals, polynomials from various variables, regression analysis of individual patients, and cluster analysis of patients group was then performed. At the cluster analysis, patients were classified into clusters by use of the individual variables as well as their composites. The calibration function was then generated for each group. By comparing the values from the noninvasive measurement with the venous plasma glucose result from the same patient, we obtained the analytical functions for each patient. This evaluation is in progress.

We performed investigation of physicochemical blood parameters obtained by developed method for patient group of 22 patients both healthy and with different pathologies to get a wide variation range of blood parameters. Obtained results were compared with results obtained with an automated multi parameter analyzers for venous specimens obtained invasively. Invasive and noninvasive measurements were temporally obtained as close to each other as possible.

Thermal and optical response of humans have been investigated. The metabolic heat generated in human body is measured based on heat diffusion principals counting non evaporative and evaporative heat losses. Non evaporative losses are evaluated from thermal conduction, convection and radiation of the body. For this purpose four temperature detectors which are used to measure surface body, ambient room and background radiation temperatures. Blood flow rate is proportional to a thermal conductivity of the skin. So it evaluated from monitoring the change in temperature for the contact duration between the contact and adjacent detectors.

Time acquisition spectral intensities of optical signals for six wavelengths and eight output bundle channels of ten patients both healthy and with different pathologies to get a wide variation range of blood parameters have been obtained. The optical wavelengths used did not necessary correspond to blood analyte infrared absorption bands. After processing of optical signals including filtering (with optimal filter design), smoothing and time domain averaging systolic and respiration bio rhythms could be evaluated. Optimal conditions for different bio rhythms evaluation depend on sensor localization on the body and distance transmitted by the radiation through the body.

Multivariate regression analysis (linear, robust, orthogonal, wavelet) was applied to compare the noninvasive method with the standard invasive reference method. Obtained noninvasive results were compared with results obtained with an automated multi parameter analyzers for venous specimens obtained invasively. Invasive and noninvasive measurements were temporally obtained as close to each other as possible. As it was mentioned above we found that optimized filtering to get the best correlation of noninvasive data with hemoglobin derivatives concentrations can be performed using wavelet of symlet type.

The pulsed temperature-modulated optical signal monitored the change in blood biochemistry. Multi wavelength spectroscopy in the spectral range of 650 – 1700 nm was used to evaluate non invasive glucose concentration. Several filtering procedures have been used to get better correlation with a standard invasive data. In general an approach was based on the assumption that the glucose concentration is the only time-dependent variable. Other factors, such as temperature, blood perfusion, tissue compression, blood oxygenation, cutaneous water, and other metabolites or medications that affect tissue blood dynamics are not considered. Spectral bands in the used wavelength range correspond to blood absorption and thus reflect hemodynamic changes in cutaneous tissue. Localized reflectance data correlate with hemoglobin concentration. Final data processing were organized under Mathlab includes multiple regression analysis (linear, robust and orthogonal). Regression analysis, depicted on a Clarke and Parkes error grids, was performed to compare the noninvasive method with the standard invasive reference method, using data points with glucose concentrations ranging from 3.0 to 22.5 mmol/L (54-405 mg/dL). The correlation coefficient (r) was 0.89.

The source-detector distances of the probe used in our study sampled cutaneous layers down to a depth of 2.5 mm. This depth includes the cutaneous vascular system (upper plexus, lower plexus and connecting arterioles, venules, and shunts), which contributes to the control of the body temperature. Cutaneous blood flow depends on temperature. Cutaneous red blood cell flux varies between comparable sites in the same individual, and 1-mm^2 areas of vascular "territories" surrounded in part by relatively avascular areas have been identified in human skin. Thus, even if the shape of the curve is tracking the change in glucose concentration, a constant-term difference may result from positioning of the illumination and detection fibers with respect to vascular territories and avascular areas in the skin, and/or with respect to pockets with different refractive index values.

Clinical testing of developed sensors is now in progress. The calibration functions are generated for different patient groups, such as healthy patients, cardiologic, pathological and diabetic patients, new born childes, elder peoples etc., to classify them into clusters by use of the individual variables as well as their composites. The calibration functions are generated for each group. Clinical changes in each patient's metabolism are also important variables that affect the values used to classify patients into clusters and would be valid as long as the metabolic condition remains unchanged.

Clinical testing of developed sensors is now in progress. Calibration functions generated for different patient groups (healthy, cardiologic, pathological, diabetic patients, new born childes, elder peoples etc.) classified into clusters for both individual variables and their composites and generated for each group. By comparing the values from the noninvasive measurement with the standard invasive reference methods from the same patient, the analytical functions for each patient are obtained. Clinical changes in patient's metabolism are also affect the values used to classify patients into clusters.

Developed non invasive multi parameter sensor is used as a prototype of point-of-care diagnostic devices for cardiologic, tumour and diabetic patients. Integrated platform for data acquisition, data processing and communication to remote networks is being developed on pocket PC.

The damage of DNA induced by UV nanosecond laser excitation at 193 nm

N.N. Vtyurina[1], S. L. Grokhovski[2], I.V. Filimonov[1], O.I. Medvedkov[3], D. Yu. Nechipurenko[1], S.A. Vasiliev[3], Yu. D. Nechipurenko[1,2]

[1] Faculty of Physics, Moscow State University, Moscow, Russia
[2] V. A. Engelhardt Institute of Molecular Biology, Russian Academy of Sciences, Moscow, Russia
[3] Fiber optics research center, Russian Academy of Sciences, Moscow, Russia

Abstract: Solutions containing dsDNA fragments were irradiated with UV laser pulses. The produced damage was investigated using polyacrylamid gel electrophoresis. The intensity of cleavage in particular phosphodiester bond after hot alkali treatment of irradiated samples was shown to depend on base pair sequence. The damage of DNA fragments has occurred predominantly in the sites containing guanine clusters.

INTRODUCTION

It is known that UV radiation might induce various types of damage in DNA. The irradiation at 193 nm is capable of producing single and double-strand breaks. In a number of works oxidative guanine lesions were analyzed at the nucleotide level as a result of DNA subjection to nanosecond ultraviolet (266 nm) laser pulses. These lesions were analyzed using polyacrylamide gel electrophoresis which has shown that the extent of guanine damages depends on the primary and secondary DNA structure. Thus, the specificity of DNA damage under the influence of laser irradiation, i.e. the dependence of damage extent on the local base pair sequence, was revealed. It is known that the chemical modification of bases in DNA which occurs upon its irradiation with UV light results in phosphodiester bond cleavage after the treatment of DNA solution with alkali.

The goal of this work is analysis of sequence specificity in cleavages of DNA fragments irradiated with ArF laser pulses at 193 nm and subsequently treated with alkali using gel electrophoresis.

MATERIALS AND METHODS

The restriction fragments of DNA were generated by digestion of plasmid pGEM7(f+) (Promega) by the corresponding restriction endonucleases. The fragments were 3'-end-labelled with $[\alpha\text{-}^{33}P]dATP$ or $[\alpha\text{-}^{33}P]dCTP$ or $[\alpha\text{-}^{32}P]dATP$ ("FGUP" Institute of reactor materials, Zarechnii, Sverdlovskaya oblast, Russia) in the presence of the unlabeled other dNTP and the Klenow fragment of *Escherichia coli* DNA-polymerase I (Sibenzyme, Novosibirsk, Russia).

10 μl of the sample were irradiated in an open test tube with various number of laser pulses of following characteristics: aperture: 9X5 mm, pulse duration: 8-10 ns, frequency of pulses: 10-20 Hz, pulse energy - 10 mJ.

After irradiation procedure the samples were heated for 10 minutes at 90° with 10 % piperidine and treated with alkali. Afterwards the samples were subjected to polyacrylamide gel electrophoresis.

Gel image was processed by SAFA package which is capable of gel's lanes alignment, calculating the overall intensity of each band and correlation of the band sequence with corresponding base pair sequence [7]. The overall intensity of each band is proportional to concentration of the corresponding fragments in solution. Thus, the values of bands intensities represent the cleavage profile of the fragment which gives one the possibility to compare the degree of cleavage in different sites of DNA.

RESULTS

In order to evaluate the optimal irradiation dose we have carried out a number of preliminary experiments. Fig.1 demonstrates the cleavage pattern obtained after irradiation of DNA fragments of known base pair sequences with various doses.

Fig.2 represents the histogram of phosphodiester bond cleavage intensities in DNA fragment irradiated with dose of 996 J (lane 12 in Fig.1). The histogram was built by correlation of band intensities values with corresponding base pair sequence.

CONCLUSIONS

It is clear that the greatest values of bands intensities correspond to enhancement of DNA cleavages in phosphodiester linkages following guanine (in a 5' to 3' direction). We note that these results can not be explained in terms of significant difference in guanine UV absorption spectrum comparing to the spectrum of adenine residues as far as no such difference have been reported. It is also worth noting that in absence of high temperature treatment of irradiated sample with alkali there was no clear sequence specificity of the cleavage patterns. Thus, the observed sequence effects on DNA cleavage patterns is actually the result of chemical modifications in guanine caused by irradiation process and revealed by alkali treatment of the samples. Our results are in line with reported data on DNA damage induced by the laser pulses at 266 nm [5].

978-1-4244-7043-3/10 $26.00 © 2010 IEEE

CAOL*2010 International Conference on Advanced Optoelectronics & Lasers, 10-14 September, 2010, Sevastopol, Ukraine

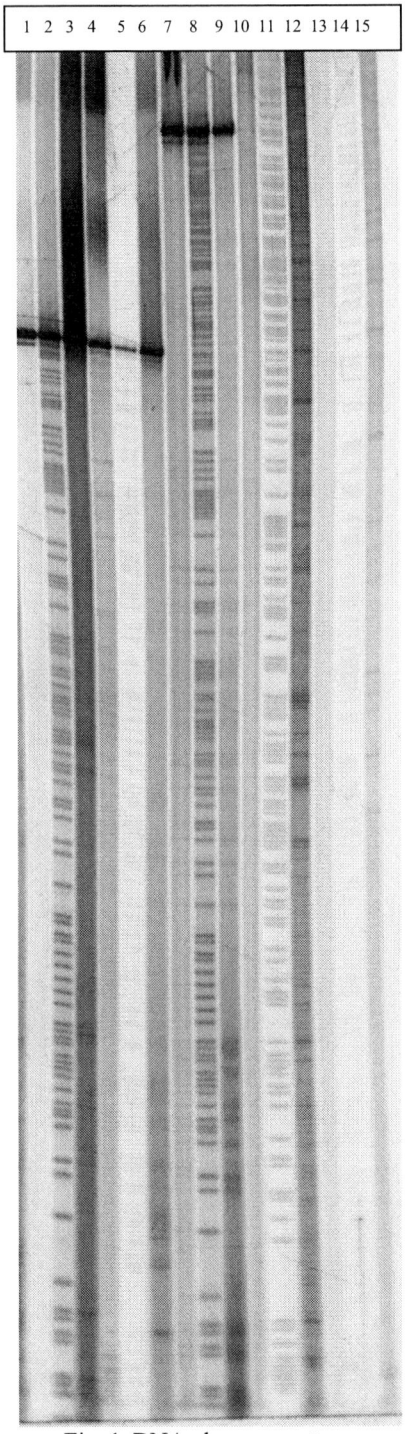

Fig. 1. DNA cleavage patters.

Lanes 1,4,7,10,13: samples without UV light treatment;
Lanes 2,5,8,11,14: chemical cleavages at the purine sites of DNA;
Lanes 3,6,9,12,15: DNA fragments irradiated with various doses (104, 102, 100, 996 and 984 Joules, respectively).

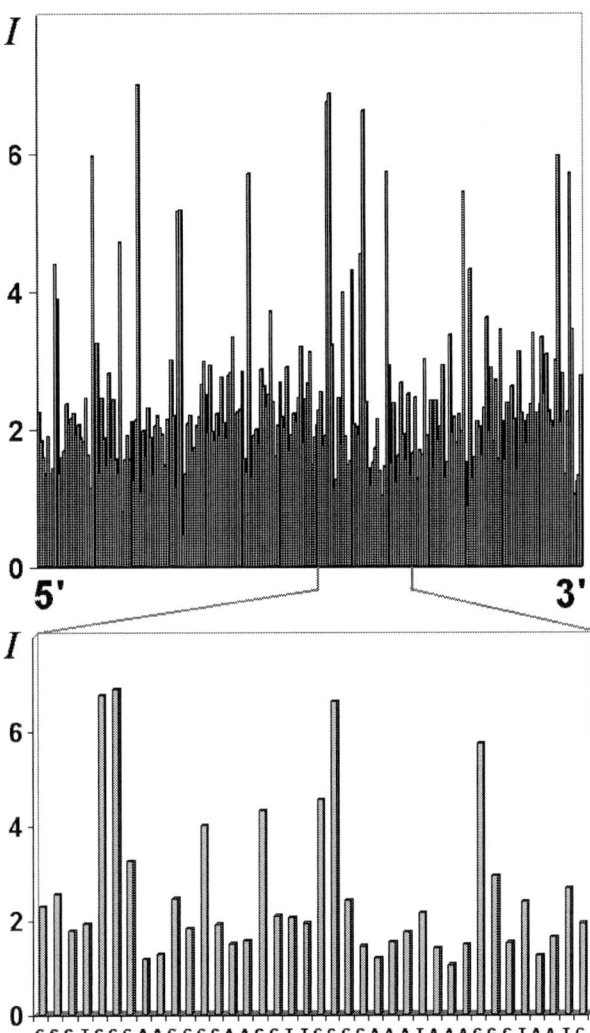

Fig. 2. The histogram of DNA cleavage intensities.
Y axes - cleavage intensities in arbitrary units,
X axes – base pair sequence.

REFERENCES

[1] K. Takakura, M. Ishikawa, K. Hidea, K. Kobayashi, A. Ito, T. Ito, "Single-strand breaks in supercoiled DNA induced by vacuum-UV radiation in aqueous solution», *Photochem. and Photobiol.*, vol. 44, pp. 397 – 400, 1986

[2] D. Schulte-Fronlinde, J. Opitz, H. Gorner and E. Bothe, «Model studies for the direct effect of high-energy irradiation on DNA. Mechanism of strand break formation induced by laser photoionization of poly U in aqueous solution», *Int. J. of Radiat. Biol.*, vol. 48, pp. 397 — 408, 1985

[3] I. E. Kochevar and L. A. Bucklcy «Photochemistry of DNA using 193 nm excimer laser radiation»,

978-1-4244-7043-3/10 $26.00 © 2010 IEEE   245

Photochem. and Photobiol., vol. 51, pp. 527 – 532, 1990

[4] P. Candeias, and S. Steenken «Ionization of purine nucleosides and nucleotides and their components by 193-nm laser photolysis in aqueous solution: model studies for oxidative damage of DNA», *J. Am. Chem. Soc.*, vol. 114, pp 699–704, 1992

[5] D. Angelov, B. Beylot and A. Spassky «Origin of the Heterogeneous Distribution of the Yield of Guanyl Radical in UV Laser Photolyzed DNA», *Biophysical J.*, vol. 88, 2005

[6] H. Sies, W. A. Schulz, Steen Steenken «Adjacent guanines as preferred sites for strand breaks in plasmid DNA irradiated with 193 nm and 248 nm UV laser light», *Photochem. and Photobiol.*, vol. 32, pp. 97-102, 1996

[7] K. Takamoto, M.R. Chance, M. Brenowitz, *Nucleic Acids Research*, vol. 32, pp. 119, 2004

Real time monitoring of micro and nano particles, blood phantoms in situ by optical micro resonance methods

V. A. Saetchnikov[1], E. A. Tcheriavskaia[1], G. Schweiger[2], A. Ostendorf[2]

[1]Belarusian State University, Minsk, Belarus
[2]Ruhr - Universitaet Bochum, Bochum, Germany

Abstract- Methods and instrumentation based on resonance frequency dependence of dielectric micro resonators on the surrounding medium is being developed as a real-time one-way disposable sensor for a number of parameters of nano particles and modeling blood in situ.

INTRODUCTION

Recently, a number of evanescent wave optical sensors have been developed and used for refractive index detection in sensing systems. Examples of such devices include the resonant mirror, metal and absorbing material-clad leaky waveguide, the differential interferometry sensor grating coupler, and the integrated optical waveguide. Though evanescent wave sensors are effective for measuring the refractive index, they typically require a relatively long physical length to achieve sufficient light-matter interaction in order to provide the necessary sensitivity. This is in contradiction to the requirements of lab-on-a-chip systems.

We show highly sensitive refractive index measurement using optical microsphere resonator as evanescent field-based sensors. The resonating nature of the cavity enhances the light that interacts with the sample volume by hundreds to thousands of times, thus reducing the required interaction length as compared with typical evanescent sensors. We discuss a variety of approaches and identify the advantages and disadvantages. We then present experimental demonstrations of approach that take advantage of extremely high Q-factors. The design uses, as the resonating cavity, a fused silica sphere, which is submerged in the sample fluid.

RING RESONATOR SENSOR

In a dielectric ring, light will circulate continually at the inner surface due to total internal reflection. These resonating modes are called the whispering gallery modes (WGMs). As with any wave guiding medium, an evanescent field extends beyond the dielectric surface and into the surrounding medium. When the refractive index changes in the surrounding medium near the surface it causes a change in the resonant condition. Thus, changes in the spectral positions of the WGMs convey quantitative and kinetic information about the refractive index of the surrounding medium. Therefore, this technique can be used to detect changes in the surrounding solution, or the binding of molecules at the surface.

Ring resonator sensors have an advantage over typical evanescent wave sensors because in a ring resonator, the light continually circulates at the surface, increasing the light-matter interaction. The quality factor, or Q-factor, is the parameter that indicates the strength of the resonance and, thus, the enhancement in light-matter interaction. Q-factors, which can result in an effective sensing length on the order of hundreds of centimeters, despite the sub - millimeter dimensions of typical microspheres. Microsphere resonators can be used as bio/chemical sensors. Captured molecules affect the effective refractive index of the WGMs, resulting in a measurable spectral shift. Low detection limits have been achieved, primarily because of the high Q-factor. Furthermore, the high Q-factor results in a very narrow mode linewidth, which improves measurement precision.

EXPERIMENTAL SETUP

We demonstrate experimentally the microsphere resonator for refractive index measurements The microspheres are created by melting the tip of a standard single mode fiber optic cable with a laser. The spheres used in this experiment are 50 - 190 micron in diameter. The light from a tunable diode laser (New Focus, 680 nm) is coupled into the microsphere through a prism. In the experiments presented here, the prism with a refractive index of 1.7 is polished at an angle of 60 degree, which is designed for optimal mode coupling.

To observe the WGM, the laser repeatedly scans across a spectral range of approximately 150 pm at a frequency of about 3 Hz. Light scattered by the microsphere is collected through a microscope by a CCD camera or photodiode and monitored with a data acquisition card and computer. When the wavelength of the tunable laser is on resonance with the WGM of the sphere, the power of the light scattered by the microsphere increases, and a spectral maximum indicating the WGM spectral position is recorded.

The microsphere is submerged into a fluidic cell and brought into contact with the prism. The well initially contains de-ionized (DI) water. To vary the refractive index, a solution of ethanol and water is incrementally added to the fluidic cell with a digital syringe. Following each injection, the WGM modes are monitored until equilibrium is reached, and then, the subsequent injection is made.

EXPERIMENTAL RESULTS

The sensitivity to refractive index changes of the microsphere sensor system is determined by monitoring the magnitude of the WGM spectral shift when a known quantity of ethanol is added to the solution in the well.. The steady-state values of WGM spectral shift for a change in refractive index exhibit excellent linearity. The expected WGM spectral shift can be computed from the theoretical relation between wavelength and refractive index, given by the asymptotic formula. The results show that the sphere has a sensitivity of approximately 30 nm/RIU. The detection limit of our system can be calculated from this sensitivity. Because the laser linewidth is extremely narrow (300 kHz), it is expected that we can resolve 1/20th to 1/50th of the WGM linewidth. For a Q-factor of $5 \cdot 10^6$, this results in a refractive index detection limit of on the order of 10^{-7} RIU, which is comparable to or even better than surface plasmon resonance or waveguide-based sensors. The change in refractive index per 1% (w/w) change in methanol is 0.00056 RIUs. This precision results in a refractive index detection limit of around . Similar results to these above have been further duplicated in our lab, which shows that these novel sensing devices can be reproduced with regularity.

The experimental demonstrations and the simulations show the potential for the excellent label-free sensors. However, it must be noted that it also suffers from some of the typical handicaps of label-free sensors. One drawback that must be countered is the temperature dependence of the resulting signal. Small temperature fluctuations result in small fluctuations of the WGM spectral position, which produces noise in the sensor signal. Temperature control is one option to combat this issue. Additionally, we have demonstrated the use of a reference channel for temperature noise subtraction.

The experimental results presented here show the excellent capability of a microsphere resonator to measure the refractive index of liquids. The microsphere can achieve similar or even better results than typical waveguide-based sensors while only consuming picoliter sample volume because of the resonating nature of the cavity. The high sensitivity and low sample consumption are excellent characteristics for lab-on-a-chip systems. However, some design concepts are in need of improvement. First, the microsphere in the experiment presented here is 100 in diameter. Although this is small compared to waveguide sensors, it is still larger than the typical desired range for advanced microfluidic systems. Second, perfect alignment of the microsphere to the light delivery system (waveguide, prism) is nontrivial, and maintaining that alignment during fluid flow is also an issue that must be addressed.

CONCLUSION

A novel emerging technique for the label-free analysis of nanoparticles including biomolecules using optical micro cavity resonance of whispering-gallery-type modes is being developed. A scheme of such a method based on microsphere matrix have been developed for ultra high resolution spectroscopy. Using a calibration principal of this scheme the method is being transformed to make further development for microbial application. Some other schemes using similar principals: stand-alone, array and matrix microsphere resonators, liquid core optical ring resonators are also being under development.

Water solutions of ethanol, HCl, glucose, vitamin C and biotin have been used to test refractive index changes by monitoring the magnitude of the whispering gallery modes spectral shift. Particular efforts were made for effective fixing of the micro spheres in the water flow, an optimal geometry for micro resonance observation and material of microsphere the most appropriate for microbial application. Optical resonance in free micro spheres from PMMA fixed in micro channels produced by photolithography has been observed under the laser power of less then 1 microwatt. Resonance shifts of C reactive protein water solutions as well as albumin solutions in pure water and with HCl modelling blood have been investigated. Introducing controlled amount of glass gel nano particles into sensor microsphere surrounding were accompanied by both correlative resonance shift (400 nm in diameter) and total reconstruct of resonance spectra (57 nm in diameter).

ACKNOLEGEMENTS

Investigations were supported by International Bureau of WTZ DLR, project BLR 08/002.

The Phenomenon of Laser-Induced Photodissosiation of Hemoglobin Complexes in Cutaneous Blood Vessels and its Biomedical Application

M.M. Asimov

Institute of Physics National Academy of Science of Belarus,
68 Nezalechnosti St., 220072 Minsk, Belarus.
E-mail: m.asimov@dragon.bas-net.by

Abstract: *New laser-optical technology based on photodissosiation of oxyhemoglobin and carboxyhemoglobin in vivo for local tissue oxygenation and increasing the efficiency of elimination of carbon monoxide poisoning effect is proposed. The results of both numerical simulations of "laser-tissue" interaction in terms of maximal effect on blood oxyhemoglobin and carboxyhemoglobin in cutaneous blood vessels and experimental investigations the efficiency of their photodissosiation are presented. It is shown that despite of similarity and overlapping of carboxyhemoglobin and oxyhemoglobin action spectrum significant difference on quantum yields of photodissociation makes possible to develop a novel optical method of elimination of carbon monoxide poisoning effect.*

INTRODUCTION

Lasers and optical technologies widely applies in different branches of modern medicine. In present, cold (low intensity or low energy) laser therapy has becoming routine [1-4] in treating a wide variety of diseases. Cold laser therapy is widely uses in dermatology for treating burns, bedsores, ulcers and anaerobic infections. Method of photodynamic therapy (PDT) is successfully applied in cancer treatment [5-7].

The biological response and therapeutic effect of laser radiation is already well established however the mechanism of this phenomenon still remains topic of discussion.

Since 1997 the hypothesis of the role of oxyhemoglobin (HbO_2) photodissociation in the biological response and the mechanism of therapeutic effect of cold laser radiation has been suggested [9-11]. It is shown that phenomenon of laser-induced photodissociation of HbO_2 gives a unique possibility in increasing the local concentration of free oxygen in tissue.

Molecular oxygen plays a vital role in the human energetic and aerobic cell metabolism is primary mechanism in energy production in tissue. Controlling this mechanism gives unique possibility of biological stimulation to reach therapeutic effect.

As it known hemoglobin (Hb) easily binds ligands of gas molecules such as oxygen (O_2) and carbon monoxide (CO).

Binding with oxygen produce oxyhemoglobin (HbO_2) which main function is transporting oxygen from lungs to tissues. Binding the ligand of CO leads to formation of carboxyhemoglobin (HbCO) complexes that decrease oxygen transport function of hemoglobin and causes poisoning effect. Normal cell metabolism required an adequate concentration of O_2 in biological tissue [8]. The problem of controlling the local tissue oxygen concentration and keeping it at the necessary for normal cell metabolism level is an actual in modern medicine.

The deficit of oxygen in tissue (hypoxia) caused complications in treatment of wide variety of diseases. The following criteria of tissue oxygen tension ($TcPO_2$) are accepted:

- $TcPO_2 > 40$ mmHg indicates the adequate concentration of O_2 in tissue for normal cell metabolism;
- $20 > TcPO_2 < 40$ mmHg – O_2 concentration is not sufficient and it may cause medical problems;
- $TcPO2 < 20$ mmHg – O_2 concentration is extremely low that leads to tissue necroses.

In clinical practice different methods to overcoming tissue hypoxia usually are used. The oldest method of oxygenation is based on hyperventilation of lung with pure oxygen at normal pressure. This method has low efficiency and not sufficient in many cases.

An alternative and more efficient method has been developed on the base of chamber technology. This method is hyperbaric oxygenation (HBO) and oxygen delivered to chamber under the pressure greater than atmospheric. The method of HBO tissue oxygenation usually takes many hours and may cause oxygen toxemia that limits it application.

In this paper new approaches both in local tissue oxygenation and elimination of poisoning effect of carbon monoxide (CO) in cutaneous blood vessels are presented. The results of selective effect of laser radiation on HbO_2 and HbCO and induced photodissociation of these hemoglobin complexes are discussed. Novel laser-optical technology in local tissue oxygenation and photodecomposition of HbCO are proposed.

LASERT-INDUCED TISSUE OXYGENATION

Amount of oxygen available for cell metabolism delivered by microcirculation is the function of:

$$\sum O_2 (TcPO_2) = f(F(HbO_2)*[O_2]),$$

were HbO_2 is the value of oxyhemoglobin arterial blood vessels and $[O_2]$ - is the concentration of oxygen released into plasma. Deterioration of the blood microcirculation required extra oxygen supply to provide the demands of cell for normal metabolism. This could be reached by in vivo laser-induced photodissosiation of HbO_2 directly at the zone were it is necessary to increase the local concentration of free molecular oxygen.

As a result we obtain total concentration of oxygen releasing in conventional way and due to laser-induced photodissosiation of HbO_2 in arterial blood:

$$\Sigma[O_2] = [O_2] + [O_2^{hv}]$$

Irradiation of cutaneous blood vessels by laser radiation induces in vivo photodissociation of HbO_2 and extraction of additional amount of molecular oxygen. This process is illustrated in fig.1.

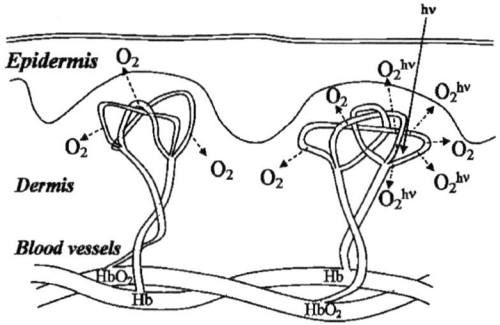

Fig.1. Optical method of additional oxygen extraction from HbO_2 and laser-induced tissue oxygenation.

The possibility of additional oxygen supply allows develop a new method of tissue hypoxia elimination that restores normal cell metabolism.

LASER-INDUCED PHOTODISSOSIATION OF CARBOXYHEMOGLOBIN

We extended our conception on carboxyhemoglobin targeting to develop new approach in reducing of CO poisoning effect. High quantum yield of photodissociation of HbCO (~98%) make this phenomenon promising for developing effective method of HbCO photo destruction and elimination of CO. Schematically it demonstrated in Fig. 2.

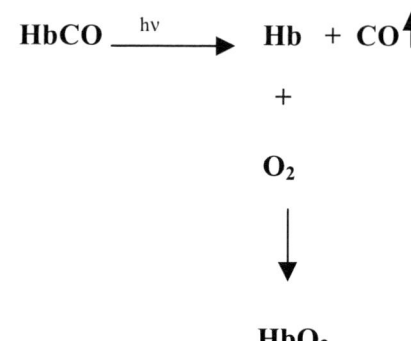

Fig.2. Illustration of HbCO photodissosiation and CO elimination with restoration of Hb oxygen transport function

Laser-induced photodecomposition of HbCO gives scientific bases in developing an effective optical method of elimination of carbon monoxide poisoning action.

RESULTS AND DICUSSION

The results of both computer modeling the "laser - Tissue" interaction in calculating action spectra of hemoglobin complexes with O_2 and CO ligands as well as experimental investigations of photodissosiation of HbO_2 and HbCO demonstrate good correlation.

Calculations demonstrate that due to optical properties of skin there is respectively narrow spectral diapason for effective photodissosiation of HbO_2 in cutaneous blood vessels. In the visible spectral range the maximum of action spectra of HbO_2 is at $\lambda = 585$ nm.

Preferable wavelength in the IR spectral region for irradiation of HbO_2 is at $\lambda = 960$ nm. It is shown that efficiency of HbO_2 photodissosiation HbO_2 at $\lambda = 585$ nm irradiation remains higher than one at 960 nm up to 2 mm depth of light penetration into tissue and about 2,5 mm they become equal.

So the wavelength $\lambda = 585$ nm may be recommended as the most efficient one for irradiation of HbO_2 in blood vessels located at the depth of tissue up to 2.5 mm while for the deeper layers of tissue the most suitable wavelength is about 960 nm.

Experimental results obtained in vivo for three voluntaries demonstrate about two time increase the local tissue O_2 concentration under the irradiation by the He-Ne laser with out put power ~1.5 mWt.

The phenomenon of laser-induced photodissosiation of HbO_2 allows understand the mechanism of biological response and therapeutic effect of laser radiation. It also gives unique possibility to develop laser-optical method of selective and local tissue oxygenation directly at the zone of irradiation.

High quantum yield of photodissociation of HbCO [12] makes this phenomenon promising for biomedical application. Significant difference in quantum yields between HbO_2 and HbCO is used for selective decomposition of latter.

The most effective ways for photodissosiation of HbCO and removal CO from blood stream are lunge and skin. In both cases the physical principle of HbCO photo dissociation remains the same, however irradiation through skin or lung tissues optical properties latter should taken account. In the case of direct elimination of HbCO in lung, one could consider that absorption spectrum may remain close to in vitro absorption spectra.

The effective wavelengths of photo dissociation should be at $\lambda_1 = 540$ and $\lambda_2 = 570$ nm. It should be noted that direct light delivery to the lung requires usage of the light guide and may do this method not convenient in clinical practice. An alternative method is to effect on HbCO in cutaneous blood vessels through human skin. In this case we must take into account optical properties of human skin in order to determine effective wavelength of penetrating radiation. This can be achieved by calculating the action spectra of HbCO in the skin.

Experimental results demonstrate that exposition at the laser wavelengths of $\lambda = 514,5$ nm the concentration of carboxyhemoglobin decreases more than 15% from initial 100% one No effect on laser wavelengths at $\lambda = 632.8$ nm was observed. Therefor for decomposition of carboxyhemoglobin the wavelengths of $\lambda = 514,5$ nm is optimal and gives possibility optically increase the rate of decomposition almost by an order of magnitude.

Proposed novel method in clinical practice could be applied by following ways depending on the severity of intoxication with CO:

- Transcutaneous irradiation of blood - at concentration of HbCO < 10%;
- Irradiation of blood directly in lung alveolus - HbCO < 30%;
- Intravenous irradiation of blood - HbCO > 30%

Significant quantum yield of photodissociation of HbCO permits us to expect high efficiency of proposed approach in clinical applications.

CONCLUSIONS

It is shown that the value of additional oxygen extracted from the oxyhemoglobin of cutaneous blood vessels depends on wavelength of low intensity laser radiation, irradiation time and the rate of the oxygen diffusion into tissue.

Direct in vivo measurements of tissue oxygen tension in different tissues are carried out. Both, the results of experimental study and computational modeling the kinetics of tissue oxygen distribution from blood plasma are presented.

It is shown that the efficiency of developed new laser-optical technology of tissue oxygenation is comparable with the method of hyperbaric oxygenation gaining at the same time advantages in selectivity and local action.

An important conclusion is drawn from the obtained results that in order to make the laser therapy really efficient one has to control the oxygen concentration in tissue keeping it at the level necessary for normal cell metabolism.

Proposed phenomenon opens new perspectives in dermatology in treating burns, bedsores, and ulcers and anaerobic infections as well as other pathologies where compensation of the tissue oxygen deficit is critical.

New approach to carbon monoxide poisoning treatment based on photodissociation of carboxyhemoglobin in blood stream of arterial and venous vessels is proposed.

Criteria of efficiency of laser-induced photodissociation of carboxyhemoglobin under direct irradiation in alveolus, transcutaneous and intravenous interaction with laser light is considered.

Phototherapy of CO poisoning in combination with oxygen hyperventilation, as well as hyperbaric oxygen therapy may become a powerful method in photomedicine.

REFERENCES

[1] S. Takas, S. Stojanovich, "Diagnostic and biostimulating lasers," Med. Pregl., vol. 51, pp. 245-249, 1998.

[2] G.D. Baxter, "Therapeutic lasers: Theory and Practice," Edinburgh; New-York, 1994.

[3] J. Tuner and L. Hode, "Laser Therapy. Clinical practice and scientific background," Prima Books AB, 2002.

[4] V.A. Mostovnikov, G.R. Mostovnikova, V.Yu. Plavski, L.G. Plavskaya, "Biophysical principles of regulatory action of low-intensity laser irradiation," *Proc. SPIE*, vol. 2728, pp. 50-62 1996.

[5] HI. Pass, "Photodynamic therapy in oncology: mechanism and clinical use," J Natl. Cancer Inst., vol. 85, pp. 443-456, 1993.

[6] H. Lui, R.R. Anderson, "Photodynamic therapy in dermatology: recent developments," Dermatol. Clin., vol 11, pp. 1-13, 1993.

[7] JS.Jr. McCaughan, "Photodynamic Therapy: a review," Drugs Aging., vol. 15, pp. 49-68, 1999.

[8] J.D. Whitney, "Physiologic Effects of Tissue Oxygenation on Wound Healing," Heart and Lung, vol. 18, pp.466-474 1989.

[9] M.M. Asimov, R.M. Asimov, A.N. Rubinov, "Application of lasers in medicine: on mechanism of biostimulation and therapeutic effect of low energy laser radiation," Proceeding of III International Conference on Laser Physics and Spectroscopy. Minsk, Belarus, pp.169-171, 1997.

[10] M.M. Asimov, R.M. Asimov, A.N. Rubinov, "Investigation of the efficiency of laser action on hemoglobin and oxyhemoglobin in the skin blood vessels,". SPIE Proceedings "Laser - Tissue Interaction," 1X. 01.27 - 01.29. 98. San Jose. CA. USA, vol. 3254, pp. 407 - 412, 1998.

[11] M.M. Asimov, R.M. Asimov, A.N. Rubinov, "Action spectra of laser radiation on hemoglobin of skin blood vessels," Journal of Applied Spectroscopy, vol. 65, pp. 877-880, 1998.

[12] R.V. Benason, E.J. Land, T.G. Truscott, "Flash Photolysis and Pulse Radiolysis. Contributions to the Chemistry of Biology and Medicine," Pergamon Press. 1983.

Some Problems of Modelling Processes of Relaxed Optics in the Regime of Saturation the Excitation

P.P.Trokhimchuck[1,2]

[1]Lesya Ukrayinka's Volyn State University, Lutsk, Ukraine
[2] Lutsk Biotechnical Institute of International Scientific Technical University, Lutsk, Ukraine

Abstract The problem of modeling processes of Relaxed Optics (RO) in the regime of saturation the excitation is discussed. Correlations between photoinized, plasmic and thermal processes are analyzed for various regimes of irradiation. The problem of creation stable structures is discussed too.

The problem of modeling processes of irreversible interactions laser irradiation with solid is very difficult problem [1-5]. The basic effects irreversible must have thermal or plasmic character [4-6]. Processes of photoionization and nonlinear optical effects are neglected. The problem of modeling and observation nonlinear effects in self-absorption range of semiconductor is very difficult problem [7]. Thus we have contradiction: photoionized nature of laser irradiation of semiconductors or another solids and only thermal or plasmic nature of irreversible relaxation first-order excitations.

Processes of very large laser pumping can cause suppression of oscillation and appearance of chaotization of laser radiation [8].

All these processes were explained with one physical-chemical point of view [2], with using elementary energetic estimations [9]. The basic idea of this method is the successive saturation of excitation of proper chemical bonds of irradiated materials [2,10]. This method allows eliminating differences in the explanation of proper experimental data.

For the sort regimes of irradiation, when irradiated time is less as relaxation time, the basic processes of irreversible changes in irradiated materials in the regime of saturation of excitations are straight processes of photoinization including multiphotonic processes of absorption. For indium antimonite most probable nonlinear processes for the regime of pulse Ruby-laser irradiation are the photon fragmentation and up-conversion absorption. First effect is basic for the excitation of first chemical bond (one photon break off ~4-5 bonds). This fact is caused grand relaxation time $\sim 10^{-7} s$. Up-conversion absorption is the result of the scattering Ruby-laser photons on excited electrons of first bond. This effect is caused break off second and third chemical bonds of $InSb$ [2,10]. For the irradiating time less as first relaxation time $\sim 10^{-7} s$ the processes of irradiated relaxation is negligible. But for the regimes of irradiation with time $\sim 10^{-3} s$ the processes of reirradiation have grand value on the

processes the formation irreversible changes in irradiated materials.

The profiles of a distribution of donor centers in $InSb$ after laser irradiation were researched by V.Bogatyryov and G.Kachurin [3]. An irradiation was created with help Ruby laser ($\lambda = 0,69\mu m$, $\tau_i = 5 - 6ms$) and series of impulses Nd:YAG laser ($\lambda = 1,06\mu m$, $\tau_i = 10ns$, frequency of repetition of impulses was $12,5Hz$). The dependence of layer concentration of electrons after Ruby-laser irradiation is represented in Fig.1 [1,3]. A value of threshold the energy of creation n-layers is equaled $\sim 5 \dfrac{J}{cm^2}$. A tendency of the saturation the layer concentration had place for the energy density $\sim 30 \dfrac{J}{cm^2}$. The melting of surface has place for this value of the irradiation.

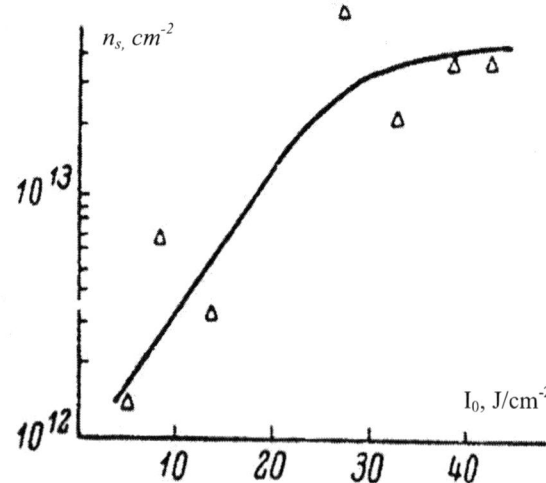

Fig. 1. Dependence of electron layer concentration from an energy density in impulse (Ruby laser irradiation)

A distributions of volume concentration of electrons in laser irradiated layers are represented in Fig.2 [1,3].

CAOL*2010 International Conference on Advanced Optoelectronics & Lasers, 10-14 September, 2010, Sevastopol, Ukraine

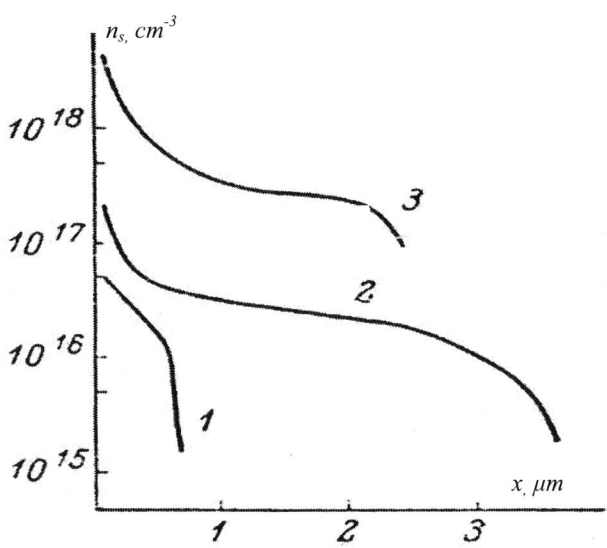

Fig. 2 Profiles of the volume distribution electrons after laser irradiation. 1, 2 – Ruby laser; 3 – YAG:Nd laser. Energy density in pulse, J/cm^2 : $1 - 5$; $2 - 40$.

Higher concentration of donor centers for more short regimes of irradiation (Fig.2) is caused of processes of reemission. For the regimes of irradiation with $\tau_i >> \tau_{r1}$ we have two types of irradiation: first order, basic, Ruby-laser irradiation with $h\nu = 1,78\,eV$, and second order reemission with $h\nu = E_g = 0,18\,eV$ for $InSb$ [1,2].

Rough estimation of effects of reemission may be made with help next formulas. The first part of reemission is equaled

$$I_{r1} = \frac{\tau_{r1}}{\tau_i} I_0 e^{-\alpha x}.$$ Let this part of absorbed radiation is reemitted. In the next time the absorbed irradiation may be represented in the next form

$$I_{r2} = \frac{\tau_{r1}}{\tau_i}\left(1 - \frac{\tau_{r1}}{\tau_i}\right)I_0 e^{-\alpha x} + \frac{\tau_{r1}}{\tau_i}\left(1 - \frac{\tau_{r1}}{\tau_i}\right)I_0 e^{-\alpha_2 x} I_0 e^{-\alpha_2 x}, \quad (1)$$

where α_1 - absorption factor of radiation with $h\nu = E_g$ (lasing effect) and α_2 - absorption factor of "blooming" radiation.

Second term in (1) is represented up-conversed absorption, which is caused irreversible changes in semiconductor. Second and third relaxation times are considerably greater as time of irradiation. Therefore second term in (1) may be represented as "irreversible" term. For the receiving number of reemission n

we must multiply second term of formula (1) on n and equate to intensity of saturation of excitation I_{sat}. Then

$$n = \frac{\tau_i I_{sat}}{\tau_r\left(1 - \dfrac{\tau_r}{\tau_i}\right)I_0}. \quad (2)$$

After substitution proper value of I_{sat} from [2] and I_0 from Fig.1 we have $n \cong 10 \div 100$.

It is very rough estimations. But experimental data of Fig.1, Fig.2 and Fig.3 [1,2,10] are certificated this hypothesis. Surface and volume concentrations donor centers in $InSb$ after irradiation of nanosecond Ruby-laser pulses (Fig.3) [1,2,10] is more in 3-4 orders as after millisecond irradiation (Fig.1 and Fig.2).

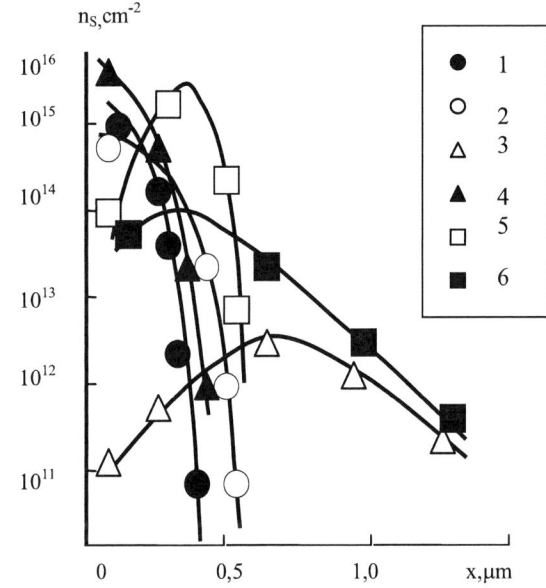

Fig.3..The profiles the distribution the layer concentration of the donor centers in inverse layers InSb and InAs after ruby laser irradiation with various density of energy (monoimpulse regime): 0,07 (1); 0,1 (2); 0,16 (3); 0,16 (4); 0,25 (5); 0,5 J·cm⁻² (6). 1-3— InSb. 4-6—InAs.

Forms of profiles of distribution of donor centers are various too. The maximum of distribution is displaced in volume for nanosecond regime of irradiation [1,2]. It is effect isn't characterized for millisecond regime of irradiation (Fig.2). Multipulse regime of irradiation of nanosecond Nd:YAG laser is analogous to Ruby-laser in millicecond regime, but for this case we have more discrete process of irradiation. The reemission is caused the decreasing concentrations of donor centers and increasing the depth of donor layers.

The honeycomb model [5] may be realized in this case too. This model may be used for modeling second order processes of plasmic and thermal relaxations, including creation

978-1-4244-7043-3/10 $26.00 © 2010 IEEE

nanostructures too after further increasing of intensity laser irradiation (Fig.4) [2]. These nanostructures (nanohills) have various heights, which are regulated interference pattern [4]. This pattern may be have various nature: interaction incident and reflected radiation [4], incident and reemitted radiation [2] and other.

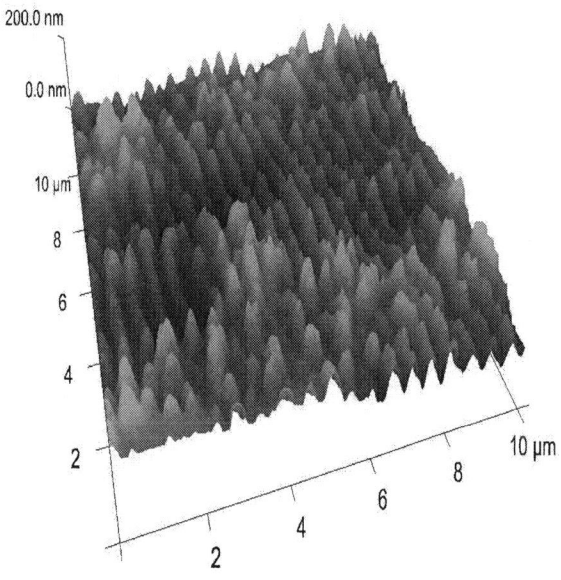

Fig.4. AFM 3D image of GaAs surface after irradiation by YAG:Nd laser at I=5.5 MW/cm^2.

Receiving structures is stable to temperatures $(375 \div 400)^0 C$ (Fig.5) [1,3].

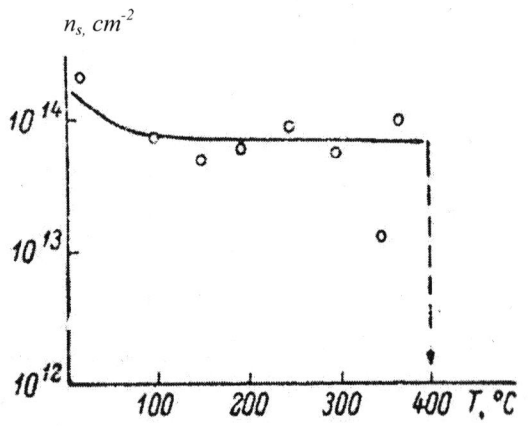

Fig.5. Thermal annealing laser induced

$n -$ layers (time of annealing $-10^3 s$.

These processes were named laser implantation [1]. It may be used for creation new technologies of optical and electronic devices. The problem of a creation three-dimensional periodical electronic structures is very important and may be have good future.

With help of these processes we can correct proper properties of materials and devices. High thermal stability of receiving donor centers on indium antimonite may be allowed to refine the basic characteristics photo electronic infrared devices. But for this we must select correct regime of irradiation.

This method allows receiving materials with properties, which can't be received with help other methods. Therefore, using of this method is expanded fundamental and applied aspects of modern laser physics and optoelectronics.

Author wishes to thank I.Khaybullin, G.Kachurin and A.Medvid' for the discussion of basic idea of this research and O.Veligurskyy for help in the preparation of this manuscript.

References

[1] P.P. Trokhimchuck. *Foundation of Relaxed Optics,* Volyn' University Press Vezha, 2006.

[2] P.P. Trokhimchuck. *Mathematical foundation of knowledge. Polymetrical doctrine,* Volyn' University Press Vezha, 2009. (In Ukrainian)

[3] V.A. Bogatyryov, G.A. Kachurin. "The creation low resistively n-layers on *InSb* with help the impulse laser irradiation", *Physics and technical of semiconductors,* vol.11, No.1,. pp.100-102, 1977. (In Russian).

[4] N.I. Koroteev, I.L. Shumay. *Physics of power laser irradiation,* Moscow Nauka Publishers, 1991.

[5] J.C. Philips. "Metastable honeycomb model of laser annealing", *Journal of Applied Physics*, vol. 52, No.12, P.7397-7402, 1981. (In Russian)

[6] V.P. Gribkovskiy. *Semiconductor lasers*, Minsk University Press. (In Russian)

[7] I.R. Shen. *Principles of Nonlinear Optics*, Moscow Nauka Publishers, 1989. (In Russian)

[8] Haken H. *Laser light dynamics*, Moscow: Mir Publishers, 1988. (In Russian)

[9] Ya.B. Zel'dovich. *Higher mathematics for beginners and its applications to physics.* Moscow Nauka Publishers, 1966. (In Russian)

[10] P. P. Trokhimchuck. *"Phase transformations and transitions in relaxed optics and its possible applications"*, *Proc. CAOL 2008*, pp.273–275 (2008)

[11] A. Medvid', I. Dmytruk, P. Onufrijevs and I. Pundyk , "Photoluminescence from nanohills formed on a surface of Ge, Si and GaAs single crystal by laser radiation*" Proc. Intern. Workshop RNAOPM-2006*, pp. 59-62 (2006)

Laser-microwave spectroscopy of Au I atoms in F rydberg state

S. F. Dyubko[1], V. A. Efremov[1], V. G. Gerasimov[1], M. N. Efimenko[1], M. P. Perepechay[1], K. B. MacAdam[2]

[1]Karazin Kharkov National University, Kharkov, Ukraine
[2]University of Kentucky, Lexington KY 40506-0055 USA

Abstract: For the first time the energy level positions of Au I atoms have been measured with high accuracy by means of laser-microwave spectroscopy in the 176-211 GHz frequency range. The new measurements cover the range of principal quantum number n=31-34 for Rydberg series $F_{5/2,7/2}$. Using our measured frequencies the quantum defect constants for this series are obtained. The description of our spectrometer, measurement technique and results are presented.

During the years 2004-2005 we were investigating microwave absorption spectrum of Au I atoms in S, P, and D Rydberg states. The results are published in Ref. [1].

Expansion of the investigation range in the region of higher orbital quantum numbers (F, G, etc.) has required significant improvement of our spectrometer. Point is that sensitivity of the resonances to the parasitic fields in such states increases very rapidly since the polarizability of the Rydberg atom is $\alpha \sim l^5$.

The recorded resonances for the transitions with $l \geq 3$ suffer from the large broadening and shift of resonance frequencies due to influence of residual (parasitic) electric fields both static and oscillating ones that penetrate inside interaction volume because of imperfect shielding as well as through output connectors.

The modernization of the experimental setup was directed first of all on the attenuation of the fields that penetrate inside the chamber (especially from the stationary units of a mobile communication). With this aim the new connectors with increased isolation of the high-frequency fields were mounted as well as the vacuum seals in the junction units of the spectrometer were changed from Teflon ones to Indium ones.

The Au atomic beam source is constructed from graphite, which is electrically heated. Sufficient intensity of the beam is obtained at source temperature about 1850 K. At such temperature a strong parasitic signal appears that is caused by thermally generated ions emitted by the source. To eliminate this disturbance, at the output of the beam source an ion trap is installed which is formed by a system of diaphragms with alternating potentials.

The ionization potential of gold is about 74409 cm^{-1} and for that reason exciting Au atoms to Rydberg nF- states requires three radiation sources, in the UV, in the visible and in the infrared range, respectively.

To transfer the atoms to nF Rydberg states the three-step excitation scheme was used:

$$6\,^2S_{1/2} \xrightarrow{\lambda_1} 6\,^2P_{1/2}^0 \xrightarrow{\lambda_2} 6\,^2D_{3/2} \xrightarrow{\lambda_3} n\,^2F_{5/2}^0 .$$

Here the laser of the first excitation step $6S_{1/2} \rightarrow 6P_{1/2}$ (coumarin-540 dye) works at fixed wavelength $\lambda = 535$ nm with subsequent frequency doubling in the BBO crystal, which was produced by United Crystals Co. (USA). The laser was pumped by the third harmonic of the Nd:YAG laser. The laser of the second step (DCM-840) with fixed wavelength $\lambda = 813$ nm with subsequent frequency doubling in the KDP crystal provided resonance pumping of the $6P_{3/2} \rightarrow 6D_{3/2}$ transition. The laser of the third step tuned in the wavelength range from 795 to 805 nm works on DCM-800 dye. Its pulse energy is 20 µJ. The second and third dye lasers are pumped by the 532-nm second harmonic of the Nd:YAG laser.

Here we present preliminary results of investigation of Rydberg F states Au I.

In the 176 to 211 GHz frequency range the frequencies of 5 resonances that belong to $nF_{5/2} \rightarrow (n+1)F_{5/2,7/2}$ type transitions in the range of principal quantum number n =31 to 34 have been assigned and measured for the first time. The results of the experiment are presented in a summary Table I.

The energy levels of Au corresponding to quantum number n, l, j may be expressed by the Rydberg-Ritz formula

$$E(n,l,j) = \frac{-R_{Au}}{n^{*2}} = \frac{-R_{Au}}{\left(n - \delta_{n,l,j}\right)^2} ,$$

where $\delta_{n,l,j} = E_{l,j} + \dfrac{A_{l,j}}{n^{*2}} + \dfrac{B_{l,j}}{n^{*4}} + \cdots$.

TABLE I
Observed resonance line centers of Au I

№	Type of transition	Frequency (MHz)
1	$33F_{5/2} \rightarrow 34F_{5/2}$	2*87913.4
2	$32F_{5/2} \rightarrow 33F_{7/2}$	2*96289.0
3	$32F_{5/2} \rightarrow 33F_{5/2}$	2*96295.7
4	$31F_{5/2} \rightarrow 32F_{7/2}$	2*105770.3
5	$31F_{5/2} \rightarrow 32F_{5/2}$	2*105778.7

The reduced-mass-corrected Rydberg constant for Au was taken as $R_{Au} = 109737.01005$ cm^{-1}. The quantum - defect constant $E_{l,j}$ for F terms has been determined: $E_{3,5/2} = 0.04679(3)$; $E_{3,7/2} = 0.04687(3)$.

In the future we are going to continue investigation of the F Rydberg states as well as Rydberg states for Au I with $l \geq 4$.

ACKNOWLEDGMENT

The research described in this publication was made possible in part by Award No. UKP1-2857-KK-07 of the U.S. Civilian Research & Development Foundation for the Independent States of the Former Soviet Union (CRDF).

REFERENCES

[1] S. F. Dyubko, V. A. Efremov, V. G. Gerasimov, K. B. MacAdam, "Millimetre-wave spectroscopy of Au I Rydberg states: S, P and D terms", *J. Phys. B: At. Mol. Opt. Phys.*, vol. 38, no. 8, pp. 1107-1118, 2005.

[2] C. M. Brown, M. L. Ginter, "Absorption spectrum of Au I between 1300 and 1900 Å", *J. Opt. Soc. Am.*, vol. 68, no. 2, pp. 243-246, 1978.

[3] J. C. Ehrhardt, S. P. Davis, "Precision wavelengths and energy levels in gold", *J. Opt. Soc. Am.*, vol. 61, no. 10, pp. 1342-1349, 1971.

[4] H. P. Look, L. M. Beaty, B. Simard, "Reassessment of the first ionization potentials of cooper, silver, and gold", *Phys. Rev. A*, vol. 59, no. 1, pp. 873-875, 1999.

Holographic sensors of glucose in model solution and serum.

V. A. Postnikov[1], A. V. Kraiski[2], T.T. Sultanov[2], V.V. Deniskin[3]

[1] Russia Institute of Physico-chemical Medicine, Moscow, Russia
[2] P.N. Lebedev Physical Institute RAS, Moscow, Russia
[3] F.V. Lukin State Research Institute of Physical Problems, Moscow, Russia

Abstract: Holographic sensors based on the hydrogels containing N-acryloyl-m-aminophenylboronic acid (N-AMB) sensitive to glucose in model solution and serum in physiological concentration (1-20 mM). The reflection (wave length max) and the shift of wave length max was influenced on the additive to polymer net. When N-acryloyl-2-glucosamine, N-AMB hydrogel was used the shift of wave length m ax was achieved 40 nm/mM when glucose concentration was increased in 0,01 M glicine solution (pH 7,4).

Holographic sensors based on the hydrogels containing affinity groups sensitive to certain components of gaseous and liquid mixtures are perspective class of sensors [1, 2]. These sensors have the following advantages: the simple qualitative visual way of the information readout, the exploitation simplicity, the reversibility, the wide spectrum of application, and the low coast at manufacturing. In same cases, for the semi-quantitative or quantitative readout a corresponding design and reading equipment are needed.

Sensor represents a thick hologram prepared in hydrogel films containing silver halogen nanocrocrystals. The photosensitized silver halogen n anocrystals were synthesized in the hydrogel matrix, that was immersed in water or acetic acid (1%) solution and irradiated by the red He-Ne laser (power - 15 mW) and developed [3]. The hydrogel consisted of three- - dimensional polymer net of acrylamide (AA), N-AMB, N,N'-- bis methylene-bis-acrylamide (bis) as crosslinking agent and N-acryloyl-2-glucosamine (GA), N-methacryloyl-lysin (Lys), 2-(dimethyl-amino)-ethylmethacrylate (DMA), acrylic acid (AC) as additive (Fig. 1). Differences in the replay of these sensors at titration of HCl solution correspond to pK of components in polymeric matrices, when the charge appears (or disappers) and matricies swelling takes place. The replay of the sensor (AA-N-AMB-GA) pointed to the fast expansion when the pH solution approaching to pK 8,6 aminophenylboronic acid and slowly contraction of the polymeric matrix, at pH more 9,2 – only the expansion. This behavior agrees with binding N-GA with tetrahedron B⁻-atoms of aminophenylboronic acid that to carry into effect of additional crosslinking. Thus, we have instruments to influence on wave length max reply of the holographic system from 450 nm (Lys) to 720 nm (DMA) at pH 7,4 (pH of serum).

All of these sensors were sensitive to glucose concentration in the model 0,1 M glycine solution (pH 7,4) of glucose(Fig.2) As seen on this figure, the shift of holographic sensor is about 32 nm to 1 mmol of glucose in the sensor of copolymer acrylamide containing 6 mol.% glucosamine, 6 mol.% boronic acid, 0,2 mol.% bis-acrylamide in glycin solution, pH 7,4.

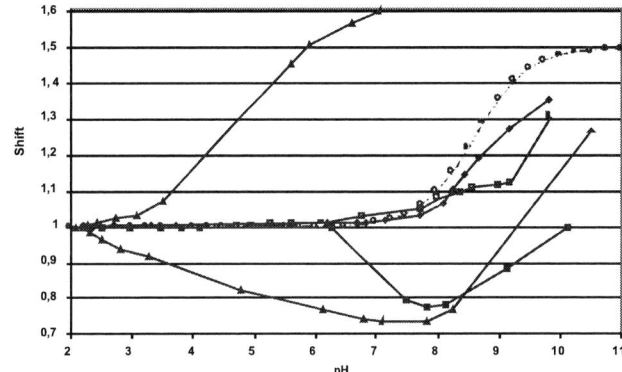

Fig. 1. The replay of sensors with additives as the function of pH in water. ▲ - AA-N-AMB-DMA-bis, △ - AA-N-AMB-Lys-bis, ■ - AA-N-AMB-AC-bis (81,4-12-3,6-3 mol.%), ◆ - AA-N-AMB-bis (85-12-3 mol.%), ◇ -Ionization N-AMB, ▫- AA-N-AMB-GA-bis (81,8-6-12-0,2 mol.%)

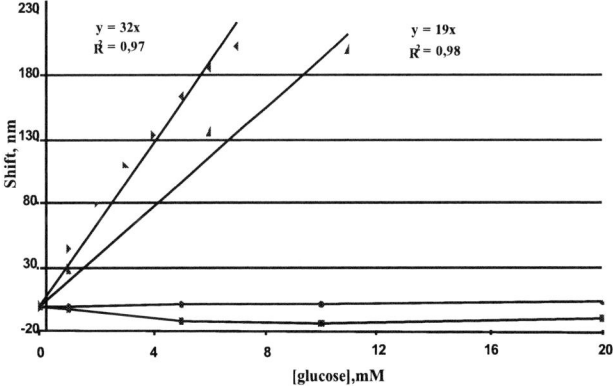

Fig. 2. The replay of sensors as the function glucose in glicine. ▫ - AA-N-AMB-DMA-bis, ◇ - AA-N-AMB-Lys-bis, (81,4-12-3,6-3, mol.%), ▲ - AA-N-AMB-bis(87-12-1, mol.%), ◆ - AA- N-AMB-GA-bis (87,98-6-6-0,2, mol.%).

The replay of sensors to glucose concentration in serum (Fig. 3) is lower but sensitive to distinguish 1 mM glucose.

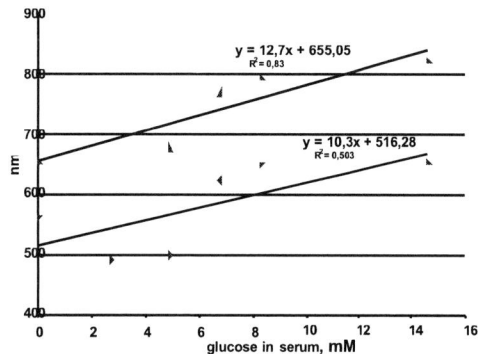

Fig. 3. The replay of sensors as the function glucose in serum.
◆- AA-N-AMB-GA-bis,(75,8-12-12-0,2), ▲ - AA-N-AMB-GA-bis, (84,5-12-3-0,5)

This research has been supporting by the Program of basic researches of Presidium RAS "Fundamental sciences for medicine".

REFERENCES

[1] X. Yang, M.-C. Lee, F. Sartain, X. Pan, and C. R. Lowe, "Complexation of l-Lactate with Boronic Acids: A Solution and Holographic Analysis", *Chem. Eur. J.*, vol.72, pp. 8491-8497, 2006.

[2] A. Horgan, A. Marshall, S. Kew, K.E.S. Dean, C. Creasey and S.Kabilan, "Crosslinking of phenylboronic acid receptors as a means of glucose selective holographic detection," *Biosensors and Bioelectronics,* vol. 21, pp. 1838–1845, 2006

[3] A.V. Kraiskii, V.A. Postnikov, T.T. Sultanov, A.V. Khamidulin "Holographic sensors for diagnostics of solution components", *Quantum Electronics*, vol 40 (2) pp. 178 – 182, 2010

Thermosensor Diagnostics (TSD) of Laser Welding Process.

A. F. Keremzhanov [1], P.P. [Arkhipov1] A. G. Lazarenko [2],
[1]ACO Ltd, Moscow, Russian Federation
[2]Kharkiv National Technical University "KhPI", Kharkiv, Ukraine

Abstract: Hardware and software means of laser welding process monitoring were developed and tested in a couple of industrial application - both with pulse and cw lasers. Factors and criteria of thermodynamic stability for metal melting and seam forming thermal fields are discussed.

The system includes IR and(or) visible sensors placed in the closest vicinity of laser focusing lens(Fig.1), control unit with PC interface (Fig. 2) and appropriate software for sensors signals (as a heat trace intensity signal function, let us call them thermosensograms) displaying, analysis and storage.

The system provides monitoring of laser welding process in real time, failures thresholds changing and their excesses indication, feedback signals generation. TSD has prospects for applications in laser fusion welding, contact welding and other kinds of welding or thermo-treatment processes[1-3]. TSD principle is illustrated by Fig. 3 for the case of pulsed laser point welding checkup.

Fig. 2. Control unit front view. Legends near buttons and LED indicators meanings (in Russian): Work Regime, Power, On /Regime, Settings, Output, Event, PC Connection.

The next PC software version provided possibility of several thermosensograms simultaneous recording and analysis with further combined signal use for errors detection. It was tested in stainless steel tubes butt welding by CO_2 Laser TL5M (Russia) in cooperation with the Institute of Laser and

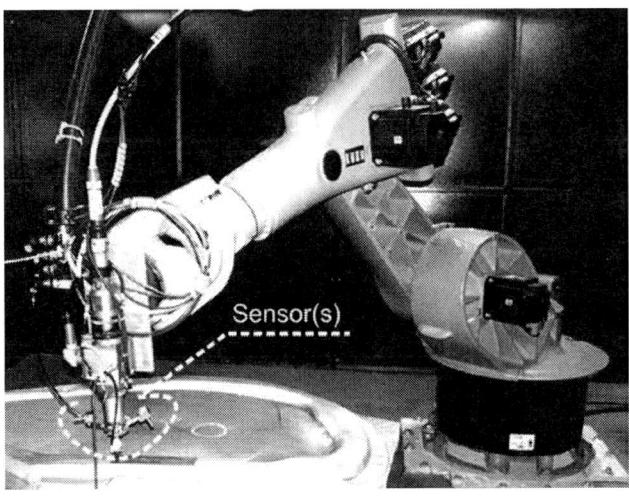

Fig. 1. IR and/or visible sensors mounted at Trumpf 4006D YAG:Nd cw laser

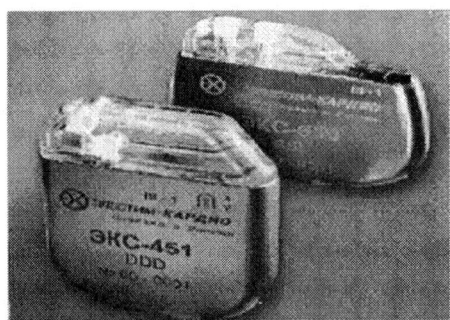

Cardiostimulators cases made of 0.6mm thikness titanium

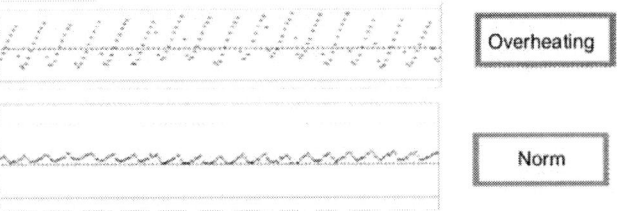

Fig. 3. Thermosensograms used during pulsing welding of titanium cases shown above

Information Technologies Problems , Russian Academy of Sciencies and "Laser Complexes Corp.", in Shatura, Moscow region (Fig. 4-6)

Defects diagnostic criteria bind the following ordinate's thresholds of Fig.4 thermosensograms to laser welding process parameters. For overheating (not considered in a given laser welding test as defect one) it is an equality or excess of signal to 30%. For burn-through – an equality or excess of signal 60% level. For poor penetration – an equality or decrease of signal lower than 17%. For time axis defect criterion (mainly for gas protection fault) - 2% of a time scale in seconds.

Separate poor-quality seams parts were fixed at laser welding of a pipe (a butt joint, welding speed was 3,5 m/min, laser power 1,5 - 2 kW, focal spot diameter - 0,2mm, for steel pipes with diameter 33mm, wall thickness 1,5mm, with a gap and flange discrepancy in limits of + 10-70 % of welded materials thickness) in real time for the following cases:

- Dot burn-through (at movement beginning, at stop of details feeding – 2nd frame, 1.2s; 3rd frame, 4.4s) and protective gas feeding suspension (1st frame - from a beginning up to 0.93s);

- poor penetration (flange discrepancy and gap exceed 50 - 70% of thickness - 3rd frame, since 8,1s).

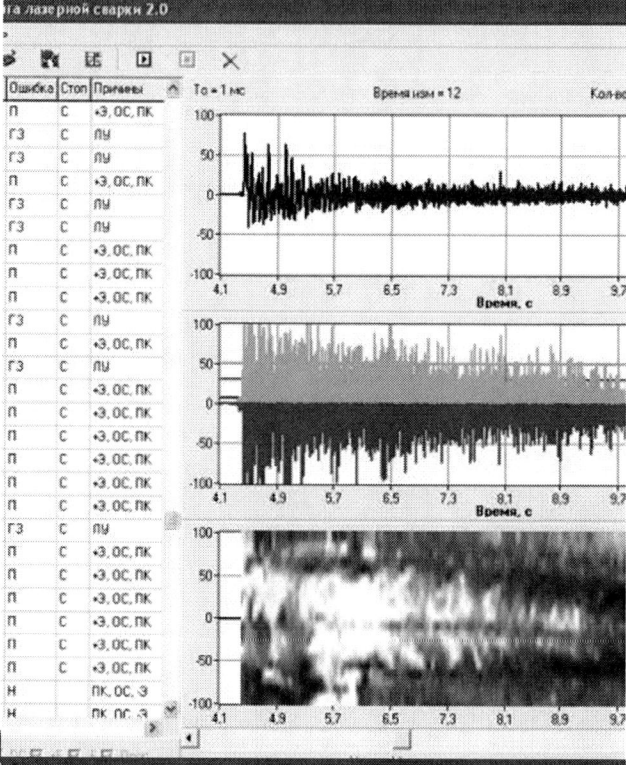

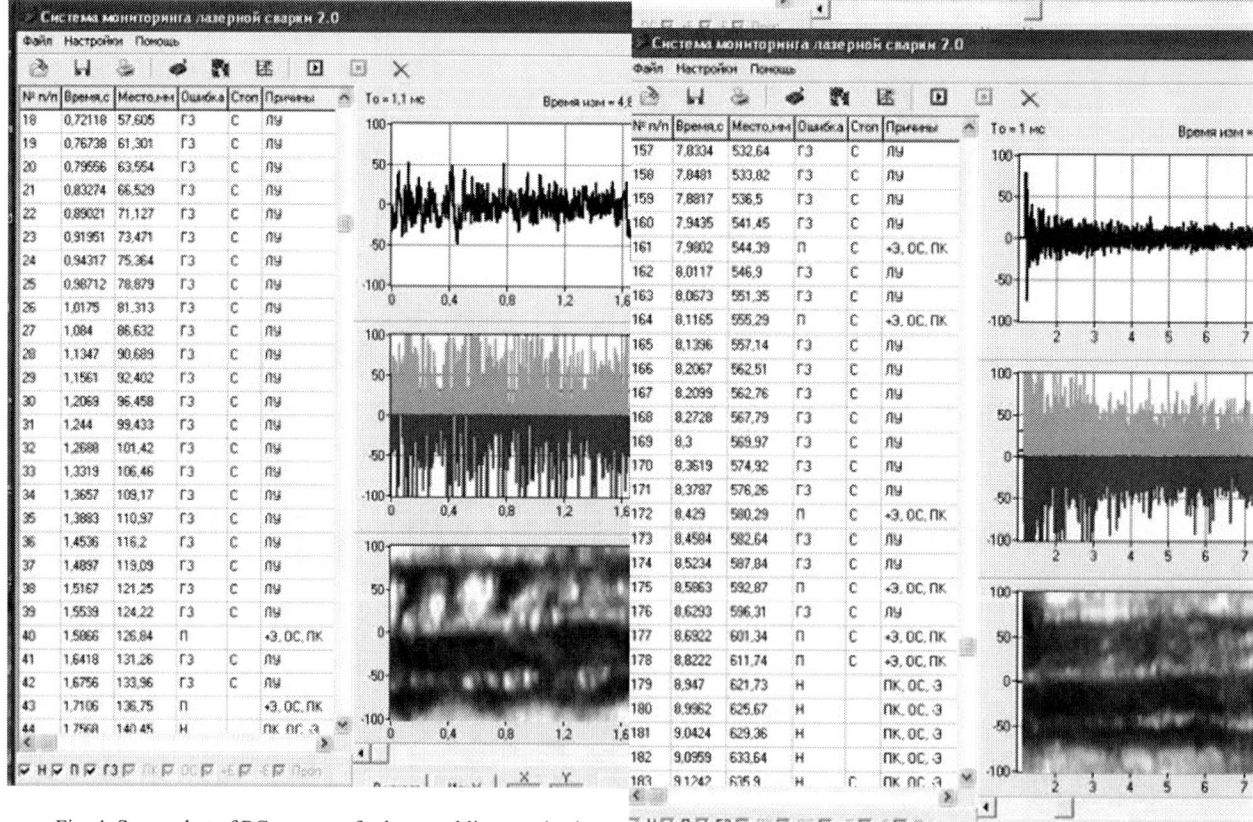

Fig. 4. Screenshot of PC program for laser welding monitoring (Russian interface, table headings mean: Time,s; Place, mm; Error, Stop, Reasons), 1st frame.

Fig. 5,6. Screenshot of PC program for laser welding monitoring (Russian interface, table headings mean: Time,s; Place, mm; Error, Stop, Reasons), 2nd (up) and 3rdframes.

Conditional overheating for the given calibration threshold in controlled laser welding process was not considered as defect. As it is shown in 1-3 frames at welding speed of 3,5m/min in power increase at 33 % (2kW - 1st frame) and normal power 1,5kW (2 - 3-rd frames) themselves do not result in burn-through.

The given laser welding optimum technological parameters achievement at normative melting-throngh is confirmed by 1st (0,96 - 1,7s) and 2nd (1,9 - 7,3s) frames of a seam surface view and appropriate laser welding thermosensograms. At thermosensograms seam signals amplitudes exceed poor penetration level of 17% and are below than burn-through level 60%. Thus, by laser welding process monitoring with the use of TSD it is possible to optimize a choice of laser welding parameters. Intentionally

The last Fig.7 shows four different seams obtained by use of cw laser Rofin-Sinar 1500 along with their thermosensograms and photos of some seam cuts – both longitudinal and transversal ones. Intentionally generated seam defects (over-heats, burn-through, pores, cracks and lack of fusion) and their correlation with thermosensograms can be seen.

REFERENCES

[1] Doc. XII-1477-97, (The 49 Annual Assemble IIW / IIS. Budapest.1996). Doc. V-1086-97, (The 50 Annual Assemble IIW / IIS. San – Francisco.1997). Doc. V-1146-99, (IIW / IIS. Paris. 1999). Doc. V-1165-00, Doc. VE-217-00, (IIW / IIS. Paris, Saarbrueken. 2000).

[2] Бутов В.Н., Архипов П.П., Дёмин Е.А., Керемжанов А.Ф. Термосенсорный контроль качества и дефектности сварных соединений по критериям термодинамической стабильности плавления металла в процессе сварки. Сварочное производство, № 11, 2006г. (Translated in Welding Journal).

[3] A method and device for laser welding diagnostics and quality control. Russian Federation Patent № 2258589, 2005. International patent publication number WO 2006 / 073334 A1, 2006.

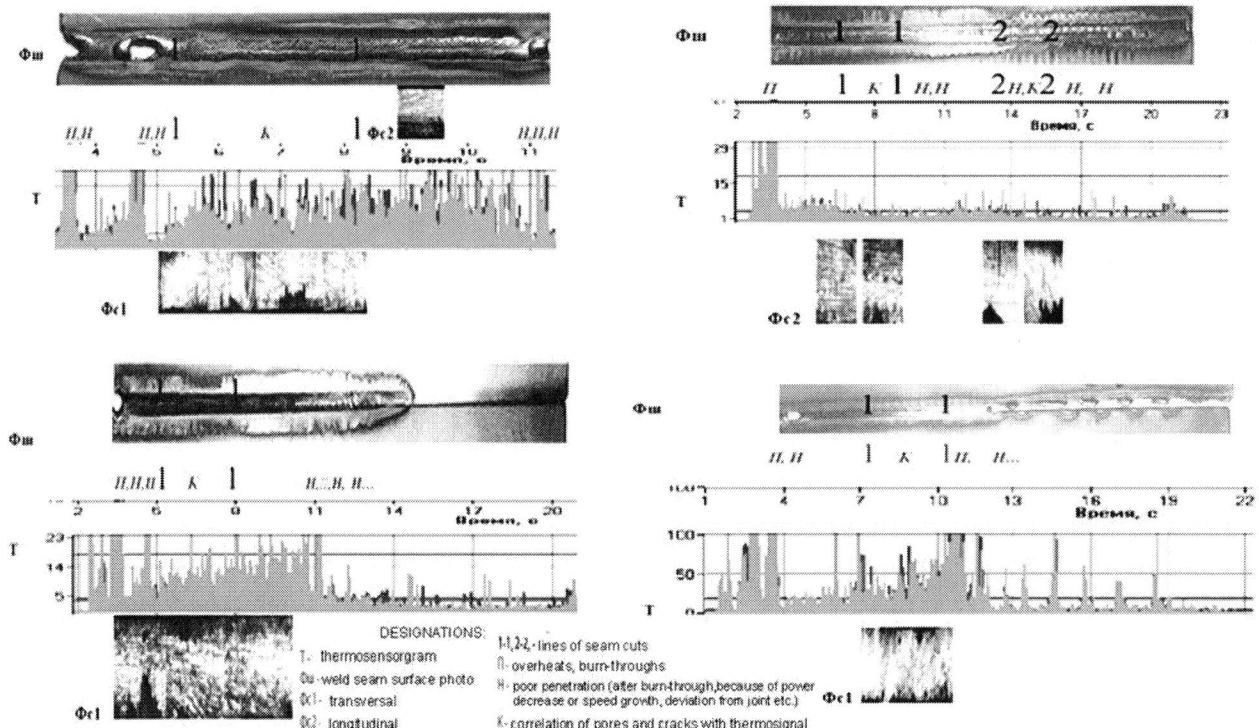

Fig.7. Examples of non-stabilities monitoring by thermosensor while cw laser Rofin-Sinar 1500 welding regime test.

About the Influence of Metallic Additives On Copper Vapor Laser Output Parameters

E. A. Svitlichniy, V. A. Kelman, Yu. V. Zhmenyak, V. V. Zvenihorodskiy, Yu. O. Shpenik

Institute of Electron Physics, Uzhhorod, Ukraine

Abstract: A universal mechanism of metal additive atoms influence on copper vapor laser parameters is proposed to interpret the peculiarities, observed at injection of cesium, silver and zinc atoms into the discharge volume. The effect of thermal instability, caused by the additives influence on the plasma processes is discussed.

For a long period of time, since 1975, investigation of metal atom additives influence on copper vapor laser (CVL) to improve the output parameters. In particular, US Patent [1] notice on the coincidence of copper atom metastable and cesium atom resonant states energies. The author supposed the effective deactivation of copper atom lower laser levels in II kind collisions with cesium atoms. Thus, cesium atoms could serve as energy acceptors. Indeed, experimental investigations [2] didn't show significant positive influence of Cs atoms additive and its influence mechanism on CVL parameters isn't so well-defined [3].

Later another additive – silver atom [4, 5] as energy donor was proposed. It's positive influence may appears in additional increasing of upper copper laser level excitation due to coincidence of copper and silver atoms resonant states energies. Experimentally this positive influence was measured for metal-halide (more effective) and "pure" copper vapor laser.

At last, zinc atom was proposed as a resonant optical influence donor [6]. The working mechanism consists in simultaneous deactivation of copper metastable states by absorption of Zn I 213.9 nm resonance line and further additional pumping of resonant states by cascade transitions. To our opinion certain positive effect of zinc additive [6, 7] as for other additives can not be interpreted only by optical processes. Thus, the ambiguous influence of different additives efficiency on CVL parameters cast doubt on the mechanisms of resonant coincidence of energies of copper and additive atoms states and explanation of a new universal influence mechanism is necessary.

Modeling of the CLV parameters behavior for a laboratory sample of CVL was performed in [8]. The output parameters were calculated for optimal and non-optimal working modes and a serious attention was taken to the consideration of the so called "thermal runaway" effect. It was shown that this effect appear at increasing the voltage on storage capacitor when inner tube wall temperature become higher than optimal ($\sim$ $1500\,^0C$) value due to increasing of electron energy losses at

elastic collisions with not merely neon atoms but also copper atoms.

The wall temperature balance at the same time becomes unstable due to the effect of singular positive feedback. Namely, the increasing of the discharge tube temperature results the increasing of copper atoms concentration. Thus, the elastic energy losses will increase with the growing of gas temperature and the tube wall temperature will also grow up. And again the copper atoms concentration also will grow up. Certainly, the origin of this phenomenon is in different energy dependence of momentum transfer cross-section for neon and copper atoms.

Agreeing with the standpoint that the increasing of copper atoms concentration over optimal value causes influence on discharge tube temperature T_w, naturally we have to discuss the possibility of such an influence mechanism at using metal atoms additives (for example, cesium, zinc or silver). Lower we want to discuss this possibility in details.

As it is well known, in the equation for heat balance of gas temperature in the laser discharge tube the process of energy transfer from electrons to the gas atoms (neon) in elastic collisions can be described by the term

$$\Delta Q_{el} = \frac{2m_e}{M} N_e N <\sigma_m v> \frac{3}{2}(T_e - T_g),$$

having the physical meaning of energy quantity, that is transferred from the electrons to atoms in cubic unit per time unit. In this equation m_e and N_e – the electron mass and concentration; M and N – gas atoms mass and concentration; T_e and T_g – electron and gas temperature; $<\sigma_m v>$ – the rate constant for the energy transfer process, obtained by averaging of the momentum transfer cross-section of electron scattering σ_m by Maxwellian distribution by velocities v. We had estimated the value of $<\sigma_m v>$ for different electron temperatures. The momentum transfer cross-section of electron scattering σ_m or (in the case of their absence) the cross-sections of elastic scattering σ_{el}, were used taken from the references, presented at Table 1 together with the $2m_e/M$ coefficients for each atom in interest.

978-1-4244-7043-3/10 $26.00 © 2010 IEEE

Table 1

Atom	M, au	$2m_e/M$	References	
			σ_m	σ_{el}
Ne	20.17	$5.4338\ 10^{-5}$	[9]	
Cu	63.54	$1.7249\ 10^{-5}$		[10]
Cs	132.9054	$0.8246\ 10^{-5}$	[11]	
Ag	107.868	$1.0160\ 10^{-5}$	[12]	
Zn	65.38	$1.6764\ 10^{-5}$	[13]	

Fig. 1 presents the energy dependence of momentum transfer cross-sections (in the case of absence – elastic scattering cross-sections) for the above mentioned metal atoms. The neon atom is characterized by a slanting increase of cross-section with maximum $\sim 4*10^{-16}$ cm^2 at energy ~ 40 eV; all the metallic atoms in interest have the cross-sections values of two order higher with maximum at very low energies. Therefore there is significant difference from the neon atom.

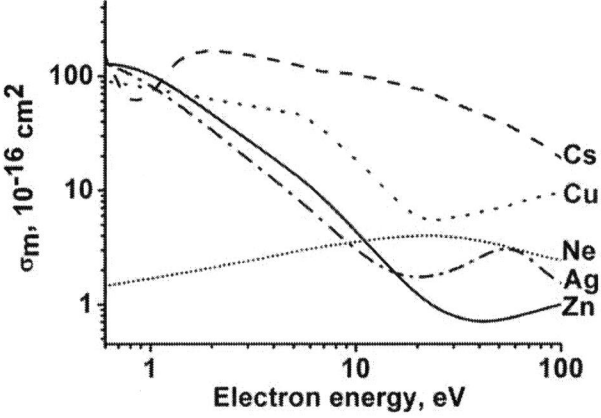

Fig. 1. Energy dependence of momentum transfer cross-sections of electron elastic scattering with neon and metal atoms (copper, cesium, zinc, silver).

The calculation of the $<\sigma_m v>$ parameter was performed following [14] by numerical integration of equation

$$<\sigma_m v> = 0.59\cdot 10^8 \int_0^\infty E^{1/2}\sigma_m f_{T_e}\, dE .$$

This equation is adapted to use the electron energy E value in eV units and the effective cross-section σ_m value in cm^2 units. The rate constant value is obtained in cm^3/s units. The Maxwellian distribution function f_{T_e} here can be presented as

$$f_{T_e} = 2(E/\pi)^{1/2}(T_e)^{-3/2}\exp(-E/T_e) .$$

Here E and T_e must be taken in eV; then f_{T_e} will be obtained in (eV)$^{-1}$ units.

The results of calculations are presented at Fig.2 as energy dependence of $\dfrac{2m_e}{M}<\sigma_m v>$ factor, which determines the efficiency of thermal transfer from electrons to different atoms. The calculations were made for the electron temperature values of 3 eV (optimal temperature for the copper vapor laser during the excitation pulse), 0.4 eV (typical for the beginning of the inter-pulse period) and 0.2 eV (typical for the ending part of the inter-pulse period). The integration interval was limited to electron energies of 0 – 100 eV, it is quite enough for our calculation accuracy of rate constants of elastic collisions thermal transfer The numerical values of the calculation results are presented at Table 2.

Table 2

Atom	$\dfrac{2m_e}{M}<\sigma_m v>$, cm^3/s		
	$T_e = 3$ eV	$T_e = 0.4$ eV	$T_e = 0.2$ eV
Ne	$1.79*10^{-12}$	$3.49*10^{-13}$	$1.95*10^{-13}$
Cu	$9.01*10^{-12}$	$4.44*10^{-12}$	$2.76*10^{-12}$
Cs	$1.26*10^{-11}$	$6.90*10^{-12}$	$5.53*10^{-12}$
Ag	$2.19*10^{-12}$	$4.53*10^{-12}$	$3.91*10^{-12}$
Zn	$4.66*10^{-12}$	$6.84*10^{-12}$	$4.67*10^{-12}$

The obtained results shows that at concentrations of copper atoms $\sim 10^{15}$ cm^{-3}, cesium additive atoms $\sim 10^{14}$ cm^{-3}, silver and zinc additive atoms $\sim 10^{15}$ cm^{-3} the amount of thermal energy obtained by each of this gas component from electron gas is comparable with the energy transfer value to the buffer gas neon atoms at $\sim 10^{16}$ cm^{-3}.

Moreover, more exact estimation of energy transfer from electrons to gas atoms needs also taking into consideration the coulomb collisions of electrons with ions. At the metal-noble gas mixtures the metal atoms are ionized mainly. To evaluate the electron-ion collisions frequency we used the equation form [15]

$$\nu_{ei} = \frac{2.8 N_e}{(1.16\cdot 10^4 T_e)^{3/2}}\ln\frac{4\cdot 10^6 T_e}{N_e^{1/3}} .$$

Here N_e must be taken in cm^{-3}, T_e in eV units. ν_{ei} will be obtained in sec^{-1}. For $T_e = 0.5$ eV, $n_e = 10^{15}$ cm^{-3} the $\nu_{ei} = 1.9*10^{10}$ sec^{-1}. For $T_e = 0.2$ eV, $n_e = 10^{12}$ cm^{-3} the $\nu_{ei} = 1.1*10^8$ sec^{-1}. For $T_e = 3$ eV, $n_e = 5*10^{13}$ cm^{-3} the $\nu_{ei} = 1.25*10^8$ sec^{-1}.

The thermal transfer to ions is of the same value as to the additive atoms (see Fig.2).

Finally, based on the performed evaluation results, we can conclude that the injection of metal atoms (Zn, Cs, Ag) additives leads to increasing of thermal transfer from electrons to the metal vapor-gas mixture. The whole amount of thermal energy, concentrated in the buffer gas and metal atoms (and ions) as kinetic energy, due to the thermal conductivity process finally will be transfer to the discharge tube inner wall and thus, its temperature T_w will increase.

Accordingly, to our opinion, by injection of metal additive atoms to the discharge volume the discharge tube temperature will grow up. It can result the following two types of influence on copper vapor laser parameters. For the laser operation mode in optimal excitation conditions the injection of additives can leads only to the worsening of lasing excitation conditions. The reason is in the decreasing of electron temperature during the excitation pulse and therefore the efficiency of upper laser levels pumping. And finally, when the electron energy will reach the lower boundary value the laser oscillation will disappear. It should be mentioned that the increasing of metal additive over optimal value always results disappearing of laser oscillation at all published investigations.

At the same time, at experimental conditions with discharge tube operating at temperature lower than optimal value, the metal atom additive injection at the beginning demonstrates a positive influence due to increasing of T_w. This increasing is limited by reaching the temperature value over optimal one and thus leads to its damping.

The proposed mechanism of metal additive atoms influence on copper vapor laser parameters is reasonably universal than the mechanisms based on resonance coincidence of energy levels of copper atom and metal additives and can interpret the peculiarities observed at experimental investigations. The obtaining of experimental proof of correctness of the proposed model needs carrying out of serious complex investigations of copper vapor laser both in optimal and non-optimal (under-heated) working modes.

REFERENCES

[1] Karras T. W., "Cesium quenched copper laser", US Patent # 3 381 107, 1974.

[2] Voronyuk L. V., "Investigation of cesium admixture influence on generation characteristics of copper vapor lasers", Ph. D. Thesis Auto-referat, Uzhhorod, 1988. 20 p.

[3] Petrash G. G., "Influence of Cs on CVL operation", Proceedings of SPIE, 2001. vol. 4747, pp. 193-197.

[4] Oouchi K., Suzuki M., Fujii K., "Green-yellow lasers in CuBr-AgBr-Ne systems operating at 40 kHz of repetition frequency", IEEE J. Quantum Electronics, 1991. vol. 27, no 11, pp. 2473-2481.

[5] Riyves R. B., Kelman V. A., Zhmenyak Yu. V., Shpenik Yu. O., Ulusova S. P., "Copper-vapour laser with silver additive", Applied Physics B. Lasers and Optics, 2005. vol. 80, pp. 865-869.

[6] Fuji K., Uno K., Tawada F., Hishida T., Nishizawa M., Suzuki M., Oouchi K., "Pulse broadening in a Ne-CuBr-Zn System by optical resonance pumping", Appl. Phys. Lett., 2002. vol. 80, no 11, pp. 1859-1861.

[7] Kelman V. A., Svitlichnyi E. A., Zhmenyak Y. V., Shpenik Y. O., "Cu-vapor laser with zinc-atom additive", Appl. Phys. B, 2009. vol. 94, pp. 301-305.

[8] Carman R. J., "Modelling of the kinetics and parametric behaviour of a copper vapour laser: Output power limitation issues", J. Appl. Phys., 1997. vol. 82, no 1, pp. 71-83.

[9] Zecca A., Karwasz G. P., Brusa R. S., "One century of experiments on electron-atom and molecule scattering: a critical review of integral cross-sections", Rivista del Nuovo Cimento, 1996. Vol. 19, no 3, p. 24.

[10] Msezane A. Z., Henry R. J. W., "Electron-impact excitation of atomic copper", Phys. Rev. A, 1986. Vol. 33, no 3, pp. 1631-1635.

[11] Zecca A., Karwasz G. P., Brusa R. S., "One century of experiments on electron-atom and molecule scattering: a critical review of integral cross-sections", Rivista del Nuovo Cimento, 1996. Vol. 19, no 3, p. 68.

[12] Toshich S. D., Kelemen V. I., Shevich D., Pejchev V., Filipovich D. M., Remeta E. Yu., Marinkovich B. P., "Elastic electron scattering by silver atom", Nucl. Instr. and Method B, 2009. vol. 267, pp. 283-287.

[13] Kelemen V., Dovhanych M., Remeta E., "Elastic electron scattering by Zn, Cd and Hg atoms in the optical potential approach", 4th Conf. on elementary processes in atomic systems (CEPAS), June 18-20, 2008. Cluj-Napoca, Romania, Book of Abstracts, pp. 76.

[14] Webb C.E., "The fundamental discharge physics of atomic gas lasers", Inst. Phys. Conf. Ser. 1976. no 29, pp. 1-28.

[15] Ginzburg V. L., "The propagation of electromagnetic waves in plasma", Moscow, Nauka, 1967. 684 p.

Influence of InGaN/GaN/Si electroluminescent heterostructure design and quantum well thickness on their luminescent and laser properties

V. N. Pavlovskii[1], E. V. Lutsenko[1], A. V. Danilchyk[1], V. Z. Zubialevich[1], A. V. Muravitskaya[1] and
G. P. Yablonskii[1], H. Kalisch[2], R. H. Jansen[2], B. Schineller[3] and M. Heuken[2,3]
[1]Stepanov Institute of Physics NAS of Belarus, Minsk, Belarus
[2]Institut für Theoretische Elektrotechnik, RWTH Aachen University, Germany
[3]AIXTRON AG, Aachen, Germany
Phone: +375-17-2949025, Fax: +375-17-2840879, e-mail: v.pavlovskii@ifanbel.bas-net.by

Abstract: Employing a strain-reducing layer stack between Si substrate and MQW layers as well as optimization of quantum well thicknesses of InGaN/GaN/Si MQW electroluminescent heterostructures led to disappearance of cracks and pinholes, to reduction of the laser threshold down to 75 kW/cm^2 at 433 nm under N$_2$ laser emission excitation.

Silicon substrates for epitaxial growth of III-nitride light emitting heterostructures have a number of advantages in comparison with the widely used sapphire and silicon carbide. These are low cost, high crystalline quality, high thermal conductivity (300 W/(m·K) instead of 40 W/(m·K) for sapphire). Additionally, Si allows the possibility of integration of GaN-based light-emitting devices with silicon electronics [1] using well-known and widely distributed silicon technologies. In the case of laser production, simple laser diode mirror creation by cleaving becomes possible ensuring low cost of final production.

Recently, Chinese Lattice Power Corporation reported that at 350 mA drive current, a 1x1 mm laboratory blue light emitting diode (LED) (453 nm wavelength) achieved more than 458 mW of output power [2]. These results show that LEDs made from III-nitrides grown on (111) Si substrates have excellent quality. Successful realization of III-nitride growth on 8 inch in diameter Si substrates will also promote industrial production of LED and laser heterostructures [3] grown on Si substrates.

Achievements in epitaxial growth of InGaN/GaN heterostructures on Si substrate make it promising for creation and mass production of not only III-nitride ultra-bright LEDs but of injection lasers as well. The first lasers made of GaN layers and InGaN/GaN multiple quantum wells (MQWs) on Si substrates were investigated in [4, 5]. Then the heterostructure design and the growth technology were improved further leading to a laser threshold reduction and to a widened spectral laser range [6, 7]. It was shown that the optical material gain of InGaN achieves a value of 11000 cm^{-1} [8].

But emission and especially laser properties of InGaN/GaN heterostructures grown on Si have not been sufficiently investigated so far. Optimized heterostructures, which are grown specially for laser diodes, should have a sophisticated design with a waveguide for optical confinement factor enhancement and for the elimination of absorption in the metal electrode at p-type region of the heterostructure. But for investigation of the laser properties, optical pumping of conventional InGaN/GaN/Si *electroluminescent* heterostructures may be used also. In this case, the influence of heterostructure design and layer thickness as well as of absorption in p- and n-type layers on radiative recombination, optical gain in InGaN QWs and laser parameters may be investigated.

In this work, emission properties of InGaN/GaN/Si electroluminescent heterostructures with different designs, quantum well thicknesses and lasing structure parameters were investigated by optical pumping.

The investigated electroluminescent MQW heterostructures were grown on Si(111) substrates in AIXTRON reactors by the MOVPE method. All samples were grown using metalorganic sources and ammonia as nitrogen source. As carrier gases, N$_2$ and H$_2$ were used. The growth parameters were optimized for obtaining the highest possible layer quality. Due to large lattice mismatch between GaN and Si, it was necessary to grow a sufficiently large sequence of strain-reducing layers prior to MQW growth. Three series of samples were investigated. The first series has an active region consisting of 5 identical InGaN/GaN QWs (with thickness of 1.6, 4.5 or 5.3 nm), two strain-reducing AlN/AlGaN stacks (with low-temperature AlN in the bottom of each stack) between the substrate and MQWs, a GaN buffer layer of different thickness (220-650 nm) and a GaN:Si$^+$/GaN:Si/GaN top stack under the MQWs (Fig. 1a). The second series contained 5 QWs (2.1 nm), one AlN/AlGaN stack, a GaN buffer (650 nm) layer, low-temperature (LT) AlN and GaN (1 μm) layer (Fig. 1b). The third series contained 5 or 10 QWs (from 1.3 to 2.5 nm), one thick AlN/AlGaN stack and GaN:Si$^+$/GaN:Si/GaN top stack under MQWs (Fig. 1c). The topmost GaN layers of the heterostructure series were as follows: GaN:Mg (200 nm) (series 1); GaN (50 nm) (series 2); GaN:Mg (170 nm) (series 3).

978-1-4244-7043-3/10 $26.00 © 2010 IEEE

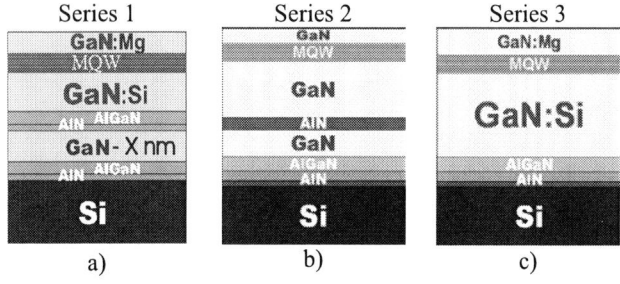

Fig. 1. Design of electroluminescent InGaN/GaN MQW heterostructures: a) with two AlN/AlGaN stacks; b) with one AlN/AlGaN stack and LT AlN layer; c) with one AlN/AlGaN stack.

Photoluminescence (PL) was excited by an emission of HeCd and N_2 lasers. Transversal pumping of the laser cavities was realized by N_2 ($\lambda=337.1$ nm) and Ti:Al_2O_3 ($\lambda=380$ nm) laser emission focused in a stripe on the sample surface perpendicular to the MQW laser cavity mirrors produced by cleaving.

We observed that an increase of the GaN layer (nearest to the substrate) thickness to 400 nm leads to a disappearance of cracks. An increase of the GaN layer thickness to 650 nm further led to the disappearance of the pinholes.

PL spectra maxima of the heterostructures with different QW thickness shifted from 415 nm to 470 nm at thickness alteration from 1.6 nm to 5.3 nm at $I_{exc}=300$ kW/cm². Here, the radiative recombination efficiency in the thinner QWs is significantly higher than that in the thicker QWs at all excitation intensities. The reduction of the PL efficiency with QW thickness rise is possibly caused by a spatial separation of carriers by internal piezoelectric fields and additionally by defect formation by means of InGaN strain relaxation in thicker QWs.

An increase of excitation level led to a short-wavelength PL band shift and a superlinear growth of PL intensity. Rise of radiative recombination efficiency in MQWs together with the blue-shift of PL band position in samples with thicker QWs evidence on the saturation of non-radiative recombination channels, on screening of piezoelectric fields, and on the filling of InGaN band tails.

Fig. 2 shows electroluminescence, photoluminescence and stimulated emission spectra of the heterostructure of series 1 with thinnest QWs without cracks and pinholes. It can be seen that the PL spectrum at $I_{exc}=1$ MW/cm² by N_2 laser emission is shifted to the short-wavelength side in comparison with the EL spectrum (considerably lower excitation level). No lasing was observed at N_2 laser excitation because of the high absorption coefficient and poor carrier transport from GaN:Mg into QWs at optical excitation above GaN band gap (3.4 eV at room temperature), caused by the low value of diffusion length in GaN:Mg. Pumping of the heterostructure by emission of the second harmonic of a tunable Ti:sapphire laser with quantum energy below the GaN:Mg band gap allowed to reach lasing but at high excitation intensity. The laser threshold was as high as 700 kW/cm² in this case. Lasing appeared on the short-

wavelength wing of the PL bands because of screening of the piezoelectric fields and the filling of band tails with a low density of states up to the energy region where the density of states is sufficient to provide such optical gain value which compensates the internal loss (absorption in GaN:Mg, GaN:Si layers and in the substrate) and the cavity mirror loss. Since the absorption of the Ti:sapphire laser emission in the MQW is not higher than 20-30% of the penetrating emission, the laser threshold value evaluated for the absorbed emission of the Ti:sapphire laser was 150 – 200 kW/cm² close to the MQW laser threshold at excitation by the N_2 laser emission after removal of a part of the GaN:Mg layer using reactive ion beam etching.

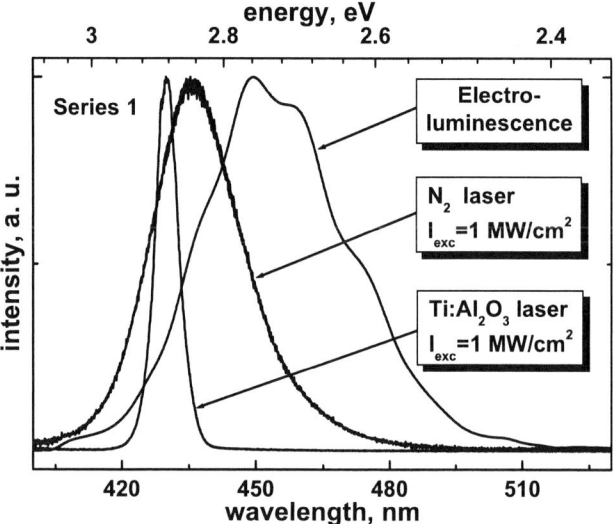

Fig. 2. Electroluminescence, photoluminescence and stimulated emission spectra from the sample surface of InGaN/GaN EL heterostructure of series 1.

Indium concentration was varied in the heterostructures of series 2 at the same thickness. An increase of In concentration led to a substantial drop of the PL intensity and to a long-wavelength shift of the PL band from 445 to 505 nm at low excitation level (0.1 W/cm²). This drop is due to defect number rise and piezoelectric field value increase. The long-wavelength shift occurs for the account of InGaN band gap reduction and piezoelectric field value rise.

All heterostructures of this series demonstrated laser action at N_2 laser excitation. This was due to efficient transport of charge carriers from the thin (50 nm) undoped top GaN layer into QWs and to a reduction of internal loss (absence of absorption in GaN:Mg and GaN:Si). A nearly linear increase of the threshold with laser wavelength was observed. A minimal laser threshold $I_{thr}=130$ kW/cm² was observed for the shortest operation wavelength of 440 nm and the largest one $I_{thr}=320$ kW/cm² was obtained for the longest operation wavelength of 465 nm. An increase of the laser threshold is due to a radiative recombination efficiency decrease with indium concentration rise.

Change of design in the heterostructures of series 3, namely use of one complex strain reducing AlN/AlGaN stack $(Al_{0.2}Ga_{0.8}N(360$ nm$)/Al_{0.3}Ga_{0.7}N(150$ nm$)/Al_{0.5}Ga_{0.5}N(150$ nm$)/AlN(300$ nm$)/)$ allowed to grow GaN:Si on it and InGaN/GaN MQWs of good quality. PL intensity of these heterostructures despite a presence of enough thick GaN:Mg layer (170 nm) was considerably higher than that of the heterostructures of previous series. Light emitting diodes were created of these heterostructures which showed good current-voltage characteristics and bright electroluminescence.

Fig. 3 presents PL and laser spectra position of series 3 heterostructures versus QW thickness. PL band position depends practically linearly on InGaN well thickness at low excitation level.

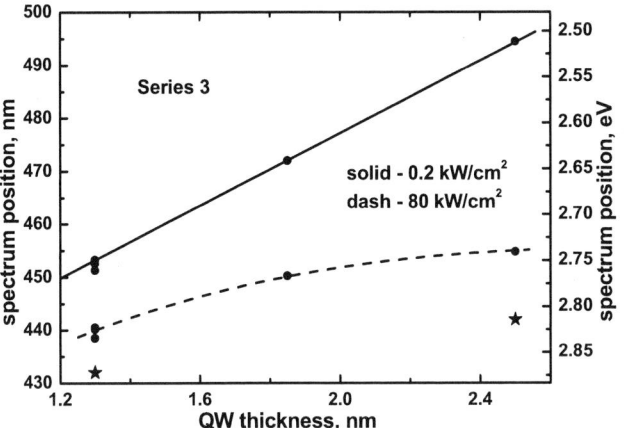

Fig. 3. PL and laser spectrum position as a function of QW thickness at different excitation levels for series 3.

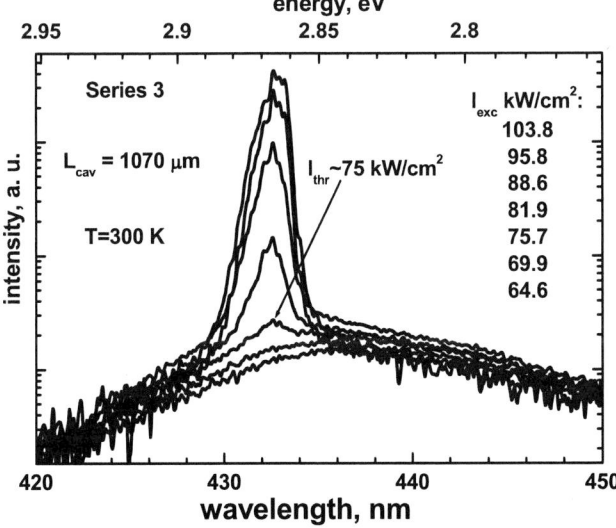

Fig. 4. Emission spectra from cavity edge as a function of excitation intensity for series 3.

The long-wavelength shift of the PL spectrum position with QW thickness rise is due to several factors: 1) an increase of the emission wavelength caused by quantum-dimensional Stark effect; 2) a reduction of dimensional quantization energy; 3) a reduction of band gap and band tail widening with In concentration increase at QW growth time rise. An increase of excitation intensity led to a short-wavelength shift of the PL spectrum as it was discussed earlier. This shift is larger for thicker QWs as it is seen from Fig. 3. An increase of the shift with QW thickness rise is due to more pronounced concentration effects such as band tail states filling and piezoelectric field screening. The laser spectra are shifted to the short-wavelength side in comparison with PL spectra at high (80 kW/cm^2) excitation level and this shift is slightly larger for thicker QWs. The lowest laser threshold was obtained for the thinnest QWs of series 3 with thickness of 1.3 nm. It was as low as 75 kW/cm^2 (Fig. 4). At the same time, the heterostructure with the thickest QWs demonstrated a laser threshold of 230 kW/cm^2 evidencing a higher crystal quality of "shorter-wavelength" InGaN.

Thus, lasing in InGaN/GaN MQW electroluminescent heterostructures with different design and QW thickness grown on silicon substrates was obtained under optical pumping in the region of 430-465 nm. It has been shown to be possible to reach a laser threshold as low as 75 kW/cm^2 at a laser wavelength of 433 nm.

REFERENCES

[1] Oussama Moutanabbir. Heterogeneous Integration of Compound Semiconductors. Annual Review of Materials Research Vol. 40 (August 2010).

[2] http://www.ledsmagazine.com/news/6/9/14

[3] http://compoundsemiconductor.net/cws/article/lab/34464

[4] G. P. Yablonskii, et al., "Luminescence and Stimulated Emission from GaN on Silicon Substrates Heterostructures", Phys. Stat. Sol. (a), vol. 192, p. 54, 2002.

[5] E. V. Lutsenko, et al.," Growth, Stimulated Emission, Photo- and Electroluminescence of InGaN/GaN EL-Test Heterostructures", Phys. Stat. Sol. (c), vol. 0, p. 272, 2002.

[6] A. L. Gurskii, et al., "High temperature operation of optically pumped InGaN/GaN MQW heterostructure lasers grown on Si substrates", Compound Semiconductors: Post-Conference Proceedings, 2003 Int. Symposium on. Vol. , Iss. p.197, 25-27 Aug. 2003.

[7] E. V. Lutsenko, et al., "Laser threshold and optical gain of blue optically pumped InGaN/GaN multiple quantum wells (MQW) grown on Si", Phys. Stat. Sol. (c), vol. 5, p. 2263, 2008.

[8] E. V. Lutsenko, et al., "High optical gain InGaN/GaN MQW electroluminescent heterostructures grown on silicon by MOCVD", Proceedings of 4th Int. Conf. on Advanced Optoelectronics and Lasers, Sept. 29 – Oct. 4, 2008, Alushta, Crimea, Ukraine, P. 360.

CAOL*2010 International Conference on Advanced Optoelectronics & Lasers, 10-14 September, 2010, Sevastopol, Ukraine

Modelling of THz-Laser Radiation Pulse Shape

V.K. Kiseliov, Senior Member, IEEE, V.P. Radionov
IRE NAS of Ukraine, 12 Ac. Proskura st., Kharkov, 61085, Ukraine
Tel: 380 57 7203335, E-mail: kiseliov @ire. kharkov.ua

Abstract: A technique for modelling the shape of the radiation pulse-pumped THz-laser is presented. The initial data to be used for modelling are as follows: the size and the shape of a discharge-current pulse; discharge current-dependent radiated power; the dependence of radiated power upon the tuning control of a resonator at different pumping currents. The application of the above technique makes it possible to produce pulses of required shape and to select laser peak-efficiency operating conditions.

In gas-discharge pulse-pumping THz-lasers the radiation pulse shape is governed by several factors. Clearly, the key factor is the discharge current pulse shape. However, even if the pumping current pulse shape is invariable, the radiation pulse form may be varied over a wide range. Specifically, one can observe the radiative pulse bifurcate to the extent that two separate pulses are generated [1]. This phenomenon occurs because of the joint impact of resonator tuning parameters, the active substance gain profile and the changes in the resonators optical length in response to the varying discharge current. The laser efficiency hinges upon the shape of discharge current and radiation pulses. Besides, the radiation pulse shape has often to be controlled in conducting different medicine and physics-related studies. In order to produce these pulses of required shape it is necessary to allow for all factors affecting the pulse shape and to be able to simulate it.

We have developed the technique for simulating the radiation pulse shape of gas-discharge pulse-pumped THz-lasers. In the course of modelling procedure the following factors are taken into account: the size and the shape of the discharge current pulse, discharge current-dependent radiated power as well as the dependence of radiated power upon the resonator tuning control at different pumping currents. The dependences of radiated power upon the discharge current and the resonator tuning are measured during pumping with continuous current. Therefore, this particular method is suitable for the gas-discharge THz-lasers that can be operated, as they are pumped both with continuous and pulse currents.

When measuring the experimental dependence of the laser radiated power upon the discharge current magnitude we adjust the resonator length for a peak radiation at each measuring point. These measurements are performed on a laser under study when it is being pumped with continuous current including the active medium parameters similar to the pulse-mode operating conditions. Fig.1 exemplifies the behavior of this dependence for a gas-discharge HCN laser with a resonator of ~1.2m long.

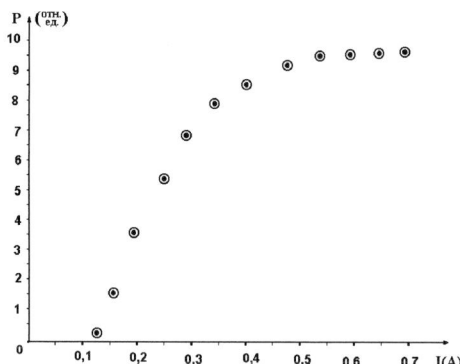

Fig. 1. The experimental dependence of the laser radiated power upon the discharge current magnitude

The resonator length correction is needed because the discharge current tends to be changed with a variation in the optical length of resonator ΔL, which is caused by the change in the refractive index of the active medium.

$$\Delta L = L_1 - L_2 = \frac{(\upsilon_1 - \upsilon_2)M}{2 f_0} = \frac{(\frac{c}{n_1} - \frac{c}{n_2})M}{2 f_0} = \frac{(n_2 - n_1)cM}{2 f_0 n_1 n_2},$$

where L_1, L_2 - are the resonant lengths of the laser resonator, which correspond to certain pumping currents; f_0 - is the center frequency of the active medium radiation line; υ_1, υ_2 - are the radiation propagation velocities on a center frequency at corresponding gas-discharge currents; n_1, n_2 - are the absolute refractive indexes of the active medium at corresponding pumping currents.

We simultaneously obtain the radiated power magnitude curves plotted against the displacement of the resonator mobile mirror. As a result, we obtain a set of dependences for each discharge current value to be measured (Fig.2). The peaks of these curves, are for the tuning to a center frequency. The resonator optical length is different for each value of the current, and this difference is around 1μm per each 0.1A.

978-1-4244-7043-3/10 $26.00 © 2010 IEEE

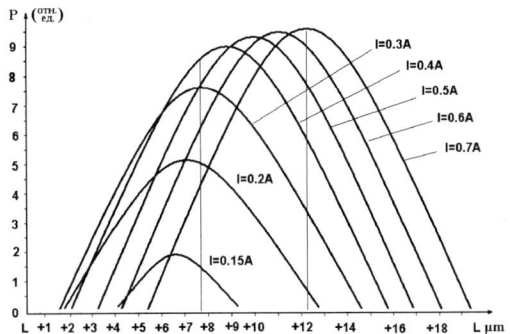

Fig. 2. A set of radiated power curves plotted against the displacement of the mobile mirror at different pumping currents.

Modelling of the shape and power of a radiation pulse is carried out in terms of a real or any arbitrary discharge-current pulse shape to be studied. It is necessary that the pulse discharge current should vary within the limits of the previously measured radiated power level as a function of the discharge current (see Fig.1.).Now take, for example, the discharge-current pulse typical for pumping the gas-discharge HCN laser with a 50-Hz continuous current (see the pulse in the top part of Fig.3). In the course of modelling (as in a real situation) the resonator tuning control has a dramatic impact upon the radiation pulse shape. In the bottom part of Fig.3a the radiation pulse is shown to be modelled when the resonator is optimally tuned at 0.7A current – the resonators mechanical length is then equal to ~L+12μm (Fig. 2). Fig.2 presents the values of radiated power for each magnitude of the discharge current. These values correspond to L+12μm and are noted in the bottom part of Fig. 3a at a respective instant of time of pulse propagation, as it is shown by thin vertical lines in the top and bottom parts of Fig. 3a.

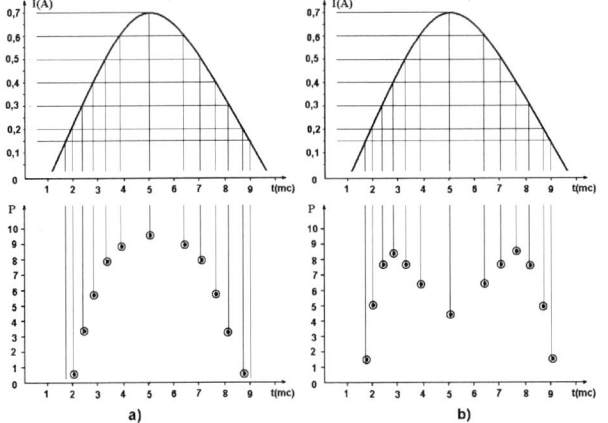

Fig. 3. The modelling radiation pulse for resonator tuning at a peak discharge current of 0.7A – (a), and for resonator tuning at a current of 0.3A –(b).

The same procedure can be used when the resonator is tuned at the current below the peak value. For instance, tuning is taken to be optimal for I=0.3A (see fig3b). With this tuning the radiated pulse bifurcation occurs. At a peak current the

resonator is found to be untuned, and this results in decreased radiated power. If the resonator is tuned for the current exceeding its peak value in a pulse, then no bifurcation occurs. One can rather observe the pulse getting narrower, and this particular effect can be easily modelled as well. The whole process is confirmed in actual practice. The pulse bifurcation effect is more pronounced with an increase in pumping current. For example, at the pulse discharge current of 1A, with the resonator being properly tuned, one can observe the radiation pulse to become fully bifurcated [1].

The real radiation pulse may differ slightly from those already modelled. This is accounted for by the presence of inhomogeneities in a gas discharge, which bring about the variations in the resonator optical length at identical discharge currents. In the modelling example we have already examined, both the time it takes for the active medium to be synthesized and the pumping rate were not allowed for. In HCN lasers this synthesis time is ~ 300ms, whereas the medium advance rate is ~ 1m/c [2]. Under certain pulse pumping conditions these factors introduce some corrections in modelling. For example, for a pulse length of less than 300 ms and several seconds of interval between them, the laser radiation is not necessarily observed. Over a discharge pulse repetition period, the active medium has no time to become synthesized in sufficient amount. Yet during the inter-pulse time the medium synthesized over the period of the previous pulse is removed. In order for the laser to operate in such a mode an additional continuous-current discharge chamber is required to bring about the active medium synthesis.

The technique described above can well be employed to model not only the shape and the radiation pulse size, but also the radiation frequency during the pulse repetition period, because the resonator optical length affects the tuning variations within the active medium radiation line. All these potential features are useful in carrying out different physics- and medicine-oriented studies. The above technique can be efficiently used to select the operating modes for peak efficiency pulsed lasers. For example, one can either calculate the laser efficiency enhancement in the change-over to the rectangular pumping pulses or calculate the required correction of the resonator length in the process of bell-shaped pumping pulse repetition.

REFERENCES

[1] V.K.Kiseliov, V.P.Radionov, "Research of effect of laser generation pulse bifurcation when HCN laser is pumped by alternating current", LFNM 2005, 15-17 September 2005, Ylta, Crimea, Ukraine p.76-79.

[2] V. K. Kiseliov, V. P. Radionov, N. F. Dakhov, "Influence of time of synthesis of the active material and intensity of his pumping on parameters terahertz gas-discharge HCN laser", ISSN 1028-821x Radiofizika i Electronika, vol.14, no1, p.p. 93-96, 2009.(in Russian).

REPARATIVE REGENERATION PROCESSES OF THYROID GLAND AND TIMUS AT EXPERIMENTAL AUTOIMMUNE THYROIDITIS AFTER LASEROTHERAPY

I. Gopkalova, V. Dubovik

V. Danilevsky Institute of Endocrine Pathology Problems of
Academy of Medical Sciences of Ukraine (IEPP)
61002, Artema Street, 10, Kharkov, Ukraine
Phone: (380) (057) 3151191, Fax: (380) (057) 7004538, e-mail: gopkalovaira@rambler.ru

Abstract: The research was showed that low-intensive laser radiation has positive influence on the leading links of AIT pathogenesis: it stops auto-immune aggression in thyroid gland and stimulates reparative regenerative processes by restitutive type, against stop of involutive processes in timus and activation of lymphocitopoesis.

Keywords: laser radiation, peparative regeneration, timus, auto-immune thyreoiditis, thyroid gland.

INTRODUCTION

During last 15 years after Chernobyl the number of people injured by chronic Hashimoto's auto-immune thyreoiditis increased 10 times in Ukraine.

The aim of the research was studying of mechanism of low-intensive laser radiation (LILR) effect on auto-immune thyreoiditis (AIT) pathogenesis.

MATERIALS AND METHODS

Research has been carried out on 50 Shinshilla rabbits with experimental AIT (EAIT). Animals were arranged in the following groups, 10 rats each: CG - controls groups, EAIT, EAIT+LT (laser therapy). Laser radiation was done by apparatus Mustang-BIO (Russia), wavelength 0,89 μm. power 10 W in pulse, frequency 80 Hz, exposition on thyroid gland (TG) proection for 30 sec, every day, during 7 days. Studying of morphofunctional state of TG rabbits with EAIT in 14 and 70 days after lasertherapy. For research of structure and function thyroid gland and timus used morphometric methods [1-5]. Was determined next parameters: of Klein index, height of follicular epithelium and coefficient of functional activity restrain, and also practically all figures of main TG structural components relative volumes (RV). In timus was determined structural components relative volumes of crust and brain substance timus and density of lymphocytes in crust and brain substance on area $4 \times 10^{-4} \, \text{mm}^2$.

RESULTS OF RESEARCH

The research was exposed normalization of Klein index, height of follicular epithelium and coefficient of functional activity restrain (Fig. 1, 2, 3), and also practically all figures of main TG structural components relative volumes (RV) through 14 and 70 days after lasertherapy.

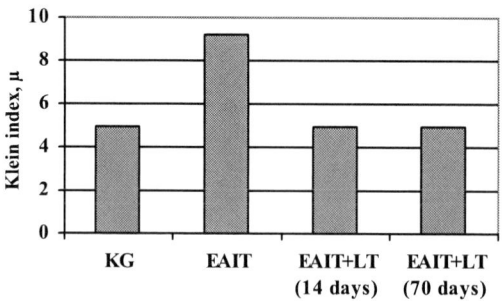

Fig. 1. Klein index.

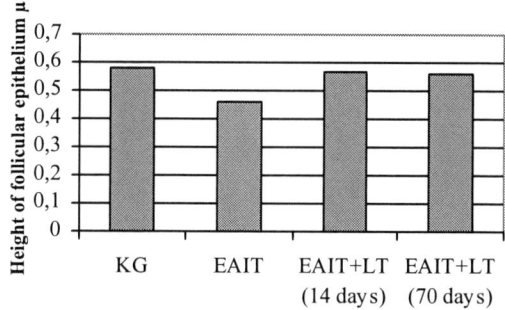

Fig.2. Height of follicular epithelium

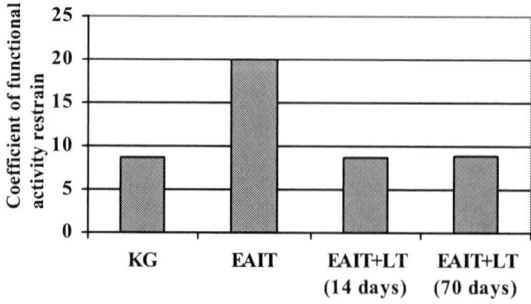

Fig. 3. Coefficient of functional activity restrain

Thus lymphoid infiltration RV decreased from (10,1±1,2)% to (0,64±0,2)%, P<0,001; stromes RV - from (15,0±1,8)% to (9,8±1,7)%, P<0,02; grease cells RV - from (6,8±0,7)% to (4,1±0,45)%. P<0,01 (Fig. 4). Simultaneously RV of follicular epithelium increased from (10,0±0,1)% to (21,8±1,9)%, P<0,001, and RV of inter-follicular epithelium increased from (4,2±0,5)% to (13,7±2,6)%, P<0,001 (Fig. 5). Experimental researches on animals with autoimmune thyroiditis model have shown, that in the rabbit thyroid gland after laser influence, compared with control bunch of animals, processes of reparative neogenesis for phylum of restitution are intensified, nonperishable downstroke of lympho-plasmocellular infiltration volume is observed

(21,8±2,3), P<0,01 and in brain substance from (14,8±2,3)% to (18,4±1,6). The attribute of stabilization of lymphocytopoesis and epitheliorhetyculopoesis are found in thymus. Thus, low-intensive laser therapy restores and stabilizes the morho-functional state as thyroid, as thymus up to the normal limits.

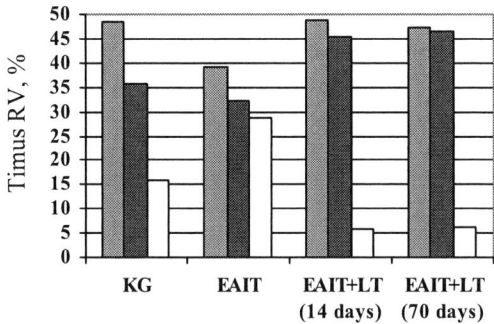

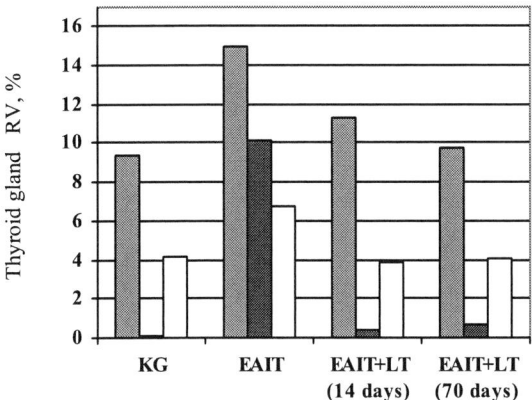

Fig. 5. Structural components relativevolumes of timus (RV).

CONCLUSION

The research was showed that low-intensive laser radiation has positive influence influence on the leading links of AIT pathogenesis: it stops auto-immune aggression in TG and stimulates reparative regenerative processes by restitutive type, against stop of involutive processes in timus and activation of lymphocitopoesis.

Fig. 4. Of main TG structural components relative volumes (RV).

REFERENCES

[1] IAC. 1522275 USSR, ISI G 09 B 23/28. «Method of modeling auto-immune thyreoiditis», Brindak O.I., Poltarak V.V., Ladogubets E.V. (USSR), No. 4017822/28-14; Appl. 22.01.86; Publ. 15.11.89, Bul. No. 42, p. 2.

[2] IAC. 1307283 USSR, ISI G 01 N 1/28. «Method of exposition of functional activity of thyroid gland», Jakovtsova A.F., Sorokina I.V., Gubina-Vakulik G.I. (USSR), No. 3855882/28-14; Appl. 17.12.84; Publ. 30.04.87, Bul. No. 16, Opening. Inven., No. 16, p. 170, 1987.

[3] IAC. 1690702 USSR, ISI A 61 B 17/00. «Method of exposition of structural components relative volumes of bodies immune and endocrine systems», Jakovtsova A.F., Sorokina I.V., Gubina-Vakulina G.I. (USSR), No. 4719016/14; Appl. 14.07.89; Publ. 15.11.91, Bul. No. 42, Opening. Inven., No. 42, p. 25, 1991.

[4] Bonashevskaja T.I., Beljaeva N.N., Kumpan N.B. «Morphofunctional of research in hygiene», M.: Medicine, p. 214, 1984.

[5] Avtandilov G.G. «Medical morphometric», M.: Medicine, p. 382, 1990.

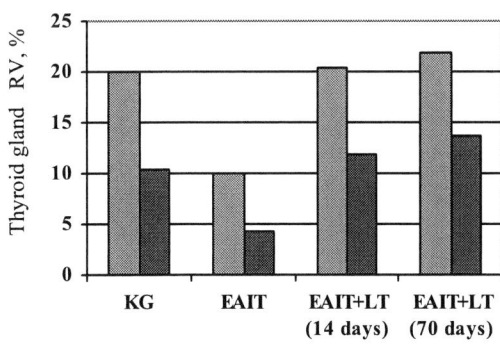

Fig. 5. Of main TG structural components relative volumes (RV).

RV of crust and brain substance increased authentically in timus up to the norm limits at the account of decreasing RV of grease cells from (28,6±2,3)% to (6,2±0,6)%, P<0,001. Density of lymphocytes in crust increased from (13,3±1,5)% to

9781424470433